WITHDRAWN
NDSU

FREGE: Logical Excavations

FREGE: Logical Excavations

G. P. BAKER & P. M. S. HACKER

Fellows of St. John's College, Oxford

Oxford University Press, New York
Basil Blackwell, Oxford
1984

Copyright © 1984 by Oxford University Press, Inc.

All rights reserved. Except for the quotation of short passages for the purposes of criticism and review, no part of this publication may be reproduced, stored in a retrieval system, or transmitted, in any form or by any means, electronic, mechanical, photocopying, recording or otherwise, without the prior permission of the publisher.

Published in the United States in 1984
by Oxford University Press, Inc.

Published in Great Britain in 1984
by Basil Blackwell Publisher,
108 Cowley Road, Oxford, OX4 1JF

Library of Congress Cataloging in Publication Data

Baker, Gordon P.
   Frege, logical excavations.
   Includes index.
   1. Frege, Gottlob, 1848–1925. I. Hacker, P.M.S.
(Peter Michael Stephan) II. Title.
B3245.F24B34 1983   160'.92'4   82-22491

ISBN 0-19-503261-6

British Library Cataloguing in Publication Data

Baker, G. P.
   Frege.
   1. Frege, Gottlob—Mathematics
   2. Mathematics—Philosophy
   I. Title   II. Hacker, P. M. S.
   510'.92'4   QA8.4

ISBN 0-631-13169-8

Printing (last digit): 9 8 7 6 5 4 3 2 1

Printed in the United States of America

For
K.J.B.
and
in memoriam
E.E.H.
(1903–83)

Denn was ein Mensch auch hat,
so sind's am Ende Gaben.

The mathematician too can wonder at the miracles (the crystal) of nature of course; but can he do so once a problem has arisen about *what* it actually is he is contemplating? Is it really possible as long as the object that he finds astonishing and gazes at with awe is *shrouded* in a philosophical fog?

I could imagine somebody might admire not only real trees, but also the shadows or reflections that they cast, taking them too for trees. But once he has told himself that these are not really trees after all and has come to be puzzled at what they are, or at how they are related to trees, his admiration will have suffered a rupture that will need healing.

WITTGENSTEIN

Preface

This book presents the fruits of three years' work on the philosophy of Frege. But the harvest gathered differs profoundly in kind and in quantity from the one originally foreseen. After completing the first volume of an analytical commentary on Wittgenstein's *Philosophical Investigations,* we looked forward to a brief change of intellectual scenery. Our study of the *Investigations* had required sketching in relevent aspects of Frege's thought and examining some of Wittgenstein's criticisms. This amounted only to a fragmentary presentation of Frege's ideas and a one-sided vision of a critical assessment of them. An appetite for a more systematic examination, together with the tendency to disagree with important aspects of current interpretations, led to the framing of a modest interim project: we resolved to spend a year writing a short introductory book for undergraduates on the general principles of Frege's philosophical logic.

As the first year of work wore on, our original plan looked increasingly unrealistic. The more closely we examined Frege's writings, the more profoundly we disagreed with standard conceptions of his theories. His famous essays in the 1890s and his last masterpiece, *The Basic Laws of Arithmetic,* no longer seemed to provide the key to his thought. On the contrary, it was that much neglected book, *Begriffsschrift,* which appeared to us to mould his subsequent work and to put his mature essays and book in the position, not of the foundation stone, but of the head of the corner. As we pored over his early theory of judgeable-content, more and more questions arose which made scant sense within the framework of ideas of late twentieth-century philosophical logic. These categories of thought proved obstacles to apprehending Frege's reasoning and to explaining the movement of his mind, the cast of the puzzles which preoccupied him, and the hue of the solutions he presented with apparently so little rationale. It slowly dawned on us that the task before us was not to sketch, for the benefit of students, a familiar edifice which was clearly visible in the sunshine and for which detailed and fairly accurate ground plans and elevations were readily available. It was rather the much more daunting task of excavating the crypt and foundations of a noble ruin, half covered with the rubble of a century, in order to lay bare its hidden structure, bring to light the principles of its construction, and evaluate its stability.

These logical excavations proved neither simple nor uncontroversial. We had to dig through the accumulated deposits of several generations to reveal the foundations of the thought of a nineteenth-century mathematical logician. This brought to the surface the articulations of a system built up in a climate of mathematical, logical, and philosophical thought which is alien to ours. To be seen aright, Frege's works must be viewed not in the perspective of a new science with neither ancestry nor precedent, nor in the light of twentieth-century categories of thought on logic, but rather in the context of the exciting expansions of nineteenth-century mathematics and, in particular, of the revolutionary applications of mathematics to logic. Just as Boole's logical system depended on a radical generalization of algebra, so Frege's rested on a novel extension of function theory. The results of these excavations and explorations throw new light on many familiar and allegedly indubitable aspects of Frege's thought.

His genius was mathematico-logical. He pushed back the frontiers of formal logic to a hitherto undreamt extent and captured terrain for the purposes of logical formalization previously altogether inaccessible. But, as our investigations proceeded, it became apparent that the philosophical foundations of his innovations were no firmer than those supporting the logical algebras admired by some of his sophisticated contemporaries. The fact that his formal calculus remains intact despite its lacking firm substrata is an important philosophical datum, which points to a multitude of morals about the nature and philosophical significance of a formal calculus. Far from Frege's philosophical thought providing the building blocks of modern philosophical logic, all its key elements are flatly inconsistent with contemporary principles of logico-philosophical construction. Our delineation of the ground plan and structure of the house that Frege built has radical implications. For we have not laboured to bring to light the defective foundations of Frege's thinking in order to glorify the quite different underpinnings of late twentieth-century logic. The point of our logical excavations is not to wallow in the wisdom of hindsight. On the contrary, it is rather to use a perspicuous representation of a great nineteenth-century logical theory as an object of comparison against which we can gauge with fresh eyes our current conceptions. The realization that modern logic initially rested on quite different foundations from those we now take for granted should lead us not to congratulate ourselves on our improved engineering, but to examine afresh our own conception of logic, and to take seriously the possibility that, though different, it may be no less flawed. The results of our excavations may be debatable or wrong in many points of detail, but unless we have stumbled on the artefact of our own *malin génie*, we can perhaps share Schliemann's belief when he discovered the ruins of Troy that the boundary between fact and myth will have to be plotted anew.

In quoting from Frege's works we use standard translations, but where these seem to us to distort the original text we modify them without specific notice. In one respect, however, we deviate systematically from very recent translations. We adhere to the traditional rendering of *'Bedeutung'* as 'reference', rather than the recent one as 'meaning'. While Frege's use of *'Bedeutung'* is by no means natural, translating it as 'meaning' produces patent nonsense (e.g. the meaning of 'Little John' met the meaning of 'Robin Hood', and the first meaning knocked the second meaning into the water!).

In analyzing Frege's formal system we alternate between employing his own notation and standard modern ones. Where aspects of his symbolism are under discussion, there is no alternative to using formulae in his concept-script. But where the topic is his conception of propositional connectives or quantifiers, the argument can often be rehearsed more clearly in modern dress.

We make no pretence to survey systematically the copious secondary writings on Frege's philosophy. We exclude from the main text explicit direct confrontation with other commentators. In footnotes we take issue with a select range of points drawn from currently influential works which purport primarily to interpret (rather than to extend) Frege's thought. Our aim in so doing is to highlight particular dangers of misinterpreting his writings or points where our account is distinctive and open to misconstruction.

While writing this book we benefited from support from institutions, friends, and colleagues. The British Academy gave us a generous grant for our research. St. John's College kindly reduced our teaching load during the final year of our labours, and provided us with facilities which we greatly appreciate. We enjoyed many stimulating discussions with Professor M.A.E. Dummett[1]; Though our disagreements were extensive, the gulf dividing us was bridged by his good humour and kindness. Dr. J. Dupré, Professor K. Fine, Dr. R. G. Frey, Dr. D. Isaacson, Dr. W. Künne, Dr. J. Raz and Mr. B. Rundle all read early drafts of chapters, and we are grateful to them for their comments, criticism, and searching questions.

St. John's College, Oxford G.P.B. and P.M.S.H.
July 1981

---

1. What instigated these exchanges was a series of papers we presented at seminars which he attended in 1978-9. Many aspects of these early drafts of ours are reflected in his recent book *The Interpretation of Frege's Philosophy*. Our preliminary ideas were often, as he intimates, the stimuli and targets of his criticisms. But since we had completed our text before his book was published, we have limited ourselves to adding a few footnotes clarifying our argument. Careful study of his new book does not in the least incline us to retract any substantial points of our interpretation of Frege. We will confront his ideas in another forum.

Contents

Abbreviations xv

PROLEGOMENA

1. Prophetic Glimmerings: The New Pythagorean 3
   1. Method in the history of philosophy 3
   2. The race for the mathematicization of logic 11
   3. Frege: half a philosopher, half a mathematician 16
   4. Hermeneutical principles for the canons 18
   5. Subliming Frege's thought 21
   6. The mirror of *our* mind 25
   7. The project 27

I YOUTHFUL INSIGHTS: THE DOCTRINE OF CONTENT

2. What Frege Inherited: Psychologism and the Theory of Content 33
   1. Conceptual content and presentation of proof 33
   2. The established tradition 37
   3. Antipsychologism: the laws of logic 41
   4. Antipsychologism: the object of judgment 46
   5. Antipsychologism: unjudgeable-contents 49
   6. A Platonist core with a Cartesian penumbra 59

3. What Frege Inherited: Language and Grammar 63
   1. Thinking and language 63
   2. Language, grammar, and logic 67
   3. The legacy of traditional logic: predicate, indicative mood, and the copula 77

4. The Formal Theory of Assertion: *Begriffsschrift*   83
   1. Behind the assertion-sign   83
   2. *Begriffsschrift*   84
   3. Problems of paraphrase   91
   4. Inconsistent demands and unintelligible stipulations   96
   5. Surveying the ruins   102

5. Judgeable-Content   104
   1. The object of assertion   104
   2. The proper objects of logic   107
   3. Frege's elucidations of the logical constants of the propositional calculus   114
   4. The propositional calculus   119
   5. Criteria of identity for judgeable-contents   125
   6. Judgeable-content and sentence-meaning   128

6. Unjudgeable-Content: The Principles   133
   1. The logical decomposition of judgeable-content   133
   2. Function/argument analysis of judgeable-contents   136
   3. Judgeable-contents as the values of functions   145
   4. Alternative analyses of judgeable-contents   154

7. Unjudgeable-Content: The Elements   164
   1. Objects: the contents of proper names   164
   2. Concepts: the contents of concept-words   170
   3. Second-level concepts: the contents of quantifiers   181
   4. Retrospect   190

8. Contextualism   194
   1. The problems of a principle   194
   2. The contextual dictum in the *Foundations*   199
   3. Beneath the *Foundations: Begriffsschrift*   205
   4. Back to the *Foundations*   215

II   THE MATURE VISION: THE DOCTRINE OF FUNCTIONS AND SENSES

9. Function and Concept   233
   1. Propaideutic   233
   2. The problem of genre-identification   236
   3. Starting afresh   241
   4. Logical doubts and difficulties   249
   5. Logic and language   260
   6. Putting the *Basic Laws* on proper *Foundations*   265
   7. New logical machinery   271
   8. Spin-off from the new technology   273

10. Sense and Reference: An Area Survey   278
    1. The rationale   278
    2. Introducing the sense/reference distinction   282
    3. Formal principles of sense and reference   290
    4. Frege's later contextualism   293

11. Sense and Reference: Digging Down   300
    1. Mode of presentation   300
    2. Poisoned springs   314
    3. Thought-constituents: intolerable tensions   322

12. Cracks in the Structure: Assertion, Truth and Thoughts   333
    1. A Fregean philosophical grammar?   333
    2. Assertion   339
    3. Truth   344
    4. Thoughts   353

PARALIPOMENA

13. The Wisdom of Hindsight: Sense, Understanding, and Truth-Conditions   365
    1. The missing keystone   365
    2. The collapse of the vault   373
    3. Understanding and thought-building-blocks   380
    4. Mistaking a reflection for a reality   386
    5. In the end lies a fresh beginning   393

    Index   399

Abbreviations

The following abbreviations are used in the text to refer to Frege's works. Unless otherwise indicated, all references are to numbered pages.

BS *Begriffsschrift, eine der arithmetischen nachgebildete Formelsprache des reinen Denkens* (Halle: L. Nebert, 1879); *Conceptual Notation, a formula language of pure thought modelled upon the formula language of arithmetic*, in *G. Frege, Conceptual Notation and related articles*, tr. and ed. with a biography and introduction by Terrell Ward Bynum (Clarendon Press, Oxford, 1972). 'BS, Preface' indicates reference to Frege's Preface; 'BS §n' indicates reference to numbered sections of *Begriffsschrift;* 'BS n' indicates reference to page number of Bynum's translation.

CN *G. Frege, Conceptual Notation and related articles*, tr. and ed. with a biography and introduction by Terrell Ward Bynum (Clarendon Press, Oxford, 1972). References are primarily to reviews of *Begriffsschrift* printed as Appendix I of this volume.

SJCN 'On the Scientific Justification of a Conceptual Notation', tr. T. W. Bynum, in CN pp. 83–9, from 'Über die wissenschaftliche Berechtigung einer Begriffsschrift', *Zeitschrift für Philosophie und philosophische Kritik*, 81 (1882): 48–56.

ACN 'On the Aim of the "Conceptual Notation"', tr. T. W. Bynum in CN pp. 90–100, from 'Über den Zweck der Begriffsschrift' *Sitzungsberichte der Jenaischen Gesellschaft für Medicin und Naturwissenschaft, Jenaische Zeitschrift fur Naturwissenschaft*, 16 (1882–3): 1–10.

FA *The Foundations of Arithmetic, a logico-mathematical enquiry into the concept of number*, tr. J. L. Austin, 2nd revised edition, (Blackwell, Oxford, 1959); *Die Grundlagen der Arithmetik, eine logisch mathematische Untersuchung über den Begriff der Zahl* (Breslau: W. Koebner, 1884). 'FA' followed by a lower case Roman numeral indicates reference to page of the Preface; otherwise all references are to numbered sections (e.g. FA §1).

BLA *The Basic Laws of Arithmetic: Exposition of the System* tr. and ed. with an introduction by Montgomery Furth (University of California Press, Berkeley and Los Angeles, 1964)—a translation of Introduction and §§1–52 of *Grundgesetze der Arithmetik, Begriffsschriftlich abgeleitet,* Band I (Jena: H. Pohle, 1893). 'BLA p.', followed by a lower case Roman numeral indicates reference to original pagination of the Introduction; otherwise all references are to numbered sections (e.g. BLA §1).

BLA ii *The Basic Laws of Arithmetic,* vol. ii, §§56–67, 86–137, 139–44, 146–7 and Appendix, in *Translations from the Philosophical Writings of Gottlob Frege,* ed. Peter Geach and Max Black (Blackwell, Oxford, 1960), translated from *Grundgesetze der Arithmetik, Begriffsschriftlich abgeleitet,* Band II (Jena: H. Pohle, 1903). All references are to numbered sections.

KS *Kleine Schriften,* ed. Ignacio Angelelli (Hildesheim: G. Olms, 1967).

FG *Gottlob Frege: On the Foundations of Geometry and Formal Theories of Arithmetic,* tr. with an introduction by Eike-Henner W. Kluge (Yale University Press, New Haven and London, 1971).

PW *Posthumous Writings,* ed. H. Hermes, F. Kambartel, F. Kaulbach, tr. P. Long, R. White (Blackwell, Oxford, 1979).

PMC *Philosophical and Mathematical Correspondence,* ed. G. Gabriel, H. Hermes, F. Kambartel, C. Thiel, A. Veraart, abridged for the English edition by B. McGuinness, tr. H. Kaal (Blackwell, Oxford, 1980).

RH 'Review of Dr. E. Husserl's *Philosophy of Arithmetic', Mind* LXXXI (1972): 321–337, tr. E. W. Kluge from *Zeitschrift für Philosophie und Philosophische Kritik,* 103 (1894): 313–32.

T 'The Thought', repr. in *Philosophical Logic* pp. 17–38, ed. P. F. Strawson (Oxford University Press, Oxford, 1967), tr. A. M. and M. Quinton from 'Der Gedanke: Eine logische Untersuchung', *Beiträge zur Philosophie des deutschen Idealismus,* 1 (1918): 58–77.

CT 'Compound Thoughts', repr. in *Essays on Frege,* pp. 537–58, ed. E. D. Klemke (University of Illinois Press, Urbana, Chicago and London, 1968), tr. R. H. Stoothoff from 'Logische Untersuchungen, Dritter Teil: Gedankengefüge', *Beiträge zur Philosophie des deutschen Idealismus,* III (1923): 36–51.

The following abbreviations all refer to papers translated by P. T. Geach or M. Black in their volume *Translations from the Philosophical Writings of Gottlob Frege* (Blackwell, Oxford, 1960).

FC 'Function and Concept', tr. of 'Funktion und Begriff', an address given to the *Jenaische Gesellschaft für Medicin und Naturwissenschaft,* January 9, 1891.

CO 'On Concept and Object', tr. of 'Über Begriff und Gegenstand', *Vierteljahrsschrift für wissenschaftliche Philosophie,* 16 (1892): 192–205.

ABBREVIATIONS

SR   'On Sense and Reference', tr. of 'Über Sinn und Bedeutung', *Zeitschrift für Philosophie und philosophische Kritik,* 100 (1892): 25–50.

SVAL   'A Critical Elucidation of Some Points in E. Schroeder's *Vorlesungen über die Algebra der Logik,* tr. of 'Kritische Beleuchtung einiger Punkte in E. Schröders Vorlesungen über die Algebra der Logik', *Archiv für systematische Philosophie,* 1 (1895): 433–56.

WF   'What is a Function?', tr. of 'Was ist eine Funktion?', *Festschrift* Ludwig Boltzmann *gewidmet zum sechzigsten Geburtstage, 20 Februar 1904* (Leipzig: A. Barth, 1904): 656–66.

N   'Negation', tr. of 'Die Verneinung: Eine logische Untersuchung', *Beiträge zur Philosophie des deutschen Idealismus,* 1 (1918): 143–57.

PROLEGOMENA

1
Prophetic Glimmerings: The New Pythagorean

1. Method in the history of philosophy

In this book we study the ideas of Gottlob Frege (1848–1925). His first masterpiece, *Begriffsschrift, eine der arithmetischen nachgebildete Formelsprache des reinen Denkens* (Conceptual Notation, a Formula Language of Pure Thought Modelled upon the Formula Language of Arithmetic), was published in 1879. In it he developed the first modern system of formal logic. His purpose in so doing was to demonstrate that mathematical induction is a special case of familiar logical forms of inference. In 1884 he published *Die Grundlagen der Arithmetik* (The Foundations of Arithmetic), in which he attempted to show that the concept of number and numbers themselves are definable in purely logical terms. The culmination of his work came with *Die Grundgesetze der Arithmetik (The Basic Laws of Arithmetic)* (vol. i in 1893, vol. ii in 1903) in which he tried to give precise formal demonstration of his logicist thesis that arithmetic is deducible from logic.

Much intellectual change can occur in a century, and the passing of so long a time should suffice to give some tolerable perspective upon Frege's thought. Indeed, it might have been so. But it was the fate of his main philosophical ideas to take root, if at all, only when planted by other hands. The bulk of his work (by no means voluminous) lay relatively unstudied, even unread, until mid-century. Many of his published papers were difficult to obtain, and his correspondence and *Nachlass* were effectively inaccessible. In recent years, however, interest in his philosophy has quickened, and publication of his surviving writings has made his thought available to a wide and receptive audience. He is now commonly viewed as the fountainhead of contemporary philosophy of language, and his ideas (or what seem to be his ideas) are held to be the essential tools for current theory-building in philosophy and linguistics. The fact that Frege's century-old ideas are seemingly also contemporaneous makes historical consciousness especially necessary for the critical analysis of his thought, even though we shall refrain from detailed historical investigations into the intellectual environment in which he worked.

History of philosophy is a first cousin of history of ideas. Like history of ideas, it *is* history. It is concerned with philosophical ideas conceived by individual thinkers at a specific time in response to questions raised in a particular cultural context. Unlike history of ideas, it is essentially, and not accidentally or intermittently, concerned with evaluating the cogency of its objects. The Scylla of the historian of philosophy is oblivion to the fact that his craft is a kind of history; his Charybdis is disregard of the fact that to practise his craft is to engage in philosophy.

Like the contemplation of past works of art, so too reflection upon past philosophical ideas is essentially dynamic, its objects being viewed as from a receding escalator in ever new and broader vistas. Though we know the Impressionists were outrageous revolutionaries in the theory and practice of painting, we can no longer *see* them as outrageous. Because we know what Mozart made of Haydn's novel musical forms, we cannot hear Haydn's works as his audience was meant to, and did, hear them. We cannot return the apples from the Tree of the Knowledge of History. The shifting viewpoint of successive generations renders the object of study essentially inconstant to view. So too with philosophical ideas. We cannot study a past philosophical system without knowledge of its subsequent fate; we cannot examine it without awareness of the ways in which we ourselves conceive of the philosophical problems it deals with. This transient perspective is unavoidable, a necessary feature of time-bound reflection upon the temporal creations of mortal creatures. It is no more a limitation than is the invisibility of the eye in the visual field. But lack of self-consciousness with respect to this necessity breeds anachronism, parochialism, and hermeneutic distortion. Lack of self-discipline, failure constantly to remember that the object of study is not contemporaneous, leads to misinterpretation of ideas, misidentification of errors. and misattributions of insights and illusions alike. An idea is not like a seed—it does not contain within itself the pattern of its future development.

How useful Frege's ideas are for modern philosophical concerns is something which we can only assess once we have identified correctly what his ideas were, what problems he tried to solve, and what success he had. Only then can we examine their fruitfulness for us and, perhaps, cast light upon the authenticity and independent worth of contemporary ideas which go by the name of 'Fregean'. Above all, we must beware of treating Frege as an absent colleague, a contemporary Fellow of Trinity on extended leave of absence. It is the exact converse of the truth to suppose that we understand him 'only in proportion as we can say, clearly and in *contemporary terms* what his problems were, which of them are still problems and what contribution ... [he] ... makes to their solution'.[1] The task of making the ideas of past thinkers intelligible to us is *not* by means of 'translating' them into modern idiom and presenting them in the guise of pre-

---

1. J. Bennett, *Kant's Analytic,* cover blurb on Kant (our italics), (Cambridge Univ. Press, Cambridge, 1966).

mature answers given in the past to questions which could only be asked in the present. For although past systems of philosophy may contain much that we can put to use, may address problems analogous to, or related in various complex ways to, our current philosophical preoccupations, the adoption of such an ahistorical point of view *ab initio* must blind us to Wölfflin's fundamental insight that 'not everything is possible in every period'. It may then, for example, seem a mere accident that Descartes or Leibniz did not solve the problem of formally representing statements involving multiple generality, and that Frege's successful resolution of the problem is to be explained simply and *solely* by reference to his (undisputed) genius.

A text is the product of intentional acts, and must be so viewed. In trying to understand a philosopher's writings our first obligation is to attempt to see the problems he addressed as he saw them. Otherwise we run the risk of reading them as inexplicably fragmentary but yet systematic contributions to late twentieth-century debates. It would, e.g., be misguided to view Frege's invention of his concept-script as an attempt to devise a formalised language which would constitute that twentieth century's philosopher's stone—the underlying (depth) grammar of natural language.[2] Equally misguided would be to conceive of his contextual dictum that a word has a meaning only in the context of a sentence as a contribution to a theory of speech-acts,[3] or as refuted by the observation that we use names vocatively or on labels.[4]

Consequently, when we examine the ideas of a past thinker we must construe his terms of art *as he* understood them, not as we currently do. It is crucial to realise that what Frege meant by 'concept' and 'object' is no more identical with what we mean by these expressions than is what Descartes meant by 'idea' identical with what we mean by it. In his use of the term 'function' Frege certainly departed radically from his contemporaries, but equally certainly, he did not construe the notion in set-theoretic terms as modern logicians do.

Just as we should beware of unconsciously projecting our concepts onto the thought of a past philosopher, so too we should exercise caution in judging that because a conclusion or a methodological principle of his chimes with our beliefs, therefore he held it for our reasons, or indeed that he understood by it what we do. Thus, e.g., Frege espoused antigeneticism in the study of concepts (FA, vif.). But his reason for this had nothing to do with the insight that 'there is no action at a distance in grammar',[5] that what concepts we have, what we understand by concept-words is something visible in what we do and say, and not (or not necessarily) in how we learnt to speak. *This* thoroughly twentieth-century thought

---

2. Cf. M.A.E. Dummett, 'Frege's Distinction between Sense and Reference', in *Truth and Other Enigmas*, p. 118f. (Duckworth, London, 1978).
3. Cf. Dummett, 'Frege's Philosophy', op.cit. p. 95, 'Frege's dictum conveys that the meaning of a word consists wholly in the contribution it makes to a precise determination of the specific linguistic act that may be effected by the utterance of each sentence in which the word may occur.'
4. P. T. Geach, *Reference and Generality*, p. 25f. (Cornell Univ. Press, Ithaca, 1962).
5. Wittgenstein, *Philosophical Grammar*, p. 81, ed. R. Rhees, tr. A.J.P. Kenny (Blackwell, Oxford, 1974), and *The Blue and Brown Books*, p. 14 (Blackwell, Oxford, 1958).

was alien to his cast of mind. *Frege's* reason for antigeneticism was wholly Platonist: concepts are objective existences. 'Often it is only after immense intellectual effort, which may have continued over centuries, that humanity at last succeeds in achieving knowledge of a concept in its pure form, in stripping off the irrelevant accretions which veil it from the eyes of the mind' (FA, vii). How this entity is initially revealed to intellectual vision is no more an account of what it is than Columbus' cartography is an accurate representation of American geography. Given *this* conception, it is surely highly misleading to suggest that 'It was Frege who first perceived ... the irrelevance of genetic questions.'[6]

In order to see a past thinker's problems as he saw them, we must investigate both what he knew, and what he did *not* know. 'Until you understand a writer's ignorance', Coleridge remarked, 'presume yourself ignorant of his understanding.' It is important not to forget the obvious: Frege had not read the *Tractatus* in the 1890s; he did not distinguish between syntax and semantics as later generations were to do (infra p. 373f.); he did not have (and could not have had) a model-theoretic conception of truth-conditions (infra p. 366ff.); he had not come across (reductive) contextual definitions (infra p. 227ff.); he did not 'adopt' bivalence (or the law of excluded middle) as an antithesis to intuitionism and anti-realism, of which he had never heard. Frege was a mathematician by training and vocation. The few philosophy courses he attended as a student at Jena and Göttingen[7] were

6. M.A.E. Dummett, *Frege: Philosophy of Language,* p. 676 (Duckworth. London, 1973); one might anyway have thought that Kant had a little to do with this 'perception', at least in respect of *a priori* concepts! And is Krug, writing shortly after Kant, irrelevant: 'It is not at all necessary to know how thoughts originate in order to find out how they must be treated in their relation to one another' (*System der theoretischen Philosophie,* Part 1, §8 n. 1 (Königsberg, 1806–10))? Bolzano (1837) likewise deserves mention: 'It would be completely wrong to believe ... that we could not ascertain at all whether one of our opinions is true as long as we do not know how it originated' (*Theory of Science* p. 10, ed. and tr. R. George (Blackwell, Oxford, 1972)). Somewhat later (1874) Lotze developed the same point: 'There are two things which I intentionally exclude from my consideration. Firstly, all enquiry into the psychological character of the growth and development of these ... ideas in our consciousness, into the order in which one of them may condition the origin of another, and into the different importance of perception of time and space in their formation ... logic is not concerned with the manner in which the elements utilized by thought come into existence, but with their value, when they have somehow or other come into existence, for the carrying out of intellectual operations.' (*Logic* I-i-§18, trs. and ed. B. Bosanquet (Clarendon Press, Oxford, 1888), cf. also III-iii-§332). Nor were these antipsychologistic stirrings confined to Germany. Jevons (1873) introduced his book with the following remarks: 'It is no part of this work to investigate the nature of the mind. People not uncommonly suppose that logic is a branch of psychology, because reasoning is a mental operation. On the same ground, however, we might argue that all the sciences are branches of psychology. As will be further explained, I adopt the opinion of Mr. Herbert Spencer, that logic is really an objective science, like mathematics or mechanics' (*The Principles of Science—A Treatise on Logic and Scientific Method,* p. 4 (Dover, New York, 1958)).
7. Bynum relates that Frege took a philosophy course under Fischer at Jena, but does not have information on its content. Since he was only there for four semesters, specialising in mathematics, it is unlikely to have been very extensive. At Göttingen (1871–73) he took one course in philosophy with Lotze, *in philosophy of religion.* (Cf. 'On the Life and Work of Gottlob Frege', p. 3f., in *G. Frege, Conceptual Notation and Related Articles,* tr. and ed. with a biography and introduction by T. W. Bynum.)

probably no more than part of the normal university experience of an educated middle-class German in the nineteenth century. His familiarity with the history of philosophy was negligible. Most of what he needed for his purposes, he picked up from an anthology. Hence to present him as consciously or intentionally generating an anti-Cartesian revolution, displacing epistemology from the foundations of philosophy,[8] is probably wrong on two counts. First, it is doubtful whether he knew that Descartes had a hierarchical conception of philosophy in which epistemology is the basis.[9] Second, it is doubtful (and no evidence has been adduced) whether he ever thought that philosophy is hierarchical, let alone that something called 'philosophy of language' is its basis.[10]

Frege did have a smattering of philosophy, and it is important to know what it was, what presuppositions he absorbed unreflectively from an existing philosophical tradition, with the bulk of which he was unfamiliar. We shall explore his tacit acceptance of a Cartesian conception of the mind, his abstractionist conception of perceptual concepts, his naïve Platonism about abstract objects. Direct evidence of positive influences of philosophical writings upon his thought is slender in the extreme,[11] and we shall not pursue that line of causal investigation. Our impres-

---

8. Dummett, *Frege: Philosophy of Language*, p. 666ff. (esp. 669), and 'Can Analytic Philosophy Be Systematic, and Ought It To Be?', in *Truth and Other Enigmas*, p. 441.
9. It is noteworthy that *Descartes* claimed that *metaphysics* was the basis of philosophy, the propaedeutic to which is logic (though not that of the Schools); cf. *The Principles of Philosophy*, in *Philosophical Works*, vol. I, p. 211, tr. and ed. E. S. Haldane and G.R.T. Ross (Dover, New York, 1955).
10. It is curious to find Dummett and Sluga (H. Sluga, *Gottlob Frege*, p. 2 (Routledge and Kegan Paul, London, 1980)) joining hands in contending that Frege was the founder of analytic philosophy, *the characteristic tenet of which is that philosophy of language is the foundation of the rest of philosophy*. If 'analytic philosophy' includes the later Wittgenstein, Ryle, and Austin among its luminaries, if analytic philosophy of law includes Hart or Kelsen, if analytic philosophy of history includes Berlin or Dray, if analytic philosophy of politics includes Nozick or Rawls, then it is *not* a characteristic tenet of the 'school'. On the contrary, it would be denied, both in theory and practice, by all these philosophers.
11. Sluga (op.cit.) makes much of the impact of Kant, Lotze, and Trendelenburg upon Frege. His method consists largely in showing affinities and convergences of superficial thought and phrase. To establish genuine causal connectedness one must show that Frege read these works, understood them in such and such ways, and that he only then *and consequently* reached such and such conclusions. Genuine and fundamental convergence of ideas must be demonstrated. It is not enough, e.g., to quote Lotze's remark that 'calculation also belongs to the logical operations' and that 'mathematics really has its proper home in logic' (H. Lotze, *Logic*, I-iii-§111) to show that Frege owes to him the idea of logicism. Did Frege read this part of the book? Is Lotze's idea really a form of logicism? (Did he think that mathematical reasoning was expressible in syllogistic or Boolean algebra? Or that numbers are definable in terms of some purely logical, non-mathematical, concept? If so, which one?) Is Lotze's trivial remark so idiosyncratic, unique, and original that Frege (a mathematician) could not have thought of it by himself, or could not have come across it elsewhere. (In fact Frege was fully aware that the primitive idea of logicism was not novel, and he refers to both Leibniz and Jevons (FA, §15) although not to Lotze. The difficulty is precisely that the idea is indeed primitive until one has at least a programme for the reductionist enterprise.) Similarly, to cite Lotze's little mathematical analogy (*Logic*, I-i-§28) for the structure of concepts (viz. that '$S = a + b + c$' does not capture the *various* ways in which *Merkmale* (characteristic marks)

sion is that what influences there are upon what is genuinely original in Frege's thought comes *largely* via reflection on the work of mathematicians and logical algebraicists.

Not only is it important to have some grasp of what a past thinker did and did not know; it is equally vital correctly to characterize both his critical targets and his positive goals. Frege's primary targets were psychologism and formalism. Two points should be noted in this matter. First, he envisaged his naïve Platonism as the only serious contender in the field once the other two adversaries were demolished. Hence Platonism, in his view, barely needed positive support. Second, his criticisms of psychologism and formalism focus almost exclusively upon those versions of the doctrines that are concerned with the *references* of expressions, not with their senses. This is particularly important with respect to Frege's criticism of psychologism (infra p. 53ff.) Unless we are clear about what he was combating, we will be prone to misconstrue his contentions; e.g., we might take his remarks on equipollence (PMC, 70f.) as an attempt to adumbrate criteria of identity for thoughts (infra p. 354ff.), or his animus towards implicit definitions as directed against contextual, paraphrastic definitions (infra p. 229).

Frege's avowed primary goal was to substantiate the logicist thesis that arithmetic is part of pure logic. Everything else he did was peripheral. Consequently *he* viewed what we judge to be his greatest achievement, i.e. his invention of concept-script, as altogether instrumental. Hence, after the collapse of his logicist programme, he took absolutely no interest in the extensive developments in formal logic which took place in the last twenty years of his life. His philosophical writings, in particular the three famous papers of the 1890s which are sometimes viewed as the foundation stones of modern philosophy of language, were wholly subordinate to his main goal. There is no evidence that his successive attempts to write a philosophical book on logic were conceived by him as a contribution to constructing a theory of meaning for a natural language, and much evidence to suggest the contrary.

Our task as *historians* of philosophy must be, as far as we are able, to render Frege intelligible to ourselves in *his* terms. We should keep our expectations under control. It is *not necessary* that a great thinker in the past should have an answer to every major question that currently preoccupies us. We must beware of converting incidental remarks into the tip of a worked-out doctrinal iceberg (e.g. Frege's footnote on proper names of people in natural language). We must be careful in extrapolating what he would have said upon subjects on which he (unaccountably!) said nothing.[12] For this may blind us to the importance of the writer's

---

    are capable of being co-ordinated in a complex concept, hence '$S = F(a,b,c,...)$' would better represent this) as the origin of Frege's function-theoretic logical system is surely far-fetched. Did a mathematician, whose speciality was function theory, need a half-baked metaphor to put him on the track of function-theoretic analysis of equations, inequations, and judgments in general?

12. Dummett argues that 'Frege himself did not make the claim that the only task of philosophy is the analysis of thought, and hence of language ... ; but by his practice in the one particular branch of philosophy in which he worked, the philosophy of mathematics, he left little doubt that that was his view' ('Can Analytic Philosophy be Systematic and Ought It To Be?' op.cit.

failure to say anything, a fact which may, like Sherlock Holmes' dog's *not* barking in the night, be of the utmost significance. Likewise it may encourage our disposition to crown our own theories with the garlands of dead heroes.

Not only must we pay careful attention to 'inexplicable' lacunae, we must also exercise care in the identification and correction of 'mistakes'. For in our excessive haste to correct what seem to us to be blunders we may, while intending to shore up the system we are scrutinizing, inadvertently undermine its foundations. An even greater danger is that our corrections may effect a fundamental transformation of the system under examination so that a new specimen is slipped under the microscope of critical assessment while the observations continue to be ascribed to the original specimen. It is commonly agreed that Frege's assimilation of sentences to proper names was a disastrous mistake. But if for him this was an essential concomitant of his application of function-theory to the analysis of judgments, then to correct this 'blunder' is to invent a novel system of logic radically different from his. Similarly, his insistence that concept-words have concepts as their reference is often thought to be unnecessary, if not indeed incomprehensible. Yet if, from his viewpoint, it was essential for his logicist project, by brushing it aside we guarantee distortion in our understanding of his thought. All agree that it is a blunder to think, as Frege did, that cogency of argument depends on the premises being both true and asserted to be true. But before rushing to correct the blunder, we should probe its grounds, for it gives us important clues against misguidedly attributing to him a modern semantic conception of validity, and may ultimately bring us to realize that he did *not* think that logic was primarily and directly an analysis of language at all.

Finally, we must beware of painting a sharply edged contour line to represent something intrinsically fuzzy. Our analysis may bring to light sharp alternatives and clear-cut options, and we may then look in vain for answers and decisions between them. This may be because the philosopher whose work we are examining did not think the matter through that far, did not see the options which we, standing on his shoulders, can perceive clearly. We shall discover that Frege was unaware of the need to choose between a compositional conception of a thought (proposition) and function/argument analysis of it; that he did not see the need to clarify the conditions of identity and difference of proof for his account of both content and sense; that he never seriously examined the question of the relation between truth and provability (and hence never gave serious thought to the possibility of verification-transcendent truth-values); that he never reflected on the relation between the obtaining of a functional relation and the requirement of *calculability* of the values of a function for its arguments. In all such cases, our preparatory sketches must be sharp and clearly defined, but the contour lines of our final representation of Frege's thought must be *sfumato*.

---

p. 442). By parity of reasoning it was his view that correct method in Ethics or Aesthetics is rigorous formalization and axiomatization, that legal philosophy ought to engage in giving canonical definitions (subject to the requirement of 'completeness') rather than describing the deliberately 'open textured' concepts characteristic of much of law (cf. H.L.A. Hart, *Concept of Law*, ch. 7 (Clarendon Press, Oxford, 1961)), and so on.

Unlike the historian of ideas, who may or may not be concerned with evaluating the ideas he recounts,[13] the historian of philosophy cannot avoid doing philosophy. It is surely an exaggeration to suggest that 'there simply are no perennial problems in philosophy: there are only individual answers to individual questions ... There is in consequence no hope of seeking ... [to] ... attempt to learn directly from the classic authors by focussing on their attempted answers to supposedly timeless questions.'[14] There may be no *perennial* questions; but then eternity is an unnecessarily long time for a historian of thought to worry about. There may be difficulties in determining identity and difference of questions; yet surely *some* of Descartes', Locke's, or Kant's questions are *very* similar to *some* of ours. It is true that classical writers were concerned with *their* questions; but it is not vain to search for similarity of questions even over a period of some centuries within a tradition, nor futile to hope to glean ideas, insights and illuminations which we may put to use. If Kant, for example, could engage in a dialogue with the Cartesians (and brilliantly refute the rationalist theory of the soul in the 'Paralogisms') then surely we can both learn from him, and engage him (cautiously) in argument. It is not impossible to identify mistaken reasoning, nor unthinkable that we might profit from knowing the *culs de sac* encountered by our predecessors. Modern philosophy differs profoundly in conception and method from the speculative activity of a Descartes or Locke, but it is its legitimate heir, not a random mutant.

Our primary question as historians of *philosophy* is not so much whether the doctrines of a past philosopher are true, but rather the prior question of whether they make sense. A philosopher, crawling along the boundaries of sense with a tape measure, does not enjoy the latitude available to a scientist. For the philosopher there is often no space between 'getting it wrong' and talking nonsense. If it is conceptual articulations that are under investigation, then their misdescription or misidentification involves traversing the limits of sense, uttering nonsense, not asserting false theories. We shall have to investigate a host of bizarre claims made by Frege: that sense *determines* reference; that concepts are functions; that thoughts refer to truth-values; etc. What these strange claims mean, whether they make any sense, will be our first (although not always last) concern.

In an elaborate philosophical construction such as Frege's, different elements are co-ordinated and articulated. An important philosophical task of the historian of philosophy is to examine the internal consistency of the system he is studying. Indeed, maybe one ought to approach a philosophical system in the *expectation* of discovering major inconsistencies. Of which great system-builder in philosophy can it be said that his thought is self-consistent? What seems to its progenitor a coherent structure may transpire to be an amalgam of incompatible constituents.

---

13. For example, a history of the idea of the sublime need not, perhaps indeed could not, evaluate it in terms of truth and falsity, correctness or incorrectness; a history of the ideas informing knight-errantry or courtly love need hardly concern itself with veracity of the ideas.
14. Q. Skinner, 'Meaning and Understanding in the History of Ideas', p. 50, in *History and Theory*, VIII (1969), an article to which we are indebted. Skinner does not here distinguish history of ideas from history of philosophy; our concern is primarily with the latter.

A crucial issue we shall have to examine is, for example, whether Frege's pair of twin ideas, function and concept as well as sense and reference, do actually cohere (as he understood them).

The primary contentions made by a philosopher must be analysed in terms of the nature and adequacy of their supporting argument. It matters little whether Copernicus hit upon the heliocentric hypothesis because he thought the sun to be a symbol of the Godhead. A scientist may hit upon the truth for the wrong reason. It is less obvious whether something analogous to this ever occurs in philosophy. Is the *content* of a philosophical thesis independent of what counts as support for it? Or is it rather that in philosophy the 'proof' shows *what* is proved? If Frege's claim that concepts are functions is held to be a profound insight, our *first* task is to examine the grounds for his contention—for what he means by it is only made clear by his supporting argument and unarticulated presuppositions.

Ultimately, as philosophers, our interest in the ideas of a past thinker must focus upon their correctness, upon the extent to which he clarified and successfully answered the questions he posed, upon the extent to which his questions are both intelligible and important, upon whether he did make perspicuous any conceptual articulations hitherto opaque and problematic. If we find that his fundamental claims, presented as insights into the true nature of things, are in fact tacit recommendations to adopt a different form of representation, we must discover what light is shed by this new creation both upon the nature of the phenomena represented and upon the structure of our own form of representation.

Of course, pursuing these philosophical goals is not truly separable from the *historical* task of accurately representing the philosophical thoughts of a past figure. It is rather the complementary aspect of the Janus-faced activity of doing history of philosophy. To achieve both is, perhaps, for the historian of philosophy, akin to an Ideal of Reason.

2. The race for the mathematicization of logic

Frege's complete formalisation of the first-order predicate calculus with identity is undoubtedly the greatest advance ever made in the subject. Yet it would be more than merely improbable to suggest that his inventions were 'born from Frege's brain unfertilized by external influences'.[15] No matter how great the originality, a seedbed must be prepared. To suggest that he 'wrote as if the world was young and the subject had just been invented'[16] is precisely to imply that it was a mere accident or deficiency of genius that prevented Plato or Aristotle from inventing function-theoretic analysis in logic.

Frege was a trained mathematician with little knowledge of philosophy. He was, however, familiar with work in mathematical logic which was being developed in the mid-nineteenth century, in particular with the Boolean logical algebra

---

15. Dummett, *Frege: Philosophy of Language*, p. xvii; cf. p. 661: 'Frege's ideas appear to have no ancestry'—presumably the same is true of Peirce or Dedekind (infra).
16. Dummett, op.cit. p. 661.

elaborated by Schröder. Similarly, throughout his mature years he kept in touch with work on the foundations of mathematics which was being done by such eminent contemporaries as Peano, Dedekind, Husserl, Cantor, and Russell. Prior to his work on *Begriffsschrift,* Frege's work was in mathematics, not philosophy, but his concerns displayed a theoretical or philosophical interest in the nature of his subject. Towards the end of his life he himself said, 'I started out from mathematics. The most pressing need, it seemed to me, was to provide this science with a better foundation. . . . The logical imperfections of language stood in the way of such investigations. I tried to overcome these obstacles with my concept-script. In this way I was led from mathematics to logic' (PW, 253). His doctoral dissertation 'On a Geometrical Representation of Imaginary Figures in a Plane' defended a Kantian conception of geometry as founded in the faculty of intuition against the objection that the geometry of imaginary figures often seems to contradict intuition. His *Habilitationsschrift* (Methods of Calculation Based upon an Amplification of the Concept of Magnitude) reflected his interest in the theory of complex functions, which he had studied with Abbe in Jena and Schering in Göttingen. In it he argued (against the Kantian conception of the syntheticity of arithmetic) that the laws of arithmetic follow from the general concept of magnitude and are independent of intuition (KS, 50f.) Themes and strategies from both these works are visible in his subsequent theory of content.

The second half of the nineteenth century was ripe for a revolution in logic, a revolution which had to come from mathematics.[17] In mid-century Boole invented a novel logical algebra, an idea quickly taken up by Jevons,[18] Venn, Huntington, and Peirce in England and America, and by Schröder in Germany. In his *Mathematical Analysis of Logic,* Boole anticipated the truth-tabular analysis of what we now call 'logical connectives', and Peirce, in 1880, anticipated Sheffer in suggesting that all Boole's 'elective functions' could be expressed by a primitive sign

17. Already in 1858 De Morgan wrote presciently, 'As joint attention to logic and mathematics increases, a logic will grow up among the mathematicians . . . This *mathematical* logic—so called *quasi lucus a non nimis lucendo*—will commend itself to the educated world by showing an actual representation of their form of thought—a representation the truth of which they recognize—instead of a mutilated and onesided fragment . . .' (Augustus De Morgan, 'On the Syllogism: III', p. 78 n., in *On the Syllogism and Other Logical Writings,* ed. P. Heath (Routledge and Kegan Paul, London, 1966)).
18. In *The Principles of Science* (1873), Jevons wrote 'we owe to profound mathematicians, such as John Herschel, Whewell, De Morgan, or Boole, the regeneration of logic in the present century. I entertain no doubt that it is in maintaining a close alliance with quantitative reasoning that we must look for further progress in our comprehension of qualitative inference' (p. 153). As Descartes had wedded algebra and geometry, Jevons remarked, so did Boole accomplish the marriage of logic and algebra (p. 635). In a burst of enthusiasm he declared 'it will probably be allowed that Boole discovered the true and general form of logic, and put the science substantially into the form which it must hold for evermore. He thus effected a reform with which there is hardly anything comparable in the history of logic between his time and the remote age of Aristotle' (p. 113). Jevons' enthusiasm was premature, since Boole's arranged marriage came to a divorce in 1879, when logic married algebra's younger and more alluring sister, function theory. But Boole's mathematicization of logic was a decisive step in the history of the subject, and was widely perceived to be so.

meaning 'neither ... nor ...'. McColl, in 1877, suggested a calculus of propositions in which the asserted principles would be implications rather than equations. In papers written between 1883 and 1885 Peirce, working in complete ignorance of Frege's labours at Jena, independently invented quantification theory. Much influenced by De Morgan and Boole, he developed a logic of relations on the basis of variable-binding quantifiers. He dropped the Boolean idea that all logical formulae must take the form of equations and inequations. His symbolism is held to be adequate for the whole of logic and identical in syntax with systems now used.[19] Dedekind, also working independently of Frege, produced in 1888 a demonstration of the validity of the principle of mathematical induction by a definition of number which has reference to all properties hereditary in a series. It was, however, Frege who made the first decisive breakthrough on a broad front in 1879 with *Begriffsschrift,* in which he applied function theory to logic, invented quantification theory, defined an ancestral of a relation, and demonstrated the reducibility of mathematical induction to laws of logic. Magnificent achievement this certainly was, but it is misleading to suggest that 'he was incapable of sailing any sea on which other ships were in sight'.[20] He may have been sailing on uncharted seas, and he was the first to sight land; but the seas were cluttered up with the ships of a few dozen other mathematicians.

This was no coincidence. The inspiration informing all these revolutionary exploratory investigations in mathematical logic was the idea of advancing the logical analysis of rigorous reasoning by applying to it techniques derived from mathematics. In particular application to the logical analysis of mathematical reasoning itself, this involved an *involution* of mathematical techniques—an application to mathematical proofs of a generalized form of a branch of abstract mathematics. Boole sought to clarify and generalize syllogistic reasoning in terms of algebraic operations on sets. Logic he regarded as the algebra of thought,[21] and he characterized abstract algebra as a 'cross-section' of rational thinking. Frege followed a similar strategy, but appealed to a more *avant-garde* branch of mathematics. He subsumed syllogistic reasoning within a logical system displaying all sound patterns of reasoning as theorems derived from a few function-theoretic axioms.

Boole's generalization of algebraic techniques was only intelligible by reference to the increasing abstraction of algebra from the entities to which it is applied, a process which took place in the nineteenth century in the hands of Peacock, Greg-

---

19. W. and M. Kneale, *The Development of Logic,* p. 431 (Clarendon Press, Oxford, 1962); see C. S. Peirce, *Collected Papers,* vol. III, p. 351ff., 393ff., 498 (Harvard Univ. Press, Cambridge, 1960).
20. Dummett, *Frege: Philosophy of Language,* p. 661.
21. G. Boole, *The Laws of Thought,* chs. I, XXII (Dover, New York, 1958). De Morgan, in 1860, remarked: 'When the ideas thrown out by Mr. Boole shall have borne their full fruit, algebra, though only founded on ideas of number in the first instance, will appear like a sectional model of the whole form of thought. Its forms, considered apart from their matter, will be seen to contain all the forms of thought in general,' in 'Logic', p. 255, repr. in *On the Syllogism and Other Logical Writings,* ed. P. Heath.

ory, De Morgan, and Sir William Rowan Hamilton. As mathematicians gradually introduced new mathematical entities, such as complex numbers and then Hamilton's hyper-complex numbers (quaternions), they were subjected to the pressures of 'the principle of the permanence of equivalent forms',[22] i.e. the desirability of retaining as much as possible of the general rules of algebra for each novel kind of number introduced. This necessitated thinking about algebra in a wholly new way. The consequent generalization and abstraction of algebra was a prerequisite within mathematics for Boole's invention of *logical* algebra, i.e. an algebra of entities which are not numbers at all.

Very similar considerations apply to Frege's invention of a function-theoretic logic. Where Boole had generalized algebra, Frege generalized the theory of functions. The precondition of this development was the growth of function theory in the first half of the nineteenth century. Mathematicians had greatly extended the application of the concept of a function by abstracting from the various features originally clustered together in the notion of a function of a real variable, and they were busy elaborating mathematical theories on these new foundations. The primitive idea of a function covered only laws of correlation of real numbers; such laws were stated by analytical formulae (e.g. '$y = x^2 + 3x + 1$'), they could be given graphical representations as smooth curves, and they indicated procedures for carrying out computations of the values of functions for arbitrary arguments. Mathematicians had gradually eased these constraints. Most notably, admissible arguments and values of functions were extended to include complex numbers (giving rise to complex analysis) and functions of real variables (giving rise to higher analysis, i.e. the study of functions of second- and higher-level). In parallel with this liberalization went further forms of emancipation. The possibility of graphical representations, even of real-valued functions, was undermined by the construction of functions with paradoxical geometrical properties (e.g. functions everywhere dense but nowhere continuous, or curves everywhere continuous but nowhere differentiable), and hence analysis was freed from any need of support from geometrical intuition. The association of functions with analytical formulae and hence with computation procedures was also severed. Indeed, one of the major landmarks of nineteenth-century mathematics is the generalization of the concept of a function accompanying Dirichlet's specification of the characteristic function on the rationals (1838) viz. the function ϕ defined by the stipulation:

$\phi(x) = 1$ if x is rational
$\phi(x) = 0$ if x is irrational.

Without this extensive mathematical development away from the traditional theory of real variables towards abstract function theory, it would have been inconceivable to erect a logical system on the idea of analyzing generalizations as the values of second-level functions whose arguments are functions which themselves

---

22. G. Peacock, 'Report on the Recent Progress and Present State of Certain Branches of Analysis', *Report of the Third Meeting of the British Association for the Advancement of Science, held in 1833*, esp. pp. 198–207.

take persons, houses, heaps of beans, etc., as arguments. There is no doubt that the key to Frege's achievement lay in his extension of function theory to logic.[23] But others were hot on the scent too. Nor is this surprising, for once the very idea of the mathematicization of logic was mooted, it was only natural to experiment with various branches of mathematics. And, as it transpired, higher analysis proved to be more fruitful than abstract algebra. Peirce in effect produced a doctrine of functions with a notation adequate for expressing all Frege's principles,[24] although he never systematized his ideas nor set out his basic principles as Frege did. Russell, in the *Principles*[25] (1903), also working independently of Frege, argued that a proposition decomposes into subject and propositional-function: the proposition is viewed as a function of the subject, the propositional-function corresponds to a sentence with a real variable, i.e. a 'dependent' entity. Of course, Frege had gone much farther than this; Russell did not realize the logical potential of introducing propositional-functions of the second-level. Yet it suffices to suggest that the extension of function-theory to logic was immanent. None of these observations are intended to belittle Frege's greatness or originality *qua* formal logician, but only to blow away the wisps of fantasy obscuring our vision of the lesser peaks in the mountain range, lest we think Mont Blanc a solitary giant in an alluvial plain. The advances Frege made were in the air in the late nineteenth century. Mathematics was both ripe for an onslaught upon logic and apparently desperately in need of a logical imprimatur. If Frege had not made the deci-

---

23. In spite of Frege's explicit declarations on this point (e.g. BS, Preface), this is not always acknowledged. Dummett isolates something different as Frege's fundamental insight, the idea of a stepwise construction of sentences incorporating logical operators (*Frege: Philosophy of Language*, p. 9ff.). Frege then devised a new notation (quantifiers and variables) 'the point [of which] was to enable the constructional history of any sentence to be determined unambiguously' (op. cit. p. 12). *If* this were a sound interpretation, then his basic idea would have been a matter of sheer inspiration—a bolt from the blue that might equally have fallen on Aristotle, Philo, Descartes, Locke, Kant, etc., and the function-theoretic notation of concept-script would be a coincidental concomitant of his basic insight. (The fact that the notion of stepwise construction has a strong appeal for us now in the syntactic and semantic analysis of language probably rests on a shift in intellectual climate between Frege's acme and the present day. It seems connected with fascination with chains of iterable operations which are the backbone of computer programming.)

    Sluga offers a similar misinterpretation: the rebuttal of Schröder's criticism of *Begriffsschrift* makes clear Frege's view that 'through the analysis of a *sentence* he has reached the notions of *function and argument* and through them in turn he has discovered ... a satisfactory account of general propositions' (H. Sluga, *Gottlob Frege* (Routledge & Kegan Paul, London, 1980), p. 93 (our italics)). As if a mathematician depended on a new vision about language to arrive at the concept of a function!

    The truth about quantification theory is not as depicted by these commentators. Their joint error is the supposition that 'thinking about *language* was essential for any real improvement in logic' (Sluga, op.cit. p. 95 (our italics)). As Frege stressed, the key to his theory is a radical generalization of function theory (FC, 21; BLA, §0)—and that inspiration has no direct connection with semantic analysis of declarative sentences.
24. Cf. W. and M. Kneale, op. cit. p. 511.
25. Russell, *The Principles of Mathematics* (2nd ed.), pp. 13, 19f., 85, 88 (Allen and Unwin, London, 1937).

sive breakthrough in 1879, others would have made it along the same lines within his lifetime (and nobody had been in a position to do so significantly earlier).

3. Frege: Half a philosopher, half a mathematician

Some of the central claims of Frege's mature theory are either singularly opaque or utterly bizarre, e.g. that concepts are functions from curious items called 'objects' to unknown entities called 'truth-values', the latter being kinds of objects; that objects further include such items as numbers, directions, differences, simultaneity of events, etc.; that sentences are proper names of the values of functions for arguments; that predicates name concepts; that the concept of a horse is not a concept but an object. To the uninitiated this galaxy of paradox and obscurity seems an impenetrable forest of shadow. The initiated are trained in the Mysteries on the correct supposition that faith comes with works. Either way it is not easy to find the correct and fruitful questions with which to confront Frege's theories, nor the best ways to press them home.

To someone accustomed to philosophical thought some of the most fundamental of Frege's claims seem to be supported by only the flimsiest of argument. Thus his contention that concepts are functions from objects (arguments) to truth-values appears to rest on nothing more than the claim that an expression such as ' $= 1$', if completed by '1^2' on the left, yields a true equation, whereas if completed by '2^2', yields a false one (FC, 28). Similarly, quite fantastical conclusions are propounded, e.g. that we *cannot* refer to concepts by the standard means which we all use so to refer, namely the expression 'the concept [of a] . . .'. (Logic will apparently seize us by the throat!) This conclusion rests on arguments which are themselves part of Frege's speculative theory, but instead of taking it as a definitive refutation of his analysis, he boldly declared that this just shows that we have a defective language in which it *looks as if* we can do something which actually we cannot! One might as well argue that our defective eyesight is to blame for the fact that it looks as if Antipodeans walk the right way up, whereas in fact they walk upside down.

Part of the explanation of this phenomenon, which we shall encounter recurrently in this study, lies in Frege's characteristically mathematical cast of mind. He was first and foremost a mathematician, thinking in a mathematical style. When an insight is achieved, his immediate instinct is towards maximal generalization. If no problems are encountered as a consequence, the generalization is acceptable. If no *contradictions* ensue, then nothing is awry. No further supporting argument is necessary; rather, the onus of proof lies with the *refutation* of a potentially fruitful generalization. Moreover, the only recognized problem is *formal*. That we would not say that a sentence is a proper name, a concept a function, a number a set, or truth an object, is as irrelevant as the fact that eighteenth-century mathematicians would not have said that the characteristic function on the rationals is a function or that the operation of differentiation is a second-level function. The general thrust of Frege's thought is determined by dogmatic adherence to Platonism. What is or is not a number is no more determined by how we use the word 'number' than what is or is not an atom is determined by how Democritus or Lucretius used the word 'atom'. What a function is is something to be *discovered*, discovered by *a priori* investigation of a broadly mathematical kind. Frege himself conceived of his introduction of the two objects the True and the False as a discovery of the same moment as that of two new chemical elements (PW, 194). Concepts are not creatures of the mind, but rather are as objective as numbers. If, in the process of generalizing an insight, contradictions are encountered, his response is typically that of a mathematician. The only options he recognized are restricting the generalization (as in his modification to Axiom V in response to Russell's paradox) or sidestepping by redefining the concept in question (as in his response to Kerry's objections to his use of 'concept').

Not only is the style of Frege's thought that of a mathematician, but also the impetus behind his thought is overtly both mathematical and formal. It is formal problems which provide the impetus for his theorizing. *Begriffsschrift* is concerned with demonstrating that mathematical induction is a special case of a purely logical law of inference. The *Foundations* is not motivated by investigating what we mean by number words, but rather by the goal of finding a purely logical definition of number. (That no one has 'consciously' meant *that* by 'number' is, from his viewpoint, irrelevant; as is the fact that our actual explanations of 'number' and of numerals take wholly different forms.) The *Basic Laws* requires discovery of a purely logical operation connecting concepts with their extensions in order to give a definitive proof of a solution to a formal problem. Its target is to show that number theory can be derived formally from pure logic.

Formal solutions are assailable *only* by formal refutation or formal improvement (BLA, pp. vi f., xxvi). Hence, Frege was overwhelmed by Russell's paradox, and presumably would have been much impressed by Gödel's incompleteness theorem. On the other hand, he would undoubtedly have dismissed (as intrusions of psychology into logic) philosophical qualms about the intelligibility of either attributing or denying knowledge of his definition of number to everyone who has mastered the use of, and hence understands, number words; or about whether ancient Greeks who counted and computed were operating his calculus, the axioms and definitions of which they did not know. He would have been altogether contemptuous of the suggestion that even if *(per impossibile)* his logicist derivation of arithmetic from logic were successfully executed, it would not show that logic provides the foundations of mathematics, but only that a new bridge has been built between two autonomous, hitherto unconnected, calculi. Such philosophical scruples are alien to his cast of mind.

Finally, his basic inspiration is mathematical. At the heart of his logic and philosophy of mathematics is a generalization of the mathematical concept of a function. From this all the rest follows; and, of course, if this is philosophically awry, all the rest does not follow! Though Frege's thought seems intricate and opaque, it in effect consists of variations on a theme, of successive terms in a formal series generated by the operation of function/argument analysis. Somebody not extensively trained in mathematics could not have thought his thoughts.

Whether or not it should be counted a virtue in a philosopher (cf. PW, 273), Frege himself was more than half a mathematician.

4. Hermeneutical principles for the canons

We must try to approach Frege's writing with fresh eyes, to forget how his ideas have been exploited in the century since the publication of *Begriffsschrift*. In particular, we must beware of reading back into his work subsequent notions[26] which, although perhaps stimulated by his insights, are nevertheless inconsistent with his thought. Our disposition to do so is a measure of our inability to read the texts save through twentieth-century spectacles. There is here a real problem of genre identification. If we read 'Function and Concept' or 'On Sense and Reference' as essays in the philosophy of language may we not be like the theatre-goer guffawing through a performance of *Othello* while under the impression that he was viewing a comedy? We can only guard ourselves against such risks by *seeing* what Frege said, not what he is typically 'known' to have said. We shall examine numerous questionable interpretations later, and these not on the periphery of his theory but at the centre. If, for example, he exposed the old incoherent notion of a grammatical predicate (or copula) as including assertoric force,[27] then why should he feel it necessary, in his concept-script, to 'dissociate assertoric force from the predicate' (PW, 184f., 198) and to speak of assertoric force in natural language as being 'bound up with the predicate' (PW, 252; cf. 129, 194, 198)? If he really 'invent[ed] the idea that the meaning of a sentence is to be identified with its truth-conditions',[28] then how could he think that '$2^2 = 4$' and '$2 + 2 = 4$' differ in sense (BLA, §2) even though both are true under the same conditions,

---

26. If function theory furnished the framework for his revolution in formal logic, and if logic is the centre of gravity of his philosophical reflections, then the anachronism most pregnant with dangers of misinterpreting Frege's thought is the substitution of the sophisticated modern conception of a function for the conception that Frege evolved from the relatively naïve notion prevalent among even the most advanced mathematicians of the 1870s. It has been our policy to consider his extension of function/argument analysis to logic exclusively in terms of reasoning about functions that would have been intelligible to him. The aim is not to mock his commitment to ideas that now seem simple-minded and indefensible, but rather to understand the integration and development of his logical system. By modern standards our account of functions is naïve, not inadvertently, but by design.
27. P. T. Geach, 'Frege', in G.E.M. Anscombe and P. T. Geach, *Three Philosophers*, p. 133 (Blackwell, Oxford, 1967). Geach connects the distinction of predication from assertion with the thesis that a thought may occur unasserted as the constituent of a molecular thought (*Logic Matters*, p. 265f. (Blackwell, Oxford, 1972)), calling the latter 'the Frege point'. This appellation seems inappropriate since that point is stressed in earlier standard works on logic (Port-Royal *Logic*, Part II, ch. ix; Mill, *System of Logic*, II-v-1; Sir William Hamilton, *Lectures on Metaphysics and Logic*, vol. III, p. 238, ed. H. L. Mansel and J. Veitch (Blackwood, Edinburgh and London, 1860); and Boole, *The Laws of Thought*, ch. XI §12). If what is distinctively attributed to Frege is instead the distinction of predication from assertion, then there is reason to doubt that Frege ever made this point clearly and consistently.
28. B. Harrison, *An Introduction to the Philosophy of Language*, p. 64f. (Macmillan, London, 1979).

viz. come what may? If he introduced the concept of sense in order to resolve the question of how it is possible to *understand* a *contingent* identity statement without knowing its truth-value,[29] then how is it that he did not so much as mention understanding in his discussion, showed no puzzlement at the possibility of understanding other kinds of statement without also knowing their reference (truth-value), and never spoke of identities as being contingent? If he only assimilated sentences to proper names of the values of functions (concepts) for arguments in the *Basic Laws*,[30] then how can it be that in *Begriffsschrift* he viewed the judgeable-content for which a sentence stands as the value of a function for an argument (BS, §10f.), and treated these entities as objects? Finally, if the dictum that a word has a meaning only in the context of a sentence really signified that we understand sentences by calculating their senses from the senses of their constituents and their mode of combinations,[31] then why did he forget to mention this until his correspondence with Jourdain in 1914, by which time he allegedly[32] had relinquished whatever primacy he had accorded to the sentence by the use of the dictum?

We shall take Frege's works seriously, not as revelation, but as philosophical argument. A mathematician moving onto philosophical terrain must surely be judged by the same philosophical canons of rigour which we would apply to philosophers ploughing similar fields. Frege accused academic logicians of standing at the gate he had opened like bewildered oxen, gaping and bellowing, but refusing stupidly to enter (PW, 186). Let us, by all means, pass through his gate and see whether what is growing in the field he cultivated is grass or lucerne.

Above all, we must avoid double standards. If we take Russell to task for saying that one cannot assert what is false,[33] we must note that Frege drifts towards similar confusion.[34] If we criticize Russell for muddles over 'assertion in a logical sense',[35] we must scrutinize Frege's remarks with equal severity.[36] If we castigate Russell for hopeless confusion of mention and use in his explanation of 'propositional function',[37] then we must similarly criticize Frege's early explanation of what a concept (function) is.[38] Many ideas that evoke contemptuous laughter in

---

29. Central State Materialists typically take Frege's example of the identity of the Morning Star with the Evening Star as a paradigm of a contingent identity.
30. As suggested by Geach (following Dummett), in 'Critical Notice of *Frege: Philosophy of Language*', Mind, LXXXV (1976), p. 438.
31. Dummett, *Frege: Philosophy of Language*, p. 4.
32. Ibid. pp. 7, 196.
33. Russell: 'It is plain that true and false propositions alike are entities of a kind, but that true propositions have a quality not belonging to false ones, a quality which, in a non-psychological sense, may be called being *asserted*'. The Principles of Mathematics, p. 35.
34. Cf. PMC, 79.
35. *Russell*, op. cit. p. 48.
36. Cf. BLA, §5; FC, 34.
37. Russell, The Principles of Mathematics, pp. 13, 19f.
38. BS, §9, Rather than praise him for having discovered the idea of a 'linguistic function' and for advocating that a linguistic function is what symbolises a numerical function or a concept (Geach, Three Philosophers, p. 143f.; infra p. 172n.).

the context of Russell's writings have exact counterparts in Frege's which are brushed under the carpet in embarrassed silence. Both responses are foolish. Errors must be acknowledged and explored, for they provide valuable clues to a philosopher's overall vision and to the structure of his thought.

Philosophers who study Frege often make light of his sundry mistakes or blunders on the grounds that in each case the general principle informing his work is correct and of deep importance. Frege's remarks on the relation between 'and' and 'but' may be flawed, but his general distinction between sense and tone is surely right! He may have slipped a little in denying that imperatives or optatives express thoughts, nevertheless his overall distinction between sense and force is surely of the utmost importance for semantic theory! Admittedly, in moments of abandon he spoke of detaching the assertoric force from the predicate, but his distinction of assertion from predication is a profound insight! Curiously enough, he thought that sentences with identical truth-conditions *but distinct proof-conditions* differ in sense, yet he is the inventor of realist semantics (as against anti-realism)! In all these cases, as we shall see, the alleged general principle cited to exonerate him from deep confusion is actually inconsistent with fundamental explicit features of Frege's thought, and in many cases the principle itself is indefensible.

Like a scientist or mathematician, Frege rarely saw any need to re-examine the foundations of his thought. He was an exemplary system-builder, and if one is ever to erect a worthy edifice one cannot constantly be digging up the foundations. This is strikingly evident from his *Nachlass,* in which it is clear that he typically re-used old manuscripts in composing new essays, often over a period spanning a decade or more.[39] This reworking of old manuscripts is not merely a literary curiosity, but is occasionally important for correct interpretation of the texts.[40] The structure that Frege erected took twenty-five years to build. His remarks about the progress of his work, his comments on alterations and modifications of design and instruments alike, are most revealing. He was remarkably self-conscious about his own development, and we should surely take his word seriously (if not, perhaps, definitively), since he is more likely to have been well-informed than those living a century later. Hence, if, in the introduction to the *Basic Laws,* where he charts his progress over the previous nine years, he does not mention abandoning contextualism, we should hesitate to ascribe to him so radical but unnoticed

39. Compare, for example, the first fragment on Logic (PW, 1–8), written sometime between 1879 and 1891, with the 1897 essay (PW, 126–51), in particular the following pages: 4/135, 5/137, 7 and 8/138, 5 and 6/142, 7/146, 5/148, 8/149. Note also the germinal idea of a generative theory of understanding in 'Logic and Mathematics' (1914) in PW, 225, and the Jourdain letter apparently of the same year (PMC, 79), which is then re-used, almost *verbatim,* in 'Compound Thoughts' in 1923 (CT, 537).
40. For example, the paragraph entitled 'Dissociating assertoric force from the predicate' (PW, 198) is apparently a reworking of PW, 185. The final sentence seems to refer to the penultimate one, and to make little sense. This appearance vanishes once one realises that two sentences about fiction have been interpolated, and hence that the reference is to the preceding remark.

a change. Similarly, if in his correspondence with Jourdain, where he corrects errors in his early works and explains changes, he nevertheless still insists on the primacy of the judgment (as in *Begriffsschrift*), we would need very weighty countervailing evidence to demonstrate abandonment of this point.

It is generally agreed that Frege's work is remarkably linear. It is the progressive working out of a single issue (namely, the reducibility of arithmetic to logic) from 1879 to 1903, followed, at the moment at which the keystone was being placed in the final arch, by effective collapse. It is important to render this singularity of vision and linearity of development surveyable. If we are to grasp Frege's thought we must obtain a clear picture of its foundations in *Begriffsschrift* (a task almost uniformly neglected by writers on his theory), and of how each successive phase is connected with its predecessor. In particular we must understand what internal pressures forced changes and innovations to the growing edifice.

Mythologizing Frege is one thing, subliming his thought is another. His system, seen through a glass pinkly, is advanced as a scientific theory, but defended as a scientific paradigm. Confrontation with the 'facts' (e.g. that we *do* refer to concepts by means of singular referring expressions; that truth and falsehood are not entities of any kind; that we do not standardly explain expressions by *Merkmal*-definitions; that concepts are not functions) is typically treated as irrelevant to the core of the theory. Either the 'facts' are reinterpreted, or the error is admitted— as a matter of detail which does not damage the basic structure. Thus, for example, it is generally accepted that sentences do not have a reference, that concept-words do not name concepts, that a concept is, if anything, the sense of a concept-word. Nevertheless, this critical consensus is not thought seriously to diminish the importance of the sense/reference distinction. Similarly, it is altogether unclear how a theory distinguishing quite generally between sense and force (hence not *Frege's*) can cope with performatives, deontic (declarative) sentences, sentences containing normative modalities, dominatives, etc. But these are just matters of detail, powerless to undermine the importance of the fundamental distinction! In this way the defence shifts from Frege's product to his instruments, from his instruments to what is held to be his programme, and from his programme back to those parts of the product, duly tidied up, which apparently vindicate the programme (while quietly disregarding the parts which do not). The phenomenon is strikingly similar to that encountered when criticising Marxist theories of history or Freudian theories of psycho-analysis. This is not surprising.

5. Subliming Frege's thought

Philosophers can be arranged along a spectrum at one end of which lie the great myth-makers, the weavers of dreams and spinners of illusion, and at the other end of which lie the shatterers of such visions, the great critical philosophers. Amongst the former are such figures as Plato and Descartes, amongst the latter Aristotle and the later Wittgenstein.[41] Frege, for more reasons than one, lies at the visionary, myth-making end of the spectrum. He is the effective[42] originator of the New Pythagoreanism—a revival of that vision according to which reality is 'made of numbers'.

Focussing on contour lines only, we can view Frege as having moulded three deeply influential ideas. First, the categories of mathematics (in particular, of higher analysis) constitute the ultimate categories of reality. Reality consists of objects and first-level functions, of first- and second-level concepts, and so on, in an ascending function-theoretic hierarchy. Second, the ultimate categories of thought are mathematical. It is not only the realm of nature which is subject to mathematicization, but also the realm of thought. What we think, and not only what we typically think about, has essentially a mathematical structure of function and argument. It is not only nature that is confined within the web of the computable, but also human thought. Indeed, the limits of possible thoughts are drawn by the type-theoretic restrictions of function-theory. Third, human language, unbeknownst to us, really has a function/argument structure. Imperfect instrument though it is, and occasionally even incoherent by reference to Frege's canons, its sentences are composed of coordinated argument-expressions and function-names which (somewhat imperfectly) mirror the functional structure of reality on the one hand, and of thoughts (Platonistically conceived as existing independently of us) on the other.

This is a breathtaking Pythagorean vision, bold and ambitious. How much more impressive it is when it transpires that Frege not only dreamt this dream, but also apparently actualized it. He invented (or discovered?) a calculus of *pure thought* whose basic form is function-theoretic. This calculus appears to constitute a structure capable of expressing human thought in all its richness.[43] Such a conception

---

41. Wittgenstein actually occupies a unique position in this respect. He had the innate temperament of a myth-maker, and produced an exemplary visionary work in his twenties. He then spent the rest of his life embarked upon the critical analysis of illusion. Just as Tolstoy was a fox driven to be a hedgehog, so Wittgenstein was a hedgehog driven to be a fox (cf. I. Berlin, 'The Hedgehog and the Fox', repr. in *Russian Thinkers,* p. 24ff. (Hogarth Press, London, 1978)).
42. In all fairness we should mention Boole's earlier remarks (1853): 'The truth that the ultimate laws of thought are mathematical in their form ... establishes a ground for some remarkable conclusions. If we directed our attention to the scientific truth alone, we might be led to infer an almost exact parallelism between the intellectual operations and the movements of external nature.' (*The Laws of Thought,* ch. XXII §6) and 'The laws of thought, in all its processes of conception and of reasoning, in all those operations of which language is the expression or instrument, are of the same kind as are the laws of the acknowledged processes of Mathematics' (ch. XXII §11).
43. Dummett, in 'Frege's Distinction between Sense and Reference' (*Truth and Other Enigmas,* pp. 116–44) remarks (p. 119): 'At least this much is not seriously challenged by anyone: that a language having the sort of syntax that Frege devised for his formalised language is adequate for the formulation of any theory of mathematics or natural science' (Could we really translate *Principia, The Origin of the Species, The Circulation of the Blood,* into Frege's concept-script without distortion?) and 'It is rather generally supposed that we shall arrive at a satisfactory syntactic analysis of natural language only by exhibiting its sentences as having an underlying (or deep) structure analogous to that of sentences of Frege's formalised language' (p. 118).

carries in its wake a host of dramatic implications. Language is, beneath its appearances, a calculus of hidden rules. According to these rules certain combinations of words make sense, represent thoughts, and others are nonsense. Vast new territories open up for philosophers (and linguists) to discover *novel* hidden forms of thought.[44] A whole galaxy of hitherto undreamt of logical forms await discovery, forms guaranteed by the canons of abstract function theory. Further, the key to the understanding of human understanding has been discovered. Speaking and understanding a language is a matter of operating a calculus. Inferring is at last recognised for what it is, namely, the operation of a handful of hitherto hidden rules. Quantification theory is really behind syllogistic, and underlies inferences with adverbs, modal operators, etc.

From these heady heights modern philosophers are prone to take ever more daring and exciting steps to peaks undreamt of by Frege. He had next to nothing to say about understanding. He never used the key he discovered to open the door to the nature of linguistic competence and creative thought.[45] As far as he was concerned the calculus he discovered mirrored reality and objective thoughts alike. But how we 'know' it, how we operate a calculus whose depth-rules are hidden from us—these are questions of which he never thought. Seen from the perspective of contemporary linguistic theorists in philosophy, Frege toiled to the top of Pisgah and was vouchsafed the vision of all the land of Gilead unto Dan, but never fully realised what he had seen. The ultimate destination which his vision enables us to reach is not the insights into logic which he revealed, but the understanding of the nature of the mind. With an unconscious irony which Hegel might have appreciated, contemporary philosophers and linguists *psychologize* Frege's calculus (or one resembling it), in the sense that according to their theories we actually know and operate it (albeit tacitly). On some views a calculus of *a priori,* indeed innate, rules *must,* in some way or other, be encoded in the brain. Otherwise, it is claimed, we would not be able to learn a language; or, according to different views, we would not be able to know (as we allegedly do) without learning, the depth-grammar of all human languages (as opposed to their superficial forms).

Since function theory, in the hands of a naïvely Platonist mathematician, is essentially concerned with laws of correlation of *entities* (mathematical objects, numbers) Frege's ideas gave a new lease of life to that most venerable and dulcet of philosophical sirens, the Augustinian picture of language. This primitive schema treats the individual words of language as names, and sentences as combinations of names. On this view, each word has a meaning which is the entity for

---

44. The contrast between (superficial) grammatical form and (real) logical form was familiar from syllogistic. Frege's innovation was not in drawing this distinction, nor in the breadth of the gulf separating grammatical from logical form (often his predecessors had purported to find far wider divergences than he did), but in the new function-theoretic forms which he 'discovered'.
45. True enough, towards the end of his life he raised the question of how it is possible to understand sentences we have never heard before (CT, 537) and in a brief paragraph sketched an answer along lines currently pursued by generative grammarians. However, this sketch actually conflicts with the thrust of Frege's theory (infra p. 383ff.) and hence must be taken with a pinch of salt.

which it stands, so that it has a meaning in virtue of its being correlated with some entity.[46] In barest outline this *Urbild* seems impossibly artless. But refined and developed into a 'scientific' theory of meaning, it assumes a new dignity. Frege first dressed it in the exotic livery of mathematical function theory. Later by 'splitting' conceptual content into sense and reference, thereby introducing a tier of sense to mediate between a sign and what it stands for, he masked any vestige of crudity or naïveté. His complex theory of sense and reference and Carnap's convoluted method of intensions and extensions are in all appearance anything but unsophisticated.

Frege not only generated this intoxicating vision, he seemingly produced the tools with which to build the theoretical structures which would realise it. Function/argument analysis provides the key to the hitherto impenetrable structure of statements involving generality, as well as providing the basic forms for the logical analysis of sentences in general. The sense/force distinction is held to be a *sine qua non* of *any* systematic theory of meaning. The categories of sense and reference provide the framework for most contemporary philosophical theorizing about language, and the notion of truth-conditional semantics, characteristic of much modern speculation, is held to be Frege's invention. The blueprint is complete, the foundation is laid, and the tools are at hand. All that is necessary is that we set to work, and complete the edifice. Philosophy, at last set upon the true path of a science by logic, has entered into an era of progress:

> there seemed to pertain to logic a peculiar depth—a universal significance. Logic lay, it seemed, at the bottom of all the sciences.[47]

And Frege seemed to lie at the bottom of logic!

This kind of vision, which goes far beyond the historical Frege, is in effect a philosophical *Weltanschauung*. It is not a discovery of a feature of reality, an empirical, scientific advance, contingent, and refutable by countervailing experience. What, after all, would its supporters accept as refuting the contention that a language has a functional structure? Is a language without such a structure (at least in its 'depth-grammar') conceivable? Is it a contingent fact that generality is expressible by second-level functions? Is it an empirical truth that concepts are functions? Is it not rather that those mesmerised by this vision confuse the *suggestion* that every sentence *can be* represented as having a functional structure with the *claim* that every sentence *has* a functional structure? Or, more clearly, they conflate the idea that every sentence *can somehow be projected* into a function/argument notation with the supposition that every sentence *is* composed of function-name and argument-expression.

The sublimed Fregean vision is a form of representation which is presented by its proponents as a discovery about the true nature of what is represented (whether

---

46. See L. Wittgenstein, *Philosophical Investigations*, §§ 1ff. (Blackwell, Oxford, 1953); cf. G. P. Baker and P.M.S. Hacker, *Wittgenstein: Understanding and Meaning*, p. 33ff. (Blackwell, Oxford, 1980).
47. Wittgenstein, *Philosophical Investigations*, §89.

language or reality in general, or both). According to Frege himself, his fundamental distinctions are founded 'deep in the nature of things' (FC, 41). The structure of natural language reflects imperfectly, and that of concept-script reflects perfectly, the objective forms of reality. This *metaphysical* conception is now out of fashion. Neo-Fregeans, on the other hand, sometimes conceive of the functional structure of language as attributable to the nature of the mind, perhaps even to the structure of the brain.[48]

In adopting a novel form of representation and presenting it as a discovery about the nature of what is represented, Frege's philosophy resembles the great metaphysical systems of the past. To cleave to such a form of representation is akin to an obsession. To bring someone to relinquish it, or to see it for what it is, is not effected by a 'refutation'. The function-theoretic conception of language might be said 'to be a superstition (*not* a mistake), itself produced by grammatical illusions'.[49] And indeed,

> the corruption of philosophy by superstition ... is ... widely spread, and does the greatest harm ...
>
> Of this kind we have among the Greeks a striking example in Pythagoras ..., another in Plato and his school ... Upon this point greatest caution should be used. For nothing is so mischievous as the apotheosis of error; and it is a very plague of the understanding for vanity to become the object of veneration.[50]

6. The mirror of *our* mind

Interest in Frege's writings quickened after 1950, and the 'Fregean vision' (as distinct from Frege's) materialised in increasingly clearer and richer forms in the 1960s and '70s. Yet philosophy had flourished happily without him for half a century. What strange movements of thought in the *Zeitgeist* suddenly transformed this obscure German logician into the Messianic figure of the alleged rebirth of philosophy, and transmuted his *recherché* doctrines into the apparent foundations of a new science? The answer to this question perhaps reveals more about us than about him. When we peer into Frege's writings we all too often miss his thought because we are diverted by apparent reflections of our own.

Twentieth-century advances in science, in particular in molecular biology and neuropsychology, have seemed to many philosophers to imply a strong reductionist thesis. To some it has suggested the culmination of the Cartesian (and anti-Aristotelian) dream of the unity of the sciences. Since Galileo the mathematicization of physics had appeared to provide the key for unlocking the secrets of nature. Late twentieth-century developments have powerfully suggested that the process would ultimately extend to all the 'human sciences'. This picture exercises

---

48. To this extent psychologism might yet win the day! Its allure is more powerful, because more elemental, than Frege imagined.
49. Wittgenstein, *Philosophical Investigations*, §110 (in a different context).
50. Bacon, *Novum Organum*, §lxv.

all the fascination of monism, coupled with the self-righteousness of the hard-headed 'realism' of the typical empiricist. It has the charm of destroying apparent prejudices, in this case alleged prejudices about humanity and its autonomy, of man as not wholly part of the order of nature, genuinely possessed of creativity and spontaneity.

Some form of reductionism in philosophy of mind meshes, to mutual advantage, with the currently popular computer analogy for understanding the operations of the brain. Where our ancestors pictured themselves in the image of God, we picture ourselves in the image of our machines. Indeed, there is here a double irony. The gods were conceived as creating us in their image, whereas we created them in ours. We now invent machines to do automatically what we do thoughtfully, and conceive of ourselves on the model of our automata.

The computer analogy in turn articulates smoothly with mind/brain identity theories, which are themselves often thought to be given support by discoveries in neurophysiology. The combination of these factors renders attractive the Fregean vision. For contemporary conceptions of a theory of meaning for a natural language typically represent language as a calculus, and depict understanding and speaking a language as operating this tacitly known calculus. What better mechanism for operating such a calculus of meaning-rules than the powerful biological computer we have between our ears?

This picture has the perverse charm of increasing the gulf between appearance and reality, a kind of charm to which philosophers are as susceptible[51] at the high intellectual plane as is the man on the Clapham omnibus at the level of Erich von Däniken. Things indeed are not as they seem to be! Language is *really* a calculus, proceeding according to intricate, adamantine, rules of which we have no (explicit) knowledge. These rules, independently of us, determine what makes sense and what does not. Unbeknownst to us, we operate this calculus of rules constantly in our daily lives. This has the enormous appeal of the uncanny—the revelation of mysteries at the heart of the mundane. It is a picture which is given further support by the supposition that the construction of a theory of meaning is *au fond* a scientific task, for is science not the last word in intellectual respectability?

Frege had no global views about the nature of philosophy. He knew virtually nothing about metaphysics, epistemology, philosophy of mind, ethics, or aesthetics. If the effect of his work has been to generate a misconception[52] of philosophy as a hierarchy of theories with philosophy of language as its foundation, at any rate this is not something he ever dreamt of. Nevertheless in the sole area where he had a passionate philosophical interest, viz. the foundations of mathematics, he had (unsurprisingly) a thoroughly scientific, cognitive, view of the subject. In

---

51. Cf. Russell: 'the point of philosophy is to start with something so simple as not to seem worth saying, and to end with something so paradoxical that no one will believe it', in 'The Philosophy of Logical Atomism', p. 193 repr. in *Logic and Knowledge,* ed. R. C. Marsh (Allen and Unwin, London, 1956).
52. Cf. Baker and Hacker, *Wittgenstein: Understanding and Meaning,* pp. 457ff. and 531ff.

his view a fundamental philosophical thesis (that arithmetic is part of logic) is capable of being definitively, formally, proved. The contention that truths of arithmetic are analytic can be rigorously demonstrated. (Whether *he* would have extrapolated this conception to other branches of philosophy is debatable.)

The extension of this picture to philosophy in general fits the temper of our times, with its adulation of science coupled with loss of confidence in humanity. The scientific conception of philosophy is, of course, no novelty. It was vigorously propounded by Russell in the first quarter of the century, but was in eclipse in the 1950s and 1960s, so much so that Ryle casually observed in 1957, 'It comes natural to us now—as it did not 30 years ago—to differentiate logic from science . . . ; it comes natural to us not to class philosophers as scientists or *a fortiori* as super-scientists'.[53] Now, unhappily, it is once more in ascendancy. Philosophy is commonly conceived to be an extension of science; it can, it is thought, produce answers of a non-trivial kind to substantive problems; its methods are scientific, for it constructs theories and postulates entities just like the advanced physical sciences. This trend fits well with contemporary philosophers' preoccupation with 'philosophical technology', i.e. the idea that formalization provides the key to the mysteries of philosophy, which in turn lends the subject an aura of 'scientific' respectability. We are all too prone today to transform natural science into a kind of sinking fund for philosophy when over impulsive Reason runs into debt.

7. The project

Fairly to represent a mythologized figure is no easy task. We must navigate between the reefs of caricaturing Frege in order thereby to capture the interest which the sublimed Fregean vision has for our contemporaries and the whirlpools of the philistinism (real or apparent) of focussing on trivial historical detail (Frege's unimportant slips, as it were) while averting our eyes from the grandiose vision he revealed. We shall try to avoid a topicality which lacks historical accuracy, as well as an accuracy which lacks interest. Consequently we shall focus sharply on the general principles informing his thought, disregarding the detail. If the former are misguided, the latter is irrelevant. Moreover, much of the supposed interesting detail is in fact imputed, and therefore depends wholly on the interpretation of underlying principles.

Our primary task is the analysis of Frege's thought. We aim to uncover its mainsprings and to illuminate its articulated components. By bringing to light the fundamental principles upon which it is constructed, we hope to make clear how this complex machinery is meant to work and what it was designed to do. Here exegetical accuracy and a sense of historical possibilities for thought are most important; we are concerned with a nineteenth-century mathematical logician, not with a late twentieth-century philosopher of language.

Frege's thought is one thing, the developed Fregean vision is another. The for-

---

53. G. Ryle, 'Review of Ludwig Wittgenstein: *Remarks on the Foundations of Mathematics*', in *Collected Papers*, Vol. 1, p. 265 (Hutchinson, London, 1971).

mer is precise enough, available in definitive form in his writings, even though surrounded with obscuring clouds of questionable interpretation. The latter is nebulous and programmatic, containing an excess of promise over performance. Its distinctions are not Frege's, but transmutations of his ideas. This firmly entrenched and widely influential cluster of doctrines stands in the way of a clear vision of his own thought. In trying to obtain such a clear viewpoint we shall frequently have to fight our way through this thicket of Neo-Fregean doctrine. It aims (unlike Frege) to produce a theory of meaning for a natural language in the belief (which he never had) of thereby rendering an account of the nature of human understanding. Its presupposition is the correctness of a calculus model of language. Its conception of the mind is that of a rule-programmed device (either neural or mental). Its tools are function/argument analysis, *a* distinction between sense and reference (which is not obviously Frege's), the separation of force from sense, truth-conditional semantics, and the principle that the sense of a sentence is derived from the senses of its constituents and their mode of composition. If we succeed in establishing its difference from Frege's own ideas, then a demonstration of its deep incoherence is an independent enterprise best deferred for a subsequent volume. Here we aim to engage Frege's doctrines in argument, not their misconceived illegitimate offspring. In doing so, we shall clarify those of his ideas which are the putative foundations of these contemporary doctrines, and we shall show that they fail to support the modern structures, either because they are not in fact in the location in which they are commonly thought to be, or because they are too weak to carry any weight. To the extent that there is a divergence, the harsh criticisms which must be rained down upon this modern set of doctrines ought by no means always to be directed at Frege. If we are removing his halo, at least we are returning his hat, which has the merit of keeping off the rain. After all, is it any less charitable to scale his theses down to his explicit goals and his stated supporting arguments than to inflate them so that a vast gulf opens between his imputed pretensions and actual achievements?

The greatness of a philosopher is not to be evaluated only by reference to the *truths* he delivered. This principle applies with special force to the great mythmakers (e.g. Plato or Descartes). Their significance lies rather in the power, clarity, and sophistication of the intellectual myth they articulate (e.g. Platonism, Cartesian dualism, Fregean Pythagoreanism), given that that myth is a perspicuous expression of a deep erroneous disposition of human thought (akin to what Kant called 'Ideas of Reason'). Once clearly articulated, such Ideas of Reason are exposed for critical evaluation. In the careful scrutiny of them we can discover what *moves* and *captivates* us. In discovering why we are so moved, we increase our own self-awareness. In disclosing what it is that captivates us, we free our thought from shackles.

We shall approach Frege as we would approach Plato or Descartes. Once having unravelled his arguments, our task is to evaluate them before the tribunal of reason. It might be said that we are restoring to Frege's statue in the philosophical pantheon its feet of clay; so doing will, at any rate, prevent spectators from thinking that a miraculous levitation is under observation. It is from pointing out his mistakes that we can learn (as from Descartes' or Plato's). It is important, and *topical,* to understand the origin, inspiration, and mainspring of the function-theoretic analysis of thought and language. Only thus can we grasp its strengths, weaknesses, and significance. It is vital, and *currently relevant,* to understand the rationale for distinguishing the assertoric force of a sentence (symbolized by a special assertion-sign in concept-script) from the object or content of assertion, in order to shed light upon the foundations of contemporary theories of meaning which claim a Fregean imprimatur. For it is upon a generalized form of such a distinction that they erect their structures. It is crucial and *urgent* to understand what Frege meant by sense and its relation to reference, in order to evaluate the claim that these are the definitive concepts in terms of which to resolve the main problems of philosophy of language.

Our project has a certain resemblance to Frege's own treatment of formalism in the philosophy of mathematics. He argued that the defects of the formalist conception of numbers become apparent only if one sticks strictly to the formalists' account and refrains from importing alien ideas actually inconsistent with formalism in defence of an allegedly formalist conception. Hence he treated the contention that numbers are signs *au pied de la lettre* and exposed its incredible consequences, whereas Thomae, who advanced this thesis, seldom reasoned in accord with it and was 'miffed when somebody else does' (FG, 123). Frege rightly persisted in his critical examination of the writings of Thomae and other formalists. We shall do likewise with respect to Frege. Our purpose is to respond to his plea for this treatment:

> I should like to hope not merely for my own sake, but also for that of science, that as many as possible try to understand my writings, my trains of thought, with equal thoroughness and with the same desire to be fair to them. (FG, 121)

We shall take him at his word.

The super-structure of Neo-Fregean thought is altogether different from its allegedly Fregean foundations. That difference is the measure of the gulf separating the historical Frege from the mythologized figure. Accordingly, our first task is to dig down through the accumulated deposits of a century of thought to reveal the fundamental layer of Frege's philosophy. We must put aside modern truth-conditional semantics, modern conceptions of formal logic, modern philosophy of language, and go back to Frege's beginning. To give an historically accurate account of his thought we must pore over *Begriffsschrift,* try to tease out from its brief introductory matter the fundamental inspiration informing his theories. Here in the theory of judgeable-content, and not in the papers of the 1890s, lies the key to his thought. Once firmly grasped, we shall use it, in Part II of this book. to unlock the secrets of his mature theory of sense and reference.

I
YOUTHFUL INSIGHTS: THE DOCTRINE OF CONTENT

2
What Frege Inherited: Psychologism and the Theory of Content

1. Conceptual content and presentation of proof

Nineteenth-century mathematics, like late nineteenth-century physics, underwent extensive and radical changes. On the one hand, doubt was thrown on the reliability of spatial intuition as a source of knowledge in real analysis, especially by the discovery of well-defined functions having paradoxical combinations of properties and thus defying representation in graphs. This led to a thorough reconstruction of analysis by such figures as Cauchy and Weierstrass. On the other hand, the development of the theory of complex numbers and the evolution of analytic geometry gave rise to proofs of theorems of Euclidean geometry in the form of calculations on complex numbers. This suggested the possibility of divorcing geometry altogether from any reliance on spatial intuition.[1] One result of this mathematical ferment was that established mathematical proofs were subjected to unprecedented critical scrutiny. Another related result was that a host of seemingly stable and secure mathematical concepts were given new and more exact definitions and were extended in unprecedented ways. A very general consequence of this destabilization of received concepts and undermining of established proofs was that the general philosophical question about the true sources of mathematical knowledge became urgent and unavoidable.

Frege's initial concern was to ascertain what are the logical foundations of arithmetic, in particular to settle the question whether complex analysis is inde-

---

1. Frege himself explored some aspects of the reduction of geometry to analysis. He claimed, for example, that 'every theorem which states a relation between lengths ... follows from the fundamental principles of analytic geometry and can be derived algebraically from the formula $r = \sqrt{(x_1 - x_0)^2 + (y_1 - y_0)^2}$ by such operations and steps as are applicable equally to complex numbers' (KS, 8).

pendent of intuition *(Anschauung)*.[2] For this purpose he considered that he had sharply to circumscribe and formalize the sound proof procedures employed by mathematicians in order to determine beyond any possible doubt the totality of premises on which the proof of any theorem of arithmetic actually depends. In his view, 'the firmest method of proof is obviously the purely logical one', since it is topic-neutral and 'based solely upon the laws on which all knowledge rests' (BS, Preface). Consequently he had to specify the laws of logic and to lay down principles of correct inference. This was his fundamental task in *Begriffsschrift*. In particular, he sought to ascertain the status of the principle of reasoning known as mathematical induction. This states that from the fact that 0 has a property F together with a demonstration that $n + 1$ has the property F if n has this property, we may correctly infer that all integers have the property F. This principle of reasoning appears to be peculiar to arithmetic, and its justification seems to turn on direct insight into the nature of the natural numbers. Therefore, to establish the independence of arithmetic from any special intuition requires providing an alternative justification for mathematical induction. The triumph of *Begriffsschrift* is the formal proof that the principle of mathematical induction can be derived from completely general laws of logic and hence independently of any considerations peculiar to natural numbers.

Begriffsschrift does not resolve the crucial epistemological question of the grounds for arithmetical knowledge. It is, however, a *sine qua non* of settling that issue. For, on Frege's view, any truths which can be established without appeal to anything empirical or *a priori* intuitable must themselves be analytic, i.e. deducible from definitions and logical laws alone (FA, §3). But determining whether all truths of arithmetic are analytic presupposes setting up a systematic method of presenting proofs which will ensure that everything necessary and nothing superfluous to a valid proof will be presented. It will then be evident whether arithmetic stands in need, at crucial points in proofs, of intuition. Proofs couched in natural language are irremediably defective for this purpose.

> A strictly defined group of modes of inference is simply not present in [ordinary] language, so that on the basis of linguistic form we cannot distinguish between a 'gapless' advance [in the argument] and an omission of connecting links. We can even say that the former almost never occurs in [ordinary] language, that it runs against the feel of language because it would involve an insufferable prolixity. (SJCN, 85)

Yet the procedures of mathematicians are hardly better. One can only be sure that there is no appeal to intuition, no employment of non-logical inference rules

> by producing a chain of deductions with no link missing, such that no step in it is taken which does not conform to some one of a small number of principles of inference recognized as purely logical. To this day, scarcely one single proof has ever been conducted on these lines; the mathematician rests content if every

---

2. His persuasion that arithmetic is independent of intuition predated his attempts to demonstrate this thesis (cf. KS, 50f.).

transition to a fresh judgment is self-evidently correct, without enquiring into the nature of this self-evidence, whether it is logical or intuitive. (FA, §90)

To remedy this defect, Frege invented his concept-script. It was designed to give a perspicuous representation of inferences, to ensure that no tacit presuppositions remain hidden, 'to be operated like a calculus by means of a small number of standard moves, so that no step is permitted which does not conform to the rules which are laid down once and for all' (FA, §91).

From Frege's point of view, the achievement of *Begriffsschrift* was twofold, technical and substantial. On the technical side, it was the first system of formal logic with the power to represent typical forms of mathematical reasoning; in particular, it formalized statements involving multiple generality. This made it a fit instrument for validation of mathematical proofs[3] as well as for the precise definition of mathematical concepts (e.g. continuity (PW, 23f.)), and it would, he hoped, enable mathematics to be erected on firm foundations. The substantive achievement was the definition of the ancestral of a relation and the consequent proof that mathematical induction is not a uniquely mathematical form of reasoning (with the status of an intuited non-logical axiom) but is rather a derived rule of inference within general logic. This encouraged him in the belief that 'Thought is in essentials the same everywhere: it is not true that there are different kinds of laws of thought to suit the different kinds of objects thought about' (FA, iii). Having proved that arithmetical reasoning employs only general laws of logic, he turned, in the *Foundations of Arithmetic,* to try to prove that the truths of arithmetic can be reduced to laws of logic by appeal to analyses of arithmetical concepts in terms of logical notions alone.

The heart of *Begriffsschrift* is then the elaboration of a notation for presenting inferences and the setting up of a formal system for rigorously testing their cogency. Frege designed his concept-script for this purpose. Negatively this calls for a principle of purity: he forswore expressing in concept-script anything 'which is without importance for the chain of inference' (BS, Preface). In this respect, he held that his notation differs greatly from natural languages. His negative principle has a positive correlate: concept-script is to give full expression to everything necessary for correct inference, thus leaving nothing to guesswork in logic (BS, §3). What his notation expresses he called 'conceptual content', and hence he declared 'I have designated by *conceptual content* that which is of sole importance for me' (BS, Preface). It is an immediate consequence of these leading remarks that an examination of the notion of conceptual content is central to the understanding of *Begriffsschrift,* and further reflection will reveal it to be no less central for interpreting the *Foundations.* Hence our primary task in the philosophical clarification of the earlier phase of Frege's thinking is the investigation and elucidation of his notion of content.

Conceptual content bifurcates into *judgeable*-content and *unjudgeable*-content.

---

3. He also hoped, at least initially, that it would prove useful in presenting proofs in kinematics, mechanics, and physics (cf. BS, Preface).

Judgeable-content is the more important notion of this pair. Logic is, on Frege's view, the science of the relations among judgeable-contents. Unjudgeable-contents are the product of splitting up judgeable-contents, and their importance in logic depends wholly on their influence on the relations among judgeable-contents of which they are components. Frege evidently supposed the distinction between judgeable- and unjudgeable-contents to be transparent (cf. BS, §2). He also assumed that very little elucidation was required to give his readers a correct understanding of the expression 'judgeable-content' (cf. PW, 185). Logicians commonly grasped roughly the same concept under such labels as 'proposition', 'thought', 'judgment', and even 'content'.[4] It was only necessary to guard against certain prevalent misunderstandings. In particular, Frege emphasized that a judgeable-content is not a mental entity and that the content of a judgment or assertion (a judgeable-content) is distinct from the acts of making a judgment and making an assertion. Hence it seems advisable to begin our investigation of conceptual content by scrutinizing the relations between judging and judgeable-content.

A judgeable-content is *what* is judged or asserted in making a judgment or assertion. But the specification of what is asserted is to be limited strictly to what is relevant to the validity of the inferences that may be drawn from an assertion. Frege assigned judgeable-contents to the sentences or utterances used to formulate assertions. He took it to be obvious that different sentences may be used to express the same judgeable-content if what is asserted in uttering them is the same (e.g. 'A hit B' and 'B was hit by A'). Clearly too, a sentence may express more than its judgeable-content, for it may suggest, intimate, presuppose, etc., a host of features concerning the speaker's or hearers' attitudes, expectations, presumed knowledge, etc., which do not affect the validity of inferences. Finally a type-sentence, isolated from a context of utterance, may evidently fail to express a (complete) judgeable-content: if it contains indexical expressions (e.g. 'He did it to him for her yesterday'), what is asserted by its utterance must be gathered from the context of its use. These obvious points together indicate the necessity of distinguishing the science of judgeable-contents from the study of sentences used to make assertions. Frege indeed held that a judgeable-content is *toto mundo* distinct from a sentence or symbol. It is an abstract object, whereas a sentence is a *perceptible* entity used as a *vehicle* for conveying a judgment; a judgeable-content is an *imperceptible* entity which is the *object* of acts of judgment; it is what is thought, judged, asserted; also what is true or false. It is thus Janus-faced, having a psychological and a logical aspect. Inasmuch as it is thought, judged, believed, or disbelieved it is connected with psychological acts and processes of human beings. Inasmuch as it is true or false it is engaged with logic. For logic is concerned with the laws of truth, i.e. the laws governing the relations between such contents, relations spelt out in the rules of valid inference.

---

4. The notion of content, though unfamiliar to the English philosophical tradition, was not invented by Frege; it was employed, for example, by Hegel, Lotze, and Schröder.

2. The established tradition

Although what we think about may be perceptible, concrete, mental, or abstract, what we think is, in Frege's view, never anything other than abstract. It is a judgeable-content, an abstract entity, which exists independently of our judging it. The sharp distinction between the act of judging and the object or content of the act, between thinking and what is thought, asserting and what is asserted, was of the utmost importance to Frege in order to insulate logic, which is concerned with contents of judgment,[5] from psychology. This 'purification' of the subject-matter of the science of valid inference was, he thought, a prerequisite for any progress in constructing an adequate logical theory.

It was commonplace to open a treatise on logic by circumscribing the subject-matter as laws of inference, thought, judgment, or assertion. Mill, for example, argued that 'Logic ... comprises the science of reasoning ... [T]o reason is simply to infer any *assertion* from *assertions* already admitted.'[6] Such introductory statements were typically soon qualified by the claim that logic is not a descriptive science describing the ways men make inferences, but rather constitutes a canon of principles of how rational beings ought to reason in pursuit of truth or knowledge.[7] So conceived, the laws of logic have as much objectivity as truths of mathematics.

Frege's conception of logic was rooted in this tradition. Inferring is 'to make a judgment because we are cognisant of other truths as providing a justification for it' (PW, 3). The task of logic is the validation of chains of inference (BS, Preface), and its goal is to set up the laws of valid inference (PW, 3). These logical laws are objective truths, wholly independent of psychology. He also accepted (PW, 128) the traditional view that logic is a normative science.

Taken with sufficient grains of salt, this body of conventional wisdom need not cause indigestion even to the most fastidious modern logicians. Of course patterns of valid argument justify taking something to be true on the grounds of other established truths; this is an analytic truth that follows from a proper (semantic) definition of 'valid argument'. Nevertheless, characterizing logic in this traditional manner both embodied and perpetuated important misconceptions about asser-

---

5. In his mature theory, Frege 'split' judgeable-content into the thought (the sense of a sentence, what is asserted in an assertion) and its truth-value (the reference of a sentence); see Part II. Here we shall occasionally shift from talking of content to talking of the thought where his anti-psychologism remained essentially the same throughout his work.
6. J. S. Mill, *System of Logic*, Intr. §2 (our italics); but compare this with Mill's agreement with Whately: 'Logic is not concerned with the actual truth either of the conclusion or of the premises, but considers only whether the one follows from the other; whether the conclusion must be true if the premises are true.' Mill, *An Examination of Sir William Hamilton's Philosophy*, p. 354, in *Collected Works*, vol. IX, ed. J. M. Robson (Univ. of Toronto Press, Toronto, 1979).
7. This qualification was also common (though not universal) amongst the psychologicians, as it was amongst their opponents. But the former, unlike the latter, did not explain the validity of the canon by reference to objective, Platonist considerations. The extensive nineteenth-century debate over whether logic is an art or a science arose out of the problem set to the psychologicians by this conception of the normativity of logic.

tion. Frege inherited several of these. That logic consists of laws for inferring assertions from assertions is open to two distinct interpretations. According to the first, permissive, interpretation, logic justifies one in drawing a conclusion from premises asserted because they are known to be true. Much of the utility of logic rests on this principle. A proof from established premises conclusively justifies the assertion of a conclusion that follows from them. This conception seems unproblematic. The second, restrictive, interpretation is, by contrast, not at all anodyne. According to it, inference can proceed *only* from judgments; the question of the validity of an argument arises *only* if its premises are *asserted* to be true. This rules out the possibility of the logical assessment of reasoning from hypotheses or from thoughts either held to be false or known to lack truth-value. Frege subscribed to this doctrine throughout his career.[8] 'Logic is concerned only with those grounds of judgment which are truths' (PW, 3); '[o]nly a thought recognized as true can be made the premise of an inference' (PW, 261); and, most strikingly, 'If a sentence uttered with assertoric force expresses a false thought, then it is logically useless and cannot strictly speaking be understood'[9] (PMC, 79; cf. also PMC, 16f.; N, 119; CT, 553).

Objections to thus restricting the domain of logic to assertions are obvious. One is that it precludes reasoning from hypotheses or conjectures; Frege *seems* to have acknowledged this possibility (BS, §2; PW, 7), but, if he did so, he never explicitly reconciled it with his official doctrine. A second objection arises from indirect proofs, where consequences are apparently drawn from false premises. Here Frege argued that appearances are deceptive; what is false or doubtful does not really occur as an independent premise in an indirect proof; but only as the antecedent of a conditional which as a whole expresses a true thought (PW, 244ff., N, 119f.).

The traditional conception of logic as the science of laws of inference was commonly involved in another dubious presupposition, namely, that inferring is a complex mental act or process involving the production or derivation of one judgment from some collection of other judgments. Although Frege avoided drawing from this idea the erroneous conclusion that these laws of inference are psychological truths, he did not challenge this conception of inference. Inferring is the *process* of making a judgment as a consequence of making other judgments (PW, 3). Since judging is a mental act, inferring must itself be a more complex *mental act*

---

8. With occasional tentative exceptions, e.g. PW, 175, sentences numbered 15 and 16.
9. Frege's conception may be contrasted with De Morgan's, which is more congenial to the modern mind: 'Logic ... [is] the examination of that part of reasoning which depends upon the manner in which inferences are formed, and the investigation of general maxims and rules for constructing arguments, so that the conclusion may contain no inaccuracy which was not previously asserted in the premises. It has so far nothing to do with the truth of the facts, opinions, or presumptions, from which an inference is derived; but simply takes care that the inference shall certainly be true, if the premises be true.... Whether the premises be true or false, is not a question of logic, but of morals, philosophy, history, or any other knowledge to which their subject matter belongs: the question of logic is, does the conclusion certainly follow if the premises be true!' De Morgan, *Formal Logic,* p. 1f. (Taylor and Walton, London, 1847).

(PW, 258).[10] This view of inferring as a psychological phenomenon helps to explain his mistaken restriction of the laws of logic to thoughts taken to be true. To judge that p is to judge it to be true that p; hence inference as a process starting from judgments must begin from thoughts taken to be true. In effect a muddle arises here from an incursion of psychology into logic that derives from a misconception of what it is to make an inference. Frege's thinking, despite his antipsychologism, was shackled by an assumption endemic among his contemporaries and predecessors.

A parallel misconception infects his conception of judgment. He inherited a Cartesian picture of judgment as an epistemically private mental act, and of assertion as an act expressing or manifesting such an act of judging. His allegiance to this picture was unswerving. To make a judgment is 'inwardly to recognize something as true' (PW, 7; cf. 139). It is a mental act or process of an agent performed at a specific time (N, 126n., 128; PW, 198) and hence belongs to the sphere of the psychological (PW, 253). Frege sided with Descartes against Hume on the issue of the voluntariness of acts of judgment. It is in principle within an agent's power to judge a thought to be true, to reject it as false, or to suspend judgment and merely entertain it as an hypothesis (BS, §§2, 4; PW, 7, 197; N, 119; FC, 34).

The act of asserting is the outer counterpart of the private act of judgment. Its paradigmatic form involves uttering a declarative sentence. Two different options now emerge in delineating the act of assertion. One is to maintain that whether an utterance of a declarative sentence constitutes an assertion depends solely on observable features surrounding its utterance. On this view, assertion is a public act, an aspect of the overt (but also intentional) behaviour of an agent. The other, antithetical, conception views asserting as a compound act consisting of uttering an appropriate sentence while judging the thought expressed by it to be true. This was the conception Frege adopted.[11] According to it, a necessary condition of a person's asserting the thought expressed by a sentence is that he judges the thought to be true. An act of assertion is the expression or externalization (*Äusserung* (PW, 2, 198)) or manifestation (*Kundgebung* (T, 22; PW, 139)) of the independent inner act of judgment. Assertion, thus conceived, is a public act only in part, i.e. the overt behaviour of uttering words in a certain form. Its other component is an inner, epistemically private act (N, 117), viz. judging. The act of judging is literally part of the act of asserting. This analysis generates obvious confusions. The most blatant are Frege's temptation to think that there is some-

---

10. A mythology of mental acts stands in the way of a correct apprehension of the ability to reason (to infer) from premises to conclusion according to rules which constitute canons of correctness. Correctly drawing conclusions from the premises of an argument is no more a mental act than is reading (cf. Wittgenstein, *Philosophical Investigations*, §156ff.). That one can read or infer *sotto voce* does not make the achievement a *mental act*.
11. There are occasional tentative moves in the first direction (PMC, 20; PW, 233) but they are never developed.

thing logically, not merely morally, awry about asserting a false thought[12] (PMC, 79), and his contention that what prevents the utterances of an actor from counting as acts of assertion is the absence of the 'requisite seriousness' backing his speech (T, 22).

Taking inference, judgment, and assertion to be mental acts created a pressure towards treating logic as a branch of psychology, to which many succumbed. There was, however, a common expedient to block this implication. This was to distinguish the object of a mental act from the act itself. Exploiting this distinction provided a non-psychological subject-matter for logic as long as the objects of judgment or assertion could be shown to be non-psychological or 'objective' (cf. Bolzano, Meinong). Even such an arch-empiricist as Mill insisted that

> An inquiry into the nature of propositions must have one of two objects: to analyse the state of mind called Belief, or to analyse what is believed. All language recognizes a difference between a doctrine or opinion, and the fact of entertaining the opinion; between assent, and what is assented to.
>
> Logic . . . has no concern with the nature of the act of judging or believing; the consideration of that act, as a phenomenon of the mind, belongs to another science. . . .
>
> In order to believe that gold is yellow, I must . . . have the idea of gold and the idea of yellow, and something having reference to these ideas must take place in my mind; but my belief has not reference to the ideas, it has reference to the things. What I believe, is a fact relating to the outward thing, gold . . . ; not a fact relating to my conception of gold. . . . [B]elieving is an act which has for its subject the facts themselves. . . .
>
> [Logic inquires then] not into Judgment, but judgments; not into the act of believing, but into the thing believed.[13]

Mill thus distinguished the act from the object of judgment, and argued that the object of judgment is an objective entity, viz. a proposition. Likewise the subject of judgment, i.e. that which we judge to be thus or otherwise, is typically (although not invariably)[14] objective, not an idea in the mind but a feature of extra-mental reality. Others combined the act/object distinction with a more strident insistence on the objectivity of the object of judgment. Bolzano, for example, introduced the *Satz-an-sich* as the object of judgment which is existence-independent of acts of judgment. It is 'any assertion [= what is asserted] that something is or is not the case, regardless whether or not somebody has put it into words, and regardless even whether or not it has been thought.'[15] In building logic on the

---

12. It is noteworthy that Frege was not the only philosopher to argue himself into such paradoxes; cf. Russell, *The Principles of Mathematics*, p. 35, on the impossibility of asserting a false proposition.
13. J. S. Mill, *System of Logic*, I-v-1. Similar antipsychologistic conceptions of the object of judgment are to be found amongst other English logicians; cf. W. S. Jevons, *The Principles of Science* ch. I and III, *passim* (1873) and later J. N. Keynes, *Studies and Exercises in Formal Logic*, 4th ed., pp. 1ff., 67 (Macmillan, London, 1906), Cf. also Peirce, *Collected Papers*, vol. II, p. 145.
14. In cases where the mind itself is the subject of a belief.
15. Bolzano, *Theory of Science*, §13.

distinction between act and object of judgment, Frege was working within an established tradition.

3. Antipsychologism: the laws of logic

Nineteenth-century German psychologism was a 'de-transcendentalized', anthropologized Kantianism. It was adamantly naturalist, with an occasional materialist, reductive bias. Where Kant had viewed the necessary forms of thought and experience as a product of the transcendental workings of the human mind, the psychologistic school dismissed Kant's transcendental psychology as unscientific mystery-mongering, and sought to explain the forms of thought and reason in terms of the empirical nature of the human mind, with occasional intimations that the ultimate explanation of the nature and structure of thought might have neurophysiological foundations. Two striking passages from Erdmann[16] indicate the character of their conception of logic:

> ... logical laws only hold within the limits of our thinking, without our being able to guarantee that this thinking might not alter in character. For it is possible that such a transformation should occur, whether affecting all or only some of these laws, since they are not all analytically derivable from one of them. It is irrelevant that this possibility is unsupported by the deliverances of our self-consciousness regarding our thinking. Though nothing presages its actualization, it remains a possibility. We can only take our thought as it now is, and are not in a position to fetter its future character to its present one.

The necessities of thought are thus viewed *sub specie humanitatis,* naturalistically not transcendentally. To be sure, we cannot think what nonlogical thought would be like, but this constraint lies within the empirical nature of our minds, not in the nature of the object of thought:

> ... we cannot help admitting that all the propositions whose contradictories we cannot envisage in thought are only necessary if we presuppose the character of our thought, as definitely given in our experience: they are not absolutely necessary, or necessary in all possible conditions. On this view our logical principles retain their necessity for our thinking, but this necessity *is not seen as absolute, but as hypothetical.* We cannot help assenting to them—such is the nature of our presentation and thinking. They are universally valid, provided our thinking remains the same. They are necessary, since to think means for us to presuppose them, as long, that is, as they express the essence of our thinking.

This psychologistic conception of logic had its supporters abroad too. Boole had taken it for granted that the laws of logic are in some sense laws of psychology. In his first great work written in 1847 he announced 'A successful attempt to express logical propositions by symbols, the laws of whose combinations *should be founded upon the laws of the mental processes which they represent,* would, so

---

16. B. Erdmann, *Logik,* quoted by Husserl, *Logical Investigations,* Vol. I, p. 162, tr. J. N. Findlay (Routledge and Kegan Paul, London, 1970).

far, be a step toward a philosophical language,[17] and in *The Laws of Thought* (1853), he declared his aim 'to investigate the fundamental laws of those operations of the mind by which reasoning is performed ... and upon this foundation to establish the science of Logic'.[18] If Boole, a mathematician, took some vague version of psychologism for granted, many of his more philosophically inclined contemporaries argued for it at length, deliberately and self-consciously following the footsteps of their much admired Continental colleagues. Mansel used as his epigraph to his *Prolegomena Logica* the remark *'La Logique n'est qu'un retour de la Psychologie sur elle-même'*, whose author, Victor Cousin, he refers to as 'one of the highest philosophical authorities of our own or any age'.[19] Cousin, 'the Great Eclectic', drew much of his inspiration from German psychologism, and Mansel similarly declared his purpose to be the 'communicating to English readers some of the most valuable results of German thought'.[20] In common with the psychologistic school, he announced that 'an enquiry into the constitution and laws of the thinking faculty, such as they are assumed by the Logician as the basis of his deductions' is 'the touchstone by which the whole truth and scientific value of Logic must ultimately be tested'.[21] Sir William Hamilton, though a critic of Cousin, was just as devoted an adherent of neo-Kantian psychologism in logic: 'Logic is the Science of the Laws of Thought as Thought',[22] i.e. the laws of the general form of thought. '[I]n so far as a form of thought is necessary, this form must be determined or necessitated by the nature of the thinking subject itself; for if it were determined by anything external to the mind, then it would not be a necessary but a merely contingent determination. The first condition, therefore, of the necessity of a form of thought is, that it is subjectively, not objectively, determined'.[23] Widespread though psychologism was, it would be altogether misguided to think that it had no opponents. Krug early in the nineteenth century, Bolzano somewhat later, and Lotze shortly before Frege were but a few of the German philosophers who took a firm antipsychologistic stand. Nor were these German objectivists or realists alone. As we have seen, Jevons and Spencer,[24] in England, had no sympathy with psychologism, in as much as it appeared to compromise the objectivity of logic and mathematics.

Frege sided firmly with the antipsychologists. The anthropologism which permeated their conception subordinated logic to psychology, rendered the laws of logic subjective and relative, and transformed the study of the laws of thought into a study of the laws of human thinking.

---

17. G. Boole, *The Mathematical Analysis of Logic,* p. 5 (Blackwell, Oxford, 1948) (our italics).
18. G. Boole, *The Laws of Thought,* ch. I, §1.
19. H. L. Mansel, *Prolegomena Logica, An Enquiry into the Psychological Character of Logical Processes,* 2nd ed., p. ix (Hammans, Oxford, 1860).
20. Ibid. p. xiii.
21. Ibid. p. iv.
22. Sir William Hamilton, *Lectures on Metaphysics and Logic,* vol. III, p. 4.
23. Ibid. p. 24f.
24. H. Spencer, *The Principles of Psychology,* 3rd ed., vol. II, ch. 3 (Williams and Norgate, London, 1881).

Logic, Frege insisted, is concerned with truth, as ethics is concerned with goodness and aesthetics with beauty (PW, 4, 128). Its concern with truth differs from that of other sciences, all of which aim to achieve the truth in their special domains, for logic studies the laws of truth. It studies the laws justifying the inferences from one true judgment to another. Truth, he stressed, is wholly independent of what is believed to be true. No matter how many people believe something to be the case, no matter what view is destined to be held true by the future community of scientists, there is no contradiction in that view being false, and its acknowledged contradictory being true.

> If being true is thus independent of being acknowledged by somebody or other, then the laws of truth are not psychological laws: they are boundary stones set in an eternal foundation, which our thought can overflow but never displace. (BLA, p. xvi)

We must sharply distinguish the descriptive laws of human thinking from the prescriptive laws of how one ought to think if one wishes to attain the truth. Psychological laws describe the nature of human thinking, but by that very token they are not laws of logic: 'The laws in accordance with which we actually draw inferences are not to be identified with the laws of valid inferences; otherwise we could never draw a wrong inference' (PW, 4).

The laws of logic are normative (BLA, p. xv); indeed logic, like ethics, is a normative science[25] (PW, 128): it spells out the rules to be followed if one is to achieve truth in inferences from true judgments. Frege conceived of laws of valid inference as anankastic rules, i.e. rules which specify the necessary (or most efficient) means for achieving a specific goal. These laws of logic are prescriptions which ought to be followed in our pursuit of truth. Just as empirical anankastic rules rest on regularities of nature, so logical anankastic rules rest on objective logical relations between judgments. These objective relations are wholly independent of how human beings think. We need not think in accord with them, just as we need not of necessity act in accord with the laws of morality. But we should do so if we are seriously concerned to attain truth.

Given the background of this realist conception of logical relations, and normative conception of the laws of logic, Frege unsurprisingly took strong exception to the relativist anthropologism of Erdmann. It would be perfectly possible, he agreed, for us to encounter beings who thought in ways flatly contradictory to the

---

25. As already noted, this conception was widespread even among psychologicians. The terminology is likewise not uncommon. Keynes opened *Studies and Exercises in Formal Logic* with the remark that logic 'is concerned with how we ought to think, and only indirectly as a means to an end with how we actually think. It may accordingly be described as a normative or regulative science'. Cf. also Peirce, who declares that 'It is pretty generally admitted that logic is a *normative* science' (*Collected Papers*, vol. I, p. 314) and attributes the neologism to the school of Schleiermacher (and elsewhere (op. cit. vol. II, p. 5n.) to Wundt in particular). The association of the (normative) laws of truth with the laws of goodness and those of beauty is ancient. The medievals conceived of *verum, bonum,* and *pulchrum* as the three 'transcendentalia' or supreme kinds of judgment, associated with the Aristotelian division of functions into theory, action, and creativity.

ways in which we think. The empirical psychological laws which describe their thinkings would differ wholly from the laws which describe our thinkings. But this does not imply that they have different, equally legitimate or valid laws of truth, as the psychologicians suggest. The psychological logician will register the different laws of thought and accept the differences as reflecting different mental constitutions. He will conclude that they have one logic, and we have another; there is no more to be said on the matter, just as ethical relativists will insist that there is no Archimedean point from which to evaluate and choose between alternative moralities. This is precisely where Frege disagreed. The impossibility of our rejecting the laws of logic as true does not indeed hinder us from imagining creatures who might do so, but it *does* hinder us in supposing that these beings are *right* to do so (BLA, p. xvii). Their thinking would appear to us as 'a hitherto unknown type of madness' (BLA, p. xvi); but if we distinguish, as we must, the laws of truth (i.e. the prescriptive laws determining how one ought to think) from descriptive laws of how and under what conditions thinkers take things to be true (i.e. descriptive laws of psychology), it will always be an open question whether a thinker's inferences are right. It will in principle be possible to determine whether the laws according to which people take things to be true accord with the laws of truth. The laws of truth are timeless and unchanging, independent of the fluctuating natures of thinkers. And it is they, and they alone, which provide the subject matter for logic.

From our point of view, Frege's case against Erdmann's conception of logical laws is remarkably weak and limited. He was content to maintain that it is in principle possible to compare laws of taking-to-be-true against the genuine laws of truth and thereby to ascertain the correctness of a set of putative logical laws. He hinted at no procedure for settling whether a codification of the laws of truth is actually correct, and he outlined no method for adjudicating between competing systems of logic (alternative logics). This lacuna would leave the determined relativist about logical laws all the latitude that he might wish for. The psychologician, like the ethical relativist, could argue that any disagreement about logical laws boils down to a discrepancy between two sets of laws of taking-to-be-true, and that Frege could give no cogent reason for enthroning his own as the sole legitimate arbiter of the correctness of inferences. The laws of truth are a merely theoretical and therefore vacuous criterion for the correctness of any entrenched principles of reasoning.

Amongst twentieth-century philosophers it was predominantly Wittgenstein who repudiated Frege's conception of the so-called 'laws of truth'. The laws of logic are not, he claimed, prescriptive, anankastic rules dictating how one ought to think, but constitutive rules, definitive of what counts as thinking; they are internally related to our concept of thinking, and not merely instrumental for correct thinking. But although his sympathies lay everywhere with a form of anthropologism, and although he found Frege's realism about logical relations to be one of the great illusions of philosophy, on the question of the possibility of discovering *contra*-logical forms of thought, he found both parties to the dispute in error. Their fundamental mistake consists in the common supposition that it is intelligible to us that creatures should *think* according to different laws of logic. 'Frege calls it "a law about what men take for true" that "It is impossible for human beings . . . to recognize an object as different from itself".'[26] But this contention, with which Erdmann agreed, is confused. This 'impossibility' is not a psychological limitation, comparable to the fact that were we differently constituted, we could have a mental image of a chiliagon, or recognize intuitively a distribution of 237 marbles on the floor (whereas, as things are we can image an octagon, or take in the fact that there are half a dozen scattered marbles without counting). It is rather that there is no such thing as recognizing an object as different from itself. The 'impossibility' is akin to the 'impossibility' of castling in draughts:[27] '[H]ere "I can't imagine the opposite" doesn't mean: my powers of imagination are unequal to the task. These words are a defence against something whose form makes it look like an empirical proposition, but which is really a grammatical one.'[28] Frege contended that if we found creatures who think according to different laws of logic we should react by saying 'we have here a hitherto unknown type of madness' (BLA, p. xvi); but, Wittgenstein comments, 'he never said what this "madness" would really be like.'[29] Frege did not describe the forms of activity in which these imagined utterances are integrated. People's agreement about accepting a structure as a proof or a valid inference consists in the fact that they use words as language, *as what we call 'language'*. 'Isn't it like this', Wittgenstein concludes,

> so long as one thinks it can't be otherwise, one draws logical conclusions. This presumably means: so long as *such-and-such is not brought into question at all*.
> The steps which are not brought in question are logical inferences. But the reason why they are not brought in question is not that they 'certainly correspond to the truth'—or something of that sort,—no, it is just this that is called 'thinking', 'speaking', 'inferring', 'arguing'. There is not any question at all here of some correspondence between what is said and reality; rather is logic *antecedent* to any such correspondence; in the same sense, that is, as that in which the establishment of a method of measurement is *antecedent* to the correctness or incorrectness of a statement of length.[30]

The laws of logic are indeed the expression of our thinking habits; they do, as the psychologicians contended, show how we think, as the laws of chess show how we play chess. But they are also an expression of the habit of *thinking,* that is, they show *what* we call 'thinking', as the laws of chess show what we call 'playing chess'.[31]

26. Wittgenstein, *Remarks on the Foundations of Mathematics*, p. 89, 3rd edition, ed. G. H. von Wright, R. Rhees, G.E.M. Anscombe, tr. G.E.M. Anscombe (Blackwell, Oxford, 1978).
27. Wittgenstein, *Zettel*, §134, ed. G.E.M. Anscombe and G. H. von Wright, tr. G.E.M. Anscombe (Blackwell, Oxford, 1967).
28. Wittgenstein, *Philosophical Investigations*, §251.
29. Wittgenstein, *Remarks on the Foundations of Mathematics*, p. 95.
30. Ibid. p. 96.
31. Ibid. p. 89.

4. Antipsychologism: the object of judgment

In Frege's view, the psychological logicians misconstrued the nature of the laws of logic because they had grossly misunderstood the nature of the *objects of judgment*. Their reasoning went roughly thus: when we judge something to be true, there must exist or subsist something which we so judge. Since our judgment may be false, i.e. may not have as its object any 'fact in the world', the object of judgment must subsist elsewhere, viz. in the mind. On Frege's view, this allowed psychological factors to contaminate the purity of logic. Psychologicians misconceived of the content of judgment (which Frege later called 'a thought') as sharing the mental character of the act of judging or thinking. Consequently, they treated the objects of judgment as ideas. But if a judgment is true, it is true quite independently of whether anyone believes it (BLA, p. xv; PW, 132f.). Indeed, that which is true, i.e. the object of thought or judgment, is true quite independently of its having been thought at all and of its having been affirmed or denied (PW, 133). Frege concluded that these platitudes excluded the possibility that what is thought or judged to be true is an idea, since ideas are subjective and hence mind-dependent.

This argument turns on the thesis that ideas are subjective entities, unlike contents of judgment. Frege called attention to three features in clarifying their subjectivity. First, they are existence-dependent upon a bearer. Ideas cannot subsist without being someone's ideas (T, 26f.). It is noteworthy (since Frege did not note it) that this feature of mental images, sensations, affections, volitions, etc. does not distinguish them from any *public* property or attribute which likewise 'needs a substance in which to inhere'. Second, ideas are *privately* owned and unshareable. I can no more have your idea than you can have my pain. It is of the essence of an idea to be had by one and only one bearer (T, 26ff.; PW, 3f.; PMC, 67). Ideas are identifiability-dependent upon their bearers. Again it is important to note, just because Frege did not, that the identifiability-dependence of 'ideas' is a feature they share with perfectly public and accessible things like smiles and sneezes. Moreover, although it precludes common ownership of a given idea, it does not exclude the possibility of qualitative identity of ideas. Third, ideas are epistemically private, only the bearer of an idea knows its nature and characteristics, 'For it is impossible to compare my sense impression with that of someone else. For that it would be necessary to bring together in one consciousness a sense impression belonging to one consciousness, with a sense impression belonging to another consciousness.' Consequently, the nature of ideas is wholly incommunicable, for when a predicate 'is supposed to characterize sense-impressions belonging to my consciousness, it is only applicable within the sphere of my consciousness' (T, 27; cf. RH, 325). This conclusion is absurd. It presupposes the possibility of a private language, which is arguably incoherent, and at the mundane level of ordinary discourse it wholly disregards the fact that we communicate information about our 'ideas' perfectly happily. We tell our doctor about our pains, our psychoanalyst about our dreams, our loved ones about our emotions.

Frege wrongly considered these three distinctions to coincide, and hence used them all to elucidate the notion of the subjectivity of ideas. In arguing against the psychologicians he gave no novel or penetrating analysis of ideas. He did not even attempt to disentangle the various threads interwoven in the classical idealists' account of thought, perception, conception and concept-acquisition, which the German psychologicians had inherited from the venerable New Way of Ideas handed down by Descartes and Locke. It is undoubtedly true that the term 'idea', reintroduced into modern philosophy by Descartes, was riddled with ambiguity and imprecision from the start. Descartes himself explicitly confessed that 'idea' sometimes signifies an *act* of the mind, sometimes the *object* of such an act.[32] Matters were, however, much worse than that.[33] When Descartes spoke of ideas, he sometimes talked of them as if they were capacities to perform certain acts, rather than acts of the mind. Equally, when considered as objects of capacities or acts, he sometimes talked of them as if they were sensations, experiences or mental images, and at other times as if they were concepts. Consequently, if any central theses of an idealist are to be subjected to criticism, the first task is to uncover whether 'idea' is being used in a reasonably precise sense, and if so, which one. It is certainly wrong to suppose that when classical idealists spoke of ideas they uniformly meant a mental object, privately owned and epistemically private. These complexities, however, did not concern Frege. His threefold characterization of ideas appeared to him to be sufficiently powerful to undermine the arguments of his immediate adversaries, the nineteenth-century German psychological logicians.

The perils inherent in his superficial analysis become evident in connection with his commitment to the doctrine of the epistemic privacy of ideas. For that is inconsistent with his own account of important aspects of language which lie outside the domain of (what is today conceived of as) semantics. Over and above the content of an expression Frege distinguished its *tone* or *colouring*,[34] e.g. its emo-

---

32. Descartes, *Meditations*, p. 138, in *Philosophical Works* tr. and ed. E. S. Haldane and G.R.T. Ross, Vol. I.
33. For detailed discussion, see A.J.P. Kenny, *Descartes*, ch. 5 (Random House, New York, 1968).
34. Neither the distinction nor the terminology was novel. The Port-Royal *Logic* distinguished within the 'signification' of a word between the 'principal idea' and the 'accessory'. The accessory ideas may be attached to the words themselves (in so far as they are excited commonly by all uses of the words) or only by the speaker (in so far as they are excited by his tone of voice, expression, gestures, etc.). Cf. Port-Royal *Logic*, Part I, ch. xiv. Boole excluded from consideration in logic those elements of speech used 'to express some emotion or state of feeling accompanying the utterance of a proposition' (*The Laws of Thought*, ch. II, §16). Lotze noted that 'the end of speech is not merely to be a brief communication of thoughts; in order to move the mind of another, to persuade, to set forth his own feeling . . . the speaker must be able to invest the content proper of his thought in manifold forms that add no material part to the logical structure of his sentence, yet throw over all its parts a peculiar *colouring* of merely psychological significance' (our italics; H. Lotze, *Microcosmus: An Essay Concerning Man and His Relation to the World*, V-iii-§4, trs. E. Hamilton and E. E. Constance Jones (Clark, Edinburgh, 1885)). In his *Logic*, Lotze remarks that 'Both poetry and eloquence aim by this method at something more than imparting ideas: they count upon the attachment to the images thus called up of feelings of pleasure or pain, of approval or disapproval, of exaltation and aversion' (Lotze, *Logic*, II-i-§155).

tive or attitudinal implications or intended consequences. Colouring, he thought, is essential to art, particularly fiction and poetry. The mechanism of colouring he explained in causal terms involving association of ideas. 'Without some affinity in human ideas art would certainly be impossible' (SR, 61). Colouring is explained by the causal powers of expressions to evoke associated ideas in reader or hearer, and it is upon this that poetic eloquence depends. Although associations may vary from person to person, complete variation would undermine the possibility of colouring altogether. 'It is indeed sometimes possible', Frege *here* insisted, 'to establish differences in the ideas, or even the sensations, of different men; but an exact comparison is not possible, because we cannot have both ideas together in the same consciousness' (SR, 60). This conclusion is doubly incoherent. It makes no sense to suppose both that a rough correspondence of ideas can be established and that it is in principle impossible to ascertain an exact correspondence. Moreover, our inability to juxtapose the different ideas in one consciousness is not a consequence of human frailty; there is simply no such thing as such a juxtaposition, just as there is no such thing as scoring goals in bridge. Hence one cannot explain the notion of an exact comparison of ideas by reference to the 'impossible possibility' of juxtaposing them in the same consciousness.[35]

Though the sword Frege forged for himself was made from impure ore mined in a fairly high-handed manner from the idealist tradition, it served him well as a weapon for assailing psychological logicians. The chink in their armour was the claim that what we judge or think is an idea, at least to the extent that it was conceived as an essentially mental item. Frege plunged his sword in at this point, and twisted the hilt with vigour. He simply drew the consequences of the subjectivity of ideas. First, the objectivity of truth is inconsistent, in his view, with truth's being a property of a mental entity which is private and subjective. Second, if the truth of a thought is independent of anyone's thinking the thought, then *what* is thought (the object of thinking) cannot be an entity existence-dependent on a thinker. Third, if what we think or judge were an idea in our minds, and if ideas are epistemically private, then there could be no communication. But we can communicate what we think, we can convey to another the object of our judgments. So what we think cannot be an idea (PW, 133f.). Fourth, if what we think were a subjective mental entity, then there could be no agreement or disagreement of judgment, no contradiction between what different people believe, no common investigation into an hypothesis. For what A would judge in judging that $2 + 2 = 4$ would concern his ideas, and what B would judge in denying that $2 + 2 = 4$ would concern *his* ideas, and there would be no common object of judgment at all (PW, 133; PMC, 80; T, 28ff.). If communication is to be possible, if common pursuit of knowledge is to be intelligible, if fruitful controversy is to be conceivable, it must be possible for different people to think the same thing. But this is

---

35. This inconsistency is arguably unimportant from Frege's point of view in as much as his purpose is largely negative, i.e. to contrast what is relevant to truth-value determination (logic) with what is irrelevant (tone). Defects in his positive account of tone do not as such affect the legitimacy of the distinction. However, whether the distinction *is* correctly drawn merits attention.

inconsistent with the claim that what we think is an idea, on the assumption that ideas are subjective, privately owned and epistemically private. Finally, truth is a sempiternal property of thoughts. What is true, viz. a thought, is true at all times. It does not become true or cease to be true. But then, Frege argued, the bearer of truth cannot be a fleeting, mental item which comes into existence and passes away according to psychological laws. It must, rather, share in the sempiternality of its most important property (T, 29, 36ff.)

These critical points rest on truisms. Truth is distinct from being believed to be true. What is true is true independently of whether anyone entertains it. We can and do convey to each other what it is that we think or judge, and it is possible for two people to entertain the same object of thought (proposition), to agree, disagree, or hypothesize it. Propositions, if true, are timelessly true. To the extent that nineteenth-century psychological logicians in Germany shared Frege's conception of the subjectivity of ideas, then the contention that the object of thought is an idea is inconsistent with these platitudes.

Frege's onslaught upon psychologicians with respect to objects of judgment is a model of an effective polemic, but not a paradigm of penetrating philosophical criticism. Far from clearing up the muddles inherent in their conception of an idea, he incorporated their conception in his criticism of their analysis of logical relations. Consequently, his attempted refutation of psychologism in logic would collapse in the face of more sophisticated forms, e.g. Carnap's methodological solipsism.[36]

5. Antipsychologism: unjudgeable-contents

Frege hammered the psychological logicians' contention that what we judge, i.e. a judgeable-content or abstract thought, is an idea. He was, however, no less concerned to destroy the philosophical theory that what we made judgments *about* must be ideas, and that what we assert, judge, or predicate about the subjects of our judgments must also be ideas. It is no less grievous an error to think that unjudgeable-contents must be ideas than it is to think that judgeable-contents are.

Frege precipitated out of assertion a Platonic entity that he viewed as the object of the mental act of judgment. His logical analysis decomposed this judgeable-content into unjudgeable-contents (BS, §2; PW, 16f.; PMC, 101) which he characterized as objects and concepts. It is unclear how exactly he intended these terms to be understood. It is even less clear whether he thought that the unjudgeable-contents into which a judgeable-content decomposes are, or are not, literally *parts* of it. In *Begriffsschrift* he multiplied confusion by suggesting that a part of the judgeable-content (or 'combination of ideas') *that there are houses* or *that there is a house* is the *idea* 'house' (BS, §2n.), and in his draft of 'On Concept and Object' he described concepts, objects, and relations as parts of a content

---

36. Cf., in particular, R. Carnap, *The Logical Structure of the World*, tr. R. A. George (Routledge and Kegan Paul, London, 1967).

(PW, 105). This tendency to invoke the part/whole terminology carries with it not only great obscurity[37] (what can be meant by suggesting that the *material object* A is *part* of the abstract judgeable-content ϕA?), but also grave risk of large-scale confusion.

The linguistic instrument of an assertion is typically a declarative sentence. It is by uttering such a sentence that we express a judgment. According to Frege, a sentence thus used stands for a judgeable-content. A sentence is composed of words. These in turn stand for unjudgeable-contents. Typically, he thought, philosophers took contents of sub-sentential expressions to be perceptible objects or properties. But, obviously, not all significant expressions fit that bill. In particular, number-words do not, since what they stand for are not objects or properties in the sensible world (FA, §§21ff., 34ff.). Philosophers standardly respond by taking the contents of such expressions to be the signs themselves, or ideas in the mind.

The first line was taken by formalists in mathematics. On Frege's construal, they did not claim that number-words and arithmetical equations had no content. This, as he saw it, would be tantamount to saying that when one judges that $2 + 2 = 4$, one does not judge anything, or that when one denies that $2 + 3 = 4$, there is nothing that one denies. This is absurd. What formalists claim, Frege contended, was that the content of arithmetical signs are the signs themselves, and hence that equations have as their contents judgments about signs. This naïve formalism he had no difficulty in refuting (FA, §§92ff.; BLA, p. xiii, and Vol. ii, §§86–137). It is, however, significant that the *Begriffsschrift* account of identities is a formalist analysis, an error which he subsequently rectified (SR, 56ff.).

The second line of argument was adopted by psychological logicians who identified unjudgeable-contents with ideas. This thesis, of course, had much wider scope than formalism, since it could be applied without manifest implausibility to most sub-sentential expressions, not merely to those not readily correlated with sensible objects or properties. Frege's primary (although not sole) target was again contemporary German psychologism as exemplified by Schloemilch (FA, §27), Husserl (RH, *passim*), and Erdmann (BLA, Preface). The application of psychologism to mathematics, Frege contended, prejudiced the objectivity or mind-independence of the entities studied by mathematicians, the objectivity of the relations holding between them, and the objectivity of the truths about them which the mathematician discovers. More generally, the uniform subjectivity of unjudgeable-contents precluded the possibility of judgments about an objective domain. Finally, the doctrine vitiated the objectivity of logic and its laws.

Frege's most general criticisms of psychologism spill over into an onslaught upon idealism in general. Although his conclusions are unexceptionable, the supporting arguments are shallow, akin to a Johnsonian foray into the refutation of idealism. 'In the end everything is drawn into the sphere of the psychological; the boundary that separates the objective and subjective fades away more and more, and even actual objects themselves are treated psychologically, as ideas. For what

---

37. This is not peculiar to Frege. Russell spoke of propositions as *containing* the entities indicated by words, cf. *The Principles of Mathematics*, p. 47.

else is *actual* but a predicate [property]? and what else are logical predicates but ideas? Thus everything drifts into idealism and from that point with perfect consistency into solipsism' (BLA, p. xix). This he found absurd. No doubt it is. But his tirade simply relies on the *assumption* that there is accessible to us an objective domain of mind-independent entities. No doubt there is. But dogmatic insistence thereon would cut little ice with such subjective idealists as Berkeley or Hume.

Frege made little attempt to argue against epistemological idealism. Only in 'The Thought' did he outline a case. First, if experience is limited to subjective perceptions, then we can never know of the existence of other minds. (This thought had crossed the minds of classical idealists!) Second, not everything can be an idea, since ideas need a bearer, viz. the self (T, 32ff.). Although Hume's theory of the self is a monumental muddle, it can hardly be thought to be refuted by such considerations. Classical idealists, about whose work Frege was singularly ignorant,[38] offered elaborate explanations of why we *think* that there is an objective domain which is the subject of most of our discourse. They would have been wholly undisturbed by the 'criticism' that their linguistic idealism and idealist metaphysics preclude reference to objective particulars as understood by Frege, or by his insistence that the self is a non-ideational object of awareness. After all, their complex metaphysical systems were designed precisely to sustain such conclusions, profoundly counter-intuitive though they may be.

Although his general case against idealism carries little philosophical weight, it is peripheral to his central campaign against the psychologicians. He conceived issues in epistemology to have no direct bearing on the proper business of logic, and therefore the task of refuting idealism was in principle irrelevant for his enterprise of codifying the laws of thought and analyzing the truths of arithmetic. In practice, too, he was concerned with establishing a very limited thesis. Just as the psychologicians believed that the objects of thought, judgment, and inference are mental entities, so too they claimed that what thoughts or judgments are *about* are likewise mental entities (ideas). In both cases the inference is invalid. Just as Frege correctly insisted on the platitude that it must be possible for different people to think, judge, or assert the same thing, so too he rightly contended that it must be possible for different people to make assertions or judgments *about* the same thing. But if the content of the subject of a simple declarative sentence is what an assertion is about, if the contents of expressions are uniformly ideas, and if ideas are essentially privately owned and cannot be shared, then this possibility is precluded.

> If every man designated something different by the name "moon", namely one of his own ideas ... an argument about the properties of the moon would be pointless: one person could equally well assert of his moon the opposite of what the other person, with equal right, said of his. If we could not grasp anything

---

38. With the exception of Mill, most of Frege's knowledge of classical empiricism is derived from Baumann's anthology *Die Lehren von Zeit, Raum und Mathematik* (Berlin, 1868); cf. FA, §5n. and *passim*.

but what was within ourselves, then a conflict of opinions [based on] a mutual understanding would be impossible . . . (BLA, p. xix).

Given that ideas are identity-dependent on their bearers, the contention that the contents of expressions are uniformly ideas makes communication and understanding impossible and thereby forecloses the very possibility of intersubjective knowledge.

Frege harped on the consequences of this form of psychologism for the possibility of the science of mathematics. '[I]f everyone had the right to understand by this name ["one"] whatever he pleased, then the same sentence about one would mean different things for different people—such sentences would have no common content' (FA, i). Consequently, 'if the number two were an idea, then it would have straight away to be private to me only. Another man's idea is, *ex vi termini,* another idea. We should then have perhaps many millions of twos. We should have to speak of my two and your two, of one two and all twos' (FA, §27). Yet in spite of this endless proliferation of objects which the psychologician considers to be the proper objects of mathematical investigation, it would be impossible to prove that the number of natural numbers is infinite (FA, §27; RH, 334)! And no shared and intersubjectively testable judgments could be formulated at all! The derivation of these consequences Frege viewed as a *reductio ad absurdum* of the psychologistic thesis that the contents of mathematical expressions are uniformly ideas.

If any of his contemporaries did argue that words must uniformly stand for ideas, he was surely right to take them to task, and his argument has considerable force. What it proves is only that not all expressions stand for ideas. It does not exclude the possibility that some expressions do stand for ideas. For the most part, Frege was perfectly content with acknowledging this possibility. He distinguished expressly between subjective and objective objects (cf. FA, §§27n., 61); he spoke of his idea or internal image of a perceptible object and distinguished this object both from the ideas of others and from the sense of an expression standing for the perceptible object (SR, 59); and he argued that everybody 'finds it necessary to recognize an inner world distinct from the outer world, a world of sense-impressions, of creations of his imagination, or sensations, of feelings and moods, a world of inclinations, wishes and decisions' (T, 26). All of this seems platitudinous. Frege, like the psychologicians, calls all of these entities (apart from decisions) 'ideas'. Hence for him, it would seem just as absurd to maintain that *no* word or expression stands for an idea as to embrace the psychologicians' claim that *every* word stands for an idea.

Nonetheless, Frege had an inclination to adhere to this very strong form of antipsychologism, and he succumbed to the temptation at least once. The source of this inclination was his tendency to think that unjudgeable-contents are *parts of* judgeable-contents. For then the objectivity of judgeable-contents apparently requires the objectivity of each and every unjudgeable-content, and this in turn entails that no judgment can be *about* any subjective entity:

Indeed would not Locke's empiricism and Berkeley's idealism, and so much that is tied up with these philosophies, have been impossible if people had distinguished adequately between thinking in the narrow sense and ideation, between the parts of a content (concepts, objects, relations) and the ideas we have? *Even if with us men thinking does not take place without ideas, still the content of a judgment is something objective, the same for everybody,* and as far as it is concerned it is neither here nor there what ideas men have when they grasp it. In any case these are subjective and will differ from one person to another. *What is here being said of the content as a whole applies also to the parts which we can distinguish within it.* (PW, 105; our italics)

The conclusion that every judgment must be about an objective entity is obviously absurd by Frege's own lights. The objectivity of unjudgeable-contents does not follow from the objectivity of judgeable-contents. Recognizing the mental nature of what a judgment is about no more entails that the judgeable-content is a mental object than does recognizing the concrete nature of what a judgment is about entail that the judgeable-content is a physical object! The part/whole analysis of judgeable-contents thus misled Frege in his polemic against psychologicians. He later sorted out this muddle. The sense/reference distinction enabled him to see clearly that what we make a judgment about is *not* part of what we judge to be true, i.e. that the reference of a proper name is never a constituent of a thought (e.g. PW, 225). His doctrine that the senses of all expressions are public and objective entities obviously carries no implication that the reference of an intelligible expression is never a subjective entity (object or concept). He could account easily for the fact that we can and do make judgments about ideas. If I judge that the pain in my knee is arthritic, what I judge is objective; my doctor may make the very same judgment. But what I make a judgment about (my pain) is subjective; and what my doctor makes a judgment about is also subjective, though it is independent of his mind. So far, so good. But Frege superimposed on this account of objective judgments about subjective objects the independent doctrine of the epistemic privacy of subjective entities. This leads to incoherence when combined with the assumption that what is epistemically private is incommunicable (cf. FA, §26). That pair of premises entails that another person cannot grasp the thought which I express in asserting that the pain in my knee is arthritic. Frege thus landed himself back in confusion about the possibility of objective thoughts about subjective entities. On the one hand, 'The being of a thought may ... be taken to be in the possibility of different thinkers' grasping the thought as one and the same thought' (N, 120). On the other hand, he acknowledged the possibility of a thought which one person alone can grasp (T, 25f.). The tension here is intolerable and the muddle unresolved.

It is striking that Frege's criticism of psychologism soldiers on unaltered even after he distinguished sense from reference. He still took the psychologicians to task for maintaining that what thoughts or judgments are *about* is uniformly ideas. He interpreted their claims as bearing only on the *references* of subsentential expressions, not on the *senses* of such expressions, although he criticized their holding that the sense of a sentence (the thought expressed by it) is an idea. He did not deepen or extend his criticism of psychologism by confronting the more plausible doctrine of representative idealism according to which it is not the reference of a name but something comparable to its sense that is alleged to be always an idea. In this form, idealism does not preclude the possibility of discourse or knowledge about public and objective entities. Frege might have thought that representative idealism would render communication and sharing of thoughts incomprehensible. This would be so if mutual understanding presupposes grasping numerically the same entity *as* the sense of any subsentential expression, since the ideas of different persons are numerically distinct (SR, 59). This defect would be remedied by his Platonist objectivism about the senses of all expressions, because that assumption does secure the bare possibility of common objects of understanding. But it would be open to a representative idealist to argue that mutual understanding does not require numerical identity of ideas, but only qualitative identity among the various ideas which different persons have which represent a single objective entity (concept or object). Frege was himself attracted by this picture of the possibility of communication. It informs his contention that the colouring or tone of expressions is roughly communicable because different persons associate approximately the same ideas with words (SR, 60f.), and it is equally essential to make sense of the claim that we can never know whether the ideas of different persons are exactly the same, e.g. whether another means the same by 'white' as I do when I use 'white' to signify a subjective sensation (FA, §26; RH, 325). But, if it makes sense to talk of qualitative identity of ideas, then a representative idealist could argue that this provides a sufficient basis for communication and mutual understanding. Against this sophisticated version of idealism Frege provided no arguments. In particular, he could not cogently argue that the fact of successful communication proves that the senses of subsentential expressions are not ideas. For if he could claim that this fact shows that identical senses are uniformly associated by speakers with each given expression,[39] the idealist could invoke the same argument to 'prove' the qualitative identity of the ideas associated with expressions. Frege simply failed to confront representative idealism. He was content to sketch in a form of Platonism about senses that is its diametrical opposite. On his view, the sense of every expression is an objective abstract entity, while on the idealists' view, the analogue of the sense of an expression is invariably a subjective mental entity.

> It might perhaps be said: Just as one man connects this idea, and another that idea, with the same word, so also one man can associate this sense and another that sense. But there still remains a difference in the mode of connection. They are *not prevented* from grasping the same sense; but they *cannot* have the same idea.... If two persons picture the same thing, each still has his own idea. (SR, 60; our italics)

39. He actually dissented from this argument, holding that identical assignments of senses to subsentential expressions are not necessary for successful communication in everyday life (SR, 58n.; PMC, 115).

Frege naively assumed that a psychologician must concede that mutual understanding presupposes the possibility of grasping numerically the same ideas, whereas that is the position that a representative idealist challenges. The sense/reference distinction makes clear the narrowness of Frege's target and the limited strength of his case.

The other main prong of his attack on this psychologicians' account of what judgments are *about* draws attention to their reliance on laws of association of ideas. Any relational judgment will express a relation among ideas, and so indeed will any atomic judgment apparently stating that an object falls under a concept. Waiving the objection that such interpretations are unintelligible or simply mistaken (BLA, p. xxi), Frege simply noted that ideas, being distinct objects of experience, can stand to each other only in contingent relations. This principle applies to *any* relation among ideas. In particular, it shows that the idea allegedly designated by the numeral '2' might so evolve that '$2 + 2 = 5$' might come to express a true relation between this idea and the idea associated with '5' (FA, §27). Similarly, laws of logic, which also must state relations among ideas, would prove to be mere contingent laws of mental associations of ideas. Frege found these implications of psychologism ridiculous, and he was content simply to state them and then propose a Platonist alternative that escaped these absurdities. This procedure is philosophically weak. An empiricist like Hume might welcome the 'accusation' that ideas are governed only by contingent laws of association; necessary truths might be declared to be trifling propositions that dress up the same idea in different verbal clothing. On the other hand, Descartes or Kant, for different reasons, would deny Frege's premise that all genuine relations among ideas must be contingent; such relations might flow from 'simple natures' or from the *a priori* structure of the intellect. Confrontation with Frege's Platonist realism about logic and mathematics demolishes none of these philosophical positions, nor does it reveal the deep sources of error and confusion that they severally manifest.

So far, so not very good. We learn little more from Frege than a handful of platitudes about assertion and the subject of assertion. He was, however, on much stronger ground when he abstained from his rather elementary forays into general epistemology and metaphysics, and concentrated on a simpler, straightforward thesis. What belongs to content is exclusively what is relevant to the validity of inferences. But what ideas, if any, different people may associate with a word is typically irrelevant to the validity of the inferences they draw from a judgment expressed by a sentence containing that word. Different men may associate different ideas with the word 'only', but that does not impugn the inference from 'Only A is F' to 'B is not F' (cf. FA, §59). With respect to some subjects of judgment we can form no idea at all, e.g. the number 0, but that does not affect the inference from 'The number of visible stars is 0' to 'The North Star is not visible'. Other subjects of judgment are such that we can form no adequate idea of them, e.g. our distance from the sun or the size of the Earth, but this does not prevent us from taking that distance or size as a fact upon which to base further inferences (FA, §59). Consequently the idea associated with a word is (at least, typically) irrelevant to its content.

> Time and time again we are led by our thought beyond the scope of our imagination, without thereby forfeiting the support we need for our inferences. Even if, as seems to be the case, it is impossible for men such as we are to think without ideas, it is still possible for their connection with what we are thinking of to be entirely superficial, arbitrary and conventional. (FA, §60)

The fundamental mistake which his empiricist antagonists in the philosophy of mathematics committed is not, in Frege's view, the assumption that words stand for something (a content). That lynchpin of the Augustinian picture of meaning was, and remained, central to Frege's vision. It is rather that his opponents held that if the content of a word is not a perceptible entity, then it must be an idea. But numbers are neither material objects (heaps or collections) nor material properties, yet they are not ideas either. We must, in his view, recognize a third class of entities, neither perceptible nor mental, but abstract, and it is such real but 'non-actual' entities that number-words have as their contents.

Frege's detailed criticisms of psychologism in logic and mathematics are more successful than his global attacks:

(i) Not only do the psychological logicians assimilate objects to ideas, they also assume that concepts are ideas, and hence mental and subjective. But there is no such thing as an object's falling under an idea. To say of an object that it falls under a concept is not formally to attribute to the object a mental property. If it does fall under that concept, its so doing is immune to the fluctuating fortunes of such mental ephemera as ideas. Since the psychologicians conceive of all unjudgeable-contents as ideas, they are forced to construe predicating a concept of an object as predicating an idea of an idea. But there is no such thing. Not only does 'A is green' not predicate green of any idea of A, since there is no such thing as a green idea, but the suggestion that ideas (e.g. the idea of green) can be asserted *of* anything is unintelligible (BLA, p. xxi).

(ii) If all contents were ideas, it would be impossible to explain how cognitively non-trivial identities are possible (RH, 327). For if the content of 'A' is a certain idea, and the content of 'B' is the same idea, then the content of 'A $\equiv$ B' cannot differ from that of 'A $\equiv$ A'.

(iii) Psychologism cannot distinguish first- from second-level concepts (BLA, xxiv f.). No amount of play with ideas can account for generalization, statements of existence or attributions of number (including 0) in count-statements. Number-words do not have ideas as their contents (FA, §27), and we cannot arrive at an idea of number by the psychologicians' favoured method of concept-formation, viz. abstraction (RH, 324, 330; cf. FA, §45). Generality, existence, and possession of number are all properties of concepts, not of objects, and to this domain abstractionism gives no access.

(iv) Psychologism fails to draw crucial distinctions which are, in Frege's view, absolutely fundamental to the development of the science of logic, and it relies upon obsolete distinctions which are an impediment to progress. In particular, psychologicians continued to employ the essentially grammatical categories of subject and predicate (PW, 142f.), failed to distinguish between concepts and objects (BLA, p. xxiv), and confused the characteristic marks of a concept with its properties (BLA, pp. xiv, xxiv).

These criticisms carry some weight. (i) We would surely agree that concepts are not peculiar mental entities, images, pictures, or other such aethereal ephemera. But it is clearly fallacious to conclude that since they are not mental, they must be abstract, sempiternal entities discovered by the mind (FA, vii). Between psycholgistic subjectivism about concepts and Fregean Platonism there has surely been an excluded middle! (ii) So too we would agree that talk of asserting or predicating ideas of ideas is *prima facie* nonsensical. But it is noteworthy that Berkeley tried to ward off a similar criticism with the remark that one must 'think with the learned and speak with the vulgar', for the term 'idea' is not commonly used to signify 'the several combinations of sensible qualitites, which are called *things:* and it is certain that any expression which varies from the familiar use of language, will seem harsh and ridiculous. But this doth not concern the truth of the proposition . . .'[40] Frege, who taught that 'All whales are mammals' is not about whales, but about the concept whale, and that 'the concept of law is open-textured' is not about a concept at all, but about an object, might be expected to take a slightly more charitable view and generous interpretation of such strategies. Merely to point out that we do not understand what it is to assert an idea of an idea, or that we can make no sense of ideas being coloured, etc. will hardly relieve the philosophical pressures which impelled great philosophers to propound such bizarre propositions. (iii) His criticism that idealists did not explain the possibility of cognitively non-trivial identities does have some historical foundation. However, an idealist is not in principle debarred from adopting a variant of Frege's own ultimate solution to this problem, arguing that ideas correspond to senses and represent (typically) objective items (as Descartes and Locke argued). On this view '$A \equiv B$' expresses the fact that two distinct ideas represent one and the same object. (iv) Finally, when it comes to Frege's critical remarks on classical and psychologistic struggles with concepts of generality, existence, and number, one must concede that he is on strong grounds. Their theories were indeed hopelessly inadequate, and the criticisms in the *Foundations* and the 'Review of Husserl' are powerful.

However, Frege's critical remarks betray that his liberation from the psychologistic inheritance is less than complete. His discussion is typically restricted to second-level concepts and abstract objects. Much of the criticism focusses upon the impossibility of obtaining ideas of numbers by abstraction. But what is less commonly noted is that the *Foundations* did *not* present abstractionism as a wholly bankrupt theory. On the contrary, he thought that numerous first-level concepts are 'arrived at' by abstraction (cf. FA, §§34, 48). Colour, weight, and hardness are cited as examples of such concepts (FA, §45); successive operations of abstraction allegedly generate series of concepts such as 'satellite of the Earth', 'satellite of a planet', 'non self-luminous heavenly body', 'heavenly body', 'body', 'object' (FA, §44). Abstraction has a legitimate place in concept-formation. The

---

40. Berkeley, *Principles of Human Knowledge,* §xxxviii.

only danger lurking here is falling into the supposition that *only* by abstraction can we arrive at a concept (FA, §§48ff.); rather, we can 'arrive at a concept equally well by starting from defining characteristics'. Far from being a criticism of abstractionism, this point is one of its standard theses. A psychological logician will allow that 'complex ideas' may be formed combinatorially from antecedently abstracted simple ideas. He will merely insist that simple ideas must be formed by abstraction. But once thus formed, complex concepts under which nothing falls can be constructed. Frege did not think that his proof that number-words are definable stands in conflict with adherence to abstractionism as *one* method of concept-formation.

Frege had no *general* criticism of the doctrine of abstraction as a theory of concept-formation, nor any *general* theory of concept-acquisition of his own with which to replace it. Nor indeed did he see the construction of such a theory to be part of his task as a logician; such investigations, in his view, belong to psychology, and can safely be disregarded by logicians. All that needs to be established is the negative point that introduction of concepts in a mathematico-logical theory need not be restricted by psychologistic accounts of concept-formation. He was oblivious to the difficulties which abstractionism encounters with a wide variety of kinds of concepts other than second-level concepts and concepts of allegedly abstract objects. He did not notice problems concerning formal concepts such as substance which even the psychologistic empiricists noted. He seemed unaware of the impossibility of giving a plausible abstractionist account of concepts of logical connectives such as conditionality or alternation. He paid no attention to the contortions necessary for psychologism to distinguish the difference between arriving by abstraction at the concept of a relation (e.g. to the left of) and arriving at the concept of its converse (viz. to the right of). He was oblivious to the difficulties abstractionism must encounter in explaining acquisition of concepts of relative properties expressed by attributive adjectives, and showed no unease over abstractionism with respect to colour concepts (in particular, for example, the difficulty of arriving at the concept of colour itself).[41]

The target of Frege's arguments was very narrow. Representative idealism lay outside the range of his fire. We would look in vain to his work for a cogent refutation of the various forms of idealism and abstractionism. Furthermore, any generalization of his criticisms will lack the power of profound philosophical insight. This is not surprising. He was too much out of sympathy with contemporary German psychologism to uncover its roots and reveal the deep temptations that had driven great geniuses from Descartes through Locke and Berkeley, to Hume, Bentham and Mill (not to mention the continental branch of the family), to embrace psychologism in one form or another. His concern was too polemical, his tone too sneering, and his vision too narrow for his criticism to be more than superficial. He was not even in a position to root out the deep errors of psychol-

---

41. For a detailed criticism of abstractionism, see P. T. Geach, *Mental Acts,* §§6-11 (Routledge and Kegan Paul, London, 1957).

ogism because of his own uncritical acceptance of some of its basic preconceptions and characteristic ideas.

6. A Platonist core with a Cartesian penumbra

Frege's endeavour was aimed at insulating logic from psychology. What he tried to do was to eliminate every vestige of mentalism from the proper objects of scientific investigation in logic. In respect of the objects of judgment and in respect of concepts too, he replaced Cartesian myths and confusions with Platonist ones. We noted that he harped on the objectivity and publicity of objects of judgment in a manner more closely akin to Bolzano's claims than to Mill's. In grasping a thought 'what is grasped, taken hold of, is already there and all we do is take possession of it ... [It] does not come into existence as a result of [any mental] activities' (PW, 137; cf. BLA, p. xxiv; T, 29n., 34f.; N, 122). Frege's insistence upon the mind-independence of *what is judged or asserted* led him to conceive of the object of judgment as a peculiar, non spatio-temporal *substance*. The content of judgment is *as objective* as the sun, only not in space and time (PW, 7); thoughts are 'like physical bodies', except that they are non-spatial and timeless (PW, 148). The process of grasping a thought is psychological, but the thought grasped is not (PW, 145). Thoughts, like numbers are real but not actual, objective but neither spatio-temporal nor perceptible (T, 20, 29, 35ff.; PW, 198).

These curious non-temporal entities nevertheless enter into temporal transactions with the sublunary world, for they are, in a wholly inexplicable manner, grasped, apprehended, communicated, affirmed or denied by human beings. These psychological acts, like all acts, occur in time, and it is in virtue of them that thoughts have causal efficacy in the actual world. For as a result of apprehending a thought thinkers act and move '[a]nd so thought can have an indirect influence on the motion of masses' (T, 38). How a supra-temporal entity can have temporal relations to actual objects is a mystery with which theologians have struggled for centuries. Frege tried rather feebly to sidestep the problem by agreeing that 'the strict timelessness is of course annulled' by a thought causally affecting a thinker through being apprehended, but tried to minimize this concession by insisting that being apprehended is an *inessential* property of a thought (T, 37): something is timeless if the changes it undergoes involve only inessential properties. This, one might feel, is to wrap up a mystery in an enigma. Being thought by a person is no more a property (whether essential or inessential) of the thought entertained than is being thought *about* by a person a property of Mont Blanc or of the number 2. If there are any abstract entities such as Fregean thoughts, they cannot enter into non-reciprocal causal relations with thinkers. This, for Frege, is tantamount to denying the reality of thoughts: 'What value could there be for us in the eternally unchangeable which could neither undergo effects nor have any effect on us? Something entirely and in every respect inactive would be unreal and non-existent for us' (T, 37). *Ipse dixit.*

As with many of his predecessors, the distinction between the act and object of judgment allowed Frege cheerfully to surrender the acts of judgment, inference, and even assertion to the sway of traditional psychologistic analysis. The objectivity of thoughts as the objects of these allegedly mental acts secured for him a beachhead for developing the science of logic independently of psychology. The Cartesian myths were left untouched within what seemed to be their proper psychological domain, while Frege, his flanks seemingly well-guarded against intrusion from psychologism, proceeded to develop his Platonist fantasies.

Frege's failure to challenge prevalent Cartesian myths about mental phenomena is a major defect in his thinking. His anti-psychologism was, paradoxically, insufficiently radical. Having isolated judgeable-contents (later thoughts), concepts, and abstract objects and secured them from psychological contamination by a wrapping of Platonism, he saw no need to carry the anti-psychologistic flag into the heartland of classical empiricism and German psychologism, the domain of the mind. This left him in the grip of a multitude of confusions. His criticism of the psychologicians' conception of the objects of judgments and of what judgments are about turn on a confused conception of subjectivity that he shared with his opponents. His ready acceptance of abstraction from experience as a legitimate source of concept-formation is difficult to reconcile with his analysis of concepts as functions whose values are always truth-values. His conception of assertion as the externalization of a mental act of judgment and his conception of inferring as essentially a transition between mental acts stood in the way of his arriving at a proper semantic account of logical validity and drove him into absurdities reflected even in the construction of his formal system of logic. His conception of understanding as a mental act directed at a Platonic object left the nature of understanding a mystery whose clarification he expressly declared not to be the proper business of a logician; in this way, he severed the internal connection between meaning and understanding and thereby debarred himself from explaining the fundamental conceptual articulations in this whole domain in terms of this crucial relationship.[42] He left himself too with a naïve and incomprehensible conception of thinking: thinking is a mental process, necessarily involving mental objects (images), in which we mysteriously come into contact with Platonic entities (thoughts or judgeable-contents). This is an exemplary instantiation of Wittgenstein's observation that 'We interpret the enigma created by our misunderstanding as the enigma of an incomprehensible process':[43]

> grasping *(Erfassen)* ... is a mental process! Yes, indeed, but it is a process which takes place on the very confines of the mental and which for that very reason cannot be completely understood from a purely psychological standpoint. For in grasping ... something comes into view whose nature is no longer mental in the proper sense, namely the thought; and this process is perhaps the most mysterious of all. But just because it is mental in character we do not need to concern ourselves with it in logic. It is enough for us that we can grasp thoughts and recognize them to be true; how this takes place is an independent question. (PW, 145)

---

42. Cf. Baker and Hacker, *Wittgenstein: Understanding and Meaning,* 664ff.
43. Wittgenstein, *Philosophical Grammar,* p. 155.

Failure to confront the Cartesian myths about the mind landed Frege in important confusions and impenetrable mysteries. In his thought there is a Cartesian conception of the mind wedded to a Platonic conception of its objects. Can we expect the offspring of such a miscegenous union to be fruitful? What we need here is a careful clarification of such concepts as understanding and the criteria of understanding, intending or meaning something by what one says, thinking, believing, judging, asserting, inferring, acquiring, and possessing concepts, and the criteria of concept-possession. To these matters Frege had nothing to contribute. But since a correct grasp of these concepts is essential for philosophy of language and philosophical logic as well as for the philosophy of mind, this fundamental weakness in his arguments against psychologism has important and widely ramifying consequences.

His naïve Platonism is equally a major defect in his antipsychologism. He erroneously assumed that the only alternative to psychologism is Platonism. If the objects of judgment are not mental entities, they must be abstract objects existing in a third world of entities which are real but not actual. Similarly, if numbers, sets, truth-values, etc. are not ideas, then they too must be Platonic objects (since they cannot themselves be symbols). The presumption that these options exhaust the possibilities is unsupported by argument and indefensible. In part it rests on Frege's unwavering adherence to the Augustinian picture, according to which expressions have significance (or content) only in virtue of standing for something. He shared this view with the psychologicians. His criticism of psychologism turns solely on arguing that psychologicians misidentified what certain expressions stood for (i.e. ideas rather than abstract entities). He was therefore not in a position to expose the deep confusions embedded in the Augustinian picture and in the conception of communication or mutual understanding as a sharing or transfer of some entity. His own Platonism clarifies nothing in this area. Indeed, it leaves communication and understanding just as mysterious and problematic as psychologism does. It transforms the task of accounting for what it is that we think and judge to be true into the task of explaining what it is for us to think what we think and to judge this to be true. If what we think is an abstract non-psychological entity, then what is involved in 'grasping', 'conceiving', 'laying hold of', 'seizing', etc. (PW, 137) this imperceptible public entity? What is it to think the sense of a sentence? Similarly, what is it to apprehend a number or a set, or to grasp the sense of a numeral or of the name of a set? Frege relegated the clarification of these mysteries to psychology (PW, 145) and thereby justified his own neglect to resolve the perplexities generated by his Platonism. His strategy here is doubly misguided. First, it offers no advance over psychologism in explaining communication and understanding. In particular, anything which could provide evidence that different persons apprehend the numerically identical abstract thought could be reinterpreted by an idealist as evidential support for their having qualitatively identical ideas. Hence what Frege's Platonism wins on the swings it loses on the roundabouts! Second, the attempt to explain communication and understanding in terms of common apprehension of abstract entities stands squarely in the way of arriving at what we now consider to be a correct logical point of view on these matters. We must look to the conventions and practices of the use of expressions to find the grounds of mutual understanding, the significance of discourse about numbers, sets, truth-values, etc., and the common content of communication. Thought, assertion, judgment, and understanding make contact in language just as do desires and their fulfilments, orders and their executions. But for clarification of this kind, Frege's Platonic entities become as irrelevant and dispensable as mental representations.[44] This strategy cuts the ground out from under the Augustinian picture, and it leads to a conception of language, thinking, and understanding that is antithetical both to psychologism and to Platonism. This is a possibility which was literally inconceivable to Frege, and its crucial importance prevents his arguments against psychologism from making any serious contribution to contemporary philosophy.

The distinctive amalgam of Cartesianism and Platonism in Frege's thinking must characterize any reasoning that could be called an extention of *his* arguments against psychologism. It would also deprive any such reasoning of real philosophical value. Both the Cartesian myths about the realm of the psychological and the Platonic myths about the realm of the logical generate deep philosophical confusions, and hence these central ingredients of Frege's thinking are themselves in dire need of philosophical investigation and clarification. The only conceivable sources of light must be alien to his framework of thought. Consequently, we must conclude that Frege's crusade against the incursions of psychology into logic is now largely obsolete. His way of drawing the distinction between logic and psychology is mistaken in detail and dangerous in its wider implications. Only somebody who shares a large measure of his Cartesian and Platonist mythology will find any seeds of the Tree of Knowledge scattered in his antipsychologistic polemic.

---

44. Unless we take the wholly unjustifiable step of identifying what Frege called 'sense' with the rules for the correct application of an expression (cf. Dummett, *Frege: Philosophy of Language,* p. 194).

3
What Frege Inherited: Language and Grammar

1. Thinking and language

Frege's 'sharp distinction' between the act of judging and the content of judgment enabled him to confine psychologism to the investigation of mental acts and processes of thinking and judging, leaving the object of thought or judgment purified of any psychological dross. It is the judgeable-content, or 'the thought', which is the concern of logic. This, as noted, he conceived of as an immaterial, non spatio-temporal entity. Thinking, in his view, is an activity of coming into intellectual contact with these entities.

For us humans, thinking necessarily involves having images (FA, §60; PW, 105, 142). Frege neither justified this supposition nor explained this necessity. Probably he thought the claim a truism and the necessity psychological. In the Kantian climate of much nineteenth-century thought this was uncontentious.[1] We might wish to quarrel even with the modest claim that having images is causally necessary for thinking, but we would surely insist that having images is logically neither necessary nor sufficient for it.

Does thinking necessarily involve language or symbols? The prevailing tradition in philosophy conceived of the relation between the psychological acts or processes of thought and their linguistic expression as *external* and contingent. Hobbes noted that

---

1. It might be illuminatingly juxtaposed with a more theory-laden passage in Lotze: 'the train of ideas alone is not *Thinking*, and does not itself discharge the offices which we require of the latter. ... the mere presence of ... images—products of the mechanical course of ideas—is not equivalent to the possession of *Concepts*, in whose form Thinking refers the manifold content to its corresponding *Universal*. For in the latter is always implied the subsidiary thought of a determining rule ...' (H. Lotze, *Microcosmus*, II-iv-§4).

the most noble and profitable invention of all other, was that of *speech,* consisting of *names* or *appellations,* and their connection; whereby men register their thoughts, recall them when they are past, and also declare them one to another....

The general use of speech is to transfer our mental discourse into verbal, or the train of our thoughts into a train of words....[2]

A decade later the Port Royal logicians expressed similar sentiments:

words are *sounds distinct and articulate, which men have taken as signs to express what passes in their mind.* And since what passes there may be reduced to *conceiving, judging, reasoning* and *disposing*..., words serve to indicate all these operations.[3]

Towards the end of the century Locke gave this conception its best known formulation in the *Essay.* Words are

immediately the signs of men's ideas, and by that means the instruments whereby men communicate their conceptions, and express to one another those thoughts and imaginations they have within their own breasts....[4]

Under the impact of Kant's complex theory of the necessary categories of thought, this naïve conception was modified. Nineteenth-century discussions manifest a movement away from the bare ideational picture adopted by most seventeenth-century thinkers, but it is altogether superficial. Hamilton, writing in 1837 and much influenced by German philosophy, suggested that

though, in general, we must hold that language, as the product and correlative of thought, must be viewed as posterior to the act of thinking itself; on the other hand, it must be admitted that we could never have risen above the very lowest degrees in the scale of thought, without the aid of signs. A sign is necessary to give stability to our intellectual progress.[5]

Mansel, his follower, was more explicit in this respect:

[An] important characteristic of all concepts is, that *they require to be fixed in a representative sign.* This characteristic cannot indeed be determined *a priori,* from the mere notion of the concept as universal, but it may be proved to be a moral certainty *a posteriori,* by the inability of which in practice every man is conscious, of advancing, without the aid of symbols, beyond the individual objects of sense or imagination....

Language, taking the word in its widest sense, is thus indispensable, not merely to the communication, but to the formation of Thought.[6]

This climate of opinion evidently prevailed among German philosophers of the

2. Hobbes, *Leviathan,* ch. IV.
3. *The Art of Thinking* (the Port Royal logic), Part II, ch. i.
4. Locke, *An Essay Concerning Human Understanding,* III-ii-6.
5. Hamilton, *Lectures on Metaphysics and Logic,* III, p. 138.
6. Mansel, *Prolegomena Logica,* p. 15 ff.

nineteenth century. Trendelenburg, writing in 1867, moved along similar well-worn tracks:

> Through the sign, ideas which would otherwise be diffuse are separated and become, as separated elements, a permanent possession which the thinker can use. Through the sign distinctions are drawn, the distinguished is fixed, and the fixed becomes capable of new connections. Through the sign the idea detaches itself from the sensory impression ... and now becomes capable of lifting itself up into the general. Thus, through the verbal sign, thinking becomes on the one hand free and, on the other, definite.[7]

Lotze, however, tended to withdraw even from this very modest association of thought and language:

> Now that man has come to use the language of sounds for the communication of his thoughts, that activity is, it is true, most clearly manifested in the forms of the parts of speech; but in itself it is not inseparably bound up with the existence of language. The development of which the ideas of the deaf and dumb are capable, though guided in the first instance by those who can speak, is enough to show that the internal work of logic is independent of linguistic expression.... No one doubts the extremely effective support which language gives to the development of thought by making the formations and transformations of ideas vividly objective to consciousness by means of sharply defined sounds ... [but] if in some languages the poverty of forms does not always allow these associations to take shape ... yet there is no doubt that the mind of those who speak them maintains the logical distinctions while forming thoughts which are vocally undistinguished. Wherever there is this inward articulation, there is thought; where it is wanted, there is no thought.[8]

Displayed against this background, Frege's conception of the matter is unsurprising. Like his predecessors he seems, at least initially, to have thought that symbols 'fix' or 'stabilize' ideas:

> if we produce the symbol of an idea which a perception has called to mind, we create in this way a firm, new focus about which ideas gather. We then select another [idea] from these in order to elicit *its* symbol. Thus we penetrate step by step into the inner world of our ideas and move about there at will, using the realm of sensibles itself to free ourselves from its constraint. Symbols have the same importance for thought that discovering how to use the wind to sail against the wind had for navigation. (SJCN, 83f.)

Likewise, he claimed that 'without symbols we would scarcely lift ourselves to conceptual thinking', for the *concept,* what similar things have in common, is 'first gained by symbolizing it' (SJCN, 84). In some places he suggested that we think *in* words or in mathematical or other symbols (SJCN, 84) or, at any rate, in some language or other (PW, 142). This commonplace supposal might seem to give

---

7. A. Trendelenburg, 'Über Leibnizes Entwurf einer allgemeinen Characteristik', *Historische Beiträge zur Philosophie,* vol. 3 (1867), p. 1, quoted and tr. by H. Sluga, *Gottlob Frege,* p. 50.
8. Lotze, *Logic,* I-i-§6.

support to his correct contention that thought, in its higher forms, is only made possible by means of language (PW, 143). Yet, though commonplace, it is confused. When I perform an elementary computation I do not think *in* numerals. When I translate Gibbon's *Decline and Fall* into German, I read his mellifluous sentences, but I do not think in Gibbonese. I *think up* German equivalents, but I do not think *in them* either. Neither what I think about (Gibbon's prose) nor what I think up (my translation) are what I think *in,* even though I may silently soloquize in English or German.[9] The issue is contentious and problematic. It calls out for philosophical elucidation. Frege provided *none*. Clearly he took it as an unproblematic psychological observation. In one striking passage, he raised the question of how it is possible for thinking to appear to be in competition with speaking. We are only imposed on by this view, he suggested, because thinking emerges in the development of an individual as 'an inaudible inner speaking'; but this, he declared, does not capture the true nature of thinking. Having this mere glimpse of the truth was all he was granted. Like Berkeley before him, who, seeing the pitfalls of language, recommended that we should think only in ideas, not in words, Frege concluded that we should disregard how thinking occurs in the consciousness of an individual and attend to its true non-linguistic nature: 'In that case we shall not derive thinking from speaking; thinking will then emerge as that which has priority and we shall not be able to blame thinking for the logical defects ... in language (PW, 270).

Enmeshed in his Platonist conception of objects of thought, he followed traditional wisdom of conceiving of language as being externally related to the thoughts it is used to express. Nor did he see any internal relation between the 'acts' of thinking, inferring, reasoning, and the *use* of language. Consequently he thought of possession of a linguistic ability as essential only for such creatures as we, with all our 'medical limitations'. *We,* because we think *in* language, must master language to think what he called 'conceptual thoughts':

> That a thought of which we are conscious is connected in our mind with some sentence or other is for us men necessary. But that does not lie in the nature of thought, but in our nature. There is no contradiction in supposing there to exist beings that can grasp the same thought as we do without needing to clothe it in a form that can be perceived by the senses. But still, for us men, there is this necessity. (PW, 269)

On this conception, the only necessity relating thinking or inferring to using language is *causal*. There is no logical necessity that the structure of a thought be reflected in the structure of a sentence expressing it. Languages, whether natural or invented, may accordingly differ in respect of their capacity accurately to mirror thought-structures in sentence-forms. These points constitute insuperable obstacles to construing Frege's philosophical remarks as deliberate contributions

---

9. G. Ryle, 'Thought and Soliloquy', pp. 41 f., in *On Thinking* (Blackwell, Oxford, 1979).

to the enterprise now called the philosophy of language.[10] His reflections on thinking, as we shall show, belong to a totally different climate of thought, and his observations about language have a radically different purpose.

2. Language, grammar, and logic

While Frege vehemently rejected psychologicians' conception of the objects of thought, he accepted their view of language as encoding something which may exist quite independently of it. They typically pictured language as encoding thoughts and ideas, conceived as psychological entities. Frege pictured it as encoding contents of judgment. The only serious alternative to psychologism about thoughts was Platonism. His naïve Platonism about judgeable-contents and, later, thoughts inclined him to view language as more or less isomorphic with entities that are language-independent without being in the least degree psychological.

Viewed from the only vantage points available to a typical nineteenth-century thinker, the supposition of an external relation between language and the thoughts or judgments it expresses was altogether plausible. One and the same judgeable-

---

10. Dummett claims that Frege reoriented modern philosophy by recognising 'that the philosophy of language is the foundation of the rest of the subject . . . because it is only by the analysis of language that we can analyse thought' (Dummett, *Truth and Other Enigmas*, p. 441f.). He allegedly established the 'proper object of philosophy . . . : namely, first, that the goal of philosophy is the analysis of the structure of *thought;* second, that the study of *thought* is to be sharply distinguished from the study of the psychological process of *thinking;* and, finally, that the only proper method for analysing thought consists in the analysis of *language*' (ibid. p. 458). The crucial third point is ascribed to Frege on the ground that he acknowledged the basic principle that 'if the philosopher attempts . . . to strip thought of its linguistic clothing and penetrate to its pure naked essence, he will merely succeed in confusing the thought with the inner accompaniments of thinking' (ibid. p. 442). Dummett here presents the thesis that the analysis of thought is the analysis of language as the only escape from the sins of psychologism, and he credits Frege with this methodological discovery. But far from intending to make this point, Frege never saw it. He deemed it a sufficient defence against psychologism to recognize thoughts (or judgeable-contents) as Platonic objects. From the fact that he denied thoughts to be mental entities we can no more legitimately conclude that he conceived the analysis of thoughts to be the analysis of declarative sentences than we can conclude from his denying that numbers are ideas that he conceived arithmetic to be the grammar of number-words.

For Frege, the relation of a thought to its linguistic expression is as that of a Platonic form to the visible shadows it casts upon the walls of Plato's cave. Such shadows are blurred, their forms distortions of the real entities they represent, although for the benighted denizens of the cave, they are the only visible representations of the objective realities that are to be studied. From the fact that Frege, like virtually all logicians, thought that language is *a* guide to the nature of thoughts, it does not follow that he conceived of logic as a study of the structure of language. From the fact that he invented a symbolic language which he conceived to be a more accurate and perspicuous representation of the structure of thoughts than natural language, it does not follow that he believed the structure of concept-script to be present in the sentences of language or in their (mythical) 'deep-structure'. He was interested in the currency of language only to the limited extent that it reflected the gold backing of Platonic objects. But to qualify as a philosopher of language one must come off the gold standard and investigate the *uses* of the paper currency of ordinary language.

content or thought may be expressed by very different symbols. Hence, Frege noted, 'the connection of a thought with one particular sentence is not a necessary one' (PW, 269). Language is a *vehicle* for the expression and communication of the thoughts we think:

> [I]t is not the sentence itself that really concerns us when we speak, but the sense or content which we associate with it and which we wish to communicate. Since the sense itself cannot be perceived by the senses, we have perforce, in order to communicate, to avail ourselves of something that can be perceived. So the sentence and its sense, the perceptible and the imperceptible, belong together. (PW, 167)

Given this loose relation between content or thought and language, it is not surprising to find that Frege emphasized the incomplete correspondence between the structure of the abstract entity which is what we think, judge, or assert, and the structure of the sentence we use to convey, express, or assert it. The idea that thought, whether conceived Platonistically or psychologistically, is only imperfectly reflected in the structures of natural language was, of course, widespread. It was indeed a corollary of conceiving the relation between thought and language as external. Mill, who also sharply distinguished the acts of thinking, believing or judging from their objects, opened *System of Logic* with the observation that

> Language is evidently, and by the admission of all philosophers, one of the principal instruments or helps of thought; and any imperfection in the instrument, or in the mode of employing it, is confessedly liable, still more than in almost any other art, to confuse and impede the process.... For a mind not previously versed in the meaning and right use of the various kinds of words, to attempt the study of methods of philosophizing, would be as if someone should attempt to become an astronomical observer, having never learned to adjust the focal distance of his optical instruments so as to see distinctly.[11]

This conception was equally well-known amongst German writers. Lotze noted that

> Speech may pass over much that thought, in order to be complete, must include; as in everyday conversation many connecting members are left to be understood by the listener, so even the typical forms of construction of a language may be an incomplete, but for all purposes sufficient, expression of the articulation of thought. It is then to make a needless demand to require that the verbal organization of discourse shall fully correspond to the logical organization of thought.[12]

Frege trod parallel paths. Language is the means whereby what is found in the

---

11. Mill, *System of Logic,* I-i-1. Hamilton had been even more emphatic: 'No tongue, how perfect soever it may appear, is a complete and perfect instrument of human thought. From its very conditions every language must be imperfect.... No language can ... be adequate to the ends for which it exists; all are imperfect, but some are far less incompetent than others' (*Lectures on Metaphysics and Logic,* IV, p. 143).
12. Lotze, *Microcosmus,* V-iii-§4.

Platonic world of sempiternal thoughts is expressed in perceptual garb. And it is precisely because the relation between thought and language is external that the defective forms of natural language stand in the way of giving a correct account of inference. He would therefore have agreed with Mill that

> logicians have generally felt that unless, in the very first stage, they removed this [linguistic] source of error; unless they taught their pupil to put away the glasses which distort the object, and to use those which are adapted to his purpose in such a manner as to assist, not perplex his vision; he would not be in a condition to practise the remaining part of their discipline with any prospect of advantage.[13]

Indeed, Frege felt himself able to go considerably farther than such familiar observations. For with the discovery of a powerful concept-script, he believed that he had, for the first time, revealed the *true* structures of thought, which in turn gave him a deeper insight into the shortcomings of language. He instanced a variety of cases where he held that the grammatical forms of sentences used to make assertions conceal the structures of the judgeable-contents expressed. So, for example, 'There are four horses in the stable' conceals the fact that a constituent of the judgeable-content asserted, namely, a number, is an *object*—a feature only transparent in the paraphrase 'The number of horses in the stable is identical with four'. Existential sentences look as if the property of existing is attributed to an object, whereas in reality statements of existence assert that a first-level concept falls under a second-level one (FA, §53; PW, 101). Such defects, in Frege's view, are repeated at the level of sentence-components. 'There is only an imperfect correspondence between the ways words are concatenated and the structure of the concepts' (PW, 12). The grammatical forms of words and phrases frequently hide the logical forms of the unjudgeable-contents they represent. Some second-level concepts appear in linguistic guise as if they were proper names of objects (e.g. 'nobody') or names of first-level concepts (e.g. 'exists'). Failure to reflect the logical type of the entity represented occasionally permits the formation of strictly illegitimate sentences. The concept of a function really must be a second-level concept, whereas in language it always appears as a first-level concept (PMC, 136), although, Frege admitted, one cannot say this in ordinary language, for the expression 'the concept of a function' does not even stand for a first-level concept, let alone a second-level one; it unavoidably stands for an *object* (PW, 177f., 250, 255).

In other ways too, language allegedly fails to fulfil the requirement of protecting our thought from logical error (SJCN, 84). First, it is rife with 'ambiguity'. Different senses are associated with one and the same proper name (SR, 58n.). The same word is often used to stand now for a concept, now for an object (SJCN, 84; cf. CO, 50; PMC, 92). This leads to fallacies of equivocation. Second, proper names and definite descriptions can be formed which denote nothing, thus depriving sentences in which they occur of any content (FA, §§95ff.) or, later, of truth-

---

13. Mill, *System of Logic*, I-i-1.

value (SR, 70). Indeed, since singular referring expressions that lack reference can be legitimately formed, ordinary language contains existential presuppositions the truth of which is a condition for the relevant sentences to have a truth-value (SR, 69). This allegedly makes the applicability of the laws of logic to the thoughts expressed dependent upon the facts! Similarly, since ill-defined concept-words occur in ordinary language, sentences lacking any content (BS, §27) or truth-value can be formed out of them. In short, ordinary language 'is not governed by logical laws in such a way that mere adherence to grammar would guarantee the formal correctness of thought-transitions *(Gedankenbewegung)*' (SJCN, 84f.). It is noteworthy that in this respect Frege's thought is inconsistent with modern semantic conceptions of validity. Since he thought that all valid inference must be from true assertions to true assertion, the occurrence of a sentence lacking a truth-value or not actually representing a judgeable-content (either because of reference-failure or because of vagueness) would, in his view, immediately invalidate an argument. Third, the diverse methods of forming complex concept-words in natural language (e.g. 'lifeboat', 'deathbed' (PW, 13)) are not systematic, nor perspicuous. The form of the symbolism (unlike the symbolism of mathematics, e.g. '$\log(\sin \pi e^2)$') does not make perspicuous the logical relations among the constituents of the complex concept, but rather leaves it a matter of guesswork (PW, 12f.). This too, Frege thought, should be remedied in a logically perfect language. He was, of course, right about the irregular and unsystematic nature of our modes of composition of complex concept-words, but wrong to think that knowledge of the correlative content has, for one who has mastered the language, anything at all to do with *guesswork*. A competent speaker of English does not have to *guess* that 'lifeboat' means 'a boat for saving lives' and that 'deathbed' does not mean 'a bed for saving deaths' any more than a mathematician has to guess the order of operations involved in calculating $\log(\sin \pi e^2)$. Finally, 'the forms in which inference is expressed are so varied, so loose and vague, that presuppositions can easily slip in unnoticed and then be overlooked when the necessary conditions for the conclusion are enumerated' (SJCN, 85). Frege gave as an example Euclid's tacit and unnoticed use of presuppositions which are specified neither in his axioms and postulates, nor in the premises of the theorems which implicitly depend on them. This too should be remedied in an ideal notation. Again, we might acknowledge Frege's point, yet feel sceptical about his assumption that a wholly perspicuous ideal notation could *guarantee* that no hidden presuppositions should occur in reasoning. While a suitably devised formal notation may (more or less) guarantee what he called 'a "gapless" advance' in the argument, just as many 'hidden presuppositions' may creep in in the multitude of unavoidably pragmatic and intuitive decisions that have to be made in *translating* a problem from natural language into the preferred concept-script.

These logical defects of natural language do not always matter. Vague concept-words are meaningless from a logical point of view, but they are used thousands of times in everyday language. Vernacular languages, after all, are not specialized instruments for conducting proofs. But when proofs are put to the test, such defects, Frege thought, do become important, and it is there that they must be prohibited (PMC, 115). Different speakers are held to associate different senses with an ordinary proper name, but as long as they thereby pick out the same referent this does not matter (SR, 58n.). Only in the theoretical structure of a demonstrative science must this be prohibited. In a perfect language *designed for such purposes,* it must not occur, since disagreement over sense allows verbal agreement to mask disagreement over what constitutes a valid proof. Similarly, every word should have the same sense in every context if proofs are to be mechanically checkable, but 'natural languages often do not satisfy this condition, and one must be content if the same word has the same sense in the same context' (SR, 58; cf. PMC, 115). The grammatical forms of ordinary language are multiple, even exuberant[14] (PMC, 68), but many of them are confusing and logically irrelevant. It is the task of logic to refine the forms for the representation of inferences by eliminating all redundancy and by accurately reflecting the true forms of thought.

In his early reflections on his invented concept-script, Frege emphasized that these shortcomings of natural language 'are rooted in a certain softness and instability of language which nevertheless is necessary for its versatility and potential for development' (SJCN, 86). He compared ordinary language to the human hand which is so infinitely dextrous and adaptable. Nevertheless, scientists find it necessary, for very specialized purposes in experiments, to build artificial hands, stiff and inflexible in certain ways, but more accurate in movement in other ways. In the same fashion, the logician who investigates the laws of truth, the mathematician who aims at complete and rigorous demonstration, and the philosopher who wishes conclusively to settle whether arithmetic is analytic, require a more specialized instrument than ordinary language.

The idea of such an ideal language was, to be sure, not a novel one. Frege paid handsome homage to Leibniz in this respect (BS, Preface; ACN, 91; PW, 9ff.). Nor was he the first to attempt to realize Leibniz' dream, for, as he recognized, the logical algebraicists were pursuing the same goal. Frege's innovation was not the notion of an improved means of representing internal relations between thoughts, nor the idea of employing distinctively mathematical techniques in devising such means, but rather in his turning to function theory rather than to algebra for inspiration.

The purpose of his invented concept-script was, 'to test in the most reliable manner the validity of a chain of reasoning and expose each presupposition which tends to creep in unnoticed' (BS, Preface). His concept-script was not conceived as an analysis of the forms of natural language, nor was it presented as 'a more perfect instrument for doing the same thing as that which we normally do by means of natural language ... [an] ideal which natural language strives after, but

---

14. Cf. Lotze: '[T]he language-forming phantasy goes beyond the needs of a commonplace thought, and produces a great number of grammatical forms and syntactical rules that with the progressive advance of reflection are gradually allowed to drop as superfluous ... (*Microcosmus*, V-iii-§4).

fails to attain'.¹⁵ On the contrary, it stands to ordinary language as the microscope to the eye. 'The latter, because of the range of its applicability and because of the ease with which it can adapt itself to the most varied circumstances, has a great superiority' (BS, Preface). Concept-script is a specialized instrument for specialized purposes in logic and mathematics. It presupposes an understanding of language and a grasp of the thoughts expressed in language, and does not investigate the nature of that language. It represents the structure of the judgeable-contents we understand, not of our understanding of them, of the thoughts we express, not of the sentences expressing them. For the science of proof, the study of valid inference, is not concerned with the structures and relations of the *expressions* for judgments or thoughts, i.e. sentences,¹⁶ but with the structures and logical relations of these objects themselves.

'It is the business of the logician to conduct an unceasing struggle against psychology and those parts of language and grammar which fail to give untrammelled expression to what is logical' (PW, 6f.). The logician must try to liberate us from the fetters of language (PW, 143, 270; PMC, 68).¹⁷ The task of the philosopher is to break the power of the word over the human mind, to free thought 'from that which only the nature of the linguistic means of expression attaches to it' (BS, Preface). To fulfil these noble roles one must investigate thoughts, not sentences, discover the laws of thoughts, not the laws of sentence-formation and transformation.

> It cannot be the task of logic to investigate language and determine what is contained in a linguistic expression. Someone who wants to learn logic from language is like an adult who wants to learn how to think from a child. When men created language, they were at a stage of childish pictorial thinking. Languages are not made so as to match logic's ruler. (PMC, 67f.)

---

15. As suggested by Dummett, 'Can analytical philosophy be systematic and ought it to be' (*Truth and Other Enigmas*, p. 441).
16. The idea that logic studies sentential structures and relations was, to typical nineteenth-century thinkers including Frege, eccentric to the point of absurdity. The great exception was Whateley (later supported by De Morgan), who contended that 'Logic is entirely conversant about language; a truth which most writers on the subject, if indeed they were fully aware of it themselves, have certainly not taken due care to impress on their readers' *Elements of Logic*, p. 56 (Oxford, 1827), and 'Logic is wholly concerned in the use of language' (op cit. p. 74; cf. De Morgan, *Formal Logic*, pp. 1, 42, 54). Since language was conceived as externally related to thought and its objects, since signs are arbitrary marks while the laws of thought were generally conceived as anything but arbitrary, since logical necessity was typically understood as either a reflection of the essential structure of the mind or as resting on inexplicable, ultimate, self-evident 'logical facts', and so wholly independent of linguistic vagaries, Whately was, on the whole, disregarded. Hamilton reacted to this linguistic conception of logic with wrath and indignation: 'Dr. Whately, in his statements relative to the object matter of Logic, is vague and obscure, erroneous and self-contradictory; and that so far from being entitled to the praise of having been the only logician who has clearly displayed the true nature of the science, on the contrary, in the exposition of this nature, he is far inferior, not only in perspicuity and precision, but in truth, to the logicians of almost every age . . .' (*Lectures on Metaphysics and Logic*, III, p. 39f.).
17. Schröder similarly thought of logical algebra as justified 'in order to free the mind from the trammels of speech' (E. Schröder, *Logik*, vol. I, p. iii).

Of course, if we follow the psychologicians in conceiving the task of logic as the investigation of how men actually think, then given that (according to Frege) they think *in* a language, we *should* have to accord great importance to language. But such an investigation is really only a branch of psychology and has nothing to do with logic (PW, 143). In logical investigation 'we need not be concerned with what linguistic usage is. Instead, we can lay down our linguistic usage in logic according to our logical needs' (PMC, 71).[18] The logician investigates the laws of logic. These do not bear the relation to thinking that the laws of grammar bear to language. The norms of grammar change as human speech patterns evolve, but the laws of logic are 'boundary stones set in an eternal foundation' (BLA, p. xvi.).

Nevertheless, Frege thought it important to recognize an approximate general correspondence between language and thought. This was a commonly accepted corollary[19] of the traditional conception of language:

> As a vehicle for the expression of thoughts, language must model itself upon what happens at the level of thought. So we may hope that we can use it as a bridge from the perceptible [i.e. sensible signs] to the imperceptible [thoughts]. Once we have come to an understanding about what happens at the linguistic level, we may find it easier to go on and apply what we have understood to what holds at the level of thought, to what is mirrored in language. (PW, 259)

---

18. Not an uncommon view in the nineteenth century; cf. De Morgan, 'Logic may take liberties with language for the expression of thought: but she must not declare her alterations to be actual parts of speech' (*On the Syllogism and Other Writings*, p. 276).
19. E.g. Boole, *The Laws of Thought*, ch. II §1: 'though in investigating the laws of signs, *a posteriori*, the immediate subject of examination is Language, with the rules which govern its use; while in making the internal processes of thought the direct object of our inquiry, we appeal in a more immediate way to our personal consciousness,—it will be found that in both cases the results obtained are formally equivalent. Nor could we easily conceive, that the unnumbered tongues and dialects of the earth should have preserved through a long succession of ages so much that is common and universal, were we not assured of the existence of some deep foundation of their agreement in the laws of the mind itself.' Of course, Boole has a psychologistic conception of thoughts, Frege a Platonistic one. But both assume that language is a guide to what lies behind it.
  Jevons, like Frege, repudiated psychologism and had a conception of meaning akin to Frege's notion of content (e.g. 'The simplest and most palpable meaning which can belong to a term consists of some single material object, such as Westminster Abbey, Stonehenge, Sirius, etc.' (*The Principles of Science*, p. 24)). He too espoused the idea of isomorphism: 'The logician then uses words and symbols as instruments of reasoning, and leaves the nature and peculiarities of language to the grammarian. But signs again must correspond to the thoughts and things expressed, in order that they shall serve their intended purpose. We may therefore say that logic treats ultimately of thoughts and things, and immediately of the signs which stand for them. Signs, thoughts and exterior objects may be regarded as parallel and analogous series of phenomena, and to treat any one of the three series is equivalent to treating either of the other series' (op. cit. p. 8f.).
  This powerful picture culminated in the rigorously developed picture theory of language in the *Tractatus*, in which the isomorphism reflects *internal relations* between language, thought, and reality. Of this Wittgenstein was later to say, 'Other illusions come from various quarters ... Thought, language, now appear to us as the unique correlate, picture, of the world. These concepts: proposition, language, thought, world, stand in line one behind the other, each equivalent to each' (*Philosophical Investigations*, §96).

Careful critical attention to language may provide valuable *clues* as to the nature of thoughts. Indeed, for us men, no alternative source of insight is available which does not presuppose this. To Kerry's criticism that he had transgressed the principle of not basing logical rules on linguistic distinctions, Frege replied, 'my own way of doing this is something that nobody can avoid who lays down such [logical] rules at all' (CO, 45). Kerry's mistakes are held to stem from *insufficient* attention to grammatical distinctions (cf. CO, 43ff.). Frege himself repeatedly treated grammar as *a* key to what is logically significant. The fact that numerals occur substantivally in sentences points towards the fundamental truth that numbers are objects (FA, §§29, 38, 57). The metaphysical distinction between concept and object is partly reflected in natural language: it is mirrored in the grammatical fact that predicates, unlike proper names, are 'unsaturated' or in need of completion (FC, 31; CO, 46f.; PW, 202), and to the good fortune of German-speaking logicians it is mirrored too in the uses of the German indefinite and definite articles[20] (CO, 45). Various reflections on language support the truth that count-statements make assertions about concepts, not about objects or agglomerations (FA, §§46, 52). Such arguments, deploying linguistic evidence in support of logical theses, are common in Frege's writings.[21]

His reflections on the relation of language to thoughts, of grammar to logic, lead to a quandary. On the one hand, sentences give the basic access to the thoughts analysed in logic, and their structures frequently provide evidence for the structures of thoughts. On the other hand, grammar is a fallible and wayward guide in logical analysis.

> [T]he use of language requires caution. We should not overlook the deep gulf that yet separates the level of language from that of thought, and which imposes certain limits on the mutual correspondence of the two levels. (PW, 259)

This attitude Frege shared with his contemporaries who likewise subscribed to the conception of language as a vehicle for thoughts, as merely externally related to the thoughts it is used to convey. Sometimes we are to trust language, sometimes not. The crucial question which now arises is how to judge in particular cases. Should we attend to the subject-predicate structure of simple declarative sentences, or should we ignore it? Should we take note of the adjectival occurrence of numerals, or should we focus only upon their substantival occurrence? If two

---

20. English-speaking logicians were equally proud of the reflective powers of their native tongue: 'Its reality [viz. the forms of a logically perfect language] and completeness will be made more apparent from the study of those forms of expression which will present themselves in subsequent applications of the present theory, viewed in more immediate comparison with that imperfect yet noble instrument of thought—the English language' (Boole, *The Laws of Thought*, ch. XI §15).
21. And not only in Frege's: 'The study of grammar ... is capable of throwing far more light on philosophical questions than is commonly supposed by philosophers. Although a grammatical distinction cannot be uncritically assumed to correspond to a genuine philosophical difference, yet the one is *prima facie* evidence of the other.... [G]rammar, thought not our master, will yet be taken as our guide' (Russell, *The Principles of Mathematics*, p. 42).

sentences differing in structure express the same thought, should we ascribe both structures to the thought? If not, how are we to choose between them? The general policy of accepting the lead of grammar but tempering this with a prophylactic dollop of scepticism contributes nothing to the resolution of these pressing questions.

Frege constructed his concept-script in order to represent perspicuously and precisely the objective structures of thoughts (cf. PW, 12). This ideal he shared with traditional syllogistic and nineteenth-century logical algebras. The means he discovered for this purpose, i.e. the notation of function theory, differed radically from anything employed previously. So, too, did the logical structures of thoughts which he claimed to discern and represent in his novel notation. But since the power of concept-script can, in his view, be no more than that of mirroring accurately objective facts about the logical structures of thoughts, the articulations of concept-script cannot be treated as the *source* of knowledge of these language-independent structures.

How then could Frege justify his account of the structures of the thoughts expressed by the use of particular declarative sentences? What grounds could he give in criticism of alternative analyses offered by other logicians? Both his Platonism about thoughts and his rigorous shunning of what he thought of as 'psychological' issues precluded giving any philosophically satisfactory answers. No doubt he supposed that thoughts had definite structures, and that these structures could be apprehended as directly as the thoughts themselves. But how we apprehend these structures must be at least as mysterious as the wholly mysterious matter of how we apprehend the thoughts (cf. PW, 145). Both mysteries, however, are of no concern to logic, belonging rather to psychology.

On Frege's behalf we might canvass the possibility of indirect justification of his analyses. Certainly he appealed to grammar both in vindicating and in criticizing proposed analyses. But without some clear standard for differentiating cases where grammar misleads from those where it hints at the truth, his reasoning is patently unsatisfactory. Equally, he attempted to vindicate his system by the correct observation that his logical notation permits the formal analysis and justification of modes of reasoning falling outside the purview of alternative logical systems. This apparently shows that he has produced a more powerful instrument for logical purposes. But it would only support the claim that it successfully mirrors the objective structure of thoughts if it were intelligible to suppose that a logical system is an inductively supported hypothesis about the nature of a transcendent domain.

We might well wonder whether Frege did not fall into the absurdity of claiming that ordinary language is misleading just in case the sentence expressing a thought diverges in form from the formula in concept-script that he would supply as its translation, and conversely, that it is perspicuous just in case there is no such divergence. Many of his criticisms of language fit this pattern, e.g. that language obscures the difference between first- and second-level concepts. He disarmingly hinted at this conception of the matter. The construction of his concept-script ensures that 'we shall have, so to speak, our very noses rubbed into the false analogies in language' (PW, 67). Since language is full of logical imperfections, logical investigations

> are especially difficult because in the very act of conducting them we are easily misled by language.... Fortunately as a result of our logical work we have acquired a yardstick by which we are apprised of these defects. (PW, 266)

If this yardstick is concept-script, then are not the 'logical defects' of language simply the *product* of adopting Frege's function/argument analysis as a norm for representing judgeable-contents or thoughts, i.e. the product of measuring sentences against formulae in concept-script? It is not that language fails to mirror exactly the objective structures of thoughts, while concept-script succeeds in this task. It is rather that one form of representation (natural language) differs profoundly from another (concept-script). Thus to differ is not to offend.

If natural language is philosophically misleading, it is not because it fails to mirror something which it is obliged to mirror. It is rather that numerous expressions in natural langauge have profoundly different uses despite superficial similarities of form as conceived by conventional systems of grammatical classification. But the fact that the grammarian's system of classification fails to capture logical differences in the uses of different kinds of expression is no defect of natural language. Nor, one might hasten to add, need it be a fault in the grammarian, who constructs grammatical systems of classification with different ends in view from the logician. That 'exists' and 'red' are conventionally classified as grammatical predicates in no way supports the criticism that natural language is defective in representing existence and redness alike as first-level concepts. What, one might ask, would English look like, if it *did* look as if 'exists' were a second-level concept? One could equally contend that since the use of 'exists' differs profoundly from that of 'red', language does *not* represent them as similar, let alone as first-level concepts. A gramophone record and a video-disc look much the same; can it then be said that technology presents them as similar? Not when one sees them in action, as it were!

Although many of Frege's criticisms of ordinary language are muddled, it is important to note that in his criticism of other systems of logic he is not challenging traditional *grammar*, but only the unwarranted reliance of conventional logical systems on grammatical categories. To characterize 'everyone' in the sentence 'Everyone breathes' as a second-level concept-word is not to challenge the thesis that it is a pronoun, syntactically parallel with proper names. Nor is the claim that 'everyone' here is logically the predicate inconsistent with the grammatical observation that 'everyone' is the subject (correctly rendered into the nominative case in Latin), 'breathes' the predicate of the sentence. While Frege misguidedly believed language to be defective in failing to be perfectly isomorphic with thoughts, it was far from his purpose to propose any reformation of traditional grammar. What *appear* to be criticisms of grammar typically turn out, on closer inspection, to have other targets. This is notably true of his discussion of the subject/predicate distinction in relation to assertion.

3. The legacy of traditional logic:
predicate, indicative mood, and the copula

Within traditional logic, the inseparable companion and somewhat uncomfortable bed-fellow of the Cartesian conception of judgment was a set of doctrines about 'forms of judgment' that emphasized certain aspects of sentences: the indicative mood, subject/predicate structure and the copula. Conventional wisdom held assertion to be a manifestation of an inner act of judging, a manifestation by means of producing a sentence of a certain *form,* namely one whose main verb is in the indicative mood. Furthermore, a judgment so expressed must be divisible into two 'terms': subject and predicate. Asserting was typically construed as predicating something, hence all assertions were thought necessarily to be of subject/predicate form. Though the predicate often has the form of a finite verb, any sentence can be paraphrased into one in which the predicate is joined to the subject-term by the copula.[22] The copula was conceived to embody the essence of assertion, for it alone could unite any pair of terms into an assertion. Traditional logic glorified its mysterious power. The copula is 'the *Form* of a proposition; it represents the Act of the Mind affirming or denying'.[23] It 'indicates the act of Judgment as by it the Predicate is affirmed or denied of the Subject'.[24] It is 'the sign denoting that there is an affirmation or denial ... the word *is* ... serves as the connecting mark between the subject and predicate, to show that one of them is affirmed of the other. ...'.[25]

Since it is obvious that not every sentence used to make an assertion consists of a subject-expression joined to a predicate-expression by a copula, traditional logic had to introduce a host of qualifications. The drift of these was to block the identification of the terms (subject and predicate) of an assertion with the parts of the uttered sentence isolated by grammar as its subject and predicate. Thus the 'logical form' of an assertion might differ greatly from the grammatical form of a sentence expressing it. Often the 'logical form' could be made grammatically perspicuous only by radical paraphrase,[26] which was held to vindicate the otherwise patently false contention that every assertion contains subject- and predicate-terms linked by a copula, and hence also the fundamental theses that asserting consists in predicating something and that assertoric force resides in the predicate, or even in the copula. So strong was this tradition that Russell felt no incongruity in using the expression 'assertion' where we should speak of a predicate. So persuasive was the common paraphrastic argument to show that every judgment contained a copula, that Russell thought it necessary to stress the role of finite verb forms in binding together the elements of a proposition into a unity.[27]

---

22. Cf., for example, Bolzano, *Theory of Science,* §127 (although he goes on to argue that the 'real' omnipresent copula is 'has').
23. Watts, *Logick,* I, 1.
24. Whately, *Elements of Logic,* II-i-2.
25. Mill, *System of Logic,* I-i-2.
26. Cf. Whately, *Elements of Logic,* p. 59ff., or Bolzano, *Theory of Science,* §127.
27. Russell, *The Principles of Mathematics,* p. 39.

A consequence of this need to distinguish grammatical from logical form was the transformation of an intelligible though over-simple doctrine into a mystery of metaphysics. The principle that every assertion is logically of subject/predicate form ceased to make any claim about sentence forms at all. The 'terms' of the judgment were taken *de re,* being conceived as entities for which sentence-constituents stood. The impact of this style of thought is evident in Russell's 'realism' about the terms of propositions, viz. things, concepts, relations, etc. Frege's early thought manifests the same doctrine, since a judgeable-content supposedly decomposes into objects and concepts.

We must therefore keep separate the three different strands in the 'traditional doctrine' of subject and predicate, viz. the grammatical distinction between parts of a sentence, the logical distinction between parts of a judgment, and the metaphysical counterpart of the logical distinction. Frege had no objection to the first distinction, for he was not concerned with grammar, but with logic. At first sight, he broke decisively with the logical tradition about forms of judgment. 'Freeing thought from that which only the nature of the linguistic means of expression attaches to it', his concept-script replaces 'the concepts of *subject* and *predicate* by *argument* and *function*' (BS, Preface). Representation of a judgment in concept-script makes no distinction between subject and predicate (BS, §3), for that distinction merely identifies what the speaker has picked out as important in his expression of a judgment, and hence is not a logical distinction. The sole vehicle of assertoric force in Frege's logical system is the assertion-sign (BS, §2), not the predicate or copula, neither of which occur. '[I]t would be best to banish the words "subject" and "predicate" from logic entirely' (PW, 120). Such remarks read like excerpts from a manifesto against traditional logical doctrines. Nevertheless, we should be wary of reading too much into them. Frege was more deeply indebted, even when criticizing traditional subject/predicate logics, to the underlying conception of assertion and judgment which informed them than he realized (or than is now recognized). Indeed, as we shall see, it was his *acceptance* of the connection between assertion and subject/predicate grammatical structure which directly motivated his introduction of the assertion-sign into concept-script. It seems that although he by-passed traditional myths about logical forms of judgment in constructing his own logical system, he by no means demolished all their foundations. Consequently, when engaged in informal philosophical analysis, he drew more from this tainted source than is immediately apparent.

Two important points stand in full view in respect of Frege's 'abstracting' from subject/predicate grammatical structure. First, the notation of his concept-script lacks any such form. With the exception of identities, his well-formed formulae are either content-letters alone (e.g. 'A', 'B', 'T'), or expressions representing the value of some function for an argument (e.g. '$\psi(A)$', '$\phi(A, B)$'), or functions of some of these formulae. None of these expressions has the form of a sentence, nor even of the mathematical surrogates for sentences (such as equations). There is here no indicative mood, no copula, and no differentiation into subject and predicate. In this respect Frege's notation stands in sharp contrast to syllogistic and to the new logical algebras, e.g. Boole's, in which assertion is expressed by the formal predicate ' = 1'. The second striking point is that Frege did not follow the standard practice of differentiating the judgments expressed by appropriate pairs of sentences differing in grammatical subject and predicate. Nothing in conceptscript corresponds to active/passive transformation of atomic sentences with transitive verbs (BS, §3). Similarly the sentences 'Some cats are pets' is held to have the same content as 'Some pets are cats' despite subject/predicate inversion (according to traditional grammar). Hence Frege identified certain judgments and inferences standardly distinguished in syllogistic and logical algebras (cf. BS, §22).

Abstracting from subject/predicate form in constructing a concept-script does not, of course, imply denying its importance in the grammar of natural language. It does not even amount to a blanket denial of its logical significance. Grammatical distinctions are partly logical, partly psychological (PW, 6). The logician's task is to sift out the logically significant aspects. Because inference can proceed only from assertions, the distinction between an asserted and an unasserted judgeable-content is crucial for logic. This in turn assigns logical importance to the subject/predicate structure of declarative sentences since *assertoric force is marked or 'carried' by the grammatical predicate*. Far from denying this traditional thought,[28] Frege constantly affirmed it (e.g. PW, 149; T, 22). '[I]t is really by using the form of an assertoric sentence that we assert truth' (PW, 129); 'assertoric force is closely bound up with the indicative mood of the verb in the main clause' (PW, 198); '[i]n language assertoric force is bound up with the predicate' (PW, 252). It is precisely because he thought that the mood of the verb embedded in the predicate carries assertoric force (he even talked of the 'assertoric force we give the word "is"' in the sentence 'It is true that 2 is prime' (PW, 194)) that he conceived of his introduction of a special assertion sign as 'dissociating the assertoric force from the predicate' (PW, 184, 185, 198). For since concept-script lacks subject/predicate form, the assertoric force, seemingly carried in natural language by the grammatical predicate or by the copula or by the copula + predicate,[29] must be detached from its normal vehicle, and carried by the new sign '⊢——'. It was no coincidence that in *Begriffsschrift* he claimed that 'the symbol ⊢—— is [the] common predicate for all judgments' in concept-script (BS, §3),[30] for it takes over the role usually discharged by the grammatical predicate.

28. As argued by P. T. Geach, 'Assertion', in *Logic Matters*, p. 265f.
29. It is noteworthy that Frege equivocated in identifying the grammatical predicate of a sentence. Sometimes he took the predicate of a simple sentence 'A is F' to be 'F'; at other times he conceived the copula to be *part* of the predicate 'is F' (rather than the particle which connects the predicate to the subject). In the draft of CO he claimed that in 'The rose is red' the grammatical predicate is 'red', but in the published version he identified it with 'is red' (PW, 96f. and note). In 'Logic' (1897) he described adjectives, without a copula, as predicates, e.g. the predicate 'true' (PW, 126), the predicates 'true', 'heavy', 'warm', 'acid', 'alkaline' (PW, 128), the predicate 'beautiful' (PW, 131). So too elsewhere (cf. PW, 233, 251f.; CO, 44). A similar vacillation is associated with his use of his technical term 'concept-word'.
30. He later dropped this point, but not the idea that the assertoric force embedded in the predicate of a declarative sentence must be disassociated from predication and embedded in the judgment-stroke.

Only if Frege thought that assertoric force *must* be marked specifically or formally can we make sense of his feeling a need for an assertion sign in concept-script, and only if he believed that in natural language it is formally marked by the predicate can we make sense of his describing his manoeuvre as 'dissociating assertoric force *from the predicate*'.

Frege's failure to repudiate this traditional dogma about assertion and predication (mood and copula) accounts for a recurrent tension in his remarks about assertion. While convinced that assertoric force was bound up with the grammatical predicate, he also noted that no syntactical characterization of a sentence can guarantee that whenever it is used it must express an assertion. For, as he recurrently stressed, declarative sentences used in fiction or drama, or used as constituents of molecular sentences, are not bearers of assertoric force. *Any* sentence can be used thus; from which one should conclude that no sign or sentence-form can guarantee assertoric force. Frege merely noted that (i) no sign in language (not even 'is true') has the sole function of indicating assertion (cf. PW, 185), and (ii) that the assertoric force that lies in the form of the declarative sentence can be removed from it by lack of 'the requisite seriousness', as in stage-assertions (T, 22).

Frege was attracted by the idea that a stage-assertion is no real assertion (T, 22; PW, 234) *because* it is not accompanied by the act of judging the expressed thought to be true (the 'requisite seriousness' is missing). The actor is supposed to do *less* than what is necessary to assert something (rather than more, viz. *acting* an act of assertion). This account is a corollary of a Cartesian conception of assertion as the expression of an inner act of judging. It is doubly misguided: even if the actor accompanied his utterance with an appropriate act of judging it would still not constitute an assertion of his, and conversely, even though the actor asserts nothing, the character he portrays does make assertions, ask questions, give orders, etc. Frege did not draw the conclusion that any sign or form indicating assertoric force is defeasible; rather, he explained the absence of assertoric force not only by 'lack of seriousness' but by the fact that the thought expressed by the actor belongs to the 'realm of fiction' (and this, presumably, is why the 'requisite seriousness' is lacking). For a sentence to 'belong to the realm of fiction' is for it to lack a truth-value (PW, 232; cf. BLA, p. xxi; N, 117; T, 34; CT, 542). Frege thus assimilated reference-failure to the use of proper names in fiction, (wrongly) presupposing that names in fiction are uniformly not correlated with anything in the 'realm of reference'. What apparently guided him here is the traditional idea that to make an assertion is to predicate something of something. Hence if there is *nothing* to predicate something of, it is *impossible* to make an assertion (and, by implication, impossible to accompany the utterance in question with an inner judgment). ' ... [I]t is certain ... that anyone who seriously took the sentence ["Odysseus was set ashore"] to be true or false would ascribe to the name "Odysseus" a reference ... ; for it is of the reference of the name that the predicate is asserted or denied. Whoever does not admit the name has reference can neither apply nor withhold the predicate' (SR, 62). This explanation is hopelessly inadequate, *inter alia* because reference-failure is not a necessary condition for failing to make an assertion in drama (cf. PW, 234).

That the conventional mark of assertoric force is unreliable further manifested itself to Frege in the fact that a subject/predicate sentence (in the indicative mood) can occur unasserted in a molecular sentence. His official doctrine (not uniformly adhered to) is that none of the sentences occurring as constituents of any truth-function of sentences is uttered individually with assertoric force. (This should have puzzled him more than it did. For if ordinary language signifies assertoric force by the copula and/or predicate, yet neither disjunct is asserted in asserting a disjunction, what part of speech in such a compound sentence carries assertoric force?) Frege's dubious doctrine[31] is no real advance over the work of Boole and Schröder in this connection. To say 'If $1 + 1 = 3$, then $1 + 2 = 4$' is not to say that $1 + 1 = 3$ or that $1 + 2 = 4$, but only that *if* $1 + 1 = 3$, then $1 + 2 = 4$. Boole and Schröder represented this aspect of assertion by attaching the formal predicate '$= 1$' to the expression for a molecular thought. Thus, for example, the assertion of a disjunction is represented by '$A + B = 1$', which neither contains nor entails '$A = 1$' or '$B = 1$'. Ordinary algebraic formation-rules guarantee that the symbol '$= 1$' can be attached only to an entire formula for a molecular judgment, not to any of its parts (e.g. '$A = 1 + B = 1$' is ill-formed). Frege's only quarrel with this is that the symbols '$= 1$' and '$= 0$' would have a double use if the formulae of logical algebra were combined with arithmetical symbols in logical representation of mathematical reasoning (ACN, 93f.), hence he chose for the same purpose the novel sign '⊢———'. Nevertheless, the insight that subject/predicate sentences can occur unasserted in this way did not lead Frege to question the dogma that the predicate (and/or copula) is the vehicle of assertoric force, but only served to increase his suspicions about the reliability of ordinary language.

It is noteworthy that Frege gave no clear explanation of *why* sentences used as disjuncts or antecedents and consequents in hypotheticals are not used to make assertions. Yet in terms of *his own* beliefs about ordinary language and about judgment, an explanation *is absolutely necessary here*. For if a speaker did judge to be true the thought expressed by a declarative sentence in an uttered disjunction, he would *ipso facto* have asserted it, since he has expressed this thought in a form appropriate for assertion (a subject/predicate sentence in the indicative mood) and also judged it to be true. Similarly, there is no *rationale* for the claim that the assertoric force carried by the indicative mood of the main clause of a sentence should extend over the thought expressed by the whole sentence rather than being restricted only to that expressed by the clause itself (PW, 198), nor is there any explanation of what carries assertoric force in molecular sentences which are not differentiated into main and subordinate clauses (e.g. disjunctions).

The upshot of careful investigation of Frege's remarks on the relations between

---

31. Dubious because of factive verbs (as Frege himself noticed (SR, 72; T, 22n.1)), e.g. 'If A acknowledges (is aware of the fact, takes into account the fact) that p, then q'; or pseudo-conditionals, e.g. 'If you want a snack, there is food in the cupboard'; etc.

grammar and logic locates his thought in the mainstream of nineteenth-century thinking. He did not demonstrate that the analysis of judgments is the analysis of sentences used to make assertions. He did not conceive of thoughts as internally related to language. His 'dissociation of assertoric force from the predicate' is not a criticism of traditional views, but an affirmation that the grammatical predicate carries assertoric force. He did not demonstrate that predication is distinct from assertion, but only stipulated that in his concept-script the force-carrying role of the grammatical predicate should be fulfilled by a different sign. Nor did he give any proper explanation of the fact that the indicative mood of the main verb in a declarative sentence is a defeasible conventional marker for the assertive use of sentences. From his point of view this was merely yet another manifestation of the logical unclarity of conventional grammar and the ambiguity of ordinary language—something to be deplored and overcome, not something meriting further investigation and careful analysis. For his failure to give due attention to our grammatical conventions concerning assertion and assertibility he paid a high price in terms of the coherence of his own formal account of assertion.

4
The Formal Theory of Assertion: *Begriffsschrift*

1. Behind the assertion-sign

One of the symbols in the lexicon of Frege's concept-script is the assertion-sign: '⊢——'. It is introduced in *Begriffsschrift* to mark the expression of a judgment (BS, §2). It is composed of two constituents, the judgment-stroke: '|' and the content-stroke '———'. The compound sign '⊢——' is used to formulate transformation-rules and to mark the axioms and theorems of his system of logic.

Frege's motives for using the assertion-sign were clearly philosophical. Contemporary formal logic did not have a standard practice of using a symbol to mark expressions of judgment (although Boole's system had such a symbol). Hence he did not simply borrow his symbolism from other logicians and then concoct a philosophical rationale for its introduction. On the contrary, he took over a host of presuppositions about the nature of inferences, judgment, and assertion which led him to think that a formal notation for inference must rigorously and consistently distinguish asserted from unasserted propositions. He expressly introduced '|' and '———' for this purpose, and he claimed that this symbolism would eliminate certain scandalous philosophical confusions.

Modern philosophers have criticized his use of the assertion-sign in formal logic. Some have thought it superfluous. Others have suggested that its use rests on a fallacy; since the validity of a proof is independent of whether the constituent formulae are expressions of assertions or merely of hypotheses, a formal logician should not introduce a symbol whose explicit purpose is to restrict attention to derivations from formulae that are used to make assertions. Frege's insistence that a logically adequate notation must contain a special sign to mark assertion has thus been repudiated. It is true that logicians still use the symbol '⊢——', but for radically different purposes. Most commonly it is a metalinguistic symbol used to indicate the syntactic property of derivability within a formal system. This is an heir to Frege's use of '⊢―――' to indicate the status of a formula as a logical law or analytic truth.

Our interest in his assertion-sign does not, however, lie in its use, justified or unjustified, in formal notations, but rather in the clouds of philosophical theory which it trails behind it. We have seen that the foundations of his reflections on assertion do not differ markedly from those of his contemporaries, and that these foundations are themselves unstable as well as shallow. Nevertheless, it is conceivable that the importance of Frege's formal theory of assertion might lie in a legitimate novel combination of well-known elements derived from tradition. To see whether this is so requires a detailed look at the introduction of the assertion-sign into concept-script.

2. *Begriffsschrift*

Begriffsschrift opens with a section entitled 'Explanation of the Symbols', the first symbol there explained being the assertion-sign. This priority was not coincidental, but rather highlights the importance that Frege attached to his introduction of this symbol and to the role he assigned it. In a jotting entitled 'What may I regard as the result of my work', dated 1906, he wrote, 'It is almost all tied up with the concept-script.... [S]trictly I should [begin] by mentioning the judgment-stroke, the dissociation of assertoric force from the predicate' (PW, 184). Later drafts of a logic book give similar prominence to this idea.

The explanation of the assertion-sign in *Begriffsschrift* must take priority over the explanations of the condition-stroke, negation-stroke, and the concavity of generality, since one of the innovations Frege claimed for his logic (PW, 11n.) was to distinguish judgments from judgeable-contents and to explain the logical connectives as names of operators on contents of judgment, not on judgments. The judgment-stroke signifies that any formula in concept-script dominated by this symbol represents a judgment. The absence of a prefixed '|' indicates the mere expression of a judgeable-content. The content-stroke introduces an expression suitable for combination with the condition-stroke or the negation-stroke. The correct employment of these symbols makes clear, for example, that the content of the antecedent of a conditional is not asserted, even though the content of the conditional as a whole is.

Because logic is conceived as setting forth canons of inference from judgments (or assertions) to judgments (assertions), whether an expression represents an assertion or not is crucial to logic.[1] On the principle that 'Everything necessary

---

1. This point, which Frege often reiterated, seems to be undermined by part of the initial explanation of the content-stroke in *Begriffsschrift*. As an example to illustrate how '――― A' produces merely an unasserted idea, he cited entertaining an hypothesis 'say, in order to derive consequences from it and to test by means of these whether the thought is correct' (BS, §2). This apparently concedes that inferences may proceed from mere hypotheses, not invariably from asserted judgeable contents. The same concession might be extracted from the later remark that, without the contrast marked by the judgment-stroke and the horizontal, 'we could not express a mere supposition [*Annahme*]—the putting of a case without a simultaneous judgment as to its

for correct inference must be expressed in full' (BS, §3), judgment or assertion must be symbolized explicitly (PMC, 79). In order to fulfil the requirement that 'nothing should be left to guesswork' in a proper concept-script, symbols must be unambiguous and context-independent. So whatever symbolizes assertion in formal proofs must do so perspicuously and invariably.

Exactly how to symbolize assertion in concept-script is primarily a theoretical question governed by principles of theoretical economy (minimizing the number of primitive symbols). Two obvious possibilities stood out for representing assertoric force in concept-script. One would take as primitive a form of representation indicating the assertoric force of a judgment, and then use a cancellation-operator to symbolize unasserted judgeable-contents; the other would take as primitive a form indicating absence of assertoric force of an expression of a judgeable-content, and then combine this with a special force-operator to symbolize assertion of that content. Frege opted for the latter, criticizing Boole for choosing the former. His main argument turns on theoretical economy (PW, 46, 52; cf. N, 130f.), but it is reinforced by other considerations. His notation complies with the principle that the structure of a symbolism should reflect the structure of the things symbolized, in particular that simple things should be represented by simple symbols, compound things by compound symbols (cf. PW, 49). Since judgment consists (according to Frege) of entertaining a content and then judging it to be true, it is fitting that this duality should be reflected in the symbolism.

Other considerations which he had for adopting his symbolism will be clarified by scrutiny of the roles he assigned to the two components of his symbol for assertion. We shall first examine the content-stroke '———'.

(i) A symbol following '———' must always have a judgeable-content (BS, §2; PW, 39). This formation rule excludes three cases: (a) prefixing '———' to a symbol with an unjudgeable-content, e.g., a symbol having the same content as 'house', in contrast to 'There are houses' (BS, §2n.); (b) prefixing '———' to any

---

arising or not' (FC, 34). The standard role of a supposition is to introduce a chain of reasoning, and therefore the need to allow the possibility of expressing mere suppositions in concept-script seems to presuppose the legitimacy of drawing conclusions from unasserted thoughts. These appearances are misleading. Frege never deviated from the dogma that inferences can be made only from premises judged or asserted to be true. This is perspicuous from his proofs in concept-script: no independent step in a proof lacks a prefixed judgment-stroke. His remarks about deriving consequences from suppositions or hypotheses have another explanation. He apparently identified making a supposition with formulating the antecedent of a conditional. To draw a conclusion from a supposition is to see that what a sentence of the form 'On the supposition that ..., (it follows that) ...' expresses is true, i.e. that a conditional is to be asserted. Anybody who supposes that something can be inferred from a thought not judged to be true 'is apparently confusing acknowledgment of the truth of a conditional compound thought with performing an inference in which the antecedent of this compound is taken for a premise' (CT, 553). This confusion allegedly besets logicians who make too much of the difference between direct and indirect proofs (PW, 244f.). As a consequence of this conception, the possibility of entertaining an hypothesis as a basis for inferences and the possibility of expressing a supposition without making a judgment as to its truth simply boil down to the possibility of formulating judgments expressed by conditionals without thereby judging what the antecedents express to be true.

symbol not belonging to the concept-script (cf. p. 112); (c) prefixing '────' to a symbol with no content, e.g., '$\frac{0}{0} = 1$'. Hence the formation rule for '────' is not purely syntactic, even when applied to arithmetical notation: if 'A' is a mathematical expression, whether '──── A' is a well-formed formula depends on the *content* of 'A', not only on its form. Although an expression of the form '├────A' might be thought to have a judgeable-content, Frege did not license prefixing it with a content-stroke; on the contrary, he must be understood tacitly to have banned as malformed any expression of the form '──── ├──── A'.

(ii) The symbols following '────' are bound together into a whole with respect to any preceding symbols, i.e., Frege assigned to the content-stroke the function of bracketing complex expressions in concept-script, thereby indicating the scopes of logical operators (cf. BS, §5, 7, 11f.).

(iii) An expression of the form '──── A' expresses 'a mere combination of ideas of which the writer does not state whether he acknowledges it to be true or not'. If, e.g., the sign '├──── A' stands for the judgment 'Opposite magnetic poles attract each other', then the sign '──── A' will 'not express this judgment, [but] it is to produce in the reader merely the idea of the mutual attraction of opposite magnetic poles' (BS, §2). '──── A' never expresses a judgment unless preceded by '|'; the absence of the judgment-stroke makes clear that no assertion is made.

(iv) A symbol of the form '──── A' can be paraphrased 'the circumstance that a'[2] or 'the proposition that a' (BS, §2). This paraphrase is meant to make it evident that '──── A' lacks assertoric force, and it seems to do so, since such noun-clauses are not instruments of assertion. Somebody who uttered the phrase 'that two is a prime number' would not normally be judged to have asserted that two is a prime.

We now turn to Frege's explanation of the second component of the assertion-sign, the judgment-stroke.

(i) If '──── A' is a well-formed formula in concept-script, then it is legitimate to prefix '|' to it provided that what '──── A' expresses is true. (It is unclear whether this is a *formation*-rule of concept script, but if it is, it is patently not purely syntactic.)

(ii) The judgment-stroke must not occur within the scope of any logical operator or predicate letter. Such formulae as '├────├──── A', 'f(├──── A)',

'┬── ├──── A
 └── ├──── B'

are ill-formed. This tacit formation rule is necessary to exclude interpreting the condition-stroke and negation-stroke as operators whose arguments are judgments.

(iii) That the judgment-stroke cannot be embedded thus is a symbolic reflec-

---

2. In this chapter we shall use 'a' and 'b' as dummy-sentences of ordinary language with the same judgeable-content as 'A' and 'B' of concept script. For the rationale of this, see p. 111ff.

tion of the stipulation that producing a well-formed expression of the form '⊢———A' is to make an assertion, and no assertion can be expressed in concept-script without the use of '⊢———'.

(iv) Since any inference is a transition from judgment(s) to judgment, any rule of inference must regulate making judgments. Hence, the judgment-stroke is indispensable for correctly representing in concept-script any formal inference-rule. Consequently, Frege used it in formulating his primary rule of inference: 'from the two judgments

$$\vdash \begin{array}{c} \text{—— A} \\ \text{—— B} \end{array}$$

and ⊢——— B the new judgment ⊢——— A follows' (BS, §6). Here '⊢——— A' is clearly meant to express a licence to make an assertion; it indicates that the judgeable-content expressed by '——— A' *is to be asserted.*

(v) Since any correct rule of inference issues a licence to make assertions, the formalization of any cogent proof in concept-script authorizes prefixing the judgment-stroke to the conclusion. Since each of the axioms (basic laws) of logic is properly prefixed with '|', every theorem of logic should be so adorned. Frege's practice conforms to this principle.

(vi) 'The judgment-stroke ... placed at the left-hand end of the content-stroke ... converts a judgeable-content into a judgment' (PW, 11n.). Conversely, omitting it at the end of the content-stroke transforms the judgment into a mere combination of ideas (BS, §2).

(vii) The distinction between '⊢——— A' and '——— A' is meant to effect 'a very clear distinction between the act of judging and the formation of a mere judgeable-content' (ACN, 94). Later Frege glossed his intention slightly differently: he meant to distinguish between acknowledging a truth and the content acknowledged to be true (BLA, p.x). According to this account, '|' signals an act of judging, whereas '———' expresses only what is judged to be true, an idea or judgeable-content.

(viii) A symbol of the form '⊢——— A' can be so paraphrased that the role of '|' is made perspicuous. Frege evidently thought that every declarative sentence could be rephrased in a form incorporating a verbal noun or noun-clause. Instead of saying 'Archimedes perished at the conquest of Syracuse', we could make the same assertion by saying 'The violent death of Archimedes at the conquest of Syracuse is a fact' (BS, §3) or 'That Archimedes died at the conquest of Syracuse is a fact' or 'That Archimedes died ... is true' or 'The circumstance that Archimedes died ... obtains *(findet statt)*' (cf. BS, §5, 7). In all of these paraphrases, the grammatical subject 'contains the whole content' of the original declarative sentence, while 'the predicate serves only to present this as a judgment' (BS, §3). Consequently such an expression exactly matches the articulation of a judgment in the symbol '⊢——— A'. The component '——— A', which exhausts the content of '⊢——— A', can be translated into a verbal noun or noun-clause (viz. 'the circumstance that *a*') and the remaining component '|' is translated into a formal predicate (viz. 'is a fact') whose sole function is apparently to effect the act of assertion. Of course, English admits the possibility of making assertions by uttering sentences not containing the formal predicate 'is a fact', and hence this paraphrase does not illuminate the stipulation for concept-script that an assertion *must* be expressed in the form '⊢─── A'. But an exact correspondence would be secured by imagining a modified form of English which required that every assertion be expressed in a sentence of the form 'The circumstance that . . . is a fact'. Then the phrase 'is a fact' would be the 'common predicate of all judgments'. The stipulations for the judgment-stroke show that concept-script is such a language (BS, §3). Frege formulated an apparently unexceptionable hypothesis according to which the role of '|' in the symbol '⊢─── A' could be made clear by paraphrase. Furthermore, this proposed paraphrase dovetails with the suggested paraphrase for the symbol '─── A' (BS, §2).

These explanations of the two constituent symbols in the assertion-sign have many different facets. They point to a number of distinct reasons that might have motivated Frege to introduce this symbol into his notation for formal logic. It may be that one of these considerations was the *primum mobile* of his innovation and that the others were rationalizations or subsidiary supports for something already firmly established. Or it may be that all of them occurred to him from the outset and jointly stimulated his inventiveness. We have no evidence to settle this question. But it is worth canvassing the different philosophical grounds associated with his use of the assertion-sign.

The claim that the contrast between '⊢─── A' and '─── A' marks the distinction between the act of judging and the content judged to be true ties the articulation of the assertion-sign to the *Grundgedanke* of Frege's antipsychologism. This fundamental distinction is both important and confusing. Indeed, one reason for its being important is that it is confusing. Certainly no one is likely to confuse the act of assertion, performed by a person at a time, with what was asserted by the person, which is neither performed nor takes place at a time. But if one is misled by 'verbs of saying', such as 'assert', 'deny', 'agree', and by psychological verbs, such as 'judge', 'believe', 'understand', as both Frege *and his psychologician opponents* were, one may come to think that what is asserted or denied, judged, believed or entertained, is an object of some kind. Then one will naturally feel impelled to characterize this *object of belief, judgment, or assertion*.[3] Since what is believed or judged may not be true, may not 'exist in reality', the psychologicians were prone to think that it must exist 'in the mind'; otherwise, when one believes falsely, it seems, there is nothing that one believes. Hence, in Frege's view, the psychologicians projected the psychological nature of the act of

---

3. Arguments against the conception of the proposition as the Platonic object of assertion, judgment, and belief have been thoroughly rehearsed since Frege, e.g. G. Ryle, 'Are There Propositions', *Collected Papers* vol. 2 pp. 12–38 (Hutchinson, London, 1971); A. N. Prior, *Objects of Thought* ch. 1–3 (Clarendon Press, Oxford, 1971); A. R. White, 'What We Believe' in *Studies in the Philosophy of Mind, American Philosophical Quarterly,* Monograph Series, ed. N. Rescher (Blackwell, Oxford, 1972); B. Rundle, *Grammar in Philosophy* ch. 7 (Clarendon Press, Oxford, 1979). Of course, to repudiate propositions thus conceived, only to conclude that what we assert, judge, or believe are sentences, is to stumble from one confusion into another.

judging onto the object of judgment, conceiving of the latter as a peculiar kind of mental entity, viz., a thought. Seeing, as had Mill, that what one judges or believes is not a psychological or subjective object, Frege thought it imperative to distinguish sharply between the psychological act of judging and the non-psychological character of the object of judging. Far from questioning the dogma that what we believe, judge, and assert is an object, he embraced it wholeheartedly, insisting only that this object is an abstract, Platonic entity. This he thought to be an unavoidable commitment if publicity, communication, and objective truth are to be rendered intelligible. In respect of the web of arguments spun by psychologicians, the distinction of the act from the object of judgment was Frege's way out.

The contrast between '⊢——— A' and '——— A' is also held to mark the distinction between a judgment and a mere judgeable-content, i.e., between an idea actually judged to be true and an idea merely entertained with suspended judgment. Frege complained that logicians generally failed to draw this distinction. Traditional logic and the logical algebras of Boole and Schröder subdivided their study into the logic of concepts (or terms), and the logic of judgments (beliefs, propositions). Within this latter branch fell the characterization of logical relations among judgments (conjunction, disjunction, conditional subordination) and the cataloguing of valid forms of inference (*modus ponendo ponens, modus tollendo ponens,* etc.). Frege thought it misleading to call this subject 'the logic of judgments', for 'one is never quite sure whether what logicians call a judgment is meant to be a thought alone or one accompanied by the judgment that it is true' (PW, 185). If 'judgment' were given the stronger reading, then the widespread thesis that a hypothetical expresses a relation between judgments would be wrong; but if it were given the weaker reading uniformly, then the claim that *modus ponendo ponens* expresses a relation between three judgments would also be false, in Frege's view. He therefore resolved to eliminate this ambiguity in the usual practice of using the term 'judgment' by restricting its application to judgeable-contents actually judged to be true. He stressed that the relation in a hypothetical is not one between judgments but one between judgeable-contents (PW, 11n.) and his notation makes clear that rules of inference such as *modus ponendo ponens* relate judgments, not mere judgeable-contents.[4] This innovation offers a gain in clarity in the informal exposition of systems of formal logic, and it might forestall certain philosophical confusions about the circularity of valid deductive reasoning. But it did not require any modifications to earlier formalizations. Boole was clear that neither disjunct in an asserted disjunction is itself asserted, and he encapsulated this principle in his algebraic notation, since it cannot be inferred from $a + b = 1$ that $a = 1$ or that $b = 1$. Despite the potential for confusion stored up in the term 'judgment', logicians generally avoided confusing themselves. Nonetheless, Frege thought their explanations of their formal systems to be scandalously inexact and misleading, like mathematicians' expositions of the theory of functions or the foundations of arithmetic.

The judgment-stroke may have been intended to be a form of punctuation to

---

4. For the first time in formal logic, he sharply distinguished axioms from rules of inference.

signal clearly that the content of a formula is being asserted. To eliminate 'guesswork' in the interpretation of formal derivations, Frege decided to lay down the convention that the expression of an asserted judgeable-content must always have the form '⊢——— A'. But his stipulation may also have rested on another foundation. As we have seen, he held that assertoric force is bound up with or carried by the predicate of a declarative sentence in a natural language. This claim might be interpreted as noting a *ground* for holding that a speaker has made an assertion: *ceteris paribus*, to utter a declarative sentence whose verb is in the indicative mood is to assert something. But Frege may have misconstrued this partial clarification of the concept of asserting something as a thesis about the *mechanism* of making an assertion. Asserting something might be thought to resemble blowing up a tank; the speaker merely launches an expression as he fires off a rocket, but if the expression, as it were, lacks explosive power, then no act of assertion is achieved. Assertoric force, on this view, must inhere in an expression if it is to be possible for anybody to use it to make an assertion. There are many suggestions of this mechanical conception of assertoric force in Frege's writings. He repeatedly claimed that *symbols* assert something (PW, 51; FC, 34n.) and that the form of a declarative sentence *contains* the act of assertion or of acknowledging the truth of a thought (BLA, p. x; T, 31; cf. N, 117) even though he openly conceded that such sentences might be deprived of assertoric force in certain circumstances. If he held that this grammatical form was the sole known *vehicle* of assertoric force, then he had an excellent reason for thinking that the expression of a judgment in concept-script presupposed the possibility of dissociating assertoric force from the predicate and loading it on to another vehicle specially built for this purpose. Since concept-script lacks formulae articulated into subject and predicate, it would be obvious that a special sign is *needed* in order to be able to assert something (cf. FC, 34). Some symbol is essential to act as the *vehicle* for assertoric force in concept-script, and the judgment-stroke is devised to serve this purpose. This interpretation makes literal sense of parts of Frege's explanations. It accounts for the claims that prefixing '|' to the symbol '——— A' *transforms* a judgeable-content into a judgment and that subtracting '|' from '⊢——— A' *transforms* the judgment into a mere combination of ideas. It also makes sense of the later claims that the symbol '⊢——— A' asserts something (FC, 34n) and that the judgment-stroke contains the act of assertion (BLA, §5). Finally, in conjunction with Frege's stipulations, it explains why presence of the symbol '|' in a proof guarantees that an assertion has been made[5] and why its absence is conclusive evidence that no assertion is made. This interpretation is speculative, to be sure, yet it coheres with Frege's arguments and it may provide the simplest explanation of the contour-lines of his reflections about assertion.

---

5. Neither of the two conditions that alone may rob a declarative sentence of assertoric force can arise in respect of the formula '⊢——— A'. Its occurrence in a subordinate clause is excluded by the formation-rules of concept script and its occurrence in fiction is of no concern to logic (PW, 198).

3. Problems of paraphrase

The primary philosophical question concerning Frege's explanation of the assertion-sign in *Begriffsschrift* and associated early writings is whether his stipulations are individually intelligible and jointly tenable. There are *prima facie* grounds for doubt. The nature and justification for the *logical* prohibition on prefixing the judgment-stroke to expressions of judgeable-contents which are false is opaque. The notion of needing to distinguish an act of judgment from its object is bewildering. The claim that the symbols '⊢——— A' and '——— A' contrast the act of judgment with its object seems difficult to amalgamate with the thesis that they contrast two acts (assertion as opposed to entertaining an idea). And the explanation that '⊢——— A' signals an *act* of judgment seems not to mesh with the use of '⊢———' in a formal proof to indicate *entitlement* to assert the content of the conclusion. The coherence of Frege's account of the assertion-sign seems dubious.

One strategy to allay these doubts would be to demonstrate that Frege's symbolism can be so paraphrased into English (or German) that its articulations are matched in a straightforward manner and his stipulations correspond (at least by and large) to evident truths about the corresponding expressions. If a systematic simple method of paraphrase were available which secured this result, the coherence of his stipulations would be as safe from criticism as the explanations of meaning which are acknowledged to be correct by competent speakers of a natural language.

Frege himself pursued this strategy in *Begriffsschrift*. He suggested a linked pair of paraphrases for '|' and '———'. According to his proposal, '⊢——— $2 + 3 = 5$' would be paraphrased as 'The circumstance that two and three make five is a fact', while '——— $2 + 3 = 5$' would correspond to the same expression shorn of the formal predicate 'is a fact'. *Apparently* he must have then thought that these paraphrases conformed to his stipulations. In particular, it seemed to him that 'the circumstance that . . .' represented the object of judgment, that such a noun clause can never be used to make an assertion, and that it expresses a mere combination of ideas. Likewise, it must have appeared to him that the formal predicate 'is a fact' contributed nothing to the judgeable-content expressed by a declarative sentence, that it is a suitable tag for theorems of logic, and that it signals that an uttered sentence is being used to express a judgment (to make an assertion). Many of these claims are problematic, as he later realized. But had he appreciated their indefensibility from the beginning, it seems incredible that he would have proposed the paraphrases given in *Begriffsschrift*.

There are obvious objections to these original paraphrastic explanations of his symbolism.

(i) The formal predicate 'is a fact' violates the main requirement laid down for the judgment-stroke. It need not invariably be used, even in the formulation of deductive reasoning, to make assertions. For, a sentence of the form 'that . . . is a fact' may be embedded in a molecular sentence as the antecedent of a conditional or one disjunct in a disjunction, and in such instances this declarative sentence would not carry assertoric force. The fact that *any* declarative sentence may be correctly used in such grammatical constructions rules out the possibility that *any* other formal predicate would exhibit the characteristic most distinctive of the judgment-stroke. Frege's failure to note this point in *Begriffsschrift* seems a mystery. More surprisingly, this counterargument scarcely makes any appearance at all (cf. PW, 251). Although he later harped upon the point that a formal predicate, in particular 'true', does not guarantee that a sentence is uttered with assertoric force, he invariably cited as counter examples the utterances of actors on the stage (PW, 129, 168, 194, 233f.; T, 21). If only for this reason he withdrew his paraphrase of '⊢─── A'.

(ii) The demise of the paraphrase 'is a fact' for the judgment-stroke was enough to kill off the paraphrase 'the circumstance that ...' for the content-stroke. But another independent objection is also fatal to this second paraphrase. At the time of writing *Begriffsschrift*, it seemed evident to Frege that a declarative sentence '*p*' as well as such noun-clauses as 'the circumstance that *p*' represent, stand for, or express the same judgeable-content. But after he distinguished sense from reference and embraced the principle that the reference of an indirect statement is the sense of the corresponding instrument of assertion, he was driven into holding that what a declarative sentence *expresses* (its sense) is what the corresponding indirect statement *designates* (its reference). Apparently the sentence '*p*' and such noun-clauses as 'the circumstance that *p*' or 'the proposition that *p*' never fulfil the same roles: they express different senses and they designate different referents. The earlier claim that they represent the same judgeable-content masks these decisive differences and renders incoherent the proposed paraphrase in *Begriffsschrift*. Frege may have concluded that the clause 'the circumstance that *a*' never expresses what is expressed by a formula of the form '─── A'.

(iii) The fact that the negation-stroke, the condition-stroke, and the concavity (the universal quantifier) are all prefixed to a content-stroke suggests another objection to Frege's paraphrase of the content-stroke as 'the circumstance that ...'. These logical constants are supposed to be *roughly* equivalent to 'not' (or 'it is not the case that ...'), 'if ... then ...', and 'everything ...'. But to paraphrase '─┬─ A' as 'It is not the case that the circumstance that *a*' or

'⊢─┬── B'
  └── A

as 'If the circumstance that *a*, then the circumstance that *b*', or '───$\overset{a}{\smile}$── $\phi(a)$' as 'Everything the circumstance that it ϕs', produces jibberish. These must be paraphrased in the first two cases so that they correspond to sentence-forming operators on sentences, and in the third so that it corresponds to a sentence-forming operator on a predicate. But the uniform translation of the content stroke into a noun-clause 'the circumstance that ...' blocks off any such appropriate paraphrase. Its acceptance would compel Frege to defend the apparently absurd conclusion that his logical constants corresponded uniformly to noun-clause forming operators on noun-clauses! And in this case they would be wholly unlike the expressions 'not', 'if ... then ...', and 'everything ...'! This asymmetry is an unavoidable and objectionable consequence of uniformly adopting Frege's paraphrase. But the alleged absurdity, though transparent for most modern logicians, need not have been conceded by him. For, his official paraphrases of his logical constants incorporate place-holders for indirect statements, not declarative sentences. The symbol

$$\vdash\!\!\begin{array}{c}\text{—— B}\\ \text{—— A}\end{array}\text{'}$$

he paraphrased as 'The case that (that a is to be affirmed and that b is to be denied) is not realized', and here the condition-stroke is presented as a noun-clause forming operator on noun-clauses! This form of explanation persisted after he distinguished sense from reference (though 'true' and 'false' then replaced 'to be affirmed' and 'to be denied'). Hence what appears to be a forceful objection delivers Frege only a glancing blow.

His verdict on the paraphrases in *Begriffsschrift* was a decisive rejection, and further reflections would lead modern logicians to confirm his finding. Since he refrained from offering any fresh paraphrases of the content-stroke and the judgment-stroke in later writings, we have now eliminated the possibility that *he* held the paraphrasing of his symbolism into a natural language to justify his stipulations about the components of the assertion-sign. Nonetheless, many modern philosophers claim that his judgment-stroke is an intelligible symbol, that he used it to mark an important distinction between sense and force, and that this distinction can be clarified by paraphrasing everyday sentences into certain canonical forms. This suggests that these theorists envisage possibilities of paraphrase not canvassed by Frege and that these paraphrases establish what his own ones failed to justify. We shall therefore explore some possibilities of paraphrase suggested by his stipulations and developed by modern logicians.

(i) It goes without saying that the articulations of the symbol '⊢——— A' are not matched if formulae of this form are paraphrased into simple declarative sentences, e.g., if '⊢——— 2 + 3 = 5' is translated into 'Two and three makes five'. For such sentences, though instruments of assertion, have no proper *part* which can be matched with the sub-formula '——— A'. Moreover, unlike the formula '⊢——— A', declarative sentences can be embedded in molecular sentences, e.g., as disjuncts, and there they lack assertoric force, whereas '⊢——— A' is invariably used to make an assertion.

(ii) Modern logicians are tempted to paraphrase the judgment-stroke by the performative verb 'assert', either as the personal performative 'I assert' or the impersonal performative 'It is (hereby) asserted'. The symbol '——— A' can then be rendered into the indirect statement 'that a'. The combination of these paraphrases gives a form of declarative sentence whose articulation matches that of the symbol '⊢——— A'. Such a sentence might plausibly be thought to express or manifest the act of judgment, to be an instrument of assertion, and to highlight the distinction between the act and the object of assertion. But they do not satisfy other requirements imposed on the symbol '⊢——— A'. In particular, such sentences can be embedded in molecular sentences; there the prefix 'I assert' or 'It is asserted' does not perform the act of assertion, and it there makes a contribution to the judgeable-content expressed. Furthermore, these prefixes have frequentative as well as performative uses, and here too they have content. Finally, it makes little sense to present chains of inference as consisting of entailments among performative utterances. A speaker can utter 'I assert that p or q' and utter 'I assert that not-p' without issuing the performative utterance 'I assert that q'. Indeed, even the claim that he did assert that q is not licensed by his pair of utterances. Consequently, explicit performatives do not provide satisfactory paraphrases of '⊢———A'.

(iii) A related proposal would be to paraphrase '⊢——— A' into a report in *oratio obliqua,* viz. into the sentence 'It has been asserted that a' or 'x has asserted that a'. Such a paraphrase matches the articulation of the symbol '⊢——— A', and it clearly distinguishes the act of assertion (designated by the verb 'assert') from the object of assertion (designated by the indirect statement). But in other respects it has glaring defects. The typical use of 'It has been asserted that a' or 'x has asserted that a' is not to make the assertion that a, but rather to refrain from making this assertion oneself while reporting another (others) to have done so. Moreover, what is asserted by uttering 'It has been asserted that a' typically differs from what is asserted by 'a', so that 'It has been asserted' contributes to the judgeable-content of an expression, unlike '|'. Sentences of the form 'It has been asserted that a' may also be embedded in molecular sentences, unlike the symbol '⊢——— A'. Finally, this paraphrase would not justify the use of the assertion-sign to mark the axioms and theorems in Frege's formalization of logic. In most respects this proposed paraphrase is completely unsatisfactory.

(iv) Another related proposal would be to introduce a normative component into the paraphrase. Instead of rendering '⊢——— A' as 'It *is* asserted that a' or 'It *has been* asserted that a', we might translate it as 'It *is to be* asserted that a'. This would account for Frege's logical ban on prefixing the judgment-stroke to judgeable-contents known to be false (cf. PW, 18, 51), and it would equally explain his prefixing this symbol to formula that he has established to express true judgeable-contents (e.g., BS, §5). The judgment-stroke would indicate not an actual *act* of assertion, but rather an *entitlement* to make an assertion. Hence its occurrence in concept-script would not contaminate logic with psychology, while its use to mark the axioms and theorems of logic would be appropriate as well as its occurrence in the formulation of rules of inference such as *modus ponendo ponens* (which express conditional entitlements to make assertions). On the other hand, this paraphrase does not square at all with the other facets of Frege's explanation of the assertion-sign. 'It is to be asserted that a' does not express or represent an *act* of assertion, and hence it does not make clear that the judgeable-content 'that a' *is* asserted by the symbol '⊢——— A'. Moreover, what is asserted by 'It is to be asserted that a' differs from what is asserted by 'a'; such a sentence is typically used either to express a norm or to make a statement about norms, and hence it differs in content from the sentence 'a' itself. The sentence 'It is to be asserted that a' can be embedded in molecular sentences, and in these cases it is not used to assert that *a*. Finally, the acceptability of this normative paraphrase would require some clarification of the nature of the normativity involved; could it be that entitlement to assert boils down to truth?[6] Once again a paraphrase with some promising features is shown to be inconsistent with many of the main features of the symbol '⊢——— A'.

Each of these proposed paraphrases captures some of the characteristics of Frege's assertion-sign, but none captures all. This is not the consequence of making an unenlightened choice. Reflection on expressions available in English reveals an impasse. The fact that the symbol '⊢——— A' is to be used invariably to make an assertion necessitates that it be paraphrased into standard instruments of assertion, viz. declarative sentences. But every declarative sentence can be so used that it fails to make an assertion, in particular because it can be embedded in molecular sentences. No expression in English has the basic features stipulated for formulae of the form '⊢——— A'.

The issue of paraphrasing the symbol '——— A' is slightly more complex. Frege held that the cogency of inference by *modus ponens*

$$\vdash\!\!\!-\!\!\!- A, \quad \vdash\!\!\!-\!\!\!\begin{array}{c}- B, \\ - A\end{array} \quad \vdash\!\!\!-\!\!\!- B$$

turns on the fact that the antecedent of the conditional (represented by '——— A') has the same content (expresses the same thought) as '⊢——— A'. It is a feature of German that the antecedent of a conditional is expressed by a subordinate clause which typically differs in word-order from the corresponding declarative sentence, and such a clause has no established use as an instrument of assertion. Hence he might have paraphrased '——— A' into the subordinate clause corresponding to the declarative sentence which has the same content as '⊢——— A' (cf. PW, 251). This need not transgress the later distinction between expressing and designating a thought. But whether it would serve Frege's purposes is unclear. Because indirect statements in German typically have the same syntactic structure as subordinate clauses, he could have incorporated this paraphrase of '——— A' as a proper part of a paraphrase for '⊢——— A', e.g. 'It is true that *a*', provided he regarded 'It is true that' as forming the expression of a thought from another expression of this thought (rather than viewing 'It is true' as generating an expression of a thought from a designation of this same thought). But even if he had accepted this premise, it would be doubtful whether uttering a subordinate clause on its own can be held to constitute the 'mere expression of a thought' or the 'entertaining of an idea without acknowledging its truth'. Hence nothing substantial would be gained by conceding the possibility of this paraphrase (into German!) of the symbol '——— A'. For it would still remain beyond the competence of paraphrase to prove the consistency of Frege's stipulations about the symbol '⊢——— A'.

---

6. Cf. Wittgenstein, *Tractatus Logico-Philosophicus*, 4.442.

4. Inconsistent demands and unintelligible stipulations

The acknowledged inadequacy of Frege's own paraphrases of the assertion-sign and the impossibility of finding any demonstrably better ones at best leaves open the question of the coherence and the intelligibility of his explanations of the components of this basic symbol in his concept-script. The situation might even be held to be far more unfavourable to his enterprise of reconstructing logic. For if it must be possible to give a clear explanation of any symbol which has a clear meaning, then the failure actually to provide explanations (in German) of his symbols is a failure to assign any meaning to them, while the stronger claim that no explanation (in German) is possible would prove that they must remain forever meaningless, i.e. that the lacuna left in Frege's account cannot be filled consistently with his descriptions of the roles of these symbols in his formalization of logic.

Frege did not conclude that his account of the assertion-sign is damned by the impossibility of providing an exact paraphrase of his symbolism into German. Instead he turned the tables on this criticism by arguing that lack of parallelism proves that German, and presumably every other natural language, is defective from a logical point of view. Is not the fact that no expressions in a natural language precisely mirror the contrast between '⊢——— A' and '——— A' in concept-script the strongest possible argument that this language is logically defective? For the distinction between judgments and mere judgeable-contents is fundamental to logic, while the instruments of assertion in natural languages (declarative sentences) often can be used indifferently whether or not the thoughts expressed are asserted. Conditionals and disjunctions, e.g., are typically expressed in English by pairs of declarative sentences although neither sentence of such pairs is used to make an assertion. This seems to be a logical defect of the language since the form of a declarative sentence seems to be meant to be the hallmark of an expression used to make an assertion. And this defect appears to be irremediable: the lack of any uniform paraphrase of '⊢——— A' and '——— A' into a natural language implies that there can be no systematic translation *within* this language which clearly differentiates expressions for asserted judgeable-contents from expressions for unasserted ones. No simple reform of a language can make it logically perspicuous in this respect. This conclusion can be reinforced, in Frege's opinion, by noting that declarative sentences have established uses other than for making assertions: in particular, they occur in performances of dramas, and there they are not used by the actors to make assertions. From these observations Frege drew the conclusion that nothing short of a novel logical symbolism which has no parallel in a natural language can suffice to give perspicuous expression to the distinction between asserted and unasserted judgeable-contents. By the absence of any uniform systematic translation between these alternative independent languages, it is not his concept-script that stands condemned, but natural languages.

This bold manoeuvre is *prima facie* dubious. Indeed, it is even uncertain what exactly is maintained in characterizing a natural language as logically defective. One interpretation might be that it is defective *tout court,* though only logicians are in a position to note this fact. Another very different interpretation is that logicians, through contemplating natural languages, reach mistaken logical theses. Sometimes Frege suggested this weaker claim, especially in observing that the occurrence of predicates, indicative verbs, and declarative sentences as clauses in conditionals or disjunctions deceived logicians into thinking that such molecular sentences express relationships between asserted judgments (cf. PW, 185). Sometimes he intimated the stronger claim; as if, e.g., the absence of an assertion-sign in a natural language left it a matter of 'guesswork' whether a sentence or clause is used to make an assertion (cf. BS, §3). The difference between these claims is crucial. The weaker is indisputable. The stronger one, however, is implausible, yet it is the one implicated in the charge that absence of any systematic paraphrase between concept-script and a natural language reveals the language to be defective.

This inference is suspect for a different reason. The charge against natural languages that predication, the indicative mood of the verb, declarative-sentence form, the occurrence of such phrases as 'it is true that ...' or 'is a fact', etc., are *unreliable* indicators of assertoric force rests on the presupposition that they are designed for the *general* purpose of marking the use of sentences to make assertions. Hence, it is inferred that they miscarry in their purpose when they occur as constituents of sentences (or clauses) within molecular sentences. But this is a travesty. What are singled out here are features of expressions that may be properly invoked to justify, *ceteris paribus,* the claim that a speaker in uttering a sentence has asserted something. Yet these features are shared by sentences and clauses within sentences the utterance of which are acknowledged by all competent speakers *not* to constitute the making of assertions. Hence it is ludicrous to characterize these features as intended uniformly to flag assertions. A logician who embraces this thesis manifestly puts an *unwarranted* interpretation on features of sentences in natural languages, and then complains that this interpretation reveals natural languages to be *misleading!* Instead, it would be preferable to backtrack. The obvious conclusion to draw from the 'data' is that a natural language does *not* single out *any* features of sentences *uniformly* to flag the making of assertions. Then Frege's argument that natural languages are defective in the matter of assertion would simply collapse.

Frege's criticism of natural languages is further defective in misallocating the burden of proof. As we earlier noted, his various stipulations about the judgment-stroke and the content-stroke are *prima facie* inconsistent. The gravamen of this charge is increased by the observation that no known expressions (of natural languages) satisfy all of his requirements, though different expressions each satisfy some of them. To brush these objections aside with disdain should make him the butt of the irony which he loosed on formalism in arithmetic. He might maintain that his invented logical symbolism has solved the problem of distinguishing judgments from mere judgeable-contents. Yet this seems precisely parallel to the formalists' assertion that 'among numbers hitherto known there is none which satisfies the simultaneous equations

$$x + 1 = 2$$
$$x + 2 = 1$$

but there is nothing to prevent us from introducing a symbol which solves the problem' (FA, §96). Frege's inference from the discrepancy between his symbols and those available in natural languages to the conclusion that natural languages are logically defective is justified only *after* he has proved that it makes sense to stipulate a pair of symbols satisfying his requirements. The need for this consistency proof is evident, and the duty to give it never discharged.

The futility of Frege's manoeuvres has even deeper grounds. His fundamental explanation of the contrast between '⊢——— A' and '——— A' is flawed by misconceptions and confusions. He invoked the distinction between asserting something and merely expressing an idea (or merely expressing a thought)—a distinction which is familiar and often used in everyday talk about speech. His special signs are meant to mark this important distinction. But he failed altogether to clarify what it is merely to express an idea or thought, and he did not explore what kinds of expression can properly be described as expressions used merely to express ideas or thoughts. The result is conceptual chaos. He sowed confusion in his own reflections on assertion, and this has borne the fruit of further confusions in more recent discussions of the contrast between sense and force.

Let us consider briefly what it means to say that somebody has 'merely expressed an idea (thought)'. This expression seems closely related to others such as 'formulated an hypothesis', 'aired a possibility', 'framed a proposition', and 'left a question open'. All of them describe acts performed in speaking. They are typically used to characterize complete utterances, not utterances held to be incomplete or in need of completion. In this respect, they are similar to such expressions as 'made a statement', 'asserted something', and 'raised a question'. It makes sense to use them to characterize utterances of declarative sentences. Finally, all of these expressions belong to the same large family: they are used to contrast and to complement one another. If it makes sense to say that a person in saying ' . . .' merely expressed the thought that . . . , then it also makes sense to say that in saying ' . . .' he asserted that . . . or left open the question whether . . . . The range of cases which may be described as somebody's merely expressing an idea is wide and not sharply bounded. But it would certainly include a policeman's speculation ('I suppose it to be possible that x is the murderer'), a scientist's hypothesis ('Smoking may cause anaemia'), and an actor's lines in a role ('Life is to steer, death is to drift'.). All of these remarks are platitudes.

The corollary of this clarification of what it is merely to express an idea (or thought) is a clarification of what expressions are typically held merely to express ideas. These are complete well-formed utterances of a language. The archetype is the declarative sentence. In many cases an expression the utterance of which on one occasion is held merely to express an idea may on other occasions be uttered to make an assertion. In all cases, the expression uttered must be one capable of being used to say something.

Against this background network of interlocking concepts, Frege's employment of the phrase 'merely expresses an idea (thought)' stands out as eccentric. Two oddities are noteworthy. First, he describes a speaker who utters the sentence 'Were π a rational number, then so too would be π^2' to make an assertion as having expressed the idea that π is rational (though without asserting this to be true). Still more bizarre is his claim that somebody who utters the sentence 'It is not the case that π is a rational number' has expressed the idea that π is rational! There is no warrant for any such applications of 'express an idea' in the explanations of this phrase which we have just canvassed. Secondly, Frege's primary application of this phrase is to speakers' utterances of clauses within molecular sentences, and hence his *paradigm* for expressions used merely to express ideas must be clauses within complex sentences (e.g., the antecedent of a hypothetical, or one disjunct of a disjunction).[7] But once again this identification is quite unwarranted by the canvassed explanation of what count as expressions used merely to express ideas. In short, the gross divergence between his applications of the phrase 'merely express an idea' and the applications authorized by a correct explanation of this everyday phrase renders unintelligible his declared intention to use the contrasting symbols '⊢——— A' and '——— A' to mark the distinction between asserting that a and merely expressing the idea (thought) that a. His content-stroke has no business to appear combined with expressions in concept-script which are the counterparts of clauses within molecular sentences. There is no question whether they are there used to make assertions, and hence none whether they are used merely to express ideas. His treating the antecedent of a hypothetical as the paradigm of an expression which merely expresses a thought is nonsense.

Frege's employment of the expression 'merely expresses an idea (thought)' cannot be defined as unorthodox, yet coherent. For to argue that he used it in a special sense would undermine the claim that he gave any explanation at all of the symbols '⊢——— A' and '——— A' in terms of the contrast between making an assertion and merely expressing an idea. And the thesis that his use of the expression is consistent is belied by his practice. His other favoured example of merely expressing an idea is utterances of actors on the stage and the writing of works of fiction. In such cases, what he called 'expressions merely expressing ideas' are declarative sentences which could be used in other circumstances to make assertions and which could intelligibly (though wrongly) be described as actually being used to make assertions in the circumstances he described. Frege had no right to treat the antecedents of hypotheticals and the utterances of actors on the stage as being on a par. His muddle is readily exposed. His saying that a clause merely expresses an idea rests on the conceptual truth that there is no such thing as using a proper part of a complex sentence to perform such speech-acts as asserting something, raising a question, canvassing a possibility, describing something, etc.

---

7. Later, and equally oddly, he billed sentence-questions in this role (N, 119, 121).

In this case, the denial that a clause makes an assertion does not leave open the possibility that it is used to perform some *other* parallel speech-act. Its utterance is a non-act (in this dimension), and it cannot be held to lack assertoric force because it has some other incompatible force (interrogative, imperative), but because it has none. The point about the utterances of actors on the stage is different in these respects. There is the possibility of an actor's uttering a declarative sentence to make an assertion even in the middle of a performance. Moreover, the claim that he is not making an assertion is accompanied by the claim that he is performing another parallel speech-act (play-acting or pretended assertion).[8] Hence his utterance can be assessed in the dimension of acts such as making an assertion, raising a question, etc., and, if it lacks assertoric force, this must be because it has some other incompatible force. To describe the actor's utterance and the antecedent of a hypothetical as being alike in respect of lacking assertoric force (or of merely expressing an idea without asserting it) involves an equivocation. One could with equal justice declare the Pythagorean theorem and the dome of St. Paul's to be alike in having no electric charge, following up this observation with the construction of a theory of electrostatics embracing theorems of Euclidean geometry! Frege's employment of the phrase 'merely expresses an idea (thought)' is manifestly inconsistent.

It is remarkable how oblivious philosophers are to this incoherence. They hold his distinguishing asserting something from merely expressing a thought to be an important insight, and they parrot with approval his examples of merely expressing a thought. They also commit the same fallacy of equivocation in treating Frege's assertion-sign (introduced to distinguish the role of a clause within a molecular sentence from the role of the entire sentence in making an assertion) as the first step in constructing a theory of force (differentiating the speech-act of assertion from other parallel speech-acts such as asking questions and issuing orders). It is important therefore to pinpoint the sources of this nearly universal muddle.

One must be the strong penchant to account for any internal relation in terms of shared entities. This was conspicuous in traditional logic. The relation of entailment between a pair of judgments of the form 'All A's are B's' and 'All B's are C's' and the conclusion 'All A's are C's' was explained as mediated through the concepts out of which these judgments are built. Similarly, the entailment of the judgment 'Q' by the pair of judgments 'P' and 'If P, then Q' was accounted for by the supposition that 'If P, then Q' expresses a relation between the judgments expressed by 'P' and 'Q'. If the judgments (propositions, thoughts) expressed by 'P' and by 'Q' were not *parts* of the judgment expressed by the conditional 'If P, then Q', then there would apparently be no connection whatever between the three judgments in a valid inference by *modus ponens*. Frege accepted this traditional idea, adding only the rider that strictly speaking it is the judgeable-contents (thoughts) expressed by 'P' and by 'Q', not judgments, which are the constituents

---

8. And the claim that the *character* is performing a speech-act (e.g. describing a landscape, stating his intentions, or expressing his discontent).

of the judgment expressed by 'if P, then Q'; the same judgeable-content (or thought) occurs now asserted in the judgment that P, now unasserted in the judgment that if P, then Q. Modern logicians follow the same path, arguing that the validity of an inference by *modus ponens* demands that the very same thought be expressed by the antecedent of the conditional and by the second premise; indeed, this rule of inference is often called 'affirming the antecedent', i.e., asserting what the antecedent expresses. It is a truism that the declarative sentence 'P' when used to make an assertion expresses a thought or proposition. But it is an illusion to conclude from this fact together with the validity of *modus ponens* that the antecedent in the asserted conditional 'if P, then Q' must express the very same proposition. Valid inference rules need no such invisible foundations in Platonic entities.

A second source of the same idea about conditionals (and other compound sentences) is a body of doctrine enshrined in traditional grammar. A declarative sentence is often there characterized as the expression of a complete thought. On this view, it is a truism that every declarative sentence expresses a thought. But main and subordinate clauses are considered to be formed out of declarative sentences by certain regular syntactical transformations, e.g., changing the tenses of verbs in indirect statements according to principles for the sequence of tenses, or altering the word order or the mood of the verb in forming certain subordinate clauses. It is entirely natural to think that such transformations do not alter what is expressed, i.e. that whatever thought a declarative sentence expressed will still be expressed by simple syntactic transformations of it. This conclusion is almost irresistible when no transformation at all is needed, as often happens in English; and it is nearly as compelling when the sole form of change is alteration of word-order, as often happens in German. How could null or minimal alterations constitute a barrier preventing a main or subordinate clause from expressing the same thought as the corresponding free-standing declarative sentence? Though attractive, this line of reasoning is subtly flawed. Expressing a thought is not something that a declarative sentence does off its own bat, as it were, but rather something that a speaker does by uttering it. The fact that he does not do so by uttering a clause belonging to a compound sentence, however closely this may resemble a declarative sentence, is no more mysterious than the fact the minimal variations in the order of proceedings or slight changes in utterances may make the difference between becoming legally married or successfully making a will and performing a ceremony with no legal consequences. The idea that a subordinate clause *must* express the same thought as the corresponding declarative sentence is as absurd as the notion that a couple must be married if they have uttered all the appropriate formulae but in a random order. In thinking that the antecedent of a conditional must express the same thought as a declarative sentence, we are impressed by a correspondence between *forms* of words, but blind to crucial differences in how they are *used*.

A third source of error is the presupposition that every significant expression expresses something. We can enquire what a declarative sentence expresses, and to answer this question will be to specify what thought it expresses. What the sentence 'Two is prime' expresses is the thought that two is prime. Since 'the thought that two is prime' answers the question 'what does "Two is prime" express?', we might well conclude, as Frege originally did, that the phrase 'The thought that two is prime' *expresses* what the sentence 'two is prime' expresses. This would be a mistake, as Frege later noted. Such an expression does not *express,* but rather *designates,* what 'Two is prime' expresses. It is not a candidate for expressing a thought at all, since it would be nonsense to describe the isolated utterance of this noun-phrase as effecting any speech-act such as asserting something, posing a question, canvassing a possibility, etc. Frege did not draw this conclusion, but instead embraced a fresh error, viz. that 'the thought that two is prime' expresses something other than what 'Two is prime' expresses (the 'indirect sense' of this sentence). This is a fragment of his account of the sense/reference distinction. He committed himself to the thesis that every significant expression whatever *expresses* something (its sense). This claim is as absurd as the parallel claim that every significant expression stands for or designates something. Although the question 'What does that express?' may intelligibly be asked about most sentences, it is nonsensical when the expression referred to is a singular designating expression, a predicate, or a clause within a molecular sentence.

Careful examination of Frege's explanations of the judgment-stroke and the content-stroke reveals that many different threads are knotted, rather than knitted, together. His conception of merely expressing an idea is thoroughly confused, and his various stipulations about his symbols apparently inconsistent. The net result is not the weaving of a rich and colourful weft upon a strong warp which depicts clearly the concept of assertion, but rather a tangle of threads emerging from a bundle of knots. Philosophers are oblivious to this partly because they mistake his purposes, partly because they share his misconceptions. Only *by disregarding Frege's explanations of his symbols* have they deluded themselves into thinking that his concept-script clarifies the concept of assertion.

5. Surveying the ruins

What is the significance of the incoherence in Frege's analysis of assertion? One reaction would be that it is of no great consequence. We can (and should!) drop his assumption that inferences must be between assertions. Systems of natural deduction are built on the notion of drawing consequences from hypotheses, and they present logical truths as the results of valid deductions in which all hypotheses have been discharged. On this improved view, any need for the assertion-sign in logic simply disappears. Consequently Frege's unfortunate explanations of the judgment-stroke and the content-stroke can be set aside as utterly irrelevant.

On the other hand this assessment might be thought grossly to underestimate the damage. Frege's formal system and his philosophical reflections in *Begriffsschrift* and *The Foundations of Arithmetic* are built on the notion of judgeable-content. But this concept is introduced through his analysis of assertion. If that is incoherent, then so too is his explanation of judgeable-content. For a judgeable-content is what is expressed by a sentence used to make an assertion *and* by a constituent clause in a sentence used to assert something; it is what remains of a judgment once assertoric force is removed. To revoke his explanation of assertoric force would leave 'judgeable-content' unexplained, while to concede the inconsistency in his use of the phrase 'to express something' would convict of incoherence his explanation of 'judgeable-content' and his presentation of the propositional calculus. The whole framework of *Begriffsschrift* would collapse.

Nor would this catastrophe extend merely to Frege's early work. *The Basic Laws of Arithmetic* and the rest of his mature writings build on the notion of a thought (the sense of a sentence). But this concept too is introduced through the analysis of assertion in a manner exactly parallel to the explanation of 'judgeable-content'. Hence the whole of his later work would also collapse in ruins. Moreover, quite independently of the doctrine of the primacy of the thought over thought-constituents (the senses of subsentential expressions), the theory of sense would be rocked to its foundations by the acknowledgment that names, concept-words, and clauses cannot correctly be said to *express* anything.

We will not attempt to adjudicate in this dispute. That would be a large undertaking, and it would require discussion of many controversial issues extremely remote from the enterprise of clarifying Frege's thinking. Instead, we will drop the tentative suggestion that the defects in his analysis of assertion may have far-reaching repercussions for the coherence of the rest of his philosophical reflections. It is far from obvious that the damage is highly localized, hence, too, not at all evident that correcting the odd 'blunder' or two (e.g. discarding the thesis that inferences must proceed from assertions) will put everything back in order. Part of the ground for this suspicion is the internal relations between assertion and the concepts of a judgeable-content and of a thought. But the other main part is the diagnosis of the source of his confusions about assertion. In particular, his Platonism seems far from a mere stylistic mannerism; it seems rather an opaque screen that prevented his properly relating the significance of speech to the *use* of expressions, and hence it sidetracked him into too exclusive a concern with forms of expressions. The great pitfall in philosophy is that *any* mistake in delineating the logical geography of concepts results in mislocating *all* of them.

Having aired these doubts, we will henceforth for the most part suspend them. There is no alternative. Were we to continue to press objections against Frege's use of the phrase 'express an idea (thought)', we would be unable to pursue the investigation of his early theory of content or his later theory of sense.

5
Judgeable-Content

1. The object of assertion

The notion of judgeable-content is the philosophical foundation-stone of the entire system of logic in *Begriffsschrift*. A judgeable-content is the conceptual content of an assertion; it is the abstract entity which remains, as it were, when the mental act of judging is subtracted from the making of an assertion. Since its essence is its potentiality to be judged to be true (or false) or to be asserted, Frege termed it '*judgeable*-content', contrasting it with *unjudgeable*-contents that are products of its logical analysis. Concept-script is expressly designed to represent judgeable-content. The judgment-stroke stands for the act of judgment, while the remainder of any well-formed expression of a judgment represents solely the object of this act, i.e. what is asserted. The notion of judgeable-content derives from the analysis of judging as a mental act directed at an abstract object.

Frege did not consider judgeable-content to be a theoretical entity postulated in the construction of logical theory, but regarded it as a genuine constituent of the act of assertion. It is as real as the mental act of judgment or the assertoric force with which a sentence is uttered. Judgeable-contents allegedly exist independently of being judged. To deny their existence would not be a misguided strategy, but an *error*. What is asserted on one occasion may be an unasserted part of what is asserted on another occasion. This point Frege later formulated in the claim that a thought may be expressed without being asserted or judged to be true. To deny the existence of contents would amount to denying that the assertion that $2 + 2 = 4$ and the assertion that if $2 + 2 = 4$ then $2^2 = 4$ had anything in common! Furthermore, we actually refer to judgeable-contents in reported speech (cf. BS, §2). In 'He asserted that $2 + 2 = 4$' the indirect statement 'that $2 + 2 = 4$' is the direct object of the verb 'assert' and designates *what* was asserted. To deny the existence of judgeable-contents would appear to be tantamount to denying that we can refer to what was asserted.

Frege's realism about judgeable-contents is of crucial importance. It underpins his repudiation of the traditional synthetic conception of judgments and his celebrated advocacy of contextualism. Traditional logic held that judgments were constructed out of independently given terms (subjects and predicates). These were apprehended prior to propositions, either directly in experience or indirectly through construction out of immediately given items. Hence the terms of propositions were logically prior to propositions, and every proposition had a unique logical analysis into its constituent terms. The cogency of any inference was derivative from the relations obtaining between the terms of the propositions of that inference. 'As opposed to this', Frege declared, 'I start out from judgments and their contents, not from concepts' (PW, 16, cf. 253). It seems as if he intended to invert the doctrine that concepts are prior to propositions: 'instead of putting a judgment together out of an individual as subject and an already previously formed concept as predicate, we do the opposite and arrive at a concept by splitting up a judgeable-content' (PW, 17). If this is meant to be a true generalization about concepts, there is a mystery about how judgeable-contents are to be apprehended independently of concepts and another about the relation of a concept to a predicate ascribing a perceptible property to an object. Ignoring these issues, Frege simply drew the consequences of his realism about judgeable-contents. Concepts are existence-dependent on judgeable-contents; they are analogous to atoms which can never be found on their own but only combined with others (PW, 17).[1] There is no such thing as 'the independent existence of concepts', but rather a concept is always to be viewed 'as having arisen by decomposition from a judgeable-content' (PMC, 101). Since a judgment is not built up out of independently given constituents, it is never the case that 'for any judgeable-content there is only one way in which it can be decomposed' (PMC, 101). Finally, since judgeable-contents are prior to concepts, the traditional doctrine that the calculus of propositions is grounded in the calculus of concepts (or classes) must be absurd (cf. PW, 184; PMC, 191 n.). Although these contentions are both controversial and in need of clarification, there is no doubt that the primacy of judgeable-content is the cornerstone of Frege's whole programme of logical analysis. The precipitation of these abstract entities out of the act of assertion or judgment is a precondition of the possibility of logic (conceived as objective non-psychological laws governing the making of correct inferences). And the further analysis of these entities, in particular their falling apart into unjudgeable-contents, must be regarded as a further analysis of *what is asserted* and only derivatively, if at all, as an analysis of anything else (e.g. of the *sentence* used to make an assertion or of the *meaning* of this sentence).

It is common to ignore the fact that Frege's original concept-script is designed to express *conceptual*-content (i.e. both judgeable- and unjudgeable-content (BS,

---

1. This remark carries no implications of any 'ontological reduction'. Nor did the antithetical doctrine of traditional logic that a proposition is constructed out of a pair of antecedently given terms.

§2)). The notation is meant to make perspicuous the cogency of inferences by clearly representing logical relations among judgeable-contents. With the exception of the judgment-stroke, every symbol in concept-script has a content. In particular, any formula replacing the dummy content-name in the schema '⊢—— A' must stand for or designate *(bedeuten)* a judgeable-content (BS, §2), and the condition-stroke

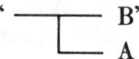

has as its content a relation between judgeable-contents, namely that if it is to be affirmed that *a*, then it is also to be affirmed that *b*. With only minimal distortion, the principle holds that conceptual content is what is symbolized by the symbols of any well-formed formula of concept-script. If we are tempted to separate off Frege's formalization of logic from his explanation of content, it is salutary to remember that he did not view logic as a game played with empty symbols, but rather as a science expressed in a concept-script by formulae whose symbols have genuine content (i.e. conceptual content).[2]

Binding the notion of content to Frege's formal system has two decisive consequences. First, it indicates the persistence of the account of content given in *Begriffsschrift* until this was expressly modified by the sense/reference distinction. The *Foundations,* being built on the formal results of *Begriffsschrift,* is demonstrably erected on the theory of conceptual content, a fact which shapes the interpretation of Frege's enterprise. Second, the fact that his symbolism is isomorphic with that of the modern predicate calculus does not prove an identity between those systems when interpreted as representations of logical relations. A claim of identity between the interpreted formal systems would have to be supported by argument showing that conceptual content can be equated with the 'semantic values' assigned to symbols in the predicate calculus, not treated as a self-evident axiom in the interpretation of Frege's logic.[3]

It is clear that the notion of judgeable-content plays a number of distinct though related roles in Frege's thought. Judgeable-contents are what is asserted or denied, what is expressed but not asserted by a constituent of a conditional or a disjunction, what is proved or disproved, what is true or false, and probably, too, what is believed or disbelieved, what is doubted or indubitable, what is known or not known, and what is understood by somebody who grasps what a speaker has asserted. Frege gave little attention to most of these matters, and he did not consider whether a single kind of abstract entity could discharge all of these functions simultaneously. Two consequences of his silence are important. First, he did not elucidate the connection of judgeable-contents with truth and falsity. Presumably he thought of truth as a property of a judgeable-content the possession of which by a particular content is external to its identity, depending on its relation to the world. The only explicit connection with truth is wholly trivial: to assert a judge-

---

2. All the basic laws of logic contain free variables, hence, according to Frege, they are generalizations.
3. Cf. Dummett, *Frege: Philosophy of Language,* pp. xiii f., 1ff.

able-content is to assert that it is true. Hence to take his explanations of the condition-stroke and the negation-stroke as clarifying relations among truth-values is to impute to Frege a doctrine identifying judgeable-contents with truth-values, and this has no warrant in his writings. Second, identifying judgeable-content with the object of assertion and contrasting this with the act of assertion indicates that the content of an utterance must be distinguished from the assertoric force with which a declarative sentence is uttered, and this separation of force from content is reflected in concept-script by the separation of the judgment-stroke from the rest of the expression of a judgment. As a consequence, any incoherence in the separation of force from content would bring down in ruins the whole of Frege's logical system interpreted as the expression of the most general laws of conceptual content. He did not think that any formal system had a scientific value independent of the cogency of the elucidation of its primitive ideas. Irredeemable defects in his explanation of conceptual content would, in his view, damn his formal logic in just the same way that the inability of mathematicians to answer the question 'What is number?' shows the precariousness of the so-called science of arithmetic.

2. The proper objects of logic

The science of valid inference, Frege thought, studies the legitimacy of transitions from one assertion to another. Acts of assertion decompose into a uniform mental act of judging directed at objects (judgeable-contents) which may vary from assertion to assertion. For logical purposes, therefore, acts of assertion are differentiated only by their objects, and consequently the cogency of an inference depends only on the relations between the relevant judgeable-contents. Judgeable-contents are the proper objects of logical investigation.

Logic can be characterized by its purpose as well as by its subject-matter. It aims to establish the laws or norms of correct thinking or judgment. It is a codification of the principles of sound inference, i.e. the rules that legitimate making one judgment or assertion on the grounds of other judgments known to be true. Logic is a normative science (PW, 128); it governs the justification for making judgments (PW, 3) just as grammar distinguishes correct from incorrect syntactical constructions (cf. PW, 6). Its laws are not generalizations about how persons actually make inferences, but rather prescriptions for how to think in order to reach the goal of truth (PW, 128; cf. BLA, p. xv). They give conditional authority to make a judgment that something is true. From these laws of sound inference can be distilled generalizations whose assertion has an unconditional justification, viz. the general laws of logic.[4] In this way,

> pure thought ... can, solely from the content that results from its own constitution bring forth judgments.... This can be compared with condensation,

---

4. The method for obtaining these laws is in effect iterated application of the deduction theorem (cf. FA, §17).

through which it is possible to transform that air that to a child's consciousness appears as nothing visible into a visible fluid that forms drops. (BS, §23)

The inference-rule of *modus ponens* serves to unfold from these general laws of logic the totality of patterns of cogent inference.

In broad outline this conception was widely shared in the nineteenth century. Logic was widely (though not universally) considered to be a normative science of the relations among entities variously termed 'propositions', 'thoughts', 'judgments', 'assertions', or 'statements' (cf. CN, 224). According to the dominant conception, logic is a *canon* of understanding and of reasoning which exhibits the principles of all logical criticism of knowledge.[5] In *avant garde* works general logic was separated into primary logic, a generalized syllogistic system built on set theory, which codified the relations among atomic or 'primary' propositions, and secondary logic, which codified the relations among molecular or 'secondary' propositions in terms of the relations among their constituent atomic propositions. Scrupulous respect for these rules of reasoning would ensure that the process of drawing inferences could never be blamed for the making of erroneous judgments. Logic provided the defence of the intellect against sin in reasoning.

It was orthodox to view every secondary proposition as stating that a certain relation holds among primary propositions (thoughts, statements, judgments).[6] A conditional, e.g., asserts that two propositions are related in a distinctive way, namely that if the first is to be asserted then so too is the second.[7] (As it were, this states the relative value of two propositions in respect of the goal of attaining the truth: the second must be prized at least as highly as the first.) The logical laws governing secondary propositions were conceived to be normative generalizations about the relations among propositions. (The directive 'From P and Q infer P' is comparable to the valuation principle 'Any two things must be at least as valuable as each separately'.)

---

5. Cf. I. Kant, *Critique of Pure Reason,* A 50ff. (B 74ff.); Boole, *The Laws of Thought,* ch. XXII, §6.
6. It had long been standard to distinguish simple from compound propositions and to consider compound propositions as containing simple ones (cf. The Port-Royal *Logic,* Part II, ch. Vff). Mill refined this doctrine, describing a conjunctive proposition as 'a mere aggregation of simple propositions' and distinguishing this kind of 'complex proposition' from an hypothetical which has the feature that the simple propositions of which it is composed 'form no part of the assertion which it conveys'. Analysing the hypothetical 'If the Koran comes from God, Mahomet is the prophet of God', he declared: 'The real subject of the predication is the entire proposition, "Mahomet is the prophet of God"; and the affirmation is, that this is a legitimate inference from the proposition "the Koran comes from God". The subject and predicate, therefore, of an hypothetical proposition are names of propositions' (*System of Logic,* I-iv-3). Boole generalized this doctrine to cover all of the traditional compound propositions: 'Every assertion that we make may be referred to one or the other of the two following kinds [primary and secondary propositions]. Either it expresses a relation among *things* or it expresses, or is equivalent to the expression of, a relation among *propositions*' (*The Laws of Thought,* ch. IV, §1). Despite differences of detail, traditional logic was committed to the idea that the logical ingredients of complex or secondary propositions were propositions.
7. Cf. Boole, *The Laws of Thought,* ch. XI, §2.

Frege did not break with either of these aspects of the conception of logic dominant in his day. He always considered logic to be a set of anankastic rules observation of which must be justified by promoting the external goal of attaining truth in making judgments. It is in this sense that he viewed logic as consisting of the laws of truth (PW, 3, 128, 252), just as he conceived of ethics as the laws of goodness or virtue.

Similarly, he did not repudiate the thesis that secondary propositions express relations between primary ones. He merely modified it, claiming that a compound judgeable-content expresses a relation among judgeable-contents. In particular, he explained his condition-stroke as the name of a relation between judgeable-contents. His explanation introduces a 'precisely defined hypothetical relation between judgeable-contents' (PW, 16) which differs from what is expressed by the conjunction 'if' through carrying no implication of causal connection (BS, §5; PW, 11n.). The condition-stroke (and any combinations of this symbol with the negation-stroke) is used to designate a new judgeable-content formed from a pair of judgeable-contents (PW, 50f.). Just as '+' names a binary operator on numbers, so the condition-stroke designates a binary operator on judgeable-contents. Because he accepted this parallelism, Frege did not criticize the symbolism of logical algebras as grounded in a fundamental error. Rather, he objected to their choice of the primitive operations (PW, 48f.), and he urged the use of non-arithmetical symbols for logical relations in order to avoid ambiguity in the applications of formal logic to mathematical proofs (PW, 13f.; ACN, 94). He accepted the general principle that molecular judgments express relations between judgeable-contents. The *raison d'être* of his concept-script is 'the perspicuous representation of logical relations by means of signs' (PW, 14), the explicit symbolization of logical relations (ACN, 94) that in mathematical proofs are expressed in natural language (SJCN, 88) and that in everyday arguments are often left to guesswork (SJCN, 85). Like other contemporary logicians, Frege sought a system of symbols for logical relations in which the fewest possible primitive concepts could be used to construct all logically significant relations among judgeable-contents (cf. BS, §7; PW, 50f.).

Do these general observations prove that Frege's logic made no substantial advance on the wisdom of his age? This would be a precipitate verdict. What they do suggest is that we should not expect to find in his work either the birth of the modern semantic conception of logical validity or a successful criticism of the traditional psychologistic conception of inference. Instead we should look for technical or mathematical improvements in executing the acknowledged business of logic—the analysis of inference viewed as transitions whose justification depends on relations among propositions, thoughts, or judgments. Although his most celebrated advance is the analysis of generalizations, he also made an important contribution[8] to the logic of secondary propositions: he gave a direct characteriza-

---

8. Another is sometimes credited to him. He is alleged to have discovered and clarified the difference between judgments and judgeable-contents, thereby resolving perennial philosophical puzzles about the 'circularity' of inferences in the form of *modus ponens* and universal instantiation (cf. Geach, 'Assertion', in *Logic Matters*, p. 255ff.; cf. 'Frege' in *Three Philosophers*, p. 133). Cer-

tion of the relations between judgeable-contents in constructing a logic of secondary propositions. Advanced logical algebras introduced the auxiliary concept of the sets of times at which propositions are true, in order to connect relations among secondary propositions with the logic of primary propositions.[9] In the analysis of inferences among compound thoughts, Frege dispensed with appeals to such sets, and he prided himself on having introduced no further auxiliary concepts. By exhibiting secondary propositions as stating relations among contents, his system brings the analysis of compound propositions into direct conformity with the then-prevalent account of the real nature of secondary propositions. His propositional calculus demonstrates that treating judgeable-contents as constituents of more complex judgeable-contents suffices without further ado to establish all of the recognized logical relations among secondary propositions. This is a formal triumph *within the framework of nineteenth-century logic.* It effects a major simplification in a theory of the relations among judgments or propositions. It is comparable to Frege's simplification of the logical analysis of proofs in arithmetic by the demonstration that mathematical induction is not an independent norm of correct reasoning whose application is limited to numbers. His propositional calculus does represent a theoretical advance on the logical algebras of secondary propositions, but not a revolution overthrowing the established order.

Like traditional logicians, Frege assigned a crucial role to abstract entities (and to mental acts). This differentiates his whole conception of logic from the modern one according to which the cogency of arguments turns solely on semantic relations among *sentences*. He meant to formulate norms for making judgments, to codify the laws of right thinking. He wished to state general principles relating judgeable-contents, not sentences used to make assertions. Having no qualms about the legitimacy of abstract entities, he aimed to make a direct assault on the task set by Leibniz, viz. *peindre non pas les paroles, mais les pensées* (PW, 13). In all these respects his ideas are remote from ours. His interpreted formal system is much more akin to traditional than to modern logic.

Not only did he make no explicit move in the direction of the modern conception, but certain of his basic ideas are incompatible with it. In particular, Frege construed the utterance of a molecular sentence as asserting that a relation holds between judgeable-contents.[10] This is dead wrong. The mistake is not one of intro-

---

tainly he harped on the distinction and stressed that secondary propositions state relations not between judgments (assertions) but between judgeable-contents (unasserted propositions). This was not, however, a fresh discovery, but the reiteration of a familiar point. Mill, e.g., analyzed conditionals as stating the 'inferribility' of one proposition from another, explicitly noting that 'what is asserted is not the truth of either of the propositions' (Mill, *A System of Logic,* I-iv-3). Similarly, the logical algebras of Boole and Schröder gave formal expression to the lack of assertoric force of the component propositions in disjunctions and hypotheticals; symbolizing a disjunction as 'A + B = 1' makes transparent, by the ordinary arithmetic of the integers modulo 2, that the derivation of 'A = 1' (and similarly of 'B = 1') is not justified.

9. The *locus classicus* of this conception is Boole, *The Laws of Thought,* ch. XI, §5ff.
10. The sense/reference distinction did not remove this confusion. The condition-stroke, e.g., stands for a function whose two arguments and whose value are truth-values (BLA, §12). If '2 > 1' asserts that 2 is greater than 1, then

ducing unnecessary abstract entities. Error would be compounded if we took such a sentence to express a relation between sentences. My assertion 'If my wife is shopping, the house will be empty' does not state anything about sentences at all. If it asserts a relation to hold, then this would be a relation between a person and a building. No minor reform will salvage Frege's fundamental conception of logical constants. To drop the idea that they name relations is to abandon treating them as functions, and this would be to knock the props from under his logical system. A revolution intervened between Frege's day and our own, the acceptance of Wittgenstein's claim that 'the "logical constants" are not representatives', i.e. that they indicate 'operations', not functions.[11]

Similarly it is inconsistent with Frege's understanding of his concept-script to construe his propositional calculus in the standard modern way. We are prone to describe his system as formulated with sentence-letters, place-holders for sentences of ordinary language. Frege held that, e.g., the formula

'———┬— A'
    └— B

might be used to express the assertion that if the moon appears in quadrature with the sun, then the moon appears as a semi-circle. But his ground for this is not that the *sentences* 'The moon appears in quadrature with the sun' and 'The moon appears as a semi-circle' may be substituted for the Greek capitals in the formulae of concept-script. In his view such expressions as

'⊢———— opposite magnetic poles attract each other'

or '⊢—┬— the moon appears as a semicircle'
     └— the moon is a quadrature with the sun

'———┬— $2 > 0$'
    └— $2 > 1$

must assert that the True stands in a certain relation to the True! Wittgenstein called attention to a consequent defect: 'if "the True" and "the False" were really objects, and were the arguments in $\sim p$, etc., then Frege's method of determining the sense of '$\sim$' would leave it absolutely undetermined' (*Tractatus*, 4.431). His argument turns on the extensionality of functions. If '$\sim$'names a genuine function whose arguments are the True and the False, then '$\sim (0 > 2)$' would have the same sense as '$\sim (1 > 2)$' (viz. each expresses the thought that the False falls under the concept of negation) since each determines the True as the value of the same function for the same argument! This is absurd. Frege obviously thought that '$\sim p$' has the same sense as '$\sim q$' if and only if 'p' has the same sense as 'q'. But in that case his explanation of the negation-stroke does indeed leave its sense absolutely undetermined.

11. Wittgenstein, *Tractatus Logico-Philosophicus*, 4.0312, 5.2341, 5.25. What are ordinarily called operations (e.g. addition or multiplication) are functions; they correlate numbers with numbers. But such entities are not meant to be covered by Wittgenstein's technical term 'operation'. He stressed that sentences are not names; they do not have references, but only senses. Since functions correlate objects with objects, the filling of the argument-places of any function-expression by anything other than the names of objects yields a nonsensical expression. It follows, therefore, that no sentence connective in an intelligible molecular sentence stands for a function. He encapsulated this thesis in the differentiation between 'operations' and 'functions'.

are malformed gibberish. *A fortiori*, he would not have considered '⊢——— A' or

'⊢┬─ A'
 └─ B

as abbreviations for such nonsense.¹² Frege conceived his concept-script as a 'formula *language* for pure thought' (BS, subtitle, our italics). Its formulae have the status of translations of asserted token-sentences of English (or German). Hence they stand to English sentences roughly as English sentences stand to German ones.¹³ Typically there is no sharing of sentence-constituents in translation, and partial translation yields nonsense belonging to no language whatever. The counterpart of substituting English sentences for letters in concept-script is the expression 'If der Mond in Quadratur steht, then erscheint er als Halbkreis'. The Greek capitals in his propositional calculus are intended as (dummy, or variable) expressions for judgeable-contents in concept-script, not as place-holders for sentences of natural language. Though '⊢——— A' may be used to express the judgment that opposite magnetic poles attract each other, neither '⊢——— A' nor 'A' is any more an abbreviation or dummy place-holder of the English sentence 'Opposite magnetic poles attract each other' than is the German sentence *'Die ungleichnamigen Magnetpole ziehen sich an'*. This conception gives a transparent rationale for a principle of purity applicable to concept-script: none of the symbols for logical constants can be combined with any expressions alien to concept-script.¹⁴ A gulf thus divides Frege's system from (at least some) modern constructions of the propositional calculus.

Admittedly, three features of his exposition mask the import of the fact that he considered concept-script to be an independent language. First, to secure generality to his elucidations of the logical constants of his system he employed free variables. Grasping the significance of these symbols therefore presupposes a correct understanding of the range of expressions that may be substituted for the free variables. There is no warrant in his exposition for taking the range of legitimate substitutions to be wider than well-formed formulae of concept-script; e.g., the elucidation of

'─┬─ A'
 └─ B

is meant to hold if '$\phi(A)$' and '$\psi(A, B)$', or '$2 + 3 = 5$' and '$3 + 4 = 7$' are substituted respectively for 'A' and 'B'. He took this limitation on substitution for

---

12. Contrast this with a randomly chosen modern explanation where '[[cobalt is present ∧ ¬ nickel is present] → a brown colour appears]' is offered as a substitution instance of the formula '[$P \land \neg Q$] → S' (W. Hodges, *Logic*, p. 116 (Penguin Books, Harmondsworth, 1977)).
13. His paradigm is algebra, which he viewed as a separate formula-language (PW, 6) distinguished from natural languages by 'directly expressing' the facts without the intervention of speech (SJCN, 88).
14. And conversely, Frege required of his logical constants that they could unambiguously and perspicuously be combined with any well-formed expressions in concept-script to express relations among judgeable-contents (SJCN, 88; ACN, 91, 97).

granted. Second, the purity of concept-script is hidden by a substantial overlap between well-formed formulae within it and those common in mathematical English (and German). Symbolic statements in mathematics, such as '2 + 3 = 5' and '$m + n = p$', appear in English mathematical texts, and they also occur as constituents of well-formed expressions in concept-script, e.g. '⊢——— 2 + 3 = 5' (PW, 11n., 18ff.; ACN, 94f.) and

$$\vdash\!\!\begin{array}{l} m + n = p \\ n + m = p \end{array}$$

(cf. PW, 18ff.). On Frege's view however, this circumstance, far from compromising the purity of concept-script, actually confirms it, since these expressions *already belong to concept-script*. Third, although he avoided combining sentences of natural language with the logical constants of his system, he did initially make use of diametrically opposite formulae, ones combining content-letters with sentence-connectives of German. This occurs in his informal elucidations of his symbolism. In explaining his primitives, he paraphrased the expression

$$\begin{array}{l} \text{———} A \\ \text{———} B \end{array}$$

as 'A *oder* B', and

$$\vdash\!\!\begin{array}{l} A \\ B \end{array}$$

'*beide*, A *und* B, *sind Tatsachen*' (BS, §7; cf. PW, 35, 48ff.)[15] No doubt, had he seen any solecism here, he would have excused it by the plea that elucidations often have to be taken with a grain of salt. Perhaps he saw nothing to discuss from habituation to similar-looking formulae in mathematics (e.g. '$x < 2$ or $x \geq 2$'). It is worth noting, however, that he later avoided parallel elucidations of his logical constants (BLA, §§6, 12). Even more circumspectly, he translated

$$\vdash\!\!\begin{array}{l} 3 > 2 \\ 2 + 3 = 5 \end{array}$$

into words as '3 is greater than 2 and the sum of 2 and 3 is 5' (BLA, §12), not as '$3 > 2$ and $2 + 3 = 5$'. Close inspection of his writing indicates the importance that he attached to viewing his concept-script as an independent language for expressing judgeable-content.

As an interpreted system, Frege's logic is the science of judgeable-content. It is firmly rooted in the concept of content, and its foundations are secure only to the extent that this concept is coherent. Only the existence of judgeable-contents and the correlations of well-formed formulae of concept-script with these entities secures content to the science of logic. Without judgeable-contents, proofs in logic

---

15. Compare too his paraphrase of the formula '⊢——— Φ (A)' by 'A has the property Φ' (BS, §10).

would be mere manipulations of symbols.[16] The proper study of logic is *judgeable-content*.

3. Frege's elucidations of the logical constants of the propositional calculus

Begriffsschrift contains the first complete axiomatization of the propositional calculus. Of course, Frege did not prove its completeness or its consistency. Indeed, the scorn that he poured on Hilbert's metalogical investigations of axiomatizations of geometry he would have turned too against metalogical proofs concerning his own logical system. The only proof of the consistency of a set of axioms is a demonstration of their joint truth, the only demonstration of their independence the compatibility of the denial of one with the joint assertion of the rest (FG, 15f., 104). Lacking any means of precisely demarcating logical laws apart from their derivation as theorems within his axiomatic system (cf. BLA, p. xvii), he was not even in a position to frame an exact question about the completeness of his axiomatization. Nonetheless he did in fact succeed in constructing what we now recognize to be the first complete axiom system of the propositional calculus, making use of only two primitives, the condition-stroke and the negation-stroke.

It is widely agreed that Frege built this system on the first truth-table definitions of these two primitives.[17] He introduced the condition-stroke by the following explanation: there are four possibilities with respect to a pair of judgeable-contents A and B, viz.

(1) A is affirmed *(bejaht)* and B is affirmed
(2) A is affirmed and B is denied *(verneint)*
(3) A is denied and B is affirmed
(4) A is denied and B is denied.

Then

$$\vdash\!\!\begin{array}{c}\rule{1em}{0.4pt}\ A\\ \rule{1em}{0.4pt}\ B\end{array}$$

stands for the judgment that *the third of these possibilities does not take place, but one of the three others does*' (BS, §5). This explanation is isomorphic with the truth-table definition of material implication and also with the later explanation of the condition-stroke (BLA, §12). Identity with a truth-table definition would be secured if 'affirmed' and 'denied' were replaced by 'true' and 'false' and if it were understood that

---

16. The later bifurcation of judgeable-content into thought and truth-value did not herald any basic alteration in his conception of logic or in the status of his concept-script. It merely transformed the normative science of judgeable-contents into the normative science of truth-values.
17. This achievement is listed in the most sober catalogues of his accomplishments; e.g. J. van Heijenoort, *From Frege to Gödel: A Source Book in Mathematical Logic, 1879–1931*, (Harvard Univ. Press, Cambridge, 1967), p. 1; M. Black, *A Companion to Wittgenstein's Tractatus* (Cambridge Univ. Press, Cambridge, 1964), p. 223.

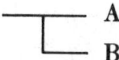

was to be asserted only if it were true. A parallel account of the negation-stroke can be extracted from Frege's explanations of how it combines with the condition-stroke. Instead of explaining that '⊢─┬─A means that "A does not take place"' (BS, §7), he could have explained that '⊢─┬─A' stands for the judgment that of the two possibilities (1) A is affirmed, (2) A is denied, the first does not take place but the second does. This too is isomorphic with a truth-table definition of negation and with Frege's later explanation of the negation-stroke; indeed, with the same provisos as before, these explanations would all be identical. Presumably the chorus of commentators who declare that he actually gave truth-table definitions in *Begriffsschrift* presuppose that these provisos are fulfilled, especially that 'affirmed' is synonymous with 'true', 'denied' with 'false'.

This interpretation carries the corollary that Frege supplied all the materials necessary to prove his axiomatization consistent and complete, although he failed to carry out the proof and wrongly denied the very possibility of doing so. He

> did not explicitly state the modern distinction between the semantic (model-theoretic) and the syntactic (proof-theoretic) treatments of the notion of logical consequence; but it is implicit in his writing. It is because of the introduction of this dual treatment that it was in Frege's writings that, so late in the history of human thought, logic came of age at last.[18]

This praise lavished on his axiomatization of the propositional calculus rests for its justification on the claim that he defined the primitive logical constants as truth-functions. But did he really do so? In respect of *Begriffsschrift*, this seems open to doubt. There he made no explicit statement that the condition-stroke and the negation-stroke are to be treated as function-names. Moreover, he made scant reference to truth and falsity; his explanations turn on whether judgeable-contents are to be affirmed or denied, and even his explanation of the judgment-stroke fails to mention truth, instead likening this symbol to the formal predicate 'is a fact' (BS, §3) and introducing it as a prefix to expressions for judgeable-contents which are to be asserted as correct *(richtig)* or which must be affirmed (BS, §5; ACN, 94, 99). (The parallel elucidation of the universal quantifier similarly avoids explicit reference to truth (BS, §11; FA, §55).) Finally, the accounts of his primitive symbols are not meant to be *definitions* at all. These signs are indefinable. They require elucidation or explanation in order that their significance be grasped properly, but these accounts do not belong to the formal system. Their role is preparatory, to enable the reader to see the truth of the axioms of logic. To ignore the distinction between definitions and elucidations would be, in Frege's view, a fundamental error (FG, 59ff.). It seems then, that he did not *define* the primitive logical constants, that he did not characterize them as names of *functions,* and that his explanations are silent about *truth.* Appearances here must be totally

---

18. Dummett, *Frege: Philosophy of Language,* p. 81

deceptive if he really did formulate truth-table definitions of the two primitives in *Begriffsschrift*. Careful scrutiny of the details of his exposition is called for.

Frege clearly thought that the condition-stroke designates a relation between judgeable-contents. Does this explicit and oft repeated claim license the conclusion that it is a *function*-name?[19] A number of considerations support a positive verdict. *Begriffsschrift* hints at the later explanation that a relation is the reference of a function-name which has two (or more) argument-places (BS, §10), and this suggests that the representation of a relation between judgeable-contents must be a function-name in concept-script. The general explanation of the concept of a function (BS, §9) actually fits the condition-stroke; for, in the expression

we can imagine '$1^2 = 9$' to be replaced by '$3 \times 7 = 21$', and '$1^4 = 81$' to be replaced by '$2 + 3 = 5$', thus treating the condition-stroke as the stable component in this formula. Frege actually availed himself of this way of viewing molecular judgments. He introduced variables whose values are judgeable-contents to express the generality of his basic laws of logic. The axiom

'⊢──┬── a'
　　└── b
　└────── a

contains two free variables,[20] and hence we have no option but to view the remainder (the array of condition- and content-strokes) as the designation of a function (BS, §9). The only escape from this conclusion would involve asserting that a symbol may have as its *content* something that has no place in the hierarchy of types of function-theory, i.e. that what it stands for is neither an object nor a function. There is no evidence that Frege ever contemplated such a *lusus naturae* at any period of his career. All of these considerations confirm that his later explicit explanation of the condition-stroke as a function-name merely develops an idea implicit in the exposition of *Begriffsschrift*.[21]

Truth is a different matter. If the condition-stroke stands for a relation between judgeable-contents, then it must be the name of a function whose arguments and values are judgeable-contents. But when Frege split judgeable-content into thought and truth-value, he explained the condition-stroke as a function whose arguments and values are truth-values. Might this not consist in making explicit something already present in *Begriffsschrift?* This inference would be fallacious. The conclusion is inconsistent with a pair of ideas informing his earlier work. The first is that the value of a function depends only on what its argument is, not on how the argument is designated. The second is that two formulae have the same

---

19. On *Frege's* conception of a function, this question is *not* trivial.
20. In exposition of Frege's logic, we follow his practice of using lower-case letters (Roman and Greek) as free variables, while upper-case letters (Greek only) are construed as dummy constants.
21. The same verdict holds in respect of the negation-stroke.

judgeable-content only if the substitution of one for the other anywhere in any cogent proof leaves the cogency of the proof unaffected (BS, §3). Suppose then that Frege had thought the condition-stroke to name a *truth*-function, i.e. a function whose arguments are truth-values. Then, if its value were a judgeable-content, any two expressions of the form

$$\vdash\!\!\!\begin{array}{c}B\\A\end{array}$$

would have identical *content* provided only that the formulae substituted for 'A' and 'B' had identical truth-values; but that would be patently absurd. Alternatively, the value of this truth-function might not be a judgeable-content, but a truth-value; but in that case the explanation in *Begriffsschrift* of the condition-stroke would leave the *content* of expressions of the form

$$\vdash\!\!\!\begin{array}{c}B\\A\end{array}$$

completely undetermined! Consequently, it is *inconsistent* with the framework of *Begriffsschrift* to view the condition-stroke and the negation-stroke as names of *functions* whose arguments are *truth*-values. In fact they are explained as designating functions whose arguments and values are judgeable-contents.

Of course, the denial that the primitive logical symbols of *Begriffsschrift* are names of truth-functions does not deny the thesis that the relations among judgeable-contents expressed in a molecular judgment are connected with dependences between the truth or falsity of one constituent content and the truth or falsity of other constituent contents. But the general idea that molecular propositions express truth-dependences is not Frege's invention. It was highlighted in Boole's calculus of secondary propositions (e.g. in the inference that $A + B = 1$ if and only if $A = 1$ and $B = 1$), and it was likewise stressed in his accompanying explanations of his system.[22] The proposition if X then Y is claimed to express a dependence of the truth of Y on the truth of X, i.e. it states that if X is true, then Y is true.[23] Not only is this idea familiar, it is wholly trivial. As Frege noted, the assertion that 2 is prime is identical with the assertion that it is true that 2 is prime, and therefore every relation among asserted judgeable-contents is *eo ipso* a relation among the truths expressed by their assertion. That trivial connection with truth would hold however Frege had explained his logical constants, provided they designated relations among judgeable-contents.

One upshot of this investigation is the realization that the explanations of the logical constants in the *Basic Laws* diverge in an important respect from the parallel explanations in *Begriffsschrift*. To a first approximation, Frege later introduced them explicitly as names of truth-functions, i.e. as names of functions

---

22. Boole, *The Laws of Thought*, ch. XI, §§2, 11.
23. Ibid., ch. XI, §12.

whose arguments and values are the two logical objects the True and the False.[24] He stipulated the references of these function-names in the orthodox manner by specifying a value for the named function for each admissible argument. This could be done in tabular form provided that the True and the False were the only admissible arguments. But, because he conceived that he had to meet the demand of completeness of definition, he had to stipulate values for any objects whatever as arguments, and hence he could not encapsulate a complete explanation in the form of a truth-table. Hence the *Basic Laws* contains no equivalent of the tabular explanations of *Begriffsschrift*. According to his mature theory, the logical constants of the propositional calculus are merely names of two[25] out of a myriad of concepts. Paradoxically the recognition of functions whose arguments are truth-values is linked with reasoning which obliterates the distinction between these functions and those whose arguments are other objects. The concept of being a prime number has just as much right to be called a truth-function as the concept referred to by the negation-stroke! Therefore, even in the *Basic Laws,* Frege failed properly to single out truth-functions, and he never gave a truth-table definition of the condition-stroke and the negation-stroke even after admitting truth-values as arguments and values of functions.

The other main upshot of this investigation is a recognition that the formalization of the propositional calculus in *Begriffsschrift* is independent of the elucidations of the primitive logical constants and the accompanying patter about constants and variables, functions and judgeable-contents. It is only in the propaedeutic of his system that Frege appealed to the conception of logical constants as function-names. This idea is not employed in the formal derivation of theorems, and the axioms are presented as indemonstrable self-evident truths, not as consequences of his explanations of the primitives. This means that the formal

---

24. In 'Introduction to Logic' (1906), Frege gave his later truth-functional explanation of the conditional (PW, 186) and contended that 'It is now almost 28 years since I gave this definition', clearly alluding to *Begriffsschrift*. If this later explanation is to be understood truth-functionally, it is untrue that Frege had already given it in 1879. It is at best isomorphic with his earlier explanation. Nor is it true that it is, by his lights, a definition, but only an elucidation.
25. Frege stipulated that the negation-stroke and the condition-stroke always occur fused with the horizontal. Our argument here treats these *compound* symbols as the logical constants of the propositional calculus and notes that these names of functions must meet the demand for completeness of definition. An alternative would be to separate out the horizontal as a distinct function-name and then to argue that the combination of the negation-stroke and the condition-stroke with horizontals legitimates *complete* definitions of these latter function-names by truth-tables. But this riposte would be futile. The fact that a composition of functions is completely defined (e.g. the function denoted by '——┬——') does not secure that *each component* function is completely defined. This demand would be met only if the negation-stroke, for example, were so defined that it admitted into its argument-place any well-formed proper name, not merely names formed with the horizontal; and for every such argument-expression, the compound symbol would have to be assigned an object as its reference. Obviously Frege's stipulations for the negation-stroke and the condition-stroke do not secure completeness of definition when the horizontal is taken to be a separate function-name. Therefore, on neither possible construal of his symbolism would truth-tables constitute *complete* definitions of his logical constants.

system can in principle be divorced from its putative foundations in the application of function theory to judgeable-contents. The same structure might be re-erected on quite different foundations. The *Basic Laws* realized this abstract possibility. Despite an essential shift in the explanations of the primitives and a reorientation about the proper subject-matter of logic, the earlier axiomatization of the propositional calculus survives with no alteration whatever. The formal system appears to be the centre of gravity in his thinking; it determines the orbits available to his more philosophical speculations. Wittgenstein later effected a similar revolution; the *Tractatus* divorced the logical constants from function theory, defined these operations by truth-tables, and thereby justified Frege's axioms for the propositional calculus without appeal to self-evidence. But the independence of Frege's system from the alleged foundations in *Begriffsschrift* and in the *Basic Laws* does not show that his calculus really stood all along on the ideas of the *Tractatus*. The possibility of moving a wooden-frame house does not rest on the fact that the house has permanent but imperceptible foundations!

4. The propositional calculus

The role of the elucidations of the primitive symbols of concept-script, in particular the sharp distinction between these informal explanations and formal definitions within concept-script, holds the key to Frege's conception of the propositional calculus. An immediate corollary of the fact that he did not formulate truth-table definitions of the logical constants is a striking divergence between his conception and the one now current among logicians.

The use to which Frege put his elucidations is straightforward. They underpin his sole explicit rule of inference[26] (the counterpart of *modus ponens* in concept-script) and his primitive laws of logic (the axioms of *Begriffsschrift*). The explanation of the conditions in which

$$\vdash \begin{array}{c} \text{---} A \\ \text{---} B \end{array}$$

is to be affirmed exclude the case that B is to be affirmed and A is to be denied. Consequently, if B is to be affirmed ($\vdash$ ─── B) and

$$\vdash \begin{array}{c} \text{---} A \\ \text{---} B \end{array}$$

is to be affirmed

$$(\vdash \begin{array}{c} \text{---} A \\ \text{---} B \end{array}),$$

then clearly A itself is to be affirmed ($\vdash$ ─── A). The explanation of the condition-stroke makes perspicuous that this rule of inference has an unqualified jus-

---

26. Frege also employed a rule of substitution, but he did not formulate it explicitly as a rule of inference in his system.

tification (BS, §6). In a parallel manner, the same explanation provides a justification for holding that any judgeable-content of the form

is to be asserted, whether A and B are severally to be asserted or denied (BS, §14). For, if A is to be denied, then

⊤── Γ
└── A

is to be asserted whatever Γ may be (and hence if Γ is replaced by

⊤── A);
└── B

while if A is to be asserted, then

⊤── A
└── B

is to be asserted, and therefore

⊤── A
 └── B
└── Γ

is to be asserted whatever Γ may be (in particular, if Γ is A). This absolute justification for asserting any judgeable-content of this form licenses treating the generalization

as a law of logic (BS, §14). The other axioms of the propositional calculus have parallel justifications.

These arguments to establish the elements of Frege's system of logic do not have the status of formal proofs from formal definitions within the system. Rather they are non-formal demonstrations based on elucidations appearing only in the propaedeutic of his axiomatization. They are comparable to the diagrams and reasoning employed to persuade a schoolchild that the parallels axiom in Euclidean geometry is a self-evident truth. Although each of the Euclidean axioms may be held to express an indemonstrable truth immediately certified by intuition of space, the ground may have to be prepared for somebody to apprehend it and to recognize its self-evidence. But on pain of compromising its status as an axiom, this conceptual technology must be distinguished from constructing a rigorous proof. Frege saw an exact analogy with geometry in respect of the nature and role of his elucidations of the logical primitives. The only difference he acknowledged is that the axioms of logic are certified directly by a 'logical source of knowledge' which is distinct from the faculty of spatial intuition that delivers the axioms of geometry (PMC, 100; PW, 269ff., 278).

The resemblance of logic to geometry extends even further according to Frege's conception of matters. Every axiom of geometry is a generalization, and hence he takes it to express a relation among concepts. The axioms, and therefore the theorems, are an unfolding of the essences of the primitive concepts of geometry (the concepts of a point and a line, the relation of a line's running through a point, etc.). Though they are truths about concepts, they are synthetic, not analytic. It would be nonsense to claim that any of them could be derived from definitions of concepts—doubly nonsensical, granted that each of them is indemonstrable and that each of the geometrical primitives is indefinable. The axioms, and therefore the theorems, of the propositional calculus are precisely similar. They unfold the natures of the primitive logical constants, the indefinable concept and relation named by the negation-stroke and the condition-stroke.[27] But it would be doubly nonsensical to assert that the axioms of logic could be derived from definitions of concepts. This would conflict with Frege's characterizing axioms as indemonstrable truths and with his taking his primitive logical constants to be indefinable. It would be equally nonsensical to derive the logical axioms from a definition of 'truth' (or 'The True') because truth is also a primitive concept in logic. Therefore the remark ' ... the laws of logic are nothing other than an unfolding of the content of the word "true"' (PW, 3) must be taken with a pinch of salt. It suggests a conception of logic parallel to Hilbert's conception of Euclidean geometry, but that was anathema to Frege, as much for logic as for geometry. He thought that logic unfolds the essence of *all* its distinctive primitive concepts (not just truth) just as geometry unfolds the natures of its particular concepts. It is the nature of the concept of negation, for example, that determines the fact that double-negation cancels out, i.e. that the generalization

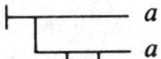

is a law of logic (BS, §18).[28] The norms of correct reasoning are no more grounded in the meaning of the word 'true' than the principles of proper conduct are in the meanings of the words 'good' or 'ought'.[29]

27. And perhaps too the concept designated by the 'horizontal'.
28. Wittgenstein called attention to this Fregean idea and marshalled arguments against it (*Remarks on the Foundations of Mathematics*, 3rd ed., Pt. I, Appx. I; *Philosophical Investigations*, §§554ff.). He considered this account of double-negation to manifest a general misconception, viz. that necessary truths flow from the natures of objects correlated with words as their meaning. A main theme of his work is the criticism of this general idea (cf. *Philosophical Grammar*, §§16ff.).
29. These claims would be true, though misleading, if the phrase 'the meaning of "true"' is equivalent to 'the concept of truth'. On that interpretation, arithmetic would be the science of the meanings of '0', '1', '2', etc. But the proper objects of arithmetic would still be numbers (Platonic entities), not symbols (linguistic entities).

For Frege, an axiomatic presentation of logic is not merely an optional style of exposition, but it reflects the essential structure of the totality of logical laws. The axioms alone secure the certainty of the theorems (provided the derivations are clearly cogent).

> The question why and with what right we acknowledge a law of logic to be true, logic can answer only by reducing it to another law of logic. Where that is not possible, logic can give no answer. (BLA, p. xvi)

This conception is a reaffirmation of the Cartesian conception that all knowledge ultimately rests on self-evident truths. Provided that each step of inference is perspicuously unassailable, then how could one conceivably arrive at a mistaken belief by reasoning from premises each of which is perspicuously true?[30] It is just this conception of logic as an axiomatic science which Wittgenstein undermined by giving truth-table definitions of the logical constants of the propositional calculus. The 'axioms' cease to be indemonstrable truths resting on a mysterious logical source of knowledge and distinguished by being self-evident; the justification of the 'theorems' is just as direct as that of the axioms, and hence none of the truths of logic are more primitive than any others.[31] Frege resisted such a model-theoretic account of logical consequence, defending to the last the conception of logic as an essentially axiomatic science.

Although he did not bring logic to its twentieth century acme, Frege's propositional calculus is a dramatic improvement on the formal systems then available. First, it is more economical than traditional accounts of the logic of secondary propositions. It is built only on judgeable-contents, thus dispensing with any auxiliary entities (especially sets of times when propositions are true). Not only is this a theoretical simplification, but also it avoids philosophical objections raised by analyses appealing to auxiliary entities.[32] Second, Frege's propositional calculus is part of a unitary logical system also embracing inferences involving generalization. Traditional logic bifurcated into two separate subsystems, primary and secondary logic; they were related only by a dubious analogy. By contrast, Frege saw that function theory is just as directly applicable to reasoning with molecular propositions as to reasoning with generalizations, and thereby he established an organic connection between two sets of laws previously held to be independent (ACN, 99; cf. PW, 14). Indeed, he drew the consequence from the primacy of judgeable-content in logical analysis that, in so far as it makes sense to speak of 'reduction', he had inverted Boole's procedure and reduced his primary propositions to his secondary ones (PW, 17f.). Finally, Frege's axiomatization of propositional logic is itself an achievement. For the first time it made the laws governing

---

30. Instances of the Sorites paradox seem at first sight to refute this principle!
31. Wittgenstein, *Tractatus Logico-Philosophicus*, 6.1ff.
32. Frege indicated the difficulty of applying Boole's account to the analysis of molecular propositions incorporating only the eternal truths of mathematics (PW, 15). His later doctrine that truth is a timeless property of thoughts would exclude the possibility of Boole's analysis of any compound thought whatever.

secondary propositions surveyable. This satisfies one of the chief requirements of understanding (FA, §5; PW, 36). Frege arrived at

> a small number of laws in which . . . the content of all the laws is included, albeit in an undeveloped state. . . . Since in view of the boundless multitude of laws that can be enunciated we cannot list them all, we cannot achieve completeness except by searching out those that, *by their power,* contain all of them. (BS, §13)

Nothing comparable was achieved until considerably later in the development of Boolean logical algebras.[33]

Against these merits of Frege's propositional calculus must be set certain fundamental weaknesses. One obvious one is the appeal to self-evidence in his selection of axioms[34] and the postulation of a 'logical source of knowledge'. Another is the conception of logical laws as anankastic rules ultimately justified by their efficacy in promoting the attainment of the external goal of the judgment. And a third is the threat of circularity implicit in an axiomatization of logical laws, for it seems as if the criterion for the correctness in the derivation of the theorems from the axioms must be conformity with the norms of logic, i.e. the theorems to be derived.[35]

On top of these general difficulties is a major problem specific to Frege's logical system. The key to his construction of the propositional calculus is the coupling of a Platonist conception of the objects of judgment with the analytical framework provided by abstract function theory. Judgeable-contents are real (abstract) objects. The cash value of this fact is that they are admissible as arguments and values of (first-level) functions. If all 'arbitrary restrictions' are lifted on what kinds of objects are admitted as arguments and values of functions, then recognition of contents as objects immediately legitimates the consideration of functions defined over contents. Frege's propositional calculus is the immediate fruit of this generalization of function theory. He explored the possibility of treating analogues of sentential-connectives as function-names defined when names of judgeable-contents are inserted in their argument-place(s), and he found the results of his exploration to be highly satisfactory. He construed molecular judgments as having contents that are the values of functions for other judgeable-contents as arguments. By implication he took 'If the fuse has not blown, the clock will be working' to have the same logical form as '$3 > 2$': both express relations between pairs of objects. But he also viewed this conditional as expressing the value of a binary function for a pair of contents as arguments; in this respect it is parallel to the symbol '$3 + 2$' expressing the result of applying the operation of addition to the number pair (3, 2). The content of a conditional sentence is thus Janus-faced. As a content of judgment, it resembles what is expressed by such sentences as '$3 > 2$'; as an object, it is comparable to what is designated by such

---

33. W. and M. Kneale, *The Development of Logic,* p. 423ff.
34. As noted by Wittgenstein, *Tractatus Logico-Philosophicus,* 6.1271.
35. One of Wittgenstein's endeavours was to avoid such circularity (ibid., 6.123, 6.1262f.).

expressions as '3 + 2'. The second aspect licenses the direct application of function theory to molecular judgments. Abstracting from the connection of judgeable-content with judgments removes any obstacle to treating sentential connectives as names for functions parallel to the symbols for elementary operations in arithmetic.

The very idea of applying function-theory to the analysis of molecular judgments involves the disastrous move[36] of construing expressions of judgeable-contents as names of objects. In particular, it demands that sentences used to express molecular judgments function logically as designations of contents.[37] Such a sentence must be exhibited as having a content which is the value of a function whose arguments are judgeable-contents. Consequently its logical significance is exhausted by its being correlated with an *entity* (the value of a function). The only question left open is what type of entity; i.e. judgeable-contents must be located in the type-hierarchy of abstract function theory. Should they be construed as objects, as first-level functions, as functions of first-level functions, etc.? Frege himself neither raised nor answered this question explicitly in constructing the propositional calculus of *Begriffsschrift*. Indeed, any formal type-theory is absent from his original exposition of the concept-script. But his conception of the matter is tolerably clear from his practice. He allowed content-names ('sentences' in concept-script) to occupy positions open to names of objects; in particular, they may flank the identity-sign (BS, §8; PW, 35f.). His formal definitions take this form (e.g. BS, §24). Any distinction in logical-type between content-names (sentences) and object-names would invalidate his procedures. It might be objected that he nowhere defined the values of functions over judgeable-contents for arguments other than judgeable-contents and thus offended against the canon of completeness of definition unless he held that judgeable-contents were type-distinct from objects. This objection has no force in interpreting *Begriffsschrift*. For there both piecemeal (BS, §8; FA, §§85, 100f.) and incomplete (BS, §27) definitions are admitted as legitimate.[38] Since partial functions are tolerated, the fact that Frege failed to define such expressions as '―― 4', '―― $\dfrac{d}{dx}$', and '⊢―― 4 / ―― 2'

cannot be cited as an argument that he took judgeable-contents to be type-distinct from objects such as numbers and geometrical points. The idea of a many-sorted domain of objects gains no support from his exposition. From his point of view to treat sentence-connectives as function-names is inseparable from treating sen-

---

36. Cf. Dummett, *Frege: Philosophy of Language*, p. 7, 184, 643ff.; Geach, 'Critical Notice of *Frege: Philosophy of Language*', p. 438, in *Mind*, LXXXV (1976).
37. Judgeable-contents are also expressed by indirect statements. The idea that these expressions function as names of objects is less immediately puzzling than the parallel thesis about sentences.
38. He also countenanced formulae in which the free variables are understood to range over limited domains of objects (e.g. FA, i f.).

tences (expressions of judgeable-contents) as names of objects. The assimilation of sentences to proper names is indispensable to *Begriffsschrift*. Far from being a late and unmotivated modification to his logic,[39] it is the foundation-stone of his propositional calculus from the very outset. Is it not incongruous to hail his logical system as a triumph when this is built on an idea commonly viewed as an unmitigated disaster?

5. Criteria of identity for judgeable-contents

The philosophical foundation of Frege's propositional calculus is the concept of a judgeable-content. But the expression 'judgeable-content' is a technical one introduced by him in the informal explanation of his logical system. How should it be understood? One view would be that no special explanation is required. The primitive terms of any theory are given meaning by the theory itself. Accordingly, the explanation of his logical theory would progressively clarify what the subject-matter of this theory really is. This was not Frege's view at all. He thought that the primitive terms of his theory should be explained. Otherwise a reader would not be in a position to see that the formulations of the first principles expressed self-evident truths.

According to his conception, however, nothing more need be said in explanation of 'judgeable-content' than is strictly necessary for the reader to apprehend the intended concept. The adequacy of explanations must be judged pragmatically. In this particular case he evidently thought that very little is quite enough. The concept indicated by 'judgeable-content' is close to the concept or range of concepts expressed by the traditional terms 'proposition', 'judgment', 'assertion', 'statement', and 'thought'. The only differences concern the explicit exclusion of certain unwanted connotations. Hence, in Frege's elucidation, *determinatio* takes the form of *negatio*. One exclusion, commonplace in works of traditional logic,[40] is features of assertions that have no relevance to inference, certainty, and truth, i.e. matters alien to the cognitive value of assertions. Among these are included manifestations of the speaker's attitudes (BS, §7), associated mental imagery (FA, §60), indications of the strength of the grounds of judgments (BS, §4), and marks of emphasis (BS, §3).[41] Another exclusion is the suggestion that a 'judgeable-content' is a mental entity. Frege stressed that it is an abstract object, comparable to a number, a direction of a line, or an extension of a concept. This was, at that time, an essential point to clarify, since some logicians treated propositions as psychological entities, while others took them to be Platonic objects. Finally, Frege introduced a further negative point of clarification: although the object of the act of judgment is always a judgeable-content, a judgeable-content may occur

---

39. Dummett, *Frege: Philosophy of Language*, pp. 7, 182ff. Sluga correctly calls attention to this important error of interpretation (*Gottlob Frege*, p. 86).
40. Lotze, *Logic*, I-i-§7, I-ii-§41, II-Prefatory remarks-§153, II-i-§165; Port-Royal *Logic*, Part I, ch. xiv; Mill, *System of Logic*, I-iv-2, II-ii-1; Boole, *The Laws of Thought*, ch. II, §§9, 16.
41. In Frege's view, this category includes such grammatical distinctions as that between subject and direct object in a sentence (BS, §3; cf. PW, 6).

as part of a content judged to be true without itself being judged to be true. A speaker may express a judgeable-content without asserting it, and on Frege's official view this occurs whenever he makes a molecular judgment. This clarifies a matter *de facto* settled in the same way in logical algebras. Having elucidated these three points, Frege believed that any sensible reader would have a full grasp of the concept of a judgeable-content.

This is the background to his explanation of identity of judgeable-content:

> ... the contents of two judgments may differ in two ways: either the consequences derivable from the first, when it is combined with certain other judgments, always follow also from the second, when it is combined with these same judgments [and conversely], or this is not the case. The two propositions 'The Greeks defeated the Persians at Plataea' and 'The Persians were defeated by the Greeks at Plataea' differ in the first way.... I call that part of the content that is the *same* in both the *conceptual content*. (BS, §3)

The argumentative context of this remark gives it a totally different significance from a modern attempt to specify the criteria of identity for an abstract object. The root of Frege's thinking is that logical relations among assertions are internal relations among their judgeable-contents. That a content is the content which it is determines that it has the relations which it has to other contents. Change in content is change in its logical relations with other contents. This has an obvious consequence for expressions of content in proofs. For if two formulae or sentences have the same content, then replacement of one by the other in any proof must leave the cogency of the proof unaltered. Conversely, if a proof is rendered invalid by such a substitution, this must reflect a difference in the contents of the formulae. These principles are mere truisms. They are restatements for linguistic expressions of the thesis of the identity of indiscernibles in respect of judgeable-contents.

In the explanation of the identity-conditions for judgeable-contents, the logical equipollence of the cited sentences is not the premise of reasoning showing them to have the same content. Rather it is the conclusion of an inference: since what is asserted by uttering one sentence is just what is asserted by uttering the other, and since the cogency of any argument turns only on what is asserted, substitution of one sentence for the other in any argument leaves the cogency of this argument unaffected. Frege took it for granted that any competent mathematician will recognize that replacement of '$A = B$' by '$B = A$' or '$3 > 2$' by '$2 < 3$' leaves the cogency (and rigour) of a proof unaffected.[42] Frege merely provided a technical terminology building on this common knowledge, not a *recherché* technique for refining these judgments in unheard of ways. Identity of content is not something to be discovered; it is a datum, the ground of all logical reflections. *A fortiori*, it is not something discoverable from the scrutiny of arrays of sentences (arguments). For if it were possible to determine the validity of arguments prior to identifying the content of constituent sentences, it would clearly be false to hold,

---

42. Indeed, he might say that it left the *proof* unaltered.

as Frege did, that cogency of proofs turns wholly and solely on judgeable-contents! His aim was not to explain how to *demonstrate* that two sentences or formulae had the same content.[43]

Modern philosophers look at these matters from a vantage point inaccessible to Frege and his contemporaries. We hold that the logical system of *Begriffsschrift* should be viewed as a theory resting on the postulation of theoretical entities termed 'judgeable-contents'. At our most tolerant, we might construe Frege's theory-building to involve writing a blank cheque; the enterprise must later be backed by a proper explanation of 'judgeable-content', incorporating criteria of identity for judgeable-contents. In a less tolerant mood, we might declare that introducing abstract objects into discourse is, *ceteris paribus,* an intellectual sin, the proper atonement for which involves performance of the ceremony of stipulating criteria of identity. In either case, we chant the slogan 'No entity without criteria of identity'. We think this (legitimate?) demand to be satisfied only if an effective procedure is specified for deciding whether two utterances of declarative sentences (or perhaps two speakers' acts of assertion) express the same or different judgeable-contents.

These modern ideas are alien to Frege's way of thinking. (One might speculate that he would have judged them to be a fresh form of intellectual barbarism, if not manifestation of a hitherto unknown form of madness!) He was light-years away from any intention to supply an operationalist criterion for determining the content-identity of different utterances or of type-sentences. His sole pronouncement on identity of content cannot be interpreted as filling this lacuna without gross distortion of his thought.

One reason for this is his pragmatic standard for the adequacy of elucidations of primitive concepts. No set of operationalist bridge-principles is necessary to secure application of logical theory. Provided that there is general agreement about what is asserted by an utterance, the theory will have all the utility that it could. Nothing would be gained by building a further theory to bridge the gap between a logical system and its application.

Second, Frege considered that judgeable-contents lie at rock-bottom in logic. His doctrine of the primacy of judgments (and later of thoughts) amounts to the claim that logical analysis begins from judgeable-contents as its data. He did not believe that a proper account of cogent inference rests on the semantic analysis of declarative *sentences*. Consequently to portray *Begriffsschrift* as adumbrating an effective procedure for determining whether or not any two sentences express the same judgeable-content is to misrepresent his conception of logic. That would open the way to going behind judgeable-contents and constructing logic on a more primitive basis. This view, in turn, has the implication that earlier he was inconsistent or failed to draw the consequences of his avowed doctrines. Surely it is

---

43. The same verdict holds in respect of his demand for criteria of identity for numbers. The *possibility* of correlating F's one-to-one with G's is not a means for determining that there are the same number of F's as G's. Indeed, the converse would be more plausible (cf. *Wittgenstein's Lectures, Cambridge 1932–1935,* ed. A. Ambrose, p. 148ff. (Blackwell, Oxford, 1979)).

better policy to accept at face value his belief that there is nothing behind his logical machinery.

Finally, the supposition that the remarks on content-identity in *Begriffsschrift* give an operational criterion of identity is ridiculous in respect of their specific content. Frege would have denied that there is a surveyable totality of cogent inferences having a given judgment as one premise (cf. BS, §13). Hence there is no such thing as verifying that a pair of assertions have the same content by inspecting every inference in which each occurs to see if it can be replaced by the other. Even a negative test for content-identity would elude our grasp. The notion of content-identity would presuppose criteria for identity and difference of proofs, and these are doubtful in respect of what counts as altering the structure of a proof and what counts as adding to its premises.[44] Furthermore, to make any sense of extracting a criterion of identity from Frege's remark, the notion of a proof as an ordered array of assertions would have to be replaced by the notion of an argument viewed as an array of sentences. That would transform his conception of logic into something altogether different.

To the extent that we find fault with his elucidation of 'judgeable-content', our thinking moves in an inertial frame distinct from his.

6. Judgeable-content and sentence-meaning

The historical question of whether Frege meant to furnish a criterion of identity for judgeable-content must be distinguished from the philosophical question of whether such a criterion could be constructed to provide a solid foundation for his logical system. That he saw no lacuna does not entail that there is none, and that he failed to fill one that we now notice does not exclude the possibility of our doing so. Most modern philosophers would characterize this as an urgent task, since until it is accomplished we do not know what Frege is talking about nor whether what he expressed in the symbolism of concept-script is true. How might we embark on this task of clarification?

A first step might be to ask of what is judgeable-content the content. It might be viewed as the content of the act of assertion whose object it is, as a command can be viewed as the content of an act of commanding. Or it might be seen as the content of a sentence used to make an assertion. According to the elucidations of *Begriffsschrift*, expressions have conceptual-content. Judgeable-content is the content of an assertoric sentence. A sentence and its passive transform are claimed to have the same content, and a nominalized sentence prefixed to 'is a fact' is held to have its entire content in the subject-term (BS, §3). Unjudgeable-contents are the contents of sub-sentential expressions in an assertoric sentence. In particular, true identity statements express the fact that two names have the same content, and formal definitions stipulate that the content of an abbreviation shall be identical with the content of a complex expression (BS, §8).

---

44. Signs of strain show in his account of definitions and in his characterization of logical laws as 'almost without content' (CT, 556).

From here one might jump straight to the conclusion that the notion of content is to be identified with the familiar notion of sentence-meaning, i.e. to argue that Frege identified what is asserted in using a sentence to make an assertion with what this sentence means.[45] One could then capitalize on familiar discussions of identity of meaning in order to supply criteria of identity for judgeable-content.

The proposal is, however, open to serious challenge. First, content is restricted, *ab initio*, to declarative sentences used to make assertions[46] (and to their appropriate parts, if they are molecular sentences). A sentence which is not assertible (or deniable) is denied any content (BS, §27). Hence non-declarative sentences such as imperatives, optatives and WH-questions lack judgeable-content,[47] although it would be absurd to deny them any meaning.[48]

Second, different tokens of a single type-sentence may differ in content. This is most apparent in the case of type-sentences incorporating personal pronouns and other indexical expressions, and in tensed sentences. 'It is raining' said now in Oxford has a distinct content from the utterance of the same sentence elsewhere or else when. But it is not *ambiguous;* variability in content is not matched by variation in sentence-meaning.

Third, Frege advanced various arguments to demonstrate that certain sentences lack judgeable-content, none of which prove any sentences to lack meaning. (i) Some sentences formulated with vague predicates lack judgeable-content. If 'A' denotes an aggregate of beans which is a borderline case for the predicate 'is a heap', then 'A is a heap' lacks content, being neither assertable nor deniable (BS, §27). Does it follow that it is *meaningless?* (ii) Reference-failure is claimed to deprive a sentence (token) of judgeable-content. If a referring-expression, e.g. 'the square root of -1', stands for nothing, then any sentence in which it occurs is 'mere printer's ink' (cf. FA, §§97, 102), lacking any content.[49] Such a sentence would undermine the cogency of any proof in which it occurred (FA, §102; cf. PW, 60). But, again, it is by no means obvious that reference failure deprives a sentence of meaning (as is evident when names in fiction are held to be cases of

---

45. Compare the parallel thesis that the sense of a sentence is an ingredient of the intuitive notion of meaning (Dummett, *Frege: Philosophy of Language,* p. 83ff., and 'Frege's Philosophy', in *Truth and Other Enigmas,* p. 103).
46. This criterion must be loosened to license assigning content to contradictions or self-evidently false utterances.
47. In his mature theory Frege denied that such sentences, with the exception of sentence-questions, express thoughts (T, 21). Sentence-questions are conceived as having content (PW, 7f.).
48. It might be regarded as a mere blunder that Frege denied judgeable-content (and later the power to express thoughts) to most non-declarative sentences (cf. Dummett, *Frege: Philosophy of Language,* p. 307f.). This, however, would distort our understanding of Frege's whole enterprise. The putative blunder rests on ideas fundamental to his philosophical reflections, viz. that inferences are confined to assertions and that judgeable-contents are explained as being the objects of assertion. Hence correcting the 'blunder' would have radical consequences for his conception of logic.
49. This idea influenced Frege even after he distinguished sense from reference. He distinguished mock thoughts from thoughts proper on the grounds of reference failure (PW, 129f.) even though his official position allowed for the possibility of sense without reference (SR, 58, 62f.; PW, 225).

reference failure). Frege was, however, deeply committed to the idea that a genuine judgment must be about something, hence to the converse principle that reference failure implies no judgment (content).⁵⁰ This notion has an impeccable philosophical pedigree in the traditional doctrine that a judgment is compounded out of subject and predicate. No subject—no judgment. (iii) Frege argued that certain modal qualifications do not affect the content expressed by a sentence.⁵¹ Prefixing 'It is necessary that' to a declarative sentence merely adds a hint about the grounds for the judgment (BS, §4). Yet this surely does not commit him to the thesis that such modal prefixes are meaningless. (iv) He also denied content to certain sentences by appeal to his own type-theory, e.g. he denied content to 'God exists' (if 'God' is construed as a proper name) on the grounds that 'exists' designates a second-level concept (FA, §53). Yet is the English sentence 'God exists' (or 'Eldorado does not exist') ungrammatical? Is it meaningless?

Fourth, Frege himself hinted at *a* distinction between sentence-meaning and judgeable-content. He claimed that two sentences may have the same conceptual content 'even if one can detect a slight difference of meaning' between them (BS, §3). Content exhausts what is logically significant (inference-relevant). He seems to have regarded the intuitive notion of meaning as containing psychological elements in addition to logical ones. The notion of content is in this respect what emerges after meaning is washed in anti-psychologistic acid.⁵²

These considerations do not settle the question of how Frege conceived of the relation between judgeable-content and sentence-meaning. (Indeed, it is doubtful whether he ever clearly and systematically confronted the issue.) But one decisive point does emerge: viz. that content can typically be assigned only to *particular tokens* of a given type-sentence. This is evident, *inter alia,* from various aspects of context-dependence of what is asserted by the use of a sentence. He indicated that the judgeable-contents of particular utterances of the sentence 'There are *n* inhabitants of Germany' will vary with the time of utterance (FA, §46). The present tense has an indexical role in such utterances, making what is asserted dependent on one feature of the context of utterance (viz. the date). Similar context-dependence of content arises for tokens of any type-sentences containing indexical expressions (e.g. 'now', 'this', 'here', 'I') or expressions having indexical uses (as

---

50. The idea resurfaces in modern writers favourably disposed towards traditional logical analysis, e.g. in Russell's analysis of singular terms (*The Principles of Mathematics,* pp. 44, 47; *Logic and Knowledge,* pp. 46f., 56; *Principia Mathematica to * 56,* p. 66f. (Cambridge Univ. Press, Cambridge, 1962), or in Strawson's claim that reference-failure renders an utterance a 'nonstarter' from the point of view of statement-making (*Introduction to Logical Theory,* pp. 174ff., 213f.).
51. Aristotle, the Port Royal logicians, and Kant all counted modalities as logically significant. But others, like Frege, repudiated this tradition. Mill considered modal terms irrelevant to the fact asserted, merely indicating the speaker's state of mind in regard to a fact (*System of Logic,* I-iv-2). Lotze argued that they indicated the grounds on which a judgment rests (*Logic,* I-ii-§§41ff.). Mansel declared the whole doctrine of modality to be 'of no consequence in formal logic' (*Prolegomena Logica,* 2nd ed., p. 231).
52. This is, of course, not to say that content is simply *part of* meaning, or an aspect of meaning—for reasons evident from the previous considerations.

Frege later acknowledged in respect of sense (BLA, p. xvi f.; T 24)).[53] Consequently content-specification must be relative to sentence-tokens, and can be extended to type-sentences only under special conditions. In thinking of judgeable-content as the content of sentences, we must bear in mind that we should understand 'sentences' to mean 'token-sentences'.

This generates a divergence between judgeable-content and what many philosophers understand by 'sentence-meaning'. For one widespread conception of sentence-meaning sharply distinguishes what a sentence means from what is stated or propounded by its use on a particular occasion. It is the latter notion, and not the former, which bears immediate kinship to Frege's judgeable-content. According to this conception of the matter, in speaking of the meaning of a sentence, as opposed to speaking of what someone meant by the utterance of a token of it, we typically speak of the type-sentence. Similarly, in explaining the meaning of a word, we give *general* directions for its use, i.e. explain the type-word. Understanding a word, knowing what it means, involves mastery of a general pattern for its use. If one knows this, one will also typically understand utterances containing tokens of it. On this view, one grasps the meaning of a typical declarative sentence if one knows the circumstances appropriate for its use to make assertions.[54] If the sentence contains indexical expressions, one must know how the context of utterance bears on the question of what assertion is made. Accordingly, the meaning of the sentence is invariant; together with the context of utterance, it determines what assertions (if any) are made by the utterance of tokens of the type-sentence.[55] Only if there is lexical or structural ambiguity have we reason to assign more than one meaning to a type-sentence, and indexical expressions are not held to manifest a radical form of lexical ambiguity. Hence the variability of content among different context-dependent tokens of a single type-sentence blocks the identification of judgeable-content with what many philosophers conceive of as sentence-meaning.[56]

---

53. Although Frege never noted it, content variability might be argued to extend to context-independent sentences containing context-independent definite descriptions whose content varies across possible worlds, since which object satisfies the description depends on the facts.
54. P. F. Strawson, 'On Referring' pp. 8f., 11 in *Logico-Linguistic Papers* (Methuen, London, 1971).
55. L. J. Lemmon, 'Sentences, Statements, Propositions', p. 95, in B. Williams and A. Montefiore ed., *British Analytic Philosophy* (Routledge and Kegan Paul, London, 1966).
56. The same argument applies to what Frege later called the thought, i.e. the sense of a sentence. There is a persistent tendency to ignore his explicit statements of the doctrine that utterances of a single type-sentence containing indexical expressions differ in *sense* in different contexts. In discussing Frege's treatment of token-reflexive expressions, Dummett observes that such expressions (e.g. 'I', 'you', 'here') 'bear a sense which provides a means of determining their reference in a systematic manner from the circumstances of utterance', concluding that 'a thought can no longer be identified with the sense of a complete assertoric sentence, when that sentence contains a token-reflexive expression' (*Frege: Philosophy of Language,* p. 383, cf. p. 400). Lemmon (loc. cit. p. 95) takes Frege to task for claiming that the sense of an expression determines its reference. His argument is that in the case of context-dependent sentences it is only the sense of the sentence *together with the context of its utterance* that determines its truth-value. This reasoning supposes that the sense of such a sentence is not context-dependent, since otherwise the

These various considerations severally and collectively reveal that judgeable-content is neither identical with, nor part of (or abstraction from), the meaning of a sentence. Much the same considerations also sharply differentiate the later notion of the sense of a sentence from that of sentence-meaning. Frege's analysis of content and subsequently of sense neither is nor was meant to be a contribution to a new subject currently referred to as 'a theory of meaning for a natural language', but rather as a contribution to a venerable subject, the science of valid inference, conceived as the study of logical relations between abstract entities (not linguistic objects such as sentences!).

The real importance of examining Frege's notion of judgeable-content does not lie in gaining a mastery of the details. There would be no advantage in rebuilding modern logic on the foundations of his system. The crucial point is to recognize how different his conception of logic is from any conception now held to be defensible. In his view, logic is something given; it unfolds the essence or nature of certain unanalyzable concepts (represented, e.g., by the condition-stroke and the negation-stroke). It is, as it were, the Euclidean geometry of the objects of judgment. These ideas are now alien. Many interpreters fail to notice them as the soil nurturing Frege's logical system. It is easy to fall prey to the illusion that the sophistication of his formal calculus and its isomorphism with standard modern logic demonstrates that his conception of logic must coincide on all essential points with our current one.[57] Serious reflection on his notion of judgeable-content is the primary means for shattering this illusion. Attention to detail is subordinate to this strategic purpose.

---

    context-dependence of its truth-value would already be fully accounted for. Kaplan makes this conception explicit in his presentation of Frege's picture of demonstratives. The sense of a demonstrative (a 'propositional component') is determined by linguistic rules, while the relation between this sense and a reference is an empirical relation, varying according to contexts of utterance and facts about the world. Sense, like 'character', is assigned to type-sentences, and therefore it is considered to be context-independent (D. Kaplan, *Demonstratives* (unpublished typescript), second draft, 1977, pp. 4, 40, 51). Such comments betray ignorance of Frege's remarks on context-dependence (especially T, 24f., BLA, p. xvif.) or misrepresent his account of the sense of sentences in a fundamental respect.

57. Conversely it seems widely thought that the history of logic prior to Frege contains nothing of interest because earlier *formalizations* of inferences seem naive and inadequate.

6
Unjudgeable-Content: The Principles

1. The logical decomposition of judgeable-content

It might appear from the general explanation of content in *Begriffsschrift* that judgeable- and unjudgeable-content are two coordinate species belonging to the genus conceptual content. But this conclusion would misrepresent Frege's whole conception of conceptual content. Judgeable-content is in his eyes the logically prior notion of the pair, unjudgeable-content the derivative concept. He subscribed to the idea of the primacy of judgeable-content in logical analysis of inferences. In *Begriffsschrift* this conception, though evident, is not stressed. Yet it informs the general explanation of what a function is (BS, §9); for there he gave an account of how to precipitate a function out of a judgeable-content. The doctrine is explicitly stated and heavily underscored in his polemical defense of *Begriffsschrift* against Schröder's criticism. Frege contrasted his own system with the bifurcation of logic in the work of Boole and Schröder into the science of the relations of concepts (i.e. of primary propositions) and the science of the relations of judgments (i.e. of secondary propositions).

> As opposed to this, I start out from judgments and their contents, and not from concepts.... I allow the formation of concepts to proceed only from judgments. If, that is, you imagine the 2 in the judgeable-content
>
> $2^4 = 16$
>
> to be replaceable by something else, by (-2) or 3 say, which may be indicated by putting an x in place of the 2:
>
> $x^4 = 16,$
>
> the judgeable-content is thus split into a constant and a variable part. The former, regarded in its own right but holding a place open for the latter, gives the concept '4th root of 16'.

> ... [I]nstead of putting a judgment together out of an individual as subject and an already previously formed concept as predicate, we do the opposite and arrive at a concept by splitting up the judgeable-content. (PW, 16f.)

For this conception of concept-formation Frege claimed substantial advantages. It circumvents some perennial logical controversies (PW, 17n**). It allows logical analysis of sophisticated mathematical and scientific concepts not amenable to explanation in terms of addition or multiplication of *Merkmale* (PW, 46; cf. BS, Preface), and 'by construing judgments as prior to concept-formation' it unifies logic, avoiding the division into the logic of primary propositions and that of secondary propositions (PW, 46). The primacy of judgeable-content in logic is thus the foundation stone of Frege's central achievements.

This idea is not an ephemera of *Begriffsschrift*. Both Frege's testimony and his practice manifest its enduring influence. The same doctrine, differently phrased, is cited as the leading principle of his life's work in logic:

> ... I begin by giving pride of place to the content of the word 'true', and then immediately go on to introduce a thought [judgeable-content] as that to which the question 'Is it true?' is in principle applicable. So I do not begin with concepts and put them together to form a thought or judgment; I come by the parts of a thought [unjudgeable-contents] by analysing the thought [judgeable-content]. This marks off my concept-script from the similar inventions of Leibniz—and his successors.... (PW, 253).

Frege applied this general conception in his various attempts to write a book on logic. The unvarying pattern is to begin with a discussion of judgeable-content (PW, 1ff.) or thoughts (PW, 128ff., 185ff., 251f.), and then to proceed to the analysis of judgments into concepts and objects (PW, 1) or into thought-constituents (PW, 224ff.; cf. BLA, §32). The direct descendant of the logical primacy of judgeable-content of *Begriffsschrift* is the doctrine of the logical primacy of the thought. A close cousin is the celebrated contextual dictum of the *Foundations:* 'Only in the context of a sentence has a word a meaning'.

Frege undoubtedly accorded priority to judgeable-content over unjudgeable-content. But the import of this claim requires some preliminary clarification:

(i) What has primacy in his logical analysis is *judgeable-content*. This must be sharply distinguished from such entities as type-sentences and the meanings of type-sentences. Logic analyzes what is asserted in using a token-sentence to make an assertion, not the vehicle of assertion or some feature of it (its meaning). In view of the important distinctions between content and standard modern conceptions of meaning, there are no grounds for identifying the doctrine of the primacy of judgeable-content in logic with the modern thesis of the primacy of the sentence in semantic theories; *a fortiori,* there are no grounds for identifying the content of an expression with the contribution to the truth-conditions of sentences in which it occurs.

(ii) The aim of analyzing judgeable-contents, not type-sentences, makes clear that logical analysis need not be fettered by the categories of grammatical analysis of sentences. In particular, it secures immunity of logical analysis from the tyranny of subject/predicate analysis in traditional grammar. It also protects Frege's theory from misguided criticism (or praise): he did not intend to criticize standard grammatical analyses of declarative sentences or to replace them by an improved account.

(iii) Frege had a strictly analytic approach to logical analysis. His sole purpose was to give a perspicuous representation of any judgeable-contents that appear as constituents of inferences. Whether or not an utterance expresses a judgeable-content has the status of a datum. This is something given independently of and prior to any theorizing about unjudgeable-contents. There is no question of appealing to an account of content to justify assigning a definite judgeable-content to some utterance not acknowledged to have a judgeable-content (i.e. whose production is not recognized to make an assertion). To this extent Frege turned his back on the notion of synthesizing judgments out of antecedently given concepts and objects. In his view the question of ascribing unjudgeable-content to a token-expression can arise only if this expression is part of an utterance actually used to make an assertion. This strictly analytic conception of content-analysis is clearly part of what he meant by his slogan 'a word has a meaning *only* in the context of a sentence'.

(iv) Since conceptual content is what is represented in the concept-script, articulations of symbols in this notation must correspond to decompositions of judgeable-contents into unjudgeable-contents (whenever they do not correspond to articulations of molecular judgeable-contents). In particular, symbols such as '$\Phi(A)$' and '$\Psi(A,B)$' are meant to represent articulations of judgeable-contents into functions and arguments. Frege even suggested that these symbols indicate the general forms of judgments expressed in natural language by the sentence-schemata 'A has the property Φ' and 'B stands in the relation Ψ to A' (BS, §10). In fact, he adopted the principle that every atomic judgeable-content can be decomposed in at least one way into a function and argument(s). Function/argument articulation is intended to be a universal norm of representation for judgeable-contents replacing the traditional pattern of subject/predicate decomposition (BS, Preface). This has immediate dramatic consequences, especially for the logical analysis of generalizations (infra, p. 181ff.).

(v) Frege held that unjudgeable-contents arise from the falling apart or decomposition *(zerfallen)* of judgeable-contents (PW, 16f.) The products of this falling apart are objects and concepts (of various levels). He likened the analysis of judgeable-contents to dividing or carving up an object (FA, §64), advocating that different concepts may be obtained from a single judgeable-content by different methods of dissection. It is uncertain how to construe these claims. Did he think that objects and concepts are literally *parts* of judgeable-contents in the same way as an exhaust-pipe is part of a car? Or did he use expressions related to the part/whole relation as metaphors, conscious always of their clashing with the literal truth that judgeable-contents are values of functions for arguments? Decision at the outset would be premature, but caution advisable.

(vi) Frege did not consider the products of analyzing judgeable-contents to be homogeneous from a logical point of view. In particular, concepts and objects are radically different from each other. They belong to different logical categories obtained by generalizing the distinction between numbers and functions, which is fundamental in function theory.

This preliminary clarification of the idea of primacy of judgeable-contents in logic is intended to forestall elementary confusion about the main lines of Frege's account of unjudgeable-contents. But to avoid basic error, it is equally necessary to bear in mind the general nature and purpose of his enterprise of content-analysis. The logical decomposition of judgeable-contents is undertaken for the strictly limited purpose of constructing a formal theory of inferential validity strong enough to encompass the modes of reasoning characteristic of advanced mathematics. Frege's direct aim is logico-mathematical. Assertions must be analysed in such a way that the logical relations among them can be exhibited within a systematic formal theory itself subject to the canons of full mathematical rigour. Although he thought that the execution of this mathematical programme for the construction of a logical system had decisive philosophical consequences (FA, §§3f.), it is essential in interpreting his account of conceptual content to remember that the controlling immediate purpose is formal; further philosophical fruits are an ornament to his theorizing in logic, but their absence would not deprive his logical system of its interest and importance.

2. Function/argument analysis of judgeable-contents

Viewed against the background of the analysis of judgments in traditional logic, Frege's account of the logical analysis of judgeable-contents can be seen to resemble its precursors in two main respects. First, what is to be analysed in logic are not declarative sentences that may function as vehicles of assertion, but rather what is asserted by their assertoric use. Second, there is a dogmatic commitment to a form of representation of what is asserted. Traditional logic had held that *every* assertion *must* be analysable into a pair of terms (subject and predicate). Frege replaced this dogma with the principle that *every* judgeable-content *must* be analysable into function and argument(s) (BS, Preface).[1] In both cases the universality claimed for a pattern of logical analysis is presented as an *a priori* truth, not as an inductive generalization.

In other respects his account differs strikingly from the traditional analysis of judgments. In jettisoning the programme of subject/predicate analysis in logic he also rejected the correlative idea that judgments are composed out of independently given entities (terms of judgments). The function/argument pattern of analysis, though just as dogmatic, is more liberal than the subject/predicate pattern. It acknowledges an indefinitely large range of forms of judgment in place of the few forms tolerated in traditional logic. Functions may have any number of

1. This is one of the cornerstones of the *Tractatus* (3.318, 4.24, 5.47). In each case, it is necessary to interpret this principle according to the author's own conception of what a function is, not according to the conception now reigning among logicians. (The differences are enormous and philosophically important.)

arguments, and they may be taken from any level of the hierarchy of functions. Frege's analysis of judgments seems to open up an infinite space of possible logical forms in contrast to the strict confines of the forms licensed by traditional logic. Many important features of his logic are immediate consequences of his conception of what a function is and of his considering his formal system to be a branch of function theory. But precisely because they were immediate consequences of his basic framework of thought, he took them for granted and did not draw explicit attention to them in the exposition of *Begriffsschrift*. Commentators, oblivious to his motivations and outlook, have therefore been prone to overlook them altogether. The clarification of *his* concept of a function is the instrument for drawing back the veil shrouding his thought from the eyes of modern philosophers. The following features of his system then stand out plainly:

(i) Since the specification of a function is intelligible and non-circular only if independently identifiable entities are thereby correlated with each other as arguments and values, the application of function/argument analysis to judgeable-contents presupposes, first, that a particular judgeable-content can be identified independently of its being the value of a particular function for a particular argument and, second, that an entity which is the argument for a function whose value is a particular judgeable-content can be identified independently of its being the entity having the property that the given function has the given content as its value when this particular entity is taken as its argument. One way to satisfy both these conditions would be to adopt the principle that every entity not itself a function is identification-independent from every function; for then every object (including every judgeable-content) could be identified directly, i.e. not as the value of any function for any argument.[2] Another way to secure independent identifiability would be to accept that every object, including judgeable-contents, which is identified as the value of some function for some argument can be identified in another manner, i.e. as the value of a different function for some other argument. Frege's position is unclear. What seems clear, however, is that he thought no object to have the property that it could be identified only as the value of a particular function for a particular argument (cf. FA, §67). Hence there is no warrant for attributing to him the principle that to identify any particular judgeable-content is to identify it as being built up in a particular way out of unjudgeable-contents. Had he adopted that principle, he would have debarred himself from the possibility of giving non-circular specifications of the functions employed in his function/argument decompositions of judgeable-contents. Like a number, any judgeable-content can be identified in indefinitely many independent ways.

2. Russell adopted this strategy in defence of his view that propositional-functions are constructed entities: 'the values of a [propositional] function are presupposed by the function, not vice versa. It is sufficiently obvious, in any particular case, that a value of a function does not presuppose the function. Thus for example the proposition "Socrates is human" can be perfectly apprehended without regarding it as a value of the function "*x* is human"' (Russell and Whitehead, *Principia Mathematica to *56*, p. 39).

(ii) Function/argument analysis interlocks with a distinction of the greatest importance for Frege's thinking. He contrasted the genuine properties of an object *(Eigenschaften)* with how we can conceive of this object *(Auffassungsweisen)*. The hallmark of properties is their objectivity: an object has whatever properties it has wholly independently of our conceiving or recognizing it to have these properties. Preëminent among the properties of an object is its construction out of constituent parts if it has any parts. The fact that an object is presented as the value of a function for a particular argument does not display its construction out of its constituents since neither a function nor an argument are typically *parts* of the value of the function for the given argument.[3] Frege apparently concluded that how an object is identified as the value of a function for an argument cannot be considered as a property of the object (FA, §67), but merely constitutes one way of conceiving of this object in relation to other objects. Objects are 'given' endowed with properties, but they are not 'given' *as* the values of functions for arguments.

The conception that the identification of objects as values of functions depends essentially on us and therefore lacks the objectivity of attributions of properties to objects seems required to explain two important aspects of Frege's account of content. First, it would justify his exclusion of the manner of determining an object *(Bestimmungsweise)* from the conceptual content of a name (BS, §8). Second, it would account for his principle that the articulation of a judgeable-content into function and argument is not an intrinsic feature of the content itself. He highlighted (and exploited) the claim that a judgeable-content can typically be viewed as differently articulated into function and argument (BS, §9; PW, 17). 'The distinction [between function and argument in most cases] has nothing to do with the conceptual content; it comes about only because we view the expression [for a judgeable-content] in a particular way' (BS, §9).[4] Only what flows from the object itself counts as a genuine property. This disqualifies as a property the identification of an object as the value of a function in all those cases in which this identification originates in us and depends on our choice.

This framework of thought seems alien and indefensible to modern philosophers. To characterize an object as the value of a function for a particular object as argument is to state a relation to hold between these two objects, and that a relation holds between one object and some other one is just as much an objective

---

3. This is an *obvious* point in function theory. No sane mathematician would argue from the fact that $4 = x^2 \rceil_2$ that the number 4 contains either the number 2 or the function x^2. (We use the notation '$f(x) \rceil_a$' to designate the value of the function $f(x)$ for the argument a. This allows a clear differentiation between the name of the value of x^2 for the argument 2 (viz. '$x^2 \rceil_2$') and the name of the value of 2^x for the argument 2 (viz. '$2^x \rceil_2$'), and it avoids various forms of ambiguity (cf. WF, 115).)

4. The only exception he envisaged consists of generalizations expressed by formulae with free variables.

property of this object as any other.[5] Hence it is mistaken to dismiss from the purview of logic the representation of an object as the value of a function on the ground that its being the value of this function for a particular argument is not an objective property of the object. Despite its unacceptability, this strategy is apparent in Frege's reasoning, and it does have important bearing on interpreting his work. It generated an uncomfortable tension. Insofar as function/argument decomposition of judgeable-contents is essential for determining the validity of inferences, the very objectivity of logic is threatened. More narrowly, the exclusion from the conceptual content of a name of the mode of determining an object (i.e. as the value of a function) requires an artificial manoeuvre to render explicable the requirement for proofs of identity-statements in mathematics (BS, §8). These tensions were major negative consequences of his conception. But it had countervailing positive implications as well. First, it is far from negative or even lukewarm about the possibility of multiple analyses of judgeable-contents: it positively encourages production of alternative analyses. There must be as many legitimate ways of decomposing a given content into function and argument(s) as there are ways of looking at a particular abstract object. The notion of a single sacrosanct *Auffassungsweise* is *prima facie* ridiculous. Multiple analyses into function and argument are not merely tolerated; they are *invited* by Frege's conception. Second, the separation of the ways of viewing an object from the objective properties of the object immediately defuses an important worry about the legitimacy of radically different decompositions of a single judgeable-content. There are logical compatibilities and incompatibilities among the properties of objects; from the fact that an object has one property it may follow that it lacks another. By contrast, the ways of viewing an object are subject to much less stringent constraints; from the fact that something may be viewed as a picture of a man striding up a hill it does not follow that it may not also be viewed as a picture of somebody sliding backwards down a hill. Ways of viewing objects, as it were, take up less logical space than properties of objects. Focussing on this matter should remove the feeling of bewilderment that might arise from Frege's claim that a single judgeable-content might now be decomposed into a second-level relation between pairs of concepts, now into an identity between objects (cf. FA, §64). Different function/argument analyses do not clash as *properties* of objects do, but only as perceived *aspects* of objects may.

(iii) According to the traditional idea that judgments are manufactured by combining terms in various sharply delimited ways, every term can occur as a constituent in many different judgments. Moreover, since terms are somehow 'given' prior to the formation of judgments, the circumscription of the terms of possible judgments together with the circumscription of the different forms of judgments suggests the doctrine that there is a determinate totality of possible judgments. Each judgment is essentially a member of a galaxy of possible judgments. Frege's thought leads by a different route to a similar conception. To view a judgeable-content as decomposed into function and argument(s) is *ipso facto* to

---

5. As Frege himself acknowledged (BS, §9)!

consider it as one of a cluster of judgeable-contents sharing constituents with the given content. Suppose, for example, that a particular judgeable-content decomposed into an object A and a first-level concept Φ. Frege considered concepts to be a species of function. It is of the essence of any function, *a fortiori* of the concept Φ, that it takes definite values for some *range* of admissible arguments. There is no such thing as a *function* completely defined by the correlation of a single object with something as its value for this argument.[6] Hence the very idea of identifying the concept Φ as a constitutent of the given judgeable-content Φ(A) presupposes the possibility of specifying which judgeable-contents are the values of this function for other arguments out of the range of objects of which A is a member. Similarly the intelligibility of singling out A as a constituent of the judgeable-content Φ(A) presupposes the possibility of specifying which object A is, i.e. the possibility of picking out some concepts under which A falls and some others under which it does not. For these reasons, the possibility of decomposing a judgeable-content into the concept Φ and the object A is inseparable from viewing the content of Φ(A) as one of a cluster of judgeable-contents Φ(B),Φ(Γ), ... and Ψ(A),X(A,B), ... This conception can be entertained in a more or less rigorous form. On one reading, it is compatible with a considerable degree of fuzziness about the boundaries of the clusters of judgments associated with a given decomposition of a judgeable-content; though, for example, any function must be defined over some range of arguments, the range for any particular function may be indeterminate. This reading would also condone partial functions and hence pairs of functions (concepts) whose ranges of arguments do not coincide. Consequently the clusters of judgments associated with the contents Φ(A), Ψ(B), X(Γ), ... might be different in shape as well as uncertain in contour. On a quite different reading, the same conception would support something more rigorous and mathematically elegant. The clusters could be made sharply bounded by the demand that every function be assigned a precise range of arguments, and they could be made symmetrical provided that every function is defined over an entire logical type of arguments. Frege's later requirement of completeness of definition (BLA, ii, §56) was a move in this direction, the seeds of which were sown in the initial conception of function/argument decomposition of judgeable-contents. The culmination of this move towards mathematical elegance and symmetry in the logical analysis of propositions came in Wittgenstein's *Tractatus*. A single proposition must be analyzed into function and argument(s);[7] such an articulation presupposes the totality of propositions containing each of the constituents in the analysis, and these in turn presuppose totalities of propositions that include every pos-

---

6. This principle is a special case of the obvious general principle that no function is uniquely characterized by specifying its values for a proper subset of its admissible arguments. The fact that $f(1) = 1$ does not discriminate between sin x, x^2, and x^3, and a partial function on a subset of the reals can typically be extended to a function defined over all real numbers in indefinitely many ways even in conformity with severe constraints (e.g. continuity or differentiability). This idea of functional underdetermination is a cornerstone of modern philosophy of science.
7. Wittgenstein, *Tractatus Logico-Philosophicus*, 4.24; cf. 3.318, 5.47.

sible constituent of any possible proposition.[8] In this way the function/argument articulation of propositions leads directly from a single object both to the totality of objects and to the totality of propositions.[9] Even to embark upon the application of function-theory to the logical analysis of judgeable-contents is to enter into the powerful gravitational field at whose centre lies the *Tractatus,* and Frege himself was pulled far in that direction by the inner logic of function theory.

(iv) Function theory comes laden with an implicit type-theory.[10] Indeed, it is linked with an open ascending hierarchy of types (numbers, first-level functions, second-level functions, ...) and with an open differentiation of types at each level (functions of one, two, three, ... arguments). Further ramifications arise from functions whose arguments are not type-homogeneous and from functions whose values are type-heterogeneous from their arguments. Finally, with this type-differentiation are associated rules prohibiting as ill-formed certain symbols constructed out of the vocabulary of function theory. All of this function theoretic type-theory must be transferred to the analysis of judgeable-contents once the decomposition is undertaken in terms of function and argument. Viewed as part of a system for *analysing* propositions, it exposes a galaxy of possibilities of which traditional logicians never even dreamt. Without the notion of second-level functions (concepts), they were not in a position to entertain the thought that count-statements ascribe second-level properties to first-level concepts. Both of Frege's principal innovations in conceptual analysis (quantification theory and the elucidation of number-words) exploited the novel possibilities opened up by function-theoretic type-theory. From the point of view of its immediate impact, his complex hierarchy of logical types more closely resembled an instrument for shaking off fetters than a straitjacket for logical analysis. His conception initially seemed a profoundly *liberating* influence, even if it has now become a prison.

(v) Abstract function-theory grants a general licence for functional abstraction. Suppose that a number n is presented as the value of the function f for the number m as argument; i.e. $n = f(m)$. Since n's being the value of f for the argument m is a matter of how we choose to view n and not an intrinsic property of the number n, and since we are free to choose how to view n, we can immediately generate

---

8. Ibid. 2.0123, 2.014, 3.311; cf. *Notebooks 1914–16*, p. 83.
9. *Tractatus,* 2.0124, 5.524.
10. Mathematicians exclude certain expressions as obviously ill-formed (and certain corresponding questions as illegitimate). Differentiation, e.g., is an operation (or function) on functions of one variable yielding as values functions of one variable; hence '$\frac{d}{dx}(\sin x)$' is well-formed, but not '$\frac{d}{dx}\left(\frac{d}{dx}\right)$' or '$x^2\big]_{d/dx}$' or '$\frac{d}{dx} + \frac{d}{dy}$'. The rigours of this type-theory are somewhat tempered by special stipulations. The expression '$\frac{d}{dx}(\pi)$' is taken to be well-formed by construing 'π' here as the designation of a constant-function (not a number), and similarly 'sin (e^x)' is construed as denoting the composition of the functions sin x and e^x, not as the value of sin x for the argument e^x (parallel to $\sin \frac{\pi}{4}$). The underlying type-theory guarantees that these notational anomalies generate no real ambiguities.

from the initial conception of n as $f(m)$ an indefinitely large array of alternative ways to view n involving functions at successively higher levels of the hierarchy of types. For instead of viewing n as the value of the first-level function f for the argument m, we may consider it as the value of a second-level function ϕ so specified that the value of ϕ for any first-level function g is the number which is the value of g for the argument m. Having defined $\phi(g) = g(m)$ for all first-level functions g, we see immediately that $\phi(f) = f(m)$; hence we can view n as the value of a second-level function for the function f as argument. In a similar way, n can be viewed as the value of a third-level function for the second-level function ϕ as argument; as the value of a fourth-level function; etc. Moreover, each of those conceptions of n has equal rights; none is *primus inter pares,* and all of them can be represented by a single formula in abstract function theory; by the definition of ϕ, e.g., the expression '$\phi(g)$' is simply equivalent to '$g(m)$', and hence '$\phi(f)$' is really just an abbreviation for '$f(m)$'. The single formula '$f(m)$' can be interpreted as including designations of functions on any (higher) level of the hierarchy of functions.[11]

This authorization of functional abstraction is transferred to logic as soon as judgeable-contents are viewed as decomposed into functions and arguments. The general welcome for alternative decompositions implicit in the doctrine of the primacy of judgeable-content is taken up by exploiting the possibility of free functional abstraction. Frege simply took this for granted without any explicit notice. In *Begriffsschrift* this is clearest in his treatment of inferences involving generality. The judgeable-content symbolized by '$\Phi(A)$' can be so viewed that either 'Φ' or 'A' is considered as a function-expression (though not designations for functions of the same type), and therefore the inference $\dfrac{(\forall \phi)\phi(A)}{\Phi(A)}$ is just as immediate a consequence of his explanation of the quantifier as is the inference $\dfrac{(\forall x)\Phi(x)}{\Phi(A)}$ (BS, §11). Indeed, Frege treated both inferences as exemplifying a single pattern in *Begriffsschrift*.[12] Free functional abstraction can take its start from formulae in his concept-script just because these belong to the notation for function theory.

11. This process of 'functional abstraction' is a movement of thought very characteristic of and widely exploited in the development of real and complex analysis. Frege's own work in function theory (KS, 51ff.) demonstrates his thorough familiarity with this mathematical technique.
12. Modern logicians note here a lack of rigorous thinking. They argue that Frege's type-theory requires his taking the expression which fills the argument-place in '$(\forall \phi)\phi(\xi)$' as the name for a second-level function, whereas 'A' in the second-inference schema is a proper name and hence must be so construed in the first-inference pattern unless the expression of the conclusions is equivocal. Hence modern logicians balk at the propriety of the expression '$(\forall \phi)\phi(A)$', and they insist that the conclusion of the first argument must be exhibited as a second-level function of the first-level function Φ. The first argument is at best elliptical according to this view. This differs radically from Frege's interpretation of his concept-script. In his view, the formula '$\Phi(A)$' is a fully adequate notation for a single judgment that can be viewed either as a first-level function whose argument is the object A or as a second-level function whose argument is a first-level function. What content is expressed by the formula '$\Phi(A)$' is independent of this difference in how it is viewed, and hence there is no case whatever for the dilemma that either '$\Phi(A)$' is equivocal or one of these inference-patterns is not immediately cogent.

Applying the concept of a function to the logical analysis of propositions involved embedding logic within the framework of thought characteristic of mathematical function-theory. This had dramatic consequences for Frege's thinking, differentiating his conception of logical analysis sharply from anything hitherto attempted.

There are also negative general features of his logic grounded in the formal character of his programme. These might be summarized in the thesis that the criteria of adequacy in the logical analysis of propositions are strictly formal. But three aspects of this claim are worthy of detailed notice because frequently overlooked.

(i) Frege did not acknowledge the propriety of any traditional metaphysical restrictions on the logical analysis of judgments. His logical categories of objects, first-level functions or concepts, second-level functions, etc., cut across all the traditional categories of metaphysics. The category of objects contains such diverse entities as persons, heaps of sand, numbers, directions, geometrical points, and judgeable-contents. Similarly, the category of first-level concepts contains such diverse things as properties, relations, concepts of general kinds, concepts of stuffs, powers, abilities, and dispositions. Thus Frege's categories unite kinds of entities normally segregated in different categories. Conversely, he drew category-distinctions among entities often classified together into single categories, e.g. the Aristotelian category of properties is subdivided into properties of objects, properties of first-level concepts, properties of second-level concepts, etc. His attitude towards objections marshalled from the side of traditional metaphysics would no doubt have been contemptuous. The traditional logical category-differentiations presumably would be declared to make no positive contributions to the function-theoretic analysis of judgeable-contents. Equally contemptuous would have been his reply to the objection that he supplied no metaphysical argument to legitimate the use of functions in the logical decomposition of judgments. His naïve Platonism left him with the impression that there was nothing to discuss. He turned his back on proving the reality of functions, concepts, and relations, just as he turned his back on certain other familiar metaphysical topics; he scorned the controversy about whether there were negative concepts (PW, 17n.), and he dismissed necessity as a concept without significance for logic (BS, §4). Perhaps he considered that he had finally attained the traditional goal of totally emancipating logic from metaphysics. But the policy of free entry for Platonist entities into logical theory has hidden costs, while the erection of a novel metaphysics on the skeleton of the type-hierarchy of function theory needs philosophical support and is open to philosophical objections.

(ii) In modern philosophy the successful analysis of concepts is generally taken to have important implications, metaphysical or epistemological, and analyses are attempted in many instances for the express purpose of carrying out some programme of 'reductionism'. If Frege were agreed to have succeeded in defining numbers as sets of concepts in the *Foundations,* many philosophers would now conclude that he had reduced our 'ontological commitments': he would have shown that only sets, but not numbers, really exist, or at least that we need not 'postulate' numbers over and above sets. Alternatively, philosophers more akin to

Russell would conclude that the achievement is an epistemological one; Frege would have proved that we are not more exposed to the risk of error by formulating assertions about numbers than we are already in making statements about sets, i.e. that numbers give no hostages to fortune over and above those already surrendered by sets.[13] Both of these conceptions are quite alien to Frege's thinking, and neither has any bearing on his programme of content-analysis. Clearly he did not consider that his definitions of numbers in the *Foundations* were necessary to prove that numbers really were objects; still less did he think that his definitions showed that only sets, not numbers, really exist. Content-analysis *chez Frege* is completely devoid of any implications of reductionism. Success is not to be measured by reduction of ontological commitments or epistemological hazards.

(iii) Logical analysis is now widely regarded as a proper part of the semantic analysis of type-sentences. This opens the door to objections that a particular logical analysis of a proposition cannot be reconciled with the structure of the sentence expressing this proposition, and it even allows the possibility of criticizing a whole logical system on the grounds that it is inconsistent with the only viable kind of semantic theory for a language. Frege would have dismissed such contentions as radically misconceived. What is analyzed in logic are judgeable-contents, not type-sentences or their meanings. Whether function/argument analysis can be applied to sentences (in syntax) or their meanings (in semantics) is irrelevant for whether it is the proper tool for analyzing judgments. He identified the content of the sentence 'Jupiter has four moons' with the content of 'Four is the number of Jupiter's moons' (FA, §57), and he argued that the second sentence indicates the truth that the number 4 is an object. He analyzed the content common to both sentences as an identity between objects. Construing this as an analysis of the first sentence, we might object first that 'four' there occurs as an adjective, not a name, and second that from the alleged equivalence in the meaning of the two sentences it cannot be inferred that the word 'four' means the same thing in both. Though well-founded, these objections do not touch Frege's enterprise, since his purpose was different from giving a syntactic or semantic analysis of either sentence. We are easily led into misunderstanding his aim for two reasons. First, finding his formal system sophisticated and up to date, we are tempted to suppose that he viewed logic as we now do; this is profoundly misguided. Second, the categories provided by function theory are so numerous and flexible that there is much greater scope for matching the articulations of a sentence in his function-theoretic representation of its content than there was in the procrustean representation of judgments in traditional logic. But that the analysis of sentences was not his target in content-analysis is conclusively proved by the fact that sentences with quite different structures are asserted to express the same judgeable-contents (e.g. PW, 16f.; cf. FA, §63f.). Content-analysis is immune to criticisms levelled against it from the direction of syntax and semantics.

---

13. Cf. the account of Occam's razor in B. Russell, 'Lectures on Logical Atomism', in R. C. Marsh (ed.), *Logic and Knowledge* (George Allen & Unwin, London, 1956), pp. 221f., 280.

3. Judgeable-contents as the values of functions

The indisputable core of Frege's logic is the decomposition or analysis of judgeable-contents into functions and arguments, the representation in concept-script of judgeable-contents as the values of functions for certain arguments. In his view, every *judgeable-content* can be taken quite literally as *the value* of a *function* for certain argument(s).

We readily overlook how radical this proposal is. We are seduced into accepting it by our familiarity with functional notation in the predicate calculus. From very early in our philosophical education we are taught to translate the statement 'Socrates is mortal' into a formula of the form '$\Phi(A)$'. Initial qualms about the intelligibility of this procedure disappear with practice, and soon we can see no reason whatever to doubt whether 'Three is prime', 'Socrates is a man', and 'Socrates is mortal' all have the same logical form. We also acquire sophisticated techniques for defusing any persisting doubts: if an ordinary mathematical function such as x^2 can be identified with a set of ordered pairs of real numbers, then the semantic value of the predicate 'x is a rational' can be identified with a set of real numbers and thus viewed as a degenerate case of a function. The better logician a philosopher is, the more imbued he is in this style of thinking and the more dulled are his sensibilities for what is *philosophically* problematic in translating sentences into the notation of function theory. The difficulty here is to resurrect the sense of wonder or bewilderment that should accompany the reception of Frege's fundamental idea. To him it must have seemed that the scales suddenly fell from his eyes; he might well have declared

> 'Then felt I like some watcher of the skies
> When a new planet swims into his ken'.

Unless we can recapture our lost innocence and experience how radical is the idea that judgeable-contents are the values of functions, we shall be oblivious to the mainspring of Frege's thinking and to something genuinely problematic in the logical tradition derived from his work.

The idea of judgeable-contents as values of functions is present already in his treatment of compound judgments (the propositional calculus). *Begriffsschrift* introduces his two logical operators (the negation-stroke and the condition-stroke) as names for functions whose arguments and values are judgeable-contents. We are prone to miss this important point. Struck by his apparently giving truth-table definitions of his two logical operators, we mistakenly jump to the conclusion that he took truth-values, not judgeable-contents, to be the arguments and values of these logical functions. There is no warrant for reading Frege's later doctrine back into the elucidations in *Begriffsschrift*. Modern logicians would doubtless not treat the logical constants of the propositional calculus as function-names at all; the truth-tables display merely truth-value dependencies between molecular propositions and their atomic constituents. But attributing this idea to Frege would be a grotesque anachronism.

Philosophers may easily fail to notice or even deliberately disregard the fact that he built his propositional calculus on the idea that judgeable-contents are the values of functions. But one can hardly miss this idea in his logical analysis of atomic propositions since there the idea occupies stage-centre. What then is so radical in Frege's conception of judgeable-contents as values of functions? We might precipitate out of this proposal two separate notions: a general extension of the concept of a function and its particular application to judgeable-contents. Our purpose in doing so is not to reconstruct the actual sequence of his thinking, but rather to crystallize the philosophical doubts engendered by his proceeding.

The strong inclination of mathematicians towards generalizing the concept of a function readily spills over into applications of the concept outside the realm of mathematical entities. Once having arrived at speaking of a function that maps every polygon onto its centre of gravity, why not continue by admitting as a function any single-valued correlation of objects with objects? We can then speak of a function that correlates each tea-cup on a table laid for tea with the saucer beneath it (cf. FA, §70), the function expressed by the phrase 'the father of x' which correlates with each human (e.g. Alexander the Great) his father (e.g. Philip of Macedon), and the function expressed by 'the capital of x' which correlates with each nation (e.g. Britain) a city (e.g. London). No obstacle seems to arise against taking any entities whatever, be they points of the compass, cabbages, colleges, or kings, as the arguments and values of functions. Only blind prejudice stands in the way of this advance of science. Of course, the idea that somebody who grasps a function can *calculate* its value for any suitable argument must be liberalized. In its place, however, we can substitute the principle that the value of a function must be *determined* for every admissible argument; or that one who grasps a function must know *how to determine* its values, provided that we admit observation and experiment as legitimate procedures, in addition to calculation, for determining the values of functions. Though apparently insignificant, this liberalization has consequences of great importance (cf. infra, p. 312ff.). But provided we set aside our qualms, the standard requirements of function theory seem to be satisfied in this wider domain of functions. In particular, the arguments and values of functions can be identified independently of their being represented as the arguments and values of any particular functions; e.g. London can be identified otherwise than as the capital of Britain, e.g. as the largest city in the country, and hence to say that 'the capital of x' names a function whose value for Great Britain as argument is London is to give a non-trivial partial explanation of a function-name. Similarly, just as argument and function are not part of the value of the function for that argument, so too from the fact that one object is the value of a function for another object as argument, it cannot be deduced that the second object is part of the first;[14] e.g., though London is the value of the function named by 'the capital of x' for the argument Britain, Britain is not part of London. Given the illusion of smoothness in extending function theory to encompass persons and

---

14. Nor can it be deduced that the argument is not part of its value! Consider, for example, the function that maps every object onto its unit set, or the function mapping each town onto the country to which it belongs.

the ordinary furniture of the world, it is scarcely surprising that such a naive extension to the domain of functions was made repeatedly and independently in the late nineteenth century.[15]

Frege undoubtedly countenanced this extension of the concept of a function. Indeed, he contributed arguments for extending the concept within mathematics (KS, 50ff.), and from the outset he acknowledged functions mapping everyday objects onto other objects (e.g. BS, §26; FA, §70). He advocated lifting any restriction on what kinds of objects may occur as arguments for first-level functions (FC, 31). Yet in one important sense this naive extension of function-theory played no significant role in his logical system. Indeed, it made no direct appearance in *Begriffsschrift*. He held that such expressions as 'London' and 'the capital of Britain', if viewed as proper names, are indistinguishable from the point of view of translating arguments into concept-script; they have the same (unjudgeable-) content, differing only in a logically irrelevant respect (*Bestimmungsweise* (BS, §8)).[16] For this reason, Frege avoided introducing material functions (i.e. func-

---

15. From one point of view this extension is difficult to pinpoint at all. From time out of mind relations or correlations have been conceived as holding between entities of any kind, mathematical and non-mathematical (e.g. between persons, as in A's being the father of B). If functions are conceived in the *modern* manner as single-valued correlations, then the extension of *this* concept of a function over arbitrary ranges of entities has been present as long as relations have been recognized. On the other hand, this concept of a function was absent in the 1870s. The idioms characteristic of function theory ('mapping', 'transformation', *'Abbildung'*) seemed to stand in the way of labelling arbitrary correlations of entities 'functions', and so too did the conceptual link between functions and *calculation*. From this point of view, what must be located is a shift in the internal constitution of the concept of a function, not merely a change in its application.

The conceptual landscape is shrouded in fog caused by familiar uses of the term 'function' in science and mathematics. In mechanics it is said that the velocity of an object in free fall is a function of the time elapsed, and in geometry mathematicians speak of linear displacements or rotations as mappings of figures in the plane onto isomorphic figures. It is unclear to what extent the notion of a function is stretched in support of such idioms. Did physicists and geometers keep clearly in mind that functions hold only between mathematical entities (e.g. numbers measuring velocity and time, or the co-ordinates of vertices of triangles)? Or did they conceive of functions as holding between vibrations, accelerations, periods of time, triangles, points, etc.? Or were they simply confused about these matters? The issue is very unclear.

From a different point of view, the extension of the concept of a function to encompass applications to arbitrary objects seems a clear and sharp change. What is required is a rigorous definition of 'function' which admits general correlations of non-mathematical entities. This seems but a continuation of the trajectory of generalization apparent in mathematical extensions of the concept of a function over complex and hyper-complex numbers or over first-order functions (in higher analysis). It might appear surprising that this was so late in being developed; Peirce credited the idea to Dedekind in 1879 (C. S. Peirce, *Collected Papers*, ed. Hartshorne and Weiss (Harvard Univ. Press, 1960), vol. III, §610). It was implicit in *Begriffsschrift* (1879), then explicit in the *Foundations* (1884) and 'Function and Concept' (1891). Similar extensions were made independently by Peano and Peirce (1885), and Russell integrated a general concept of a function into *The Principles of Mathematics* (1903).

16. Nonetheless, he did not deny that the complexity of what he later considered to be complex proper names may have significance for the cogency of inferences; on the contrary, he exploited the complexity of such symbols as '2^4' in the inequation '$2^4 > 15$' and the symbol '$\Phi(A)$' in his analysis of the continuity of the real function $\Phi(x)$ at the point A (PW, 24). There is no incon-

tions whose values are objects other than judgeable-contents) into his formula language for pure thought. Although we might require them to capture the logical significance of complex proper names in concept-script, he perceived no parallel need. What we regard as material functions he either ignored (as in 'the capital of Britain') or absorbed into functions whose values are judgeable-contents (e.g. 'the centre of mass of the solar system has no acceleration' is construed as stating that the solar system falls under a concept (BS, §9)). He introduced symbols for material functions only in the application of concept-script to the formalizations of judgments about material functions, especially in the formalizations of theorems and definitions in real analysis (PW, 24ff.). Hence *this* extension of function theory is not itself the mainspring of Frege's revolution in logic.

What distinguishes his work is a more subtle and sophisticated extension of the concept of a function. This finds clear expression in his concept-script. The name for a material function, whether mathematical or not, can occur only as a proper part of an expression fit for making a judgment. Mathematicians would use expressions such as '$2 = f(1)$', '$x^2 > 0$', or '$m + n = n + m$' for stating truths, particular or general, but never expressions such as '$f(1)$', '$g(x)$', or '$\frac{d}{dx}(fx)$'. If we defined $f(x) = y$ as holding if and only if y is the capital of x, and $g(x) = y$ as holding if and only if y is the father of x, the naïve extension of function theory would license using such expressions as '$f(\text{Britain}) = \text{London}$' or '$g(\text{Socrates})$ was born before 450 B.C.' for making assertions; but it would not license treating any expression of the form '$f(\text{Britain})$' or '$g(\text{Socrates})$' as alone expressing a thought. By contrast, Frege did employ expressions of the forms '$\Phi(A)$' and '$\Psi(A,B)$' as expressions for a judgeable-content. What informed this innovation was his discerning novel entities (judgeable-contents) and his inclusion of these entities within the category of *objects*. Only this claim can bridge the gap between the naive extension of function theory (viz. lifting all restrictions on what objects are admissible as arguments and values of functions) and his representation of judgeable-contents as the values of functions. The decisive move in constructing his logical system is to treat judgeable-contents as objects, and thereby to authorize taking them to be the values of functions.

---

sistency here with his denial of the relevance of *Bestimmungsweise* to logic. Whether a complex expression is treated as a proper name from a logical point of view depends on how the judgeable-content expressed by the sentence in which it occurs is viewed as decomposing into function and argument. Provided this judgeable-content is viewed as the ascription of a property to the object corresponding to the complex proper name (e.g. to the number 16 for which '2^4' stands), then the mode of determination of this object and the corresponding articulation of the complex proper name are logically without significance. On the other hand, the complexity of the 'complex proper name' does have logical significance provided that the judgeable-content is so viewed that it does *not* ascribe a property to the object corresponding to this name (e.g. if '$2^4 > 15$' is construed as the ascription of the property of having a fourth power which is greater than 15 to the number 2). Frege would fall into inconsistency only if he were simultaneously to view an expression as logically a proper name and to treat how the corresponding object is determined as logically significant. In fact, he assigned logical significance to the complexity of an expression when and only when he did not view it as standing for an object.

In a sense this description of Frege's innovation is anachronistic. It introduces a distinction where he would have noticed none. It separates off the issue of his 'extending his ontology' (to include judgeable-contents) from the matter of his extending the concept of a function to admit any object whatever as an argument or a value of a suitably defined function. Since he did not regard the delimitation of the category of objects as a pragmatic matter, he would have found the distinction unintelligible. Whether judgeable-contents are objects or not would not be an issue of *deciding* how to *speak,* but of *recognizing* a matter of *fact.* Moreover, the arguments for acknowledging the existence of judgeable-contents at all would be arguments for regarding them as objects (rather than as another kind of entity from the function-theoretic hierarchy of types); in particular, they must be objects if they are to be what indirect statements stand for. Consequently, in Frege's view, the extension of functions over all objects whatever leaves nothing to discuss about admitting judgeable-contents as values of functions provided the fact that judgeable-contents are objects is recognized. Of course, the precise nature of these objects may be somewhat unclear, but this can be left to be discovered later. Frege's free-and-easy Platonism would mask from his eyes the radical character of his generalization of function-theory over judgeable-contents. The point of making a distinction here is to bring clearly into view something of the utmost importance that he failed even to notice.

The seemingly anodyne proposal to admit judgeable-contents as the value of functions is in fact fraught with philosophical difficulties in the light of *his* conception of what a function is. Three stand out from the very first, and they call into question the intelligibility of Frege's whole logical system.

(i) It must be possible to specify a function; in other words, it must be possible to give an explanation of any function-name. In mathematics the standard form is an open equation or analytical formula; e.g. the function-name '$f(x)$' might be defined by the formula '$f(x) = e^x + \log x + 1$'. In some cases it might consist of a list of values for arguments together with a similarity rider; e.g. '$1 + 1 = 2, 1 + 2 = 3, 2 + 2 = 4, 1 + 3 = 4$, and so on' would be a (somewhat unsophisticated) explanation of the binary function of addition. In similar ways it is straightforward to satisfy the demand that function-names be explicable in respect of the naive extension of function theory; e.g., if we abbreviate 'the capital of x' by the expression '$f(x)$', then we could specify the function $f(x)$ by the stipulation (analysis) '$f(x) = $ the city which is the seat of the government of (the country) x' or by the open list, 'f(Britain) = London, f(France) = Paris, f(Spain) = Madrid, etc.'. Apart from oddity of notation, there seems no ground for rejecting these as unintelligible and unserviceable explanations. In every case specification of a function requires the use of identities: either an open identity (with a free variable) or a list of identities. An explanation must make clear what the value of a function is for any admissible argument, and hence it must provide a means for determining whether an arbitrary object is identical with the object which is the value of the given function for a given argument. Specifications of functions make use of identities with the goal of clarifying the conditions under which identities hold. This feature alone seems an insuperable hurdle to introducing functions whose values are judgeable-contents. The expression of a judgeable-content is a sentence (or the mathematical analogue of a sentence, especially an equation or inequality). Consequently, the specification of a function whose value is a judgeable-content must take the form of one or more identities in which a *sentence* flanks the identity-sign on one side. But any such formula is literally nonsense. There is no intelligible use for the identity-sign allowing a sentence to occur on either or both flanks. Such formulae as '$(3 = 1 + 2) = 4$' or $(3 > 2) = (2^2 = 4)$', *a fortiori* such formulae as '$\Psi(2) = (2 > 1)$' or '$\Phi(2) = 2$ is prime', are plain gibberish.[17] They violate the syntax of identity statements. Consequently there is no such thing as an explanation of a function-name whose values are judgeable-contents, and correlatively there is no such question as whether any given function might take a particular judgeable-content (e.g. the one expressed by '$2^2 = 4$') as its value. We have no way to break into the magic circle of functions whose values are judgeable-contents. Nothing counts as an explanation of such a function. Consequently, unless we credit ourselves with a capacity to grasp truths that are ineffable, there is no such thing as understanding the phrase 'a function whose values are judgeable-contents'. In the attempt to extend the concept of a function over judgeable-contents, Frege runs up against the limits of logical space.

(ii) Specification of a function requires a non-circular specification of its values for admissible arguments. This presupposes the possibility of identifying any entity declared to be the value of the function $f(x)$ for the argument A independently of its description as the value of $f(x)$ for A as argument. This requirement is readily satisfied in mathematics; e.g. the number 4 can be identified otherwise than as the square of 2, and hence '$4 = x^2 \rbrack_2$' is a non-circular specification of

---

17. Frege might have protested at this objection. Although we do not usually regard such formulae or corresponding sentences in natural language as well-formed, his account of identity reveals that they are unexceptionable, since they assert a coincidence in the content of pairs of sentences. Does this counterargument secure the intelligibility of his specifications of concepts? It would make the metalinguistic interpretation of identities essential to the construction of his logical system, with the consequence that the problem would re-emerge with his later denial that identities make assertions about symbols. Moreover, not every identity is well-formed even on his early view: he did not allow function-names to flank the identity sign except in combination with signs designating or indicating their arguments. Some rationale is therefore required to allow sentences to occupy positions open to some expressions (e.g. numerals or proper names of persons) but not to others (e.g. function-names or concept-words). The only obvious one is that sentences are themselves really names of objects, and then this would be another commitment of Frege's counter-argument. Finally, we might consider something to be awry about an analysis of identity-statements that assigns sense to patent nonsense. Surely the sentence 'If an opponent lays a chase of less than half a yard, then the only possibility of winning the point is to hit a force into the dedans is identical with the rules of real tennis are more complex than those of lawn tennis' makes no sense, whatever Frege might say to the contrary. Would it be reasonable to brush this point aside with the claim that the expression 'is identical with' cannot be employed to formulate genuine identity-statements? Even if his explanations of concepts are not plainly gibberish on his view, plain gibberish lies not far off.

the value of x^2 for 2 as argument. The naive generalization of function theory, as noted, raises no general difficulties in this respect; e.g., since London can be identified otherwise than as the capital of Britain, the identity 'London = f(Britain)' can serve as part of a non-circular specification of the material function corresponding to the phrase 'the capital of x'. What is crucial to non-circular specification of functions is not merely the availability of different names for each entity picked out as the value of a function for a given argument. This requirement could be met by arbitrary stipulation, i.e. by introducing for each symbol '$f(A)$' a simple symbol (say, 'B') which abbreviates it. That would achieve nothing except to mask circularity in function-specifications: removal of the abbreviation by substituting the *definiens* for the *definiendum* would make the hidden circularity perspicuous. Non-circular specification makes a stronger demand: every entity specified as a value of the function f must be identifiable in some essentially independent way. There must be associated with it some mode of determination *(Bestimmungsweise)* independent of the function f. The hallmark of independence is the need for a *proof* that two different modes of determination yield the same result. (This is why definitional abbreviations are useless. Substitution of *definiens* for *definiendum* is not a step of inference, but simply an alteration in the way of presenting a single judgeable-content.) Mathematics contains many different *independent* specifications of numbers and functions. That '2 + 2', '4', '2^2', and '7 − 3' all designate the same number requires a series of proofs (cf. FA, §6); and similarly proofs are needed that 'x^2', '$\frac{d}{dx}\left(\frac{x^3}{3}\right)$', and '$\int_0^x 2y\,dy$' all designate the same first-level function. Alternative names for an entity guarantee the possibility of non-circular specifications of functions with this entity as a value only if these names are associated with different modes of determination (cf. BS, §8).

It is far from obvious that there are any parallel *independent* specifications of judgeable-contents in concept-script.[18] Although Frege acknowledged the possibility of alternative decompositions of judgeable-contents and although he could admit distinct formulae in concept-script to represent these different decompositions, the formulae so related would typically be linked by explicit definitions. If, e.g., we consider the content symbolized by '$\Phi(A,B)$', we can make clear that we view this as a function of A alone by applying the definition '$\Psi(x) = \Phi(x,B)$' (cf. BS, §9). But this furnishes only an apparently independent specification of the content $\Phi(A,B)$, since the linking definition shows that '$\Psi(A)$' simply abbreviates

---

18. Frege might have thought that this requirement is sometimes satisfied. In particular, the pair of formulae '$a // b$' and '$Da = Db$' (translating 'The direction of the line a is identical with the direction of the line b') appear to us to be independent specifications of what he considered to be a single judgeable-content (FA, §64). But this would not show that the *general* requirement of independent specifiability of the values of functions can be met in respect of judgeable-contents. Moreover, even this specific instance is problematic. The supposition that these formulae are independent on the model of '$x^2\Big]_2$' and '$2(\sin x + \cos 2x)\Big]_{\pi/2}$' yields an antinomy about the need for a step of proof to justify a transition from '$a // b$' to '$Da = Db$' (or vice versa). Frege evidently saw no genuine step of inference here (cf. KS, 1).

the formula '$\Phi(A,B)$'. The same conclusion holds of abstraction of complex first-level functions (BS, §9) and of higher order functions (BS, §10). Though we can view the content expressed by '$\Psi(A)$' as articulated into a second-level function whose argument is $\Psi(\xi)$ and rewrite this decomposition according to the definition '$\mathfrak{F}(\phi) = \phi(A)$', '$\mathfrak{F}(\Psi)$' would be merely an abbreviation for '$\Psi(A)$'. So far was Frege from thinking that there were essentially independent specifications of judgeable-contents in concept-script that he canvassed the idea of there being in concept-script a single formula representing simultaneously all possible analyses of any one judgeable-content (cf. PMC, 67). Any alternative formulation would presumably be derived from this canonical formula by functional abstraction and definitional abbreviation.

This interpretation can be substantiated by reflection on the role of concept-script. Frege's aim is to incorporate into this symbolism all and only what is relevant to the cogency of inferences. Relative to a given argument, this demands a single representation of every sentence that expresses the same judgment. Otherwise an unjustified transition would occur in the course of a proof. Moreover, unless a single formula in concept-script can encompass all of the logical powers of a given judgment, there will be a difficulty in relating the judgments common to different inferences. The verdict that two different formulae not related by definitional abbreviation express the same content might rest on intuition, in which case the formalization of proof would be incomplete and the rigour of proof imperfect. Alternatively a non-trivial proof would be necessary to justify the transition from one formula to the other, but this would be inconsistent with the hypothesis that they were logically equipollent (BS, §3). It is of the very essence of the concept-script that there should be no analogue of non-trivial proofs of numerical identities in the case of different symbolizations of conceptual-contents. Hence the possibility of essentially independent representations of a single judgeable-content is not available in concept-script. This forecloses the possibility of independent specifications in concept-script of the judgeable-content which is specified as being the value of a function for certain arguments. Consequently the function/argument articulation given in a complete logical analysis of a judgeable-content must be internal to the identity of this particular judgeable-content. Here Frege's thinking runs aground on the logical prerequisites for applying function-theory to the logical analysis of judgeable-contents.

Does this argument, perhaps, pin too much on the issue of the possibility of essentially different representations of a single content *in concept-script*? Could it not be circumvented by noting the role of ordinary sentences in the expression of judgments? Surely we could introduce a function whose values are judgeable-contents with the aid of such sentences. Suppose we let 'A' stand for Socrates, 'B' for Plato, 'Γ' for Aristotle, ..., and we specify the function $\Phi(\xi)$ by the stipulations

$\Phi(A)$ = Socrates is mortal
$\Phi(B)$ = Plato is mortal
$\Phi(\Gamma)$ = Aristotle is mortal
⋮

Provided these identities are acknowledged to be intelligible at all, does the suspicion of circularity not fall by the wayside? The procedure has no resemblance to the illegitimate stipulations 'the value of $f(x)$ for the argument 1 is $f(1)$, for the argument 2 is $f(2)$ ...'! True enough, but what does it achieve? Frege regarded concept-script as a self-contained language, parallel to German, English, French, etc. Consequently, the proposed stipulations have the status of translations into another language, not transformations within a single one. This renders them useless for the purpose of generating alternative specifications of a single judgeable-content. The possibility of identifying the number 4 in essentially different ways is not established by noting that 'four', 'vier', 'quatre', 'quatro', etc. designate the same thing. By parity of reasoning, the possibility of translating '$\Phi(A)$' as 'Socrates is mortal' shows nothing more than the possibility of translating the latter sentence as *'Sokrates ist sterblich'*. Frege's only recourse would be to claim that we have some direct apprehension of judgeable-contents. This would replace an antinomy with a mystery. Moreover, it would war against some of his fundamental ideas. Postulating direct apprehension of judgeable-contents as unarticulated or as articulated otherwise than in their canonical forms in concept-script would be as inadvisable a strategy as that of attempting to behead a hydra.

(iii) The hierarchy of types in abstract function theory is taken by Frege to be comprehensive: it pigeon-holes every entity relevant to logic. In later writings this idea finds expression in two ways. First, everything which is not a function is declared to be an object (e.g. BLA, §2). Second, it is a matter of negligence if every first-level function is not defined over every object; nothing is essentially altered in the nature of a given function by extending it over the totality of objects (e.g. by extending addition from numbers to heavenly bodies (FC, 33)). These doctrines combine to yield a noteworthy account of assertion or judgment; since the sense of a sentence, not being a function, must be an object, and since the concept expressed by 'A judged ...' takes such objects (thoughts) as arguments, it must be a mere accident that we do not speak of judging (and asserting) objects other than thoughts. This claim abrogates the internal relations between thoughts and assertion. It thereby distorts the concepts of judgment and assertion, as well as related concepts such as thinking, believing, and inferring. Although the later doctrine of completeness of definition of concept-words is not present in *Begriffsschrift*, the seed of the later misconception of assertion is already present there. For it is essential to Frege's programme of reconstructing logic that judgeable-contents be objects. To segregate judgeable-contents into a distinct logical category would be tantamount to undermining this foundation of his thought. If failure to make such a category distinction is an unmitigated disaster in logic, then it is a defect present from the outset in Frege's thinking.

Consider first-level functions whose values are judgeable-contents (concepts in the *Foundations*). Admissible arguments are taken from a variety of traditional categories: e.g. numbers (BS, §§23ff. cf. §5), sets (BS, §9), and physical objects (BS, §§2, 27). Hence concepts share arguments with material functions, mathematical and non-mathematical. Suppose judgeable-contents were held to constitute a distinct logical category from objects such as numbers, physical objects, and ideas; i.e. suppose that it made no sense to say of judgeable-contents what can be said of those other objects, and vice versa. Then, of course, judgeable-contents would be sharply segregated from those other objects in respect of assertion and judgment, for only of contents would it make sense to say that they could be asserted. But the price of this conceptual purity would be astronomic. It would debar identifying concepts with *functions*. Just as a rigid category-theory for objects induces category-differences among properties and relations, so too it would induce category-differences among 'functions'. Of a material function it makes no sense to assert that its value is an object of assertion, judgment, belief, etc. Yet it is of the essence of a concept that it does make sense to assert that its value is such an object. Consequently, concepts must be *toto mundo* distinct from functions proper on the supposition that judgeable-contents constitute a logically isolated category of objects. This would reduce the claim that judgeable-contents are to be articulated into *functions* and arguments to the status of a bad pun or an uncashable metaphor. That is the very antithesis of Frege's intention. He aimed to develop logic as a branch of *function* theory and to build his logical system on the foundation-stone of the concept of a function. Any defense of his thinking which undermined the literal truth of these claims would be ludicrous as an interpretation of his work.

What we have exposed is a tension between analyzing judgeable-contents as the values of functions and according a unique logical status to them as the objects of assertion (also a unique logical status to sentences as vehicles of assertion). Frege's headache is worthy of attention because yet again it arises from his bumping against the limits of logical space.

4. Alternative analyses of judgeable-contents

Part of Frege's formal logical system is isomorphic with the first-order predicate calculus with identity, and in addition it contains an attempt to formalize second-order quantification. The structure of the system, together with the elucidations of the primitive logical operators, make it tempting to ascribe to Frege a conception of logic consonant in fundamentals with the conception of sophisticated modern logicians.

The key to the modern analysis of the notion of logical consequence is that validity of arguments is determined by the semantic *analysis* of the constituent *sentences*. Assuming for simplicity that a particular unambiguous sentence is context-independent, logical semantics will reveal its inferential-powers in the context of any putative argument in which it may occur. Of course, relative to determining the validity of a particular argument, a full semantic analysis may be unnecessary, a partial one sufficient. If, for example, we consider the argument schema $\frac{\text{3 is greater than 2}}{\text{Something is greater than 2}}$ we see that the semantic structure of the predicate 'is greater than 2' plays no role here; relative to this argument-form, the complex predicate can be treated as if it were semantically simple. Nonetheless, it is assumed that there is a single full semantic analysis of such a sentence from which its full inferential-powers will be apparent. Though, for example, '3 is greater than

2' may be analyzed in certain contexts as exemplifying the form $\Phi(A)$ and in others as exemplifying the form $\Phi(B)$, it may always be analyzed for logical purposes as having the form $\Psi(A,B)$. Partial analyses are derived from such an underlying full analysis by abstracting from part of the semantic structure of the sentence. Hence the set of admissible semantic analyses of a sentence has the structure of a partial ordering with a *supremum*. This idea provides the backbone of modern logical semantics, though the skeleton is hidden beneath the flesh of qualifications and addenda necessary to the attempt to analyze context-dependent sentences, sentences containing terms susceptible of further semantic analysis, etc.

To what extent did Frege share this conception? There are at least three considerations indicating that he did not.

(i) The foundation of the modern conception of validity is a particular account of what an argument is. An argument is a sequence of declarative sentences, one of which is singled out as the conclusion; the argument is valid provided that the conclusion cannot be false unless at least one of the premises is false. According to these explanations, certain sequences of sentences turn out to be valid arguments that non-logicians would commonly not acknowledge to be *arguments* at all. Such 'degenerate' argument-forms are exemplified by the schemata $\frac{P}{P}$ and $\frac{P \& \neg P}{Q}$ also perhaps $\frac{P}{\neg\neg P}$, $\frac{\neg\neg P}{P}$, $\frac{P \& Q}{Q \& P}$, and $\frac{P \to Q}{\neg Q \to \neg P}$. Frege in this respect lined up with the philosophically unsophisticated (such as Aristotle)[19] not with modern logicians. He held that there was no such thing as an *inference* of the form $\frac{P}{P}$, since an inference consists in a transition from one judgeable-content to another one (cf. PW, 3). A mathematician who writes '3 > 2' on one line of a proof, then '2 < 3' on the next, has made no inference; he has simply replaced one expression of a judgment by another for the very same judgment. The diametrically opposite degenerate valid argument-form $\frac{P \& \neg P}{Q}$, Frege also refused to recognize as corresponding to any inference, since an inference must start from a judgeable-content actually judged to be true. (This is clear from his account of indirect proof (N, 119f; PW, 244ff.).) No inference, *a fortiori* no *valid* inference. This verdict surprisingly holds for certain pairs of formulae in concept-script which can be interderived within his own logical system: in particular, he later expressly stated that the formulae in each of the pairs P and $\neg\neg P$, $P \& Q$ and $Q \& P$, and $P \to Q$ and $\neg Q \to \neg P$, expressed the same thought (CT, 548, 541f., 553f.). This thesis has the bizarre implication that certain correctly executed complex derivations in concept-script do not constitute inferences at all! Indeed, it even implies that the application of *modus ponens* to certain axioms of concept-script does not constitute making an inference. These observations support the conclusion that Frege did not have a purely semantic conception of the cogency of inferences. Since his verdicts about cogency of arguments do not square with

---

19. Cf. *Posterior Analytics*, I, 2.

the modern ones, it would be remarkable if the grounds for his verdicts were identical with the grounds for ours.[20]

(ii) Frege held that the primary task of the logician is to determine the logical structure of thoughts, and he insisted that there was only an approximate and external correspondence between the logical structures of thoughts and the grammatical structures of declarative sentences. This makes altogether implausible the thesis that he considered any form of *sentence*-analysis to be fundamental for determining the validity of *inferences* among *thoughts*. *Prima facie* doubts can be sharpened. From the beginning Frege noted that different sentences may express the same judgeable-content. This holds both of sentences of natural language and of the analogues of sentences in concept-script. Thus active/passive transforms have the same content (BS, §3). So too do the sentences 'Jupiter has four moons' and 'Four is the number of the moons of Jupiter' (FA, §57); '2 is a fourth root of 16', 'the individual 2 falls under the concept "4th root of 16"', and '4 is a logarithm of 16 to the base 2' (PW, 16f.); 'Jesus is a man' and 'Jesus falls under the concept man' (cf. CO, 47); 'all mammals are land-dwellers' and 'the concept *mammal* is subordinate to the concept *land-dweller*' (CO, 48; cf. FA, §47); and 'Frederick the Great won the battle of Rossbach', and 'It is true that Frederick the Great won the battle of Rossbach', and 'The victory of Frederick the Great at the battle of Rossbach is a fact' (PW, 141; BS, §3). Similarly, there are different expressions in concept-script having identical judgeable-contents; e.g. '3 > 2' and '2 < 3' (cf. PW, 101), or '4 = $\log_2 16$', '$x^4 \big]_2 = 16$', '$2^x \big]_4 = 16$', and '$x^y \big]_{x=2, y=4} = 16$' (cf. PW, 16f.). In many of these cases it is not plausible to suggest that the identity of judgeable-content expressed by each of several sentences can be traced back to any single *structure* visible in the *sentences* of each set. Intuitively many of these sentences differ in structure, and modern semantic analysis accords with these intuitions since it often assigns different structures to sentences that Frege declared to express a single judgeable-content. We thus face a dilemma: either Frege made numerous mistakes in giving instances of different sentences which have the same judgeable-content, or he did not ground his judgments of content-identity on what we now recognize to be identity in the semantic analysis of sentences. Given his background conception of the relation of thought to language, it seems more reasonable, as well as more charitable, to conclude that his logical analysis of inferences is fundamentally different from our semantical analysis of arguments than to conclude that he made repeated blunders in assigning semantic structures to sentences.

(iii) The application of function theory to the analysis of judgeable-content demands the possibility that each content can be analyzed in more than one way

---

20. By contrast, his verdicts do resemble some found among traditional logicians. Mill, e.g., argued that so-called 'conversion' in syllogistic logic is not a genuine form of inference at all; 'No A's are B's' and 'No B's are A's' are two ways of stating the same proposition (Mill, *A System of Logic*, II-i-2).

as the value of a function for some argument(s), since it must be possible to specify the value of a given function for an argument independently of its being the value of this particular function for this particular argument. It is clear that partial analyses of a judgeable-content derived by abstracting from a putative complete analysis would not suffice for this purpose. The possibility of radically different logical decompositions of any judgeable-content is a presupposition of the application of function-theory to the logical analysis of judgments.

Indeed, function theory suggests an even stronger thesis: for any entity that can be represented as the value of a function for some argument(s), there must be indefinitely many ways that it can be picked out as the value of functions for arguments. Compare, for example, the integer 4; it can be represented as the value of indefinitely many functions of one real variable (e.g. $x^2 \rfloor_2$ or $x^3 \rfloor_1 + 3$), as the value of indefinitely many functions of two real variables (e.g. $x + y \rfloor_{2,2}$ or $x \cos y \rfloor_{4,0}$), and as the value of indefinitely many second-level functions (e.g. if $\mathfrak{F}(f(x)) = \frac{d}{dx}(f(x)) \rfloor_{x=1}$, then $4 = \mathfrak{F}(x^4)$). No *a priori* limits can be imposed on the level of a function whose value may be the integer 4, nor on the number of its arguments. There seems no rationale for denying that the possibilities of the logical analysis of any single judgeable-content are similarly *wide open*. By stipulation, it seems, we could introduce a first-level concept Φ whose value for the argument 2 will be a given judgeable-content; or a second-level concept $\mathfrak{F}$ whose value for the concept *prime* is the same judgeable-content, etc. (Or would a dearth of concepts prevent such stipulation?) It seems therefore an integral part of function/argument analysis of judgeable-contents that there in principle be indefinitely many different analyses of any single judgeable-content, even indefinitely many involving entities on a single level of the function hierarchy. This conception conflicts with the guiding idea of the semantic analysis of sentences: on any given level of functional abstraction, any unambiguous sentence must have a single complete analysis together with a definite number of partial analyses derived from the complete analysis by neglecting part of the complete structure. Moreover, function theory gives no warrant for regarding as fundamental or privileged any particular representation of an entity as the value of some function for some argument(s). There is, for example, equality of esteem among the different possible ways of designating the integer 4. The parallel doctrine for logic would exclude treating any one representation of a judgeable-content as fundamental. Frege subscribed to this principle: 'I do not believe that for any judgeable-content there is only one way in which it can be decomposed, or that one of these possible ways can ... claim objective preeminence' (PMC, 101).[21] Semantic analysis of sen-

---

21. The word 'always' *(immer)* has been omitted in this quotation. It is wholly unclear what grounds Frege might have had for envisaging exceptions to the principle. Perhaps he had in mind decompositions represented in concept-script by formulae containing free variables (cf. BS, §9).

tences does not conform to this principle. It treats the full analysis of a sentence as taking precedence over partial analyses, and it takes an analysis at some one level to be fundamental, those at higher levels of functional abstraction to be derivative.[22] In a number of basic respects the grounding of Frege's logical analysis in function theory is antithetical to the framework of thought informing the modern semantic analysis of sentences.

These three arguments constitute a powerful case against supposing that his logical analyses of judgments and inferences rest on the semantic analysis of sentences and arguments. This amounts to an important, though negative, clarification of his thinking. But this conception of the possibility of alternative analyses of judgeable-contents merits positive clarification. We need to make clear what he thought to be involved in his fundamental insight that judgeable-contents are logically prior to unjudgeable-contents (PW, 16f., cf. 253). And we should scrutinize what he thought to be the implications of function/argument analysis in logic. Without this knowledge we run the risk, in following his lead, of walking nonchalantly across a minefield into perdition, or alternatively of trampling on nuggets of gold which he has scattered in our path.

Frege's conception of the possibility of alternative analyses of judgeable-contents arises directly from function theory. The fundamental idea is that an entity may be designated as the value of different functions for different arguments. He simply transferred this principle to the logical analysis of judgeable-contents. *Ceteris paribus*, the possibilities for analyzing a single judgeable-content as the value of functions for suitable arguments are wide open. What must guide a logician in choosing a particular representation among a myriad of possible ones is his aim to give a perspicuous and economical representation of the inferential-powers of a given judgeable-content. Frege apparently thought that a single formula in concept-script could typically be chosen to encapsulate the totality of the inferential-powers of a given judgeable-content (cf. PMC, 67), though he offered no support for this traditional belief.

Three aspects of this conception of logical analysis are obvious, and prominent in Frege's thoughts.

(i) The idea that a judgment is the value of many different functions contrasts with the traditional conception that judgments are logically composite, built up from independent self-subsistent concepts (and objects). Frege took exception to this synthetic conception of judgments in part because it wrongly excluded the possibility of radically different decompositions of a single judgment.

(ii) That the value of a mathematical function is a particular number is an objective fact; e.g., it is a matter of common knowledge that 4 is the value of x^2 for the argument 2. But whether somebody regards 4 as the value of x^2 for the

---

22. Dummett elaborated this doctrine in response to our harping on the importance of alternative decompositions of judgeable-contents in the interpretation of Frege's logic. He now projects this view onto Frege, claiming that Frege drew a clear distinction between the 'analysis' of a thought into its 'constituents' and its 'decomposition' into its 'components' (Dummett, *The Interpretation of Frege's Philosophy* (Duckworth, London, 1981), p. 271ff.).

argument 2 or in some other way is a matter of arbitrary choice, and to say of somebody that he conceives of 4 in a particular way is apparently to describe his state of mind. Hence Frege was inclined to regard the way of conceiving a number or the mode of determining it (as the value of a function) as something subjective. This conception he transferred to function/argument analysis of judgeable-contents. That a given content is the value of a concept for a particular object is an objective fact, basic to the very possibility of the science of logic. But how a given judgment is viewed as decomposed into function and argument is typically an arbitrary matter of interpretation (*eine Sache der Auffassung* (BS, §9)). 'The distinction [between function and argument] has nothing to do with the conceptual content [of a formula in concept-script]' (BS, §9). Frege held this principle to be without exceptions 'as long as function and argument are completely determinate', i.e. as long as a formula contains no free variables (cf. BS, §1).

> But, if the argument becomes *indeterminate* ... , then the distinction between function and argument takes on a significance for content *(eine inhaltliche Bedeutung)*.... (T)hrough the opposition between the *determinate* and the *indeterminate* ... , the whole is decomposed into *function* and *argument* according to its content and not merely according to the point of view adopted *(nicht nur in der Auffassung)*. (BS, §9)

With the sole exception of judgeable-contents expressed by formulae containing free variables,[23] how a given content is conceived to be articulated into function and argument, i.e. how it is conceived or determined as the value of a function, is nothing intrinsic to the content, but rather is something optional and apparently subjective. It should, as a special case of a mode of determining an object, be irrelevant to the logical analysis of judgments in the expression of which names have their customary designations (cf. BS, §8). This generated a deep tension in Frege's fundamental idea of building logic on the basis of function/argument decomposition of judgeable-contents!

(iii) Formal definitions impose constraints on logical analysis. This idea derives from the use of definitions in mathematical proofs. If, for example, logarithms are defined by the equivalence $a = \log_b c$ if and only if $b^a = c$, then any rigorous proof requiring analysis of any premise of the form $a = \log_b c$ must proceed via a step eliminating the *definiendum* in favour of the *definiens*. Such a formal definition is treated as introducing a mere abbreviation for the *definiens,* and therefore the first step in analyzing the *definiendum* is compulsory, not optional. Frege generalized this conception to formal definitions in logic. A definition introduces by

---

23. The rationale for this exception is now transparent. In a formula containing a free variable, there actually occurs an expression (the variable) which is replaceable by different expressions, and therefore there is no option to viewing this expression as replaceable by other expressions, i.e. it must be considered to be the argument-name, and the other (relatively determinate) expression the function-name (BS, §9). This seems to generate a tension with the thesis that the same judgeable-content can be expressed in concept-script by a formula containing a variable bound by a quantifier, since this formula calls for an antithetical articulation into function and argument. But such asymmetry is typical of content analysis (infra, p. 227f.).

stipulation of content-identity a previously unused simple symbol as an abbreviation for some complex formula (BS, §24). A definition has a special function in respect of proofs. It licenses (indeed compels) the substitution of the *definiens* for the *definiendum*, thus unpacking or making apparent what was previously 'put into' the *definiendum* (BS, §24). (The special role of definitions with respect to proofs highlights the connection between content and demonstration or inference. Definiteness of content is a presupposition of the objectivity of questions about the cogency of proofs.) If a symbol is introduced by a formal definition, the fact that it designates an entity in a particular way (as the value of a function) seems to be an altogether objective feature of it, and hence there seems pressure towards adopting the principle that in this special case the way of regarding (or the mode of determining) an entity is part of its content. Frege did not draw this conclusion, but this idea too set up a tension in his thinking which he later partly resolved.

Function theory does not merely tolerate alternative analyses of entities as the values of functions. It positively requires this possibility, and it constantly exploits such alternative analyses in demonstrations. This makes it highly likely that alternative analyses would play a prominent role in Frege's extension of function theory to the logical analysis of judgeable-contents. This expectation is substantiated both by his general comments and by his actual practice. 'In general a content can be analysed in a number of ways' (PW, 107); 'I do not believe that for any judgeable-content there is only one way in which it can be decomposed' (PMC, 101). And he retained the principle after the distinction of sense from reference; '... a thought can be split up in many ways ...' (CO, 49; cf. PW, 201f.). To the extent that different decompositions are merely different ways of regarding or viewing a judgeable-content, each one is logically on a par with every other one, just as the different aspects perceived in the duck-rabbit drawing enjoy parity of esteem. No such decomposition can claim 'objective preeminence' (PMC, 101). The content $\Phi(A)$ is just as much the value of a second-level concept for the concept $\Phi(x)$ as argument as it is the value of $\Phi(x)$ for the object A as argument (BS, §10). And the same principle of parity is reiterated after Frege distinguished sense from reference (e.g. PW, 187, 201f.; BLA, §22).

Frege constantly exploited this general licence to give alternative decompositions of judgeable-contents. Many of his resulting manoeuvres seem so bizarre to anybody educated in the modern conception of logical analysis that they have been misunderstood or ignored altogether. *Begriffsschrift* highlighted the generation of alternative decompositions by the technique of functional abstraction. A content can be viewed now as decomposing into a first-level relation with two arguments, now as consisting of a first-level concept of one argument (BS, §9); or now as consisting of a first-level concept Φ and an object, now as composed of a second-level concept applied to the concept Φ (BS, §10); or now as consisting of a logical operation ⊢⌐ applied to a pair of judgeable-contents $\Phi(A)$ and $\Psi(A)$, now as composed of a complex concept

$$\begin{array}{c} \vdash \Psi(x) \\ \llcorner \Phi(x) \end{array}$$

applied to the object A (BS, §9). In all these cases functional abstraction brings about a metamorphosis in the decomposition of a judgeable-content. Such alternative analyses involve entities from different levels in the hierarchy of types in function theory. Functional abstraction is a powerful tool for generating alternative analyses. In fact, it opens up unlimited vistas of possible analyses exploiting ascent to ever high levels of concepts: just as we can view $\Phi(A)$ as containing a second-level concept applied to the concept Φ, so too we can view it as containing a third-level concept applied to that second-level concept, as containing a fourth-level concept applied to this third-level concept, etc.

On the other hand, functional abstraction is not the sole tool for generating alternative analyses. Frege also exploited methods of effecting descent in the levels of entity involved in decomposing a given judgeable-content. What he did was to follow the embryonic mathematical procedure of introducing equivalence classes to represent the relations among concepts in favourable cases as relations among objects. The paradigm for this procedure is the introduction of points at infinity into projective geometry or directions into Euclidean geometry (cf. KS, 1). Because the relation of parallelism between pairs of lines is reflexive, symmetric, and transitive, the set of lines can be partitioned into disjoint sets of lines each of which is parallel to every other one in the same set. Consequently, there is a one-to-one correspondence between concepts of the form ' . . . is parallel with a' and sets of mutually parallel lines. This allows the representation of certain second-level relations among concepts (e.g. the one expressed by 'Whatever is parallel with a is parallel with b') by relations among objects (e.g. the one expressed by 'The direction of a is identical with the direction of b' or 'The intersection of a with the line at infinity is identical with the intersection of b with the line at infinity').

Frege canvassed a similar manoeuvre in his analysis of statements of arithmetic. Just as the fact that the judgeable-content expressed by 'line a is parallel to line b' can be taken as an identity suggests the possibility of decomposing it into an identity of directions (FA, §64), so the characteristics of the content expressed by 'The F's are in one-to-one correlation with the G's' ('$F(\rho) \breve{\bar{\rho}} G(\rho)$')[24] suggest the possibility of carving up the content in a new way to yield an identity between numbers (FA, §63). Frege's aim is to find a judgeable-content which can be identified independently of its decomposition into an identity between numbers and which can therefore be employed to define expressions of the form 'the number of F's' in virtue of its alternative decomposition as an identity between numbers. Starting from the judgeable-content $F(\rho) \breve{\bar{\rho}} G(\rho)$, he 'replaces' the relation of one-to-one correspondence by the relation of identity, 'removes what is specific' to the former relation and 'divides it between F and G', thus arriving at the new concept of Number (cf. FA, §64). His putative definition of 'the number of F's' (or '$N_\tau F_\tau$') turns on the possibility of carving up the content $F(\rho) \breve{\bar{\rho}} G(\rho)$, considered as a second-level relation between first-level concepts, in the alternative way $N_\tau F_\tau = N_\tau G_\tau$, i.e. as a first-level relation between two objects. Even though he rejected this definition, the fact that Frege seriously canvassed it (and probably

---

24. Frege's notation: cf. M. Schirn, *Studien zu Frege*, vol. I, p. 95 (Frommann-Holzboog, Stuttgart, 1976).

employed it in the first attempt to execute his logicist programme (cf. KS, 51)) demonstrates that he admitted the possibility of descent in the level of entities introduced in the decomposition of a given judgeable-content. This idea recurs later. In particular, he thought that the judgment expressed by the sentence 'There is at least one square root of 4', though standardly analyzed into the assertion that the first-level concept *square root of 4* falls under a second-level concept, could also be analyzed into the assertion that an object (designated by 'the concept *square root of 4*') falls under a first-level concept (viz. the concept *concept that is realized*) (PW, 110). The *Basic Laws* advanced the general principle that second-level concepts can be represented by first-level concepts (BLA, §25), introducing a special logical operation into concept-script to effect this reduction (BLA, §§34f.). This operation, together with functional abstraction, guaranteed that a judgeable-content (or thought), decomposed in some way into entities from the hierarchy of function-types, could typically be decomposed in other ways into entities of higher or lower function-types. The range of possible logical decompositions of a given judgeable-content is far wider than the range of modern semantic analyses of a single sentence expressing this content.

Frege's liberalism in admitting alternative analyses of particular judgeable-contents has important implications for the interpretation of his work. Some of these seem absurd or unintelligible to modern logicians who view their enterprise and their symbolic notations from a different angle. It is worth briefly surveying them:

(i) Symbols in concept-script are not *typically* type-specific. Either of the Greek capitals in '$\Phi(A)$' may be regarded as designating a function (BS, §10). In *Begriffsschrift* the type-indeterminacy of such symbols is radical: the argument-pattern $\frac{(\forall \phi)\phi(A)}{\Phi(A)}$ is treated as a substitution-instance of the pattern $\frac{(\forall x)\Phi(x)}{\Phi(A)}$ (BS, §28).

(ii) Mathematical abstraction does not come to an end when a judgeable-content is translated into concept-script. Rather, it *begins* with formulae of concept-script, since from a given formula we typically can (by functional abstraction) read off alternative decompositions of the judgeable-content expressed.

(iii) There need be no correspondence between the logical decomposition of a judgeable-content and the structure of a sentence in natural language which expresses this content. Admittedly, Frege remarked that 'In general a content can be analysed in a number of ways and language seeks to provide for this by having at its disposal different expressions for the same content' (PW, 107). But he would not have conceded that the range of alternative analyses of a judgment should be circumscribed by the possibilities of perspicuously representing different function/argument decompositions in a natural language, e.g. German. First, he held that natural languages were defective in lacking forms clearly designating higher-level functions. Second, he would have thought that the issue of how a content may be articulated can be settled *a priori,* whereas what means there are for expressing such articulations in German is contingent.

(iv) Alternative analyses do not exhibit simple patterns of mutual exclusion. That different incommensurable analyses of content are compatible with each other is an immediate corollary of considering judgeable-contents as values of different functions for arguments. Though equipollent, no single one enjoys primacy. This would have seemed odd to his contemporaries because of their synthetic conception of judgment as built up out of antecedently given terms in a definite manner. It seems equally curious to a modern logician, since he would see each declarative sentence as having a single complete semantic analysis. This is an important measure of how different Frege's conception of logic was from the one now current (and wrongly ascribed to him).

(v) A single judgeable-content may have more than one translation in concept-script. This is a corollary of the observation that different formulae may have the same content. The pairs of formulae 'A & B' and 'B & A', 'A' and '¬¬A' and 'A → B' and '¬B → ¬A' have the same content. For a time, Frege probably thought that the formulae '$F(\rho) \breve{p} G(\rho)$' and '$N_r F_r = N_r G_r$' had the same content.[25] In none of these cases does the verdict of content-identity rest on a formal definition which reveals one of the pair to be an abbreviation of the other. Rather, it depends on direct insight into the content expressed by each formula. Content-identity is self-evident in these cases (cf. CT, 543). Given the identity in content of formulae of each pair, the inference-patterns $\dfrac{A \& B}{A}$ and $\dfrac{B \& A}{A}$ are not to be differentiated at all (contrary to modern practice). Similarly the inference-pattern $\dfrac{\Phi(A) \quad (A = B)}{\Phi(B)}$ is indistinguishable from $\dfrac{\Phi(A) \quad B = A}{\Phi(B)}$, nor is *modus ponendo ponens* distinct from *modus tollendo tollens*. By modern canons of rigour, Frege's rules of derivation are defective in formulation.

Our task here has been to clarify his conception of the possibilities of alternative logical analyses of judgeable-contents, especially by drawing out ingredient principles of which he was clearly aware and to which he appealed in philosophical argument. It is a measure of a deep failure of comprehension of his work among modern philosophers that the issue of alternative decompositions has passed unremarked, *a fortiori* unanalyzed.[26] Lack of understanding of this matter ramifies into mistaken conceptions of concept-script and into radically confused interpretations of the celebrated contextual dictum 'A word has meaning only in the context of a sentence'. To dismiss as mere slips Frege's comments about carving up single judgeable-contents in fundamentally different ways is to throw away the key to any interpretation of his work that he would have regarded as intelligible.

---

25. He indicated a belief that the whole content of arithmetic is contained in the general science of magnitude, and that the concept of magnitude can be extracted from the conditions for identity of magnitude (KS, 51). This suggests that he intended to carry out his logicist programme on the foundation of the equivalence $F(\rho) \breve{p} G(\rho) = (N_r F_r = N_r G_r)$. Assailed later by certain logical doubts and difficulties, he evidently struggled to overcome them and persevere with his original programme (as evidenced by some lost papers, viz. N47 in Scholz's catalogue on the *Nachlass*, cf. Schirn, *Studien zu Frege*, vol. I, p. 95). This would account for Frege's comment that he 'was long reluctant to recognize ranges of values, ... but ... saw no other possibility of placing arithmetic on a logical foundation' (PMC, 140f.).
26. In response to our raising this issue, Dummett has now addressed himself to alternative analysis (*The Interpretation of Frege's Philosophy*, ch. 15).

7
Unjudgeable-Content: The Elements

1. Objects: the contents of proper names

The previous chapter aimed at establishing the broad outlines of content-analysis. It unfolded the complex implications of Frege's conception of what it is to apply function theory to the analysis of judgeable-contents. We must now consider the products resulting from the application of this programme by scrutinizing the categories of the entities arising out of his decompositions of judgeable-contents.

The categorial framework for classifying unjudgeable-contents is obvious from his determination to articulate judgeable-contents into functions and arguments for the purpose of the logical analysis of inference. In his terminology, functions whose values are judgeable-contents are 'concepts'. If the category of objects embraces everything that is not a concept (cf. FC, 32), then the decomposition of any judgeable-content into function and argument(s) will produce at least one concept and possibly, though not invariably, some object(s). Since the generalization of the notion of a function from mathematics to logic carries in its wake the type-hierarchy of function theory, concepts will be subdivided into distinct logical categories according to the number of their arguments and the levels of their admissible arguments. Unjudgeable-contents will always be concepts (of various levels) or objects. Because objects enjoy pride of place in the order of explanation of the notion of a concept (function), an investigation of unjudgeable-contents may best start from the category of objects.

Elucidation of Frege's conception of the role of objects as unjudgeable-contents requires bearing in mind various important general principles easily lost sight of under the pressure of competing modern ideas. First, objects can be characterized as unjudgeable-contents only relatively to particular decompositions of particular judgeable-contents. (A persisting concrete object is not a self-subsistent unjudgeable-content, but only, as it were, the permanent possibility of an unjudgeable-content.) Second, the distinction between function and argument in decomposing a judgeable-content is a matter of how the judgeable-content is regarded, i.e. a matter of interpretation *(Auffassung)*. Third, every judgeable-content can be decomposed in fundamentally different ways by appeal to functional abstraction; no decomposition enjoys the privilege of monopoly. In particular, any judgeable-content which can be regarded as a first-level function whose argument is an object can also be viewed as a second-level function whose argument is the previously discerned first-level function. To depart from these principles would be to misrepresent Frege's thinking completely.

This framework of ideas explains away a major puzzle about the symbolism of concept-script. He evidently considered the distinction between objects and concepts to be fundamental in the doctrine of content. He criticized logicians, in particular Boole, for conflating concepts with objects by failing to distinguish the subsumption of an object under a concept from the subordination of one concept to another (BS, §9; PW, 18; PMC, 101; cf. infra, p. 184f.), and he highlighted the concept/object distinction as a crucial methodological principle in the *Foundations*. Yet there is no differentiation of dummy letters for unjudgeable-contents in concept-script. Greek capitals are used alike for the functions and the arguments into which arbitrary judgeable-contents are decomposed. How can an allegedly perspicuous logical notation fail to mark a basic logical distinction? The answer is straightforward. We interpret the symbolism of Frege's extension of function theory in the same way that mathematicians commonly construe the notation used in real analysis. We read the complex symbol '$\Phi(A)$' as the value of the function Φ for the argument A, never as the value of the function A for the argument Φ. By doing so, we overlook an important and explicit point of divergence between Frege's concept of a function and the 'far more restricted' concept of a function employed by mathematicians in real analysis (BS, §10). Typically a mathematician would distinguish type-symbols into number-words and function-names; he would take '2', '½', 'π', 'e', etc., to designate numbers, not functions, and he would take 'sin', '( )^2', '$\sqrt{}$', '+', etc., to stand for functions or operations. Hence, he would declare 'sin $\frac{\pi}{2}$' to designate the number 1 in a unique way, viz. as the value of the function sin x for the argument $\frac{\pi}{2}$; he would deny that 'sin $\frac{\pi}{2}$' represents 1 as the value of a second-level function whose argument is the function sin x. This restriction is what Frege explicitly lifted. By regarding the expression 'sin' as replaceable by other function-names in 'sin $\frac{\pi}{2}$', i.e. by viewing 'sin $\frac{\pi}{2}$' as one of the set of formulae 'cos $\frac{\pi}{2}$', 'ctn $\frac{\pi}{2}$', '$\left(\frac{\pi}{2}\right)^2$', '$e^{\pi/2}$', etc., we automatically view 'sin $\frac{\pi}{2}$' as presenting the value of a second-level function, and presumably we must regard the symbol '$\frac{\pi}{2}$' as standing for this second-level function since 'sin' is already known to stand for a first-level function. On Frege's view, these alternative ways of viewing the symbol 'sin $\frac{\pi}{2}$' have equal rights, and so too do the corresponding ways of viewing the symbol '$\frac{\pi}{2}$'. This cannot properly be characterized (on the basis of syntax) as logically a proper name *as opposed to* a function-name. From a logical point of view, the classification of symbols in function theory into number-words and function-names is held to depend on how they are regarded. In abstraction from some particular *Auffassungsweise*, the symbols of real analysis should be seen to be type-ambiguous.

In generalizing the notion of a function, Frege appealed to this radical conception that function/argument articulation is a matter of *Auffassung*. If the symbol 'Φ' in the formula '$\Phi(A)$' is regarded as being replaceable by other symbols 'Ψ', 'X', etc., then this is to view 'Φ' as the argument-expression in '$\Phi(A)$', i.e. it is to see the content expressed by '$\Phi(A)$' as the value of a function whose argument is designated by 'Φ'. Even if the judgeable-content $\Phi(A)$ is first described as ascribing a property to an object named by 'A', it can with equal propriety be characterized as stating that the concept Φ falls under a second-level concept; the content $\Phi(A)$, as it were, contains a second-level concept (cf. BLA, §22). On this second interpretation, Frege apparently concluded through an argument by elimination, the symbol 'A' must itself be viewed as the name of a second-level concept.[1] The essential arbitrariness of function/argument decomposition of judgeable-contents must be reflected in the type-ambiguity of the symbols for unjudgeable-contents in concept-script, and this excludes the possibility of any syntactic distinction between symbols for objects and symbols for concepts in concept-script.

The systematic ambiguity of the symbols in concept-script for unjudgeable-contents is not a mere idiosyncrasy, the excision of which would leave Frege's logic intact. It is related to two leading ideas. The first is the license for unrestricted functional abstraction that lies at the heart of his analysis of generality (infra. p. 182ff.). Function/argument analysis cannot be divorced from the hierarchy of types embedded in function theory. Consequently to apprehend the *immediate* inference $\frac{\Phi(A)}{(\exists \phi)\phi(A)}$ as *self-evidently* cogent (cf. BS, §11) requires construing Φ as the argument of a second-level function common to this pair of judgeable-contents, while construing the *immediate* inference $\frac{\Phi(A)}{(\exists x)\Phi(x)}$ as *self-evidently* cogent

---

1. One might argue that he should have concluded not that 'A' is the name of this second-level concept, but that '(A)' is such a name. Even had he done so, type-ambiguity of symbols would re-emerge in respect of decompositions into functions of higher-level. In any case, the proposal would make an anachronistic distinction: Frege was evidently understood to have treated the symbols in concept-script as being interchangeable with respect to function/argument analysis of judgeable-contents (CN, 215).

presupposes the different construal of Φ as a first-level function whose argument is an object. Unless what is expressed by the single formula 'Φ(A)' can, with equal propriety and equal immediacy, be viewed in *both* ways, there is either a lacuna or an error in Frege's basic account of the logic of generality. The second point is connected with the first. Since a single judgeable-content can be regarded as decomposing into function and argument in different ways involving functions from indefinitely many different levels of the hierarchy of types of function, the imposition of the requirement that symbols for unjudgeable-contents in concept-script be type-specific would demand that every judgeable-content could be represented by indefinitely many *different* formulae. But what would now justify the substitution of one such formula for another in a formal proof in concept-script? If such a transition rested on an appeal to the intuition that both formulae had the same content, the contention that concept-script makes possible a *mechanical* check on the gaplessness of proofs would be sabotaged. But if such a transition were justified by appeal to a formal definition, then the claim that definitions introduce abbreviations which are convenient but dispensable in principle from proofs would be undermined. Frege's whole conception of concept-script as embodying all and only what is relevant for the cogency of inference necessitates that a single formula can represent all of the possible decompositions of a single judgeable-content into function and argument. To refuse to countenance the type-ambiguity of symbols in logic would be to demand that his logical system be rebuilt from scratch.

The explanation of what it is to consider an expression to be a function-name from a logical point of view seems to be as immediately applicable to subsentential expressions in declarative sentences of ordinary language as to the symbols of concept-script. Hence it is natural to extend the thesis that expressions for unjudgeable-contents are systematically type-ambiguous from concept-script to language in general.[2]

To regard part of a sentence as invariant and the remaining expressions as replaceable by others (in comformity with syntax) is to view the constant expression as a function-name and the other expressions as standing for arguments. This licenses treating what are ordinarily called 'proper names' as function-names; e.g., if we regard everything apart from 'Hector' as replaceable in the sentence 'The father of Hector ransomed his son's body', then 'Hector' is *ipso facto* viewed as a function-name (for a concept of second-level). Consequently, if we think of a proper name as an expression designating an object and a concept-word as an expression standing for a concept, then the differentiation of proper names from

---

2. Of course this extension would be blocked unless logic may legitimately dissociate the content of an expression from the pattern of its use, the practice of explaining how it is to be used, and the criteria for its correct use. Otherwise such features about the word 'Aristotle' could be invoked in order to deny that 'Aristotle' is just as appropriately called a second level concept word as a proper name in the declarative sentence 'Aristotle admired Plato's Socratic dialogues'. Frege apparently thought that the proper drawing of the boundary between logic and psychology justifies logicians in abstracting from criteria of understanding and modes of explaining expressions in natural language.

concept-words is not a matter of syntax at all, but rather a matter of how the sentence (and the content expressed by it) is viewed. Only relative to a particular way of regarding a judgeable-content can any expression be classified as a proper name. Hence it is not in the least surprising that *Begriffsschrift* follows no uniform procedure in translating sentences with singular referring expressions into concept-script, sometimes ignoring a complex designator, sometimes representing it by a single letter in concept-script (cf. BS, §9).

The upshot of Frege's reasoning is not a repudiation of the thesis that the content of a proper name is an object (the object named). Properly understood, that is a truism. For what it is to view an expression as a proper name from a logical point of view is to view it as standing for an object. Frege embroidered on the familiar 'Fido'—Fido model of proper names. It would be mistaken, however, to suppose that he must have arrived at his conception of a proper name by independently reflecting on the use of expressions in natural languages or by accepting some prevailing conception from contemporary grammarians. Function theory itself gives a transparent rationale for the contention that the content of a proper name is simply the object named, i.e. that a full specification of the content of an expression viewed as a proper name involves nothing more than its correlation with an object. Within the parameters of function theory, the only alternative is that the content of a name may also depend on how the object is determined as the value of some function for suitable arguments. But it is a cardinal principle of function theory that the value of a function for an argument depends only on what the argument is, not as how it is designated as the value of another function (or operation). This carries over to the functions employed in function/argument decompositions of judgeable-contents. In particular, if the argument is taken to be an object, then the judgeable-content expressed by the formula on this interpretation must be independent of how this object is presented as the value of any other functions.[3] Leibniz's Law expresses this fundamental extensionality of functions. So too does the exclusion of the 'mode of determination' *(Bestimmungsweise)* of an object from the content of a name (BS, §8) provided that by 'mode of determination' is meant the presentation of an object as the value of a function for an argument. It would be self-contradictory to assert both that a formula presents a judgeable-content as the value of a first-level concept for an object as argument and that how the name in this sentence determines this object is relevant to the judgeable-content expressed. If Frege's conception of the content of an expression regarded as a proper name is fundamentally erroneous, his mistake is not a gratuitous blunder but rather a manifestation of a basic aspect of function/argument analysis.

The application of function theory to the analysis of judgeable contents carries with it a general commitment to purify the unjudgeable-content of any expression taken to designate an argument of any contamination from how this unjudgeable-

---

3. Any apparent violation of this norm must be explained away by arguing that what appears to be the argument of a non-extensional function is not its real argument (e.g. BS, §8 on identity and SR, 69ff. on *oratio obliqua*).

content is itself determined as the value of a function. In particular, this demands the exclusion of *Bestimmungsweise* from the content of a proper name. To retreat from this position would necessitate a fundamental revision to Frege's general conception of logic. Yet this idea generates important tensions. First, it gives rise to an antinomy about identity and difference of judgeable-contents. For surely how an object is designated may matter to the cogency of a proof and therefore be relevant to the content expressed by a sentence. It is transparent, e.g., that '2.2 is a composite number' expresses a true judgment, but not that '4 is a composite number' does. But may both sentences not be viewed as ascribing the property of being composite to the same number? And must they not have the same content on this view? On the other hand, does the asymmetry in what would be required to prove these two sentences (or their formal analogues) not show that they differ in judgeable-content? Does identity and difference of judgeable-content depend on one's point of view? Would this not undermine any right to speak of *the* judgeable-content expressed by a particular utterance of a sentence? These difficulties seem immediate and acute. Similarly, Frege's conception of the content of proper names might seem to conflict with his presuppositions that declarative sentences are names of judgeable-contents and that function/argument decomposition of judgeable-contents is crucially significant for logic. Conflict, however, is not immediate. The logical significance of the mode of determination of an object is denied only when the object itself is taken to be the argument of a function. This is manifestly not true of the principal function/argument articulation of the judgeable-content asserted by the use of a declarative sentence (e.g. the articulation visible in '⊢——— $\Phi(A)$'). But inconsistency does break out in any case where a judgeable-content occurs as the argument of a logical function, e.g. the function expressed by the condition-stroke. For then the decompositions of the constituent judgeable-contents into function and argument would officially not be available for the purpose of the functional abstraction required for Frege's quantification theory.

The product of these investigations of his early notion of a proper name is a recognition that we must be circumspect in interpreting his undoubted belief that the content of a proper name is the object named. Two crucial points separate his conception from what might seem the natural interpretation of this principle. First, the classification of an expression as a proper name from a logical point of view depends on how it is viewed. Consequently, there is no set of type-expressions in any language which can be identified as uniformly playing the role of proper names for the purpose of logical analysis, and therefore the logical category of proper names cannot in principle be isolated by any syntactic criteria. Second, the notion of an object is presupposed in the identification of expressions regarded from a logical point of view as proper names.[4] Relative to a particular decomposition of a judgeable-content, an expression in a declarative sentence is construed as a proper name if and only if the corresponding constituent of the judgeable-

---

4. Geach rightly calls attention to this principle in examining Frege's claim that number-words are proper names (Geach, 'Frege', in *Three Philosophers*, p. 136).

content expressed is an object. Hence the syntactic category of referring expressions cannot serve as the key to elucidating the category of objects.[5]

From the perspective of modern logic, and *apparently* also from that of Frege's mature writings,[6] the dependence of the logical classification of expressions on our way of viewing what a sentence expresses seems absurd. Yet, as already noted, this is an altogether natural extrapolation of reflection on designators in real analysis. Similarly, his characterization of what it is to view an expression as a proper name and his exclusion of the mode of determination of an object from the content of an expression so viewed simply recapitulates one facet of his initial commitment to function/argument analysis of judgeable-contents. The deep aspect of his account of proper names is apt to escape our notice, and we are likely to confuse his early conception of proper names with Mill's contention that ordinary proper names lack connotation. The systematic ambiguity of the symbols of concept-script makes the possibilities for alternative analyses of judgeable-contents integral to the representation of each single judgeable-content in concept-script.

2. Concepts: the contents of concept-words

For the purpose of logical analysis of judgeable-contents, the super-category of concepts is correlative to the category of objects. Frege's notion of a concept is the foil for his notion of an object. This pair of notions is indispensable for articulating or 'decomposing' judgeable-contents into functions and arguments in order to clarify the inferential connections among judgments. Any admissible decomposition must exhibit at least one concept among the constituents of a judgeable-content. The construction of his entire logical system rests foursquare on the legitimacy of his notion of a concept.

From the outset Frege treated concepts as species of functions. This is evident from the fact that the decomposition of a judgeable-content into concept and object is an analysis into function and argument (PW, 16f.) Indeed, in *Begriffsschrift*, he applied the term 'function' itself to what he later called concepts (BS, §§9f.; cf. FA, §70). Concepts are simply functions whose values are judgeable-contents (cf. BS, Preface). Logical analysis of judgeable-contents into functions and arguments represents judgeable-contents as the values of functions for particular arguments. Hence judgeable-contents are expressed in concept-

---

5. Dummett contends that Frege (at least in his mature theory) did try to elucidate the ontological categories of objects and concepts by reference to the linguistic categories of proper names and predicates, and therefore he lavishes attention on what Frege would have offered as a syntactic characterization of proper names and on what principles Frege might have used to avoid genuine objects from being drowned in a sea of pseudo-entities corresponding to expressions meeting these putative syntactic criteria (Frege, *Philosophy of Language*, p. 56ff). Unless Frege later gave up some of the fundamental ideas of function theoretic analysis that informed *Begriffsschrift*, Dummett's labours have no bearing on the elucidation of Frege's thought.
6. Though not of *all* of these writings! In particular, he claimed that $\Phi(2)$ contains a second-level concept (BLA, §22), which seems at odds with the thesis that numerals are proper names of numbers.

script by formulae such as '⸺ Φ(A)' and '⸺ Ψ(A,B)'. This conception of judgeable-contents as the values of functions presupposes a radical generalization of the notion of a function borrowed from analysis.

Begriffsschrift gives scant explanation of what a concept is, but it is clear from Frege's employment of this notion there and in the *Foundations* what form a fuller explanation would have taken. The notion of a concept (or the generalized logical conception of a function) is derived from the then current mathematical notion of a function by a combination of extension and restriction. First, the concept of a first-level function of n arguments is extended by lifting any restriction on what objects may occur as its arguments or values; in particular, *judgeable-contents* are admitted as values for such functions. Simultaneously, all restrictions are lifted on what operations may be employed for the purpose of constructing analytical formulae specifying first-level functions; in particular, expressions such as '$>$', '$<$', '$=$' may be used to construct binary first-level functions (over numbers) the values of which are judgeable-contents (conventionally expressed by such formulae as '$3 > 2$' or '$4 = 2^2$') (cf. FC, 28; PW, 24ff.). This extension of the notion of first-level functions is parallelled by corresponding extensions of the notions of higher-level functions, especially of second-level functions. The product is the full hierarchy of entities familiar from abstract function-theory. Every function has a unique place in this hierarchy determined by the levels and number of its arguments. Concepts can now be singled out by imposing a restriction on this generalized notion of a function: concepts are functions whose values are uniformly *judgeable-contents*.

This explanation of concepts has three noteworthy features. First, it guarantees that the full hierarchy of types of functions yielding numbers as values is matched by a full hierarchy of types of concepts. Concepts must be stratified into levels and further subdivided according to the number of their arguments. In particular, function theory hands the discerning logician second-level concepts as a ready-made tool of analysis. Second, his explanation of what concepts are makes clear why Frege thought a proper conception of concepts and of concept-formation to be essential for executing correct function/argument analyses of judgeable-contents (BS, Preface). Limitation to the standardly recognized forms of concept-formation (by abstraction and *Merkmal*-definition) would have made his logical analysis of generalizations impossible. Third, his early conception of concepts as functions is distinct from his later explanation of concepts as functions whose values are truth-values (FC, 30). For it is *judgeable-contents* that are the values of concepts in *Begriffsschrift* (BS, Preface) and the *Foundations* (FA, §§66n.,70), and judgeable-contents certainly cannot be equated with truth-values (as is clear from BS, §3). The account of concepts as functions antedates any discussion of truth-values as objects.

Deferring consideration of higher-level concepts, we focus now on first-level concepts. This fundamental category, like that of objects, is not spotlighted by a typographical distinction in concept-script. Greek capitals are construed both as object names and as names for concepts of various levels (BS, §§9f.). No syntactic distinction in concept-script corresponds to the basic distinction of concept from object (FA, x).

Frege's guiding idea here parallels his account of the content of a name; the content of a concept-word is the concept designated by the symbol. The primacy of the notion of designation *(bezeichnen)* in this explanation is important. '... [I]t will not do to call a general concept-word the name of a thing.... The business of a general concept-word is precisely to *designate* a concept' (FA, §51, our italics). This idea remained prominent in his later thought. In his view, mathematicians transgressed the distinction between a sign *(Zeichen)* and what it signifies *(Bezeichnetes)*, thereby conflating a function (the entity designated by a function-name) with a mathematical expression (the function-name itself (FC, 21ff.; WF, 113)). That Frege had the same idea earlier is obscured by his conflation of mention and use in the explanation of the notion of a logical function (BS, §9) and a concept or relation (FA, §§66n., 70). Though meriting censure for being logically slipshod, these passages are no ground for denying that he then upheld the doctrine that the content of a concept-word is the concept designated, still less for affirming that he then advanced the thesis that the content of a concept-word is an expression. These interpretations would lead to the conclusion that he barred his own way to giving any *function*/argument analysis of *judgeable-contents* in *Begriffsschrift!* and to the conclusion that a count-statement predicates a property of a *symbol* according to the analysis advanced in the *Foundations!* There is no room for doubt that he thought the content of a concept-word to be the concept designated, and that he considered concepts to be functions whose values are judgeable-contents.[7] In respect of assigning unjudgeable-contents to subsentential expressions, he subscribed to the Augustinian picture of language.

His doctrine that the content of a concept-word is the concept designated must be understood in a way parallel to the correlative doctrine that the content of a proper name is the object named. Both are subject to the same general constraints and provisos. The most important of these is that the logical categorization of any expression as a concept-word for a concept of a particular level is a matter, at least typically, of how a judgeable-content is viewed as decomposed into unjudge-

---

7. On the strength of BS, §9, Geach credits Frege with inventing the notion of a 'linguistic function', i.e. of a function whose arguments and values are expressions. According to this function-theoretic formulation of syntax, a predicate is a linguistic function whose argument is a proper name and whose value is a sentence. Frege was allegedly groping towards the idea that only linguistic functions can symbolize functions, or that the proper sign of a function is itself a function and not a quotable expression. This interpretation presupposes that Frege considered it the primary business of a logician to analyze sentence-structures (which he did not), that he would not countenance isolating parts of sentences as names of concepts or functions (which he did), and that the value of a syntactic predicate (viz. a sentence) can always be identified otherwise than as the value of a particular linguistic function for a particular name as argument (which is problematic unless formal syntax also incorporates a non-function theoretic characterization of predicates). Moreover, Geach ignores the restriction of the account of BS, §9 to judgeable-contents expressed in concept-script! What Frege in fact did in this passage is to explain how to apply the mathematical conception of functional abstraction to his generalized concept of a function, and his account suffers from the typical mathematician's vice of failing clearly to distinguish between signs and what they signify. (Cf. P. T. Geach, 'Frege'; in G.E.M. Anscombe and P. T. Geach, *Three Philosophers*, p. 143ff.).

able-contents. This depends on one's point of view *(Auffassung)*. In respect of concept-script, because any symbol with determinate content can be taken to be held constant while other such symbols are replaced, it can always be viewed as a function-name standing for a concept of an appropriate level, and it can be so viewed in indefinitely many different ways. As in the parallel case of proper names, there seems no obstacle in principle to applying this reasoning to any symbols expressing judgeable-contents, in particular to declarative sentences of natural languages. But then any expression in such a sentence which contributes to the judgeable-content expressed can immediately be viewed in indefinitely many different ways as designating a concept. In distinguishing proper names and concept-words of different types, the logician is engaged in an enterprise of classification in principle radically unlike the grammarian's syntactic categorization of type-expressions.

This divergence between logical classification and syntax makes clear that the logical categorization of expressions is derivative from independently given ontological categories. It is only in virtue of our viewing an expression as standing for an nth-level concept that we characterize it as an nth-level concept-word, just as what it is to view an expression as a proper name is to regard it as the name of an object. Precisely because there is no independent criterion for identifying n^{th}-level concept-words, the principle that the content of an nth-level concept word is an nth-level concept is a platitude. On his view, no conceivable refinement of syntax holds the key to ontology.

Although his conception makes it a truism that any expression in natural language which is used to stand for a concept is a concept-word from a logical point of view, his notion of a concept as a function whose values are judgeable-contents makes it far from obvious that there are *any* expressions in any natural language which are concept-words at all. Most logicians isolated the term 'wise' in the declarative sentence 'Socrates is wise' as the predicate, concept-word, or property-word, and many contended that it stands for a property, universal, or concept which can alternatively be designated by the abstract noun 'wisdom'. But such logicians did not consider the relevant Platonic entity to be any species of function. Did Frege suppose that direct insight into the essential nature of these tradition-hallowed entities confirms that they are really functions? Or did he consider that this conclusion can be derived from reflecting on perspicuous features of concept-words? The only reasoning which his texts even hint at, viz. the inference from the fact that 'Socrates' may significantly be replaced by 'Plato', 'Aristotle', etc., in the sentence 'Socrates is wise' to the conclusion that 'wise' may be viewed as a *function*-name, scarcely gives the impression of being free of gaps! At the least it rests on the presuppositions that judgeable-contents are genuine objects and that any object can be considered as the value of a function for some argument.

These presuppositions generate serious, though seldom noticed, tensions in Frege's thinking. The most damaging consequence is that they exclude in principle the possibility that a judgeable-content should have a unique analysis. This conclusion is certified by the limitless possibilities of abstracting higher-order functions from a judgeable-content articulated into function and argument. But more surprisingly, it holds even if the logical types of the components are taken to be predetermined; and this makes it in principle impossible that a sentence expressing a judgeable-content should have a unique translation into concept-script. The reasoning is straightforward and compelling. The fundamental ideas of function theory license multiple distinct translations of any judgeable-content that can be represented as the value of some (first-level) concept for an object as argument. Consider a judgeable-content expressed by the formula '$\Phi(A)$'. From the fact that a function is underdetermined by one of its values for a single argument we must conclude that the identity of the concept Φ cannot be recovered simply from knowledge that Φ takes the given content $\Phi(A)$ as its value for A as argument. In addition, it must be specified what values Φ takes for other admissible objects as arguments. Specifying the content of 'Φ' is inseparable from this wider undertaking. From the requirement that the value of a function for an argument must be specifiable independently its being the value of this particular function for this particular argument, we can infer that $\Phi(A)$ must be identifiable, independently of the function Φ. But seemingly nothing could prevent our introducing a concept Φ' by the following stipulation:

$$
\begin{aligned}
&(\ \Phi'(A) = \Phi(A)\ )\\
&(\ \ )\\
&(\ \Phi'(B) = \Phi(\Gamma)\ )\\
&(\ \ ) \text{ where } A, B, \Gamma \text{ are distinct objects.}\\
&(\ \Phi'(\Gamma) = \Phi(B)\ )\\
&(\ \ )\\
&(\text{otherwise } \Phi'(x) = \Phi(x).)
\end{aligned}
$$

Ex hypothesei, Φ' has the same judgeable-content as its value for the argument A that Φ has for the same argument. Therefore, on pain of violating fundamental norms of function theory, the possibility of translating a given judgeable-content into the formula '$\Phi(A)$' guarantees the possibility of translating it into other formulae of the same form, especially into '$\Phi'(A)$' (but *pari passu* into '$\Psi(B)$'). Indeed, it must be legitimate to translate what is expressed by '$\Phi(A)$' into any formula that presents the same judgeable-content as the value of any function differing from Φ only in the values assigned to objects other than A as arguments. This has the dramatic consequence that even identifying a judgeable-content as the value of some function for a particular object as its argument does not suffice to determine which function has the given content for the given argument. The object and the judgeable-content fail to determine the remaining 'constituent' of the judgeable-content, viz. the concept. (This is an important asymmetry with the relation of part to whole: subtraction of a part from a whole yields a unique remainder (cf. SR, 65).) It is uncertain to what extent Frege was aware of this aspect of the underdetermination of a function by any specification of its values for a proper subset of its admissible arguments (cf. especially FA, §70). But this fundamental principle of function theory excludes the very possibility of a unique analysis of any judgeable-content at any level of functional abstraction, and hence it renders incoherent the ideal of a canonical representation of any judgeable-content in concept-script.

The other major tension in Frege's position centres on the intelligibility of the claim that he explained or clarified the notion of a concept. With what justification can we construe any of his observations about what he called 'concepts' and 'concept-words' as a contribution to a philosophical investigation of the concept of a concept or the notion of a concept-word in natural language? Since his classification of an expression as a concept-word is not purely a matter of syntax, but rests on how one chooses to view a judgeable-content and the declarative sentence expressing this content, his use of 'concept-word' is obviously idiosyncratic from the point of view of traditional grammar and logic. But since the identification of a concept-word is in his view parasitic on regarding it as standing for an identifiable concept, the central problem is the extent to which his remarks constitute a clarification of our concept of a concept. Even superficial reflection should make us deeply sceptical about this claim. Our ways of speaking about concepts employ none of the idioms standard in discussions of functions, and what can intelligibly be said about concepts (e.g. that they can be grasped or understood, that they may be more or less vague) cannot intelligibly be said of functions. However unclear we might be about how to explain what 'concept' means, we would never give (or accept as correct) the explanation that a concept is a function whose value is a judgeable-content. Surely Frege's identifying concepts as species of *functions* cuts off what 'concept' means in his writings from our concept of a concept. He seems simply to have redefined a word already in use and misleadingly employed it according to his own non-standard explanation. He, of course, did not draw this conclusion. He saw himself in the rôle of a scientist making fresh discoveries and developing novel theories. He had finally revealed what concepts really are. Their true nature is not transparent, and any reluctance that we might display about accepting his explanation would merely testify to our lack of enlightenment.[8] His account of concepts, so conceived, stands to the everyday notion of concepts in the same relation as crystallography stands to the everyday conception of crystals. It allegedly improves and refines our prescientific knowledge. This raises a deep question about whether there is any analogue of scientific discovery or scientific theory in the philosophical investigation of concepts.

In his eyes, two ideas stood out as major advances in his scientific investigation of concepts: a standard of adequacy for definitions of concept-words and a liberalized conception of modes of concept-formation. Both ideas have proved profoundly influential. Indeed, both have been canonized in modern philosophical semantics.

First, function theory provides a norm for the specification of functions that serves as a paradigm for logically adequate analysis or definition of concepts. The core of a correct analysis is the employment of analytic formulae containing free

---

8. Russell conceded that in defining the number 2 as a class of classes 'we cannot avoid what must at first sight seem a paradox. . . . We naturally think that the class of couples . . . is something different from the number 2.' But he recommended the definition as the proper foundation for a science of arithmetic, adding that the impression of paradox 'will soon wear off'! (Russell, *Introduction to Mathematical Philosophy*, p. 18 (Allen and Unwin, London, 1918)).

variables. (The general case would involve a partition of the domain of arguments together with a specification of the values of the function for arguments in each subset either by use of an analytic formula or by complete enumeration of cases.) Once the domain of operations used to construct functions is suitably expanded, the analogue of an analytic formula in mathematics, e.g.

$$`f(x) = \frac{1}{x} + (\sin x)\, e^{x}\text{'}$$

is the type of formula used by Frege for justifying substitutions in concept script (e.g. '$f(x) =$

$$\begin{array}{c} g(x) \\ h(x) \end{array}\text{'}$$

(BS, §22)). Similar formulae incorporating free variables ranging over functions are used in analysis to define higher-level functions: e.g.

$$\mathfrak{F}(f) = \left.\frac{d}{dx}f(x)\right|_{x=0} \quad \text{or} \quad \nabla^2 f = \frac{\partial^2}{\partial x^2}f + \frac{\partial^2}{\partial y^2}f + \frac{\partial^2}{\partial z^2}f.$$

Frege made free use of such formulae in giving formal definitions of second-level concepts: e.g. (BS, §31)

$$`|\!\!\vdash\!\!\text{------} (\forall x)(\forall y)(f(y,x) \to (\forall z)(f(y,z) \to x = z)) \ = \ \underset{\epsilon}{I}^{\delta} f(\delta,\epsilon)\text{'}.$$

This generalization of functional definition by means of analytic formulae containing free variables is appropriate for introducing abbreviations to facilitate the writing out of complete proofs in concept-script (BS, §24).

But the procedure has far greater interest and wider implications too. If it is possible to identify the content of the defined formal expression with the content of some expression in natural language, then the analytic formula can be described as a precise and rigorous definition of this expression or as an analysis of the designated concept. Frege made claims of this kind. The quoted formal definition is offered as an analysis of the mathematical concept of being single-valued *(eindeutig)*. It sets a standard of clarity, precision, and rigour that allegedly cannot be equalled by 'informal explanations' of this familiar concept. Perhaps a better instance is the concept of continuity for a real function of one variable. Mathematicians had for some time operated with a 'common sense idea of continuity' explained in terms of absence of gaps or jumps in the graph of a function. They had gradually improved the concept by developing informal elucidations employing the notion of a limit. But they did not know what the term 'continuity' *really means* until Frege showed how to construct the formal definition of this concept out of the materials of the predicate calculus (PW, 24).[9] Here a formula

---

9. This argument is paraphrased from G. H. Hardy, *A Course of Pure Mathematics*, p. 216f., 10th ed. (Cambridge University Press, Cambridge, 1958).

in concept-script is claimed to specify the content of the concept-word 'continuous':[10] in effect it gives a definition of this term, viz.

$$\Phi \text{ is (everywhere) continuous} = (\forall x)(\forall y)(y > 0 \rightarrow$$
$$\sim (\forall z)(z > 0 \rightarrow \sim (\forall w)(|w| \leq z \rightarrow$$
$$|\Phi(x + w) - \Phi(x)| \leq y))).$$

This idea is of colossal importance. It gives rise to the expectation that the logician can make discoveries about the contents of expressions, about the true nature of concepts. It supports the contention that a system of formal logic (concept-script) furnishes the solvents necessary for analyzing complex concepts into their atoms. It is the very backbone of Frege's logicist programme of analyzing arithmetic. And finally, it supplies a novel norm for the adequacy of explanations of concept-words: anything other than an open generalization will be contemptuously dismissed as defective from a rigorous logical point of view. Philosophical thinking about concepts is reoriented, diverted into a path undreamt of in discussions of definitions in traditional logic.

Frege gave no general argument to establish the possibility of identifying the content of a concept-word from a natural language with the content of a formula in concept-script. Evidently he saw nothing here in need of justification. More strikingly, he made no case for identifying the contents of particular concept-words with the contents of particular formulae in concept-script. Far from thinking that argument is necessary here, he apparently thought it to be impossible. Identity of content in such cases is simply seen to obtain by the eye of the mind (cf. FA, vii); given the necessary reflective insight into the concepts expressed by our concept-words, it must be self-evident (cf. PW, 209ff.).

Against this background Frege saw no obstacle to subsuming traditional forms of definition under the umbrella of function-specification by analytic formulae in concept-script. A standard definition *per genus et differentiam* is clearly translated into a formal definition of the form '| |—— $\Psi(x) = \Phi_1(x) \& \Phi_2(x)$'. The more liberal pattern of definition discussed in logical algebra is just as easily provided for; here the translation is a formal definition of the form '| |—— $\Psi(x) = \Phi_1(x) \& \Phi_2(x) \& \ldots \ldots \& \Phi_n(x)$' or '| |—— $\Psi(x) = \Phi_1(x) \vee \Phi_2(x) \vee \ldots \Phi_m(x)$', where $\Phi_1(x), \Phi_2(x), \ldots$ are the defining characteristics *(Merkmale)* of the concept $\Psi(x)$. The possibilities of concept-analysis that Frege envisaged thus encompass all the forms that had received the cachet of philosophical respectability. Therefore there seems nothing to lose and much to gain by adopting the more liberal norm that a concept can legitimately be analyzed by a formal definition in which the *definiens* may be any suitable well-formed formula of concept-script.[11] This conception of definition is now so commonplace that it is diffi-

---

10. Frege himself did not write out any such identity, still less regard such an identity as a formal definition. No such formula mixing everyday expressions with symbols of concept-script is well-formed in his view.
11. This account of the analysis of concepts has important philosophical weaknesses. It leaves intact the cluster of misconceptions surrounding the notion of 'indefinable' or 'primitive' concepts because it embodies too restrictive a notion of explanations of meaning. On the other hand, it is

cult for us to recapture the exhilaration which Frege must have experienced on discovery of this powerful new technique of concept-analysis.

The second major advance which he saw himself as achieving in the scientific investigation of concepts is a liberalized conception of legitimate modes of concept-formation. The variety of possibilities for abstracting functions from expressions for judgeable-contents in concept-script discloses a whole galaxy of hitherto unsuspected kinds of concept-formation! Logic had previously been shackled by the dogmatic assumption that the only acceptable form of concept-analysis was to exhibit a concept as the sum of its characteristic marks (*Merkmale* (BS, Preface; FA, §53), i.e. to break down a concept into the constituent concepts (*Teilbegriffe* (FC, 4)). This had had a host of catastrophic consequences. It had restricted the category of concept-words to linguistic ready-mades (cf. PW, 17n.), especially individual common nouns, adjectives, and verbs, and to simple combinations of such expressions (e.g. 'equilateral rectilinear triangle'). It had fostered fruitless controversy: for example, logicians had often disputed whether there were negative concepts corresponding to such predicates as 'not triangular' or 'non-triangle' (PW, 17n.). It had contributed to neglect in logic of the formal properties of relations (cf. FA, §70). It had blocked the logical analysis of sophisticated mathematical inferences, e.g. arguments in function theory turning on the concept of continuity (cf. FA, §88). Finally, it had supported the mistaken idea that analytic truths must by their very nature be trivial or self-evident.

> If we represent the concepts (or their extensions) by figures or areas in a plane, then the concept defined by a simple list of characteristics corresponds to the area common to all the areas representing the defining characteristics; it is enclosed by segments of their boundary lines. With a definition like this, therefore, what we do ... is to use the lines already given in a new way for the purpose of demarcating an area. Nothing essentially new, however, emerges in the process.... [Hence, in making inferences from such definitions we are] simply taking out of the box again what we have just put into it. (FA, §88; cf. PW, 32ff.)

Commitment to this form of analysis of concepts into constituent *Merkmale* was, in Frege's eyes, a fundamental defect of traditional logic. He liberated logicians from their self-imposed imprisonment, using the tool of functional abstraction to strike off their fetters. Announcing the achievements of *Begriffsschrift*, he made the modest claim, 'It is easy to see how regarding a content as the value of a function for an argument leads to the formation of concepts' (BS, Preface). Later, as he sharpened his criticisms of other logicians, he boasted more of his own success. He alone was able to produce precise formal definitions of basic mathematical concepts, e.g. continuity of real functions (PW, 24) and limits (PW, 25). In such analyses of concepts

---

also too liberal because it acknowledges as legitimate definitions of concepts stipulations which do not specify *concepts* at all (infra, p. 253f.). A symptom of this error is the impossibility of finding coherent paraphrases for certain expressions introduced into concept-script as function-names, e.g. the horizontal (supra. p. 91ff.).

> there is no question there of using the boundary lines of concepts we already have to form the boundaries of the new ones. Rather, totally new boundary lines are drawn by such definitions—and these are the scientifically fruitful ones. (PW, 34)

This is of decisive importance for the thesis that arithmetic is analytic. For in the case of this really fruitful type of definition,

> what we shall be able to infer from it, cannot be inspected in advance; here, we are not simply taking out of the box again what we have just put into it. The conclusions we draw from it extend our knowledge.... The truth is that they are contained in the definitions, but as plants are contained in their seeds, not as beams are contained in a house. (FA, §88)

The fact that philosophers had too narrow a conception of the possibilities of concept-formation shielded from them the very possibility of proving that the truths of arithmetic are all analytic. Only by widening the domain of forms of acceptable concept-formation did Frege create the logical space necessary for the execution of his central logicist programme.

He viewed his conception of concepts as pregnant with dramatic implications for logical or philosophical concept-analysis. It seems immediately vindicated by its wide applicability and general fruitfulness. On top of all these advantages, it also effects a theoretical simplification in the analysis and formalization of sound inferences. The introduction of functions into the analysis of judgeable-contents frees logic from the necessity of introducing auxiliary entities (the extensions of concepts). This is a major theoretic simplification of formal logic. Frege appealed directly to a form of inference basic in function theory: instantiation. An open equation uses free variables to express generality. Consequently, any appropriate substitution of symbols for these variables must yield a truth provided the generalization is true. A scheme such as

$$\frac{(x + 1)^2 = x^2 + 2x + 1}{(10^{10} + 1)^2 = (10^{10})^2 + 2.10^{10} + 1}$$

is an immediate inference. Frege simply generalized this pattern of inference in extending the notion of a function into logic. The following patterns of inference he treated as primitive and as basic to his system of logic:[12]

$$\frac{\Phi(x)}{\Phi(A)} \quad \text{and} \quad \frac{\phi(A)}{\Phi(A)}.$$

What justifies these forms of inference is not any truth of set theory but rather the very idea of concepts as kinds of function.

Frege gathered in a rich harvest from the germinal idea that concepts are functions whose values are judgeable-contents. He stored away a subtle and elaborate formal theory that has served as the seed bank for much philosophy over the last

---

12. Here we follow Frege's practice of using lower-case letters as free variables, upper-case letters as dummy constants.

hundred years. But the value of these derivative crops is as doubtful as the worth of his original seed. The very idea that concepts are functions and judgeable-contents the values of functions raises deep questions about its intelligibility and coherence. On top of these fundamental doubts, there is another layer of residual worries deserving at least brief mention:

(i) The contention that concepts (for which concept-words stand) are uniformly objective and public entities does not square readily with the idea that concept-words may stand for mind-dependent and epistemically private properties, qualities, or relations.

(ii) The idea that a concept-word lacks any content if it is necessarily all-embracing or necessarily empty (cf. FA, §29) seems *prima facie* inconsistent with assigning content to an expression defined in terms of a self-contradictory predicate such as the number 0 (FA, §74).

(iii) The extensional equivalence of functions typically requires non-trivial mathematical proof, just like the identity of objects differently determined as the values of functions. The analogue for concepts would be a typical need for non-trivial proofs of concept-identity. It would seem that we can no more easily dispense with different concept-words in concept-script having identical contents than we can with different names having identical content. Yet this would generate a paradox parallel with the antinomy of identity (BS, §8), and presumably its resolution would demand a parallel bifurcation in the content of concept-words. *Mirabile dictu,* Frege did not expressly mention this matter.[13]

(iv) In one respect, Frege's concepts more closely resemble paradigmatic mathematical functions than typical material functions (e.g. the function correlating each person with his father or each city with its population). What judgeable-content is correlated with an object by a first-level concept does not depend on matters of fact, and hence is not something beholden to experiments or observations. By contrast, the value of typical material functions must be determined empirically; it is 'calculable' only from the argument together with a complete description of the relevant facts. Though the consequence is that concepts preserve one of the characteristic features of functions in mathematics, the price seems considerable. The truth or falsity of a judgeable-content does often depend on matters of fact. (In particular, it is possible to grasp a judgeable-content without knowing its truth-value (PW, 7f.).) The analogue at the level of concepts is that the extension of a concept often depends on matter of fact, and even when it can be settled *a priori* (e.g. the extension of $x \neq x$ (FA, §74)) it is only indirectly related to the concept itself (cf. FA, §53). In particular, it is possible to grasp a concept without knowing its extension, and identity of extension does not guarantee concept-identity (cf. FA, §49; PW, 18).[14] The external relation of concepts

---

13. Perhaps it was meant to be covered by BS, §8, on the assumption that unjudgeable-contents are *logically* homogeneous.
14. In distinguishing between the content *(Inhalt)* of a concept-word and its extension *(Umfang)*, Frege applied traditional terminology in a standard way. Cf. E. Schröder, *Der Operationskreis des Logikkalkuls* (Leipzig, 1877), p. 2.

to their extensions threatens to block the execution of Frege's logicist programme once it is conceded that extensions of concepts must be introduced in the analysis of numerals (FA, §68).

The principle that the content of a concept-word is the concept designated by this symbol simply recapitulates Frege's initial commitment to function/argument analysis of judgeable-contents. The apparent advantages and the less apparent difficulties clustering around this principle are the immediate corollaries of adopting his norm of representation of judgments.

3. Second-level concepts: the contents of quantifiers

The crowning achievement of *Begriffsschrift* is the construction of the predicate calculus with identity (or quantification theory). This opened up the possibility of exhibiting mathematical induction as a theorem derived from general logical laws, and it established the framework in which Frege undertook to analyze numbers and the concept of natural number in the *Foundations*. Quantification theory is the foundation of his logicism. It was itself the direct product of his employing abstract function theory for the logical analysis of judgeable-contents. Once possessed of a solution to the long-standing problem of constructing a logical system sufficient for analyzing inferences involving judgments with multiple generality, Frege must have felt that the key to his success (analysis into functions and arguments) was indispensable in logic, and he must have been tempted, moreover, to tackle any residual logical quandaries by further application of abstract function theory.

It is important not to exaggerate the achievement of *Begriffsschrift*. Frege did not *discover* the general logic of generality, nor even the logic of multiple generality. At least we should avoid this description if it carries the implications that no earlier thinker succeeded in detecting basic fallacies in arguments involving multiple generality, or that none succeeded in carrying out intricate and sophisticated reasoning in this genre. Many philosophers had detected and explained the fallacy of the form of argument now represented by the scheme $\frac{(\forall x)(\exists y) R xy}{(\exists y)(\forall x) R xy}$ without the benefit of quantification theory. Similarly, mathematicians made subtle distinctions between concepts differentiated only by quantifier shifts (e.g. between convergence and uniform convergence of infinite series, or between connectedness and local connectness of point-sets), and they even proved complex theorems involving such concepts (e.g. that a function may be everywhere continuous and nowhere differentiable). It was only because such reasoning was tolerably well understood and well developed that Frege was able to judge that his logical system encapsulated the general logic of generality. He did not discover a whole network of hitherto unrecognized forms of reasoning. (Indeed, it is doubtful whether that description even makes sense.)

What he did accomplish was a *systematization* of the logic of generality. He discovered, or invented, a formal system, the *formal* logic of generality. (Some would conclude that he thereby discovered the *mechanism* of inferences involving generality, a mechanism previously operated, though unconsciously, by any person who had intelligently engaged in such inferences.) This advance allows a transformation of the presentation of arguments, especially of proofs in advanced mathematics. Hitherto reasoning turning on multiple generality could be presented only informally. Some stages of inference could be expressed only with an admixture of everyday expressions (pidgin-mathematics, as it were): to unpack the premise that $f(x)$ is a continuous function, a mathematician would have to write: 'At every point x, for every positive number y, $f(x) = \lim_{y \to o} f(x + y)$ and $f(x) = \lim_{y \to o} f(x - y)$'. Moreover, there was no mechanical procedure for deciding whether inferences between propositions so formulated were sound; appeal to intuition and a mathematician's understanding of these formulae was the sole support for the claim that a putative proof was both rigorous and cogent. Frege thought that the invention of his concept-script allowed mathematicians to escape from this unsatisfactory situation, and hence he believed that he provided the means to build mathematics more rigorously on firmer foundations.

The guiding idea of his systematization of the logic of generality grew out of function theory. It is that quantifiers are concept-words designating second-level concepts. This idea has two main ingredients, each of which is a fundamental aspect of function/argument analysis of judgeable-contents. The first is the very idea of a second-level concept, i.e. of a function whose arguments are concepts and whose values are judgeable-contents. But this idea is an immediate consequence of conceiving of first-level concepts as functions, since the full hierarchy of types of functions is an integral part of function theory. The possibility of taking any expressions as designating second-level concepts could not cross the mind of anybody unfamiliar with function theory, but it could not fail to occur to somebody educated in this branch of mathematics. Though commonly viewed as a set of constraints, the hierarchy of types of functions may serve in this way to open up new vistas in theory-construction. The second ingredient is the notion of functional abstraction applied to function-theoretic expressions. For if second-level concepts are to be exploited to generate a wholly general logic of generality, a vast stock of first-level concepts will be required to function as the arguments of second-level concepts. Because of its limited conception of concept-formation, traditional logic would be unable to provide an adequate stock of concepts. It is for this reason that Frege insisted on the possibility of a very wide range of forms of concept-formation. What supplied the galaxy of concepts he required is the process of functional abstraction basic to function theory. The germ of his thought is apparent from his own example: we can regard the formula '$2^4 = 16$' in various ways, e.g. as stating the value of the function x^4 for the argument 2, or the value of 2^x for the argument 4, or even the value of $\log_2 y$ for the argument 16 (PW, 16f.). Similarly, we can view it as stating the value of the second-level function, defined by $F(f(\xi)) = f(2)$, for the argument x^4 (cf. BLA, §22). Frege generalized this idea by applying it to any formula expressed in concept-script. Starting from the formula '$\Psi(A,B)$' (BS, §9), we can abstract one concept whose sole argument is A, another whose sole argument is B, and even a second-level concept whose sole argument is the relation Ψ (cf. BS, §10). Equally, strikingly, we can abstract from the formula '$\Phi(A) \rightarrow \Psi(A)$' (representing the judgment that if internal forces alone act on the solar system, then the solar system has no acceleration) a concept whose sole argument is A or a relation both of whose arguments are the object A (BS, §9). This is the model for forming concepts of unlimited degree of complexity, in the specification of which any logical operators, whether quantifiers or propositional connectives, may occur (PW, 34). Liberality in modes of concept-formation obviates the danger of any shortage of first-level concepts in constructing a formal logic of generality.

Given this background of thought, the idea of incorporating second-level concepts in the analysis of general judgments is immediately compelling. Several routes converge on the same proposal. First, the standard mathematical notation for generality employs letters or free variables: e.g., where the domain is understood to be real numbers, the formula '$x^2 \geq 0$' expresses a true judgment. But similar formulae are also used as designators of functions (e.g. 'x^2' or 'sin x'). Once the notion of function is extended to cover Fregean concepts, such a formula as '$x^2 \geq 0$' counts as a concept-word (function-name). This makes it plausible to treat free-variable formulae expressing generalizations as assertions about concepts; indeed the only problem appears to be to prevent concept-words and generalizations in concept-script from collapsing together. If, for example, '$x^2 \geq 0$' designates a concept, then how can this formula also be used to express a judgment about this concept (PW, 16)? The sense of paradox, however acute, would not be lessened by the claim that the formula '$x^2 \geq 0$' asserts nothing about the concept designated by the self-same formula! Second, the standard alternative mathematical notation for generality, for similar reasons, highlights concept-designators; e.g., in the formulations 'For every real number x, $x^2 \geq 0$' or 'Every real number x is such that $x^2 \geq 0$'. What follows the preamble is what Frege would identify as a concept-word (viz. '$x^2 \geq 0$'). Consequently, it is wholly natural to canvass the possibility of so analyzing the judgment that this apparent reference to a concept is taken to be genuine. Type-theory leaves open only two straightforward possibilities: either the verbal preamble designates an object (i.e. has the role of a name) or it designates a concept whose arguments are first-level concepts. The first option is not intuitively appealing: the expression 'for every real number x' and paradigm names (e.g. the numeral '2') play different roles in the formation of complex mathematical expressions. With a paradigmatic name, it makes sense to ask *what* object it names, i.e. *what* its content is, but this question is anomalous when applied to the expression 'for every number x'. Some mistakenly treat the question as having sense, and then supply the answer 'an indefinite number'. But Frege remarked on the absurdity of this response (FA, §47). Are we really to suppose that the universe of objects, in addition to definite numbers such as 0, 1, 2, 3, ..., contains indefinite numbers (WF, 110; PW, 160f.)? These considerations constitute a *prima facie* case against construing as names such expressions as 'for every real number x', hence a *prima facie* case for taking them to be designations of second-level concepts. A third line of argument buttresses this conclusion. Frege's stipulation that only names with content are legitimate in concept-script carries the corollary that existential generalization holds without restriction: i.e. if we read '$\Phi(A)$' as 'A has the property Φ' (BS, §10), then the schema $\dfrac{\text{A has the property } \Phi}{\text{Something has the property } \Phi}$ is a valid form of inference. This imposes a necessary condition on names in concept-script. This condition is satisfied in the case of the expression 'for every number x', since 'Something has the property Φ' can be correctly inferred from 'For every number x, x has the property Φ'. So far, so good. But disaster follows hard on the heels of this success. The expression 'for no number x' seems to have the same logical status as 'for every number x'. Even were it insisted that 'for no number x has x the property Φ' was an abbreviation for 'for every number x, x lacks the property Φ', i.e. 'for every number $x, \sim\Phi(x)$', Frege's admission of the possibility of functional abstraction for concept-script licenses singling out the concept of Φ as one component of this judgeable-content. But it would be fallacious to infer 'Something has the property of Φ' from the premise 'For no number x has x the property Φ', and hence the concept Φ must be treated as the argument of a second-level concept in the judgment that no number has the property Φ. On pain of contradiction, this judgeable-content cannot be analyzed into a first-level concept Φ whose argument is an object. Symmetry requires that the same verdict be passed on what is expressed by the phrase 'for every number x'. Once Frege conceived the plan of carrying out the logical analysis of judgments with the apparatus of function theory, there were several obvious considerations to lead him to the conception that generalization involved second-level concepts. His arriving at that conclusion required no fresh fundamental insights.

Traditional logic may also have prompted him to the same conclusion. It was standard to describe generalizations as expressing relations between concepts; thus 'All whales are mammals' expresses the subordination of the concept of a whale to the concept of a mammal (hence a relation between two concepts). With a minor qualification Frege supported this contention (FA, §47). From the viewpoint of logical tradition, the novelty of his formal logic of generality is his treating 'degenerate' generalizations as the basic ones in logic and in his demoting paradigmatic 'complete' generalizations to a derivative status.[15] Traditional logic had held that every universal statement contains two terms; hence the paradigm for universal generalization is the scheme 'All A's are B's'. Apparent exceptions were assimilated to this pattern. The judgment expressed by 'Everything is extended' appears to lack a subject term, and hence to be degenerate; orthodoxy was saved by introducing a dummy subject such as 'thing' or 'object', and then by paraphrasing the sentence into 'Every object is extended'. Frege stood this conception on its head. He singled out the degenerate cases as fundamental, symbolizing them in concept script either in the form $\Phi(x)$ or in the form $(\forall x)\Phi(x)$.[16] Sub-

---

15. Another novelty is Frege's sharp distinction between subordination of one concept to another and the subsumption of an object under a concept (PW, 18).
16. In effect, he treated such expressions as 'thing', 'object', 'concept', etc. as formal concepts, i.e. as not translated into concept-words in concept-script (cf. PW, 63), but as expressed by different styles of variables in quantified formula (cf. *Tractatus*, 4.126ff.).

ordination of concepts is simply a special case of judgeable-contents with this form; for such judgments are expressed in concept-script by formulae of the form '$\Phi(x) \to \Psi(x)$' (BS, §12), and functional abstraction licenses considering judgments of this form to exemplify the form '$\Omega(x)$' (cf. BS, §9). In Frege's view traditional logic had gone astray by mistaking what is logically complex for something logically simple (cf. PW, 49).

Given the framework of abstract function theory, it was natural for him to arrive at the conception of generalizations as incorporating second-level concepts. But whether or not this idea is sound and fruitful is an independent question. Frege contemplated his creation and found it to be good. It passes the minimum test of accounting for the validity of those inferences formalized in syllogistic, the logic of 'primary propositions' (Boole). Both universal and singular judgments, whether positive or negative, can obviously be expressed in concept-script, and the inference from 'For all x, x has the property Φ' to 'A has the property Φ' can be expressed as a case of function-theoretic instantiation, viz. $\dfrac{\Phi(x)}{\Phi(A)}$. Singular judgments, however, pose a problem; for, as long as only free variables are available for expressing generalizations, there is no way to translate 'For some x, x has the property Φ', i.e. 'Not: for all x, x lacks the property Φ', since this judgeable-content is evidently not to be represented as '$\Phi(x)$' or as '$\sim\Phi(x)$'. But the solution is immediate. What Frege required is an explicit concept-word to express generality; say the symbol '$\mathfrak{F}$'. Then, provided that a formula expresses a judgeable-content if and only if its negation does (FA, §29; PW 7f.), functional abstraction delivers a concept $\sim\Phi$ corresponding to the concept Φ. Consequently, particular judgments will take the form '$\sim(\mathfrak{F}\Phi)$' or '$\sim(\mathfrak{F}(\sim\Phi))$', in contrast to universal judgments which exemplify the forms '$\mathfrak{F}\Phi$' and '$\mathfrak{F}(\sim\Phi)$'. The difference between formulae '$\sim(\mathfrak{F}\Phi)$' and '$\mathfrak{F}(\sim\Phi)$', which we now describe as a distinction of scope, turns simply on differences of what the functions $\sim$ and $\mathfrak{F}$ take as arguments. In the first case, the content expressed is the value of the operation of negation applied to the content that is the value of $\mathfrak{F}$ for the argument Φ, while in the second the content is the value of the function $\mathfrak{F}$ for the first-level concept which is the value of the operation of negation applied to the argument Φ. Since in general the order of operations in function theory matters to the result,[17] there is no reason to identify these two contents. The fundamental inference-rule for quantifiers can be reduced to the pair $\dfrac{\Phi(x)}{\mathfrak{F}\Phi}$ and $\dfrac{\mathfrak{F}\Phi}{\Phi(A)}$, and the inference-patterns sanctioned in syllogistic can be derived from these simple rules. The basic idea that logical operators have scopes is an integral part of the very idea that they designate functions.

---

17. The non-commutativity of the operation of composition of functions is an idea basic to function theory. E.g., the two functions compounded out of $\sin x$ and $2x$ (viz. $2\sin x$ and $\sin 2x$) differ from each other, and similarly, if $\mathfrak{F}(f(x)) = f\left(\dfrac{1}{x}\right)$, then typically $\dfrac{d}{dx}\mathfrak{F}(g(x))$ differs from $\mathfrak{F}\dfrac{d}{dx}(g(x))$.

In fact, Frege's conception of generality has far more to recommend it than its capturing the logical relations sanctioned in syllogistic. Three further achievements stand to its credit.

(i) His system displays fundamental connections between inferences involving generality and those that do not. Whereas traditional logicians had bifurcated logic by segregating primary propositions (generalizations) from secondary propositions (molecular propositions), he employed functional abstraction to mediate the ascent from atomic and molecular judgeable-contents to the general logic of generality and thereby set up a simple and organic relation between the two separated parts of logic (PW, 18).

(ii) His treatment of generality extends the range of inferences amenable to formalization. He succeeded in decomposing standardly recognized sound inferences involving generality into iterations of a few basic inference-patterns (officially, according to *Begriffsschrift*, to a single inference-pattern (BS, §22)). This achievement required something slightly more than the bare idea that generalizations contain second-level concepts. Even introducing an explicit concept-word '$\mathfrak{F}$' into concept-script would not alone suffice to distinguish different judgeable-contents incorporating multiple generality. For this would not differentiate the true generalization about natural numbers that there is no greatest number from the false generalization that there is no smallest number; both would be translated into concept-script by the formula $\mathfrak{F}(\sim\mathfrak{F}\sim(\Psi))$. The problem to be resolved is to indicate which occurrence of '$\mathfrak{F}$' is related to the first argument of the function Ψ and which to the second argument. For this purpose Frege borrowed from the calculus the practice of using variables to *index* second-level functions; the paradigm here is the notation '$\frac{\partial}{\partial x}$' and '$\frac{\partial}{\partial y}$' for indicating the first and second partial derivatives of a real function of two arguments. He combined variable-indexed quantifiers with designations for concepts that include free variables; hence, in place of the ambiguous formula of the form '$\mathfrak{F}(\sim\mathfrak{F}\sim\Psi)$', he wrote formulas of the form '$(\forall x)(\sim(\forall y)\sim(x > y))$' and '$(\forall y)(\sim(\forall x)\sim(x > y))$', which manifestly differ in structure. This left one further matter to be clarified; the occurrence of a formula with free variables within a formula of concept-script makes it seem as if the whole formula must contain a free variable (and hence designate a function). This appearance is illusory. The illusion can be exploded by reflecting on *variable-binding* operations familiar from the calculus. A formula such as '$\int_0^1 f(x)dx$' or '$\left[\frac{d}{dx}f(x)\right]_{x=0}$' is not a function-designator, but rather a complex designator of a number—a number which in principle can be designated by an expression containing no variable (e.g. a numeral). The variables appearing in such formulae are bound, not free, and hence they are merely *apparent* variables, not real ones.[18] Quantifiers are thus variable-indexed variable-binding oper-

---

18. This suggestive terminology was introduced by Peano and then given prominence by Russell (*Principles of Mathematics*, p. 13).

ators applied to concept-words and designating second-level concepts. This conception suffices for the formalization of the general logic of generality.

(iii) In conjunction with the concept of identity, Frege's logical analysis of generalizations can be extended to yield a logical formalization of count statements, i.e. statements of the form 'There are n F's' (FA, §§55ff.). This account would proceed on familiar lines:

There are no F's $\quad\sim (\exists x) Fx$
There is (exactly) one F $\quad(\exists x)(Fx \ \& \ (\forall y)(Fy \rightarrow y = x))$
There are (exactly) two F's $\quad(\exists x_1)(\exists x_2)(Fx_1 \ \& \ Fx_2 \ \& \ x_1 \neq x_2$
$\quad\& \ (\forall y)(Fy \rightarrow y = x_1 \lor y = x_2))$
$\quad\vdots$

Functional abstraction would license distinguishing the concept F in each of these judgeable-contents as the argument and treating the remainders as second-level concepts which take the concept F as argument. (Numerically definite quantifiers, '$(\exists_1 x)$', '$(\exists_2 x)$', '$(\exists_3 x)$', could be introduced to designate these second-level concepts.) This analysis would provide cogent backing for the central thesis that the content of a count-statement is an assertion about a concept (FA, §§46, 57). Frege himself gave an account approximating to the modern conception of numerically definite quantifiers (FA, §55). But the *Foundations* no more succeeded in developing the modern semantic analysis of assertions of applied arithmetic than *Begriffsschrift* did in giving the modern semantics of quantified assertions. The abstraction of numerically definite quantifiers from quantified formulae including identity presupposes that identities express judgments about objects, whereas Frege had a strong propensity to view them as expressing judgments about symbols (cf. FA, §57; infra, p. 220f.). Though he broke new ground in analyzing count-statements, he cannot be judged successfully to have analyzed statements of applied arithmetic.

Abstract function theory proved to be a powerful tool of logical analysis. In Frege's hands it advanced the frontiers of formal logic by providing a novel model for the formalization of hitherto intractable judgments and inferences. Yet quantification theory has a price. He overlooked this in the exhilaration of technical triumph, while we do so through an inclination to believe that quantification theory simply describes the depth structure of language, if not the logical structure of the world. There are at least three aspects of quantification theory that should have seemed problematic from Frege's point of view.

(i) The quantifier-notation of concept-script is introduced with an elucidation (not a formal (semantic) definition): '⊢—$\mathfrak{a}$—$\Phi(\mathfrak{a})$ ... stands for the judgment that, whatever we may take for its argument, the function is a fact' (BS, §11: italics removed). This could be rephrased: every judgeable-content of the form $\Phi(A)$ is to be affirmed. This seems to most modern logicians a somewhat careless formulation of his later elucidation of the quantifier, namely that '$(\forall x) \Phi(x)$' denotes the True if the value of the function $\Phi(\xi)$ is the True for every object as argument, and otherwise it is the False (BLA, §8). This interpretation is as dubious as the parallel claim that *Begriffsschrift* gives truth-table definitions of the two primitive propositional connectives. Both are open to similar counter-arguments. Were the alleged elucidation Frege's, it would not serve to specify what second-level concept the universal quantifier designates. Specification of the content of '$(\forall x)$' requires determining what *judgeable-content* is expressed by each formula of the form '$(\forall x)\Phi(x)$' where '$\Phi(x)$' designates a first-level concept. But the elucidation as interpreted explains only what *truth-value* '$(\forall x)\Phi(x)$' has for each concept-word '$\Phi(x)$', and only then on the supposition that the truth-value of each formula of the form '$\Phi(A)$' is known (although the truth-values of these formulae are not yet part of their content). If '$(\forall x)$' is a second-level concept-word, this explanation would leave its content completely undetermined. This objection can even be formalized. Suppose that the concepts Φ and Ψ are coextensional but differ in content; this is a possibility that Frege clearly envisaged. Then for any object x, $\Phi(x)$ is true if and only if $\Psi(x)$ is true. Therefore, if Frege's elucidation were considered to determine the *content* of '$(\forall x)\Phi(x)$', we should conclude that the content of '$(\forall x)\Phi(x)$' is identical with the content of '$(\forall x)\Psi(x)$'. Yet this is absurd, for the inference-pattern $\dfrac{(\forall x)\ \Phi(x)}{\Phi(A)}$ is obviously cogent, whereas the corresponding pattern $\dfrac{(\forall x)\ \Psi(x)}{\Phi(A)}$ is fallacious, becoming cogent only with an added premise, viz. $(\forall x)(\Phi(x) \rightarrow \Psi(x))$. By Frege's explanation of content (BS, §3), this inferential asymmetry demonstrates a difference in content between '$(\forall x)\ \Phi(x)$' and '$(\forall x)\ \Psi(x)$' which the modern interpretation of his elucidation of '$(\forall x)$' fails to account for. (This argument shows that the sense of '$(\forall x)$' is left indeterminate by the elucidation in the *Basic Laws*, although the reference of '$(\forall x)$' is uniquely specified.)

(ii) Though the universal quantifier is officially taken to designate a second-level concept whose arguments are first-level functions of a single argument, Frege employed it in ways inconsistent with this categorization. Its use often offends against a strict interpretation of the canons of function-theoretic type-theory, namely the requirement that any properly defined function must be determinate in respect of the number and types of its admissible arguments.[19] These features characterize the type to which any given function belongs. Any attempt to pigeonhole the content of Frege's universal quantifier in the hierarchy of function-types is doomed to frustration. First, *Begriffsschrift* treated the inference-pattern $\dfrac{(\forall \phi)\phi(A)}{\Phi(A)}$ as a *substitution-instance* of the pattern $\dfrac{(\forall x)\ \Psi(x)}{\Psi(A)}$ (BS, §28). Hence admissible argument-expressions for the quantifier include names of second-level concepts, so that its content must be now a second-level concept, now a third-level one. Secondly, functional abstraction licenses decomposing the judgeable-content $(\forall x)\ (\forall y)\Phi(x,y)$ into a second-level concept $(\forall x)$ whose argument is a first-level

---

19. This renders chemical metaphors immediately attractive. Functions (and hence concepts) may be said to have determinate valences, and hence comparable to a chemical radical (cf. PW, 17; also C. S. Peirce, *Collected Papers*, vol. III, pp. 262f., 295ff.).

concept $(\forall y)\Phi(x,y)$. But further functional abstraction allows viewing this concept as itself decomposable into a second-level concept $(\forall y)$ applied to a first-level relation $\Phi(x,y)$. Therefore admissible argument-expressions for the quantifier are not homogeneous; its content must now be a second-level concept whose arguments are first-level concepts, now a second-level concept whose arguments are first-level relations of two arguments. In respect of the hierarchy of types of abstract function theory, the universal quantifier designates a type-promiscuous entity. If it were a logical requirement that every well-defined function belongs to a single type, then the quantifier would be ill-defined, or at least its content would vary from context to context.[20] In fact, Frege would probably not have drawn this conclusion. Like modern logicians, he would have noted a unifying pattern among the diverse uses of the quantifier. A single algorithm subsumes all of them: a quantifier of level m maps a concept with n arguments of level $m - 1$ into a concept of $n - 1$ arguments of level $m - 1$.[21] By appeal to the intentional conception of function-identity common in mathematicians,[22] he would have concluded that the universal quantifier always designates the same function (concept), even though the arguments of this function are not type-homogeneous.[23] (This solution to this dilemma about quantifiers, and other logical operators, leaves unresolved a tension between two different criteria for function-identity in Frege's logical system.)

(iii) He tended to consider concepts to be the designations of grammatical predicates. If in a simple atomic sentence such as '2 is prime' the content of the numeral is the number 2 and the content of the sentence is identical with the content of the formula '$\Phi(2)$', then by elimination the content of the predicate 'prime' must be the concept Φ. Guided by this idea, he paraphrased the formula '$\Phi(A)$' into the sentence 'A has the property Φ' (BS, §10). Later this doctrine became more explicit; a concept is identified as the reference of a grammatical predicate (CO, 47f.), and the unsaturatedness of the predicate is held to mirror the unsaturatedness of the concept itself (CO, 50) and also that of the sense of the concept-word (CO, 54). Sometimes the unsaturatedness of a concept-word is not apparent in a particular sentence. In such cases it can be made perspicuous

---

20. So too would the logical constants of the propositional calculus. They can be used to construct first-level concepts out of first-level concepts, second-level concepts out of second-level concepts, etc.
21. For simplicity, a judgeable-content is treated here as a concept (of level $m - 1$) with 0 arguments.
22. Co-ordinate geometry employs a single 'projection function' that maps any n-tuple of reals $\langle x_1, x_2, \ldots, x_{n-1}, x_n \rangle$ onto the $(n - 1)$-tuple $\langle x_1, x_2, \ldots, x_{n-1} \rangle$, and real analysis makes use of a single operation of partial differentiation with respect to x which is applied to any real function of two or more variables. In such cases, identity of an algorithm is taken to license identity of a function in spite of type-differences among its admissible arguments and values. Here the criteria for intensional identity of functions are less restrictive than the criteria of extensional identity.
23. The same argument could be applied to neutralize the objection that the numerals '0', '1', '2', ... must have different meanings when used to count objects and when used to count first-level concepts. Numerically definite quantifiers can be treated on a par with the existential and universal quantifiers.

by a paraphrase; e.g. the predicative nature of the concept *mammal* which occurs in the judgment expressed by 'All mammals have red blood' appears clearly in the alternative formulation 'Whatever is a mammal has red blood' (CO, 47; cf. FA, §47). It is a striking fact about the counterpart of the universal quantifier in natural languages that it fails to pass either of these tests for concept-words. In the sentence 'Everything is extended', the term 'everything' appears as the grammatical subject, not as the predicate. Other expressions of generality ('everyone', 'something', 'nothing') resemble paradigmatic names far more closely than pardigmatic predicates or concept-words.[24] Moreover there is no natural direct paraphrase of sentences in which such expressions occur as grammatical subjects that has the effect of making their predicative nature perspicuous. The most plausible paraphrases (e.g. 'It is true of everything that if it is a mammal it is warmblooded') still exemplify expressions of generality as proper *parts* of grammatical predicates; and the availability of parallel paraphrases for names seems also to stress the difference of expressions of generality from predicates. The best escape from this apparent dilemma appeals to the content-equivalence of, e.g., 'Something is a whale' with 'Whales exist'. This suggests reducing all generalizations to the form of sentences whose predicate is 'exist'; e.g. 'All whales are mammals' could be paraphrased by 'Whales which are not mammals do not exist'. Nevertheless the fact that the counterparts of quantifiers are *not* predicates and *cannot* be directly paraphrased into predicates is difficult to reconcile with Frege's general conception of *concepts* as the contents of predicates. Moreover, it is far from perspicuous that the term 'exist', though clearly a predicate, designates a second-level concept rather than a first-level one. All of this awkwardness would no doubt have been dismissed with the remark that here one of the principal logical defects of language comes to the fore. This response seems shallow, for what is called into question is the very intelligibility of employing second-level concepts in the analysis of generalization, on the supposition that language reflects the predicative nature of concepts. (The fact that Frege thought the key to quantification theory to be the idea of second-level concepts and the nature of these concepts not to be perspicuous in natural language seems sufficient to refute the claim that he derived the basic idea of quantification theory from the inspection of the structure of sentences expressing generality.)

In constructing his general logic of generality Frege navigated in unchartered waters with aplomb, and he returned to a hero's welcome bearing considerable treasures. He seems to have been oblivious to the many dangers of the deep which threatened his enterprise with shipwreck.

4. Retrospect

This analytical survey of unjudgeable-contents must consider one last issue. This is the crucial ambivalence about whether the relation of a judgeable-content to

---

24. This fact is responsible for a perennial form of philosophical joke familiar from such diverse sources as *The Odyssey* and *Through the Looking Glass*.

the unjudgeable-contents cited in its logical analysis is the relation of a whole to its constituent parts. Frege described judgeable-contents as decomposing or falling apart *(zerfallen)* into unjudgeable-contents. He even said that they could be differently carved up *(zerspalten)* into concepts and objects (FA, §64). This suggests that he viewed the relation of unjudgeable-contents and judgeable-contents as that of parts to wholes, just as he so regarded the relation of certain constituent judgeable-contents to a compound judgeable-content (cf. PW, 36). Did he take this part/whole relation to be a literal truth? Or rather a suggestive metaphor, a mere *façon de parler?* He vacillated about this. On the one hand, he used terminology related to part/whole analysis without explicit warning that it should not be construed literally. Moreover, the implication of objectivity in part/whole analysis would be a powerful antidote to the unwanted implications of subjectivity attendant on his official doctrine that the decomposition of judgeable-contents is a mere matter of how we choose to regard them. On the other hand, the fundamental idea of function/argument analysis of judgeable-contents is deeply inconsistent with part/whole analysis: neither the argument nor the function itself can typically be said to be parts of the value of a function for an argument. Moreover, the basic principles of part/whole analysis are inconsistent with prominent theses advanced by Frege. Parts must be homogeneous with each other and with the whole of which they are parts (cf. N, 131f.); nuts, bolts, tyres, windows, etc., are parts of cars, but not the colour of the paintwork nor the texture of the upholstery. If judgeable-contents are objects, then it is nonsense to call a concept *part* of a judgeable-content, although every decomposition must contain at least one concept of some level (cf. PW, 192f.). Moreover, no judgeable-content can be constituted of entities that are all type-homogeneous since its constituents would not then hold together (cf. CO, 54). A whole is greater than its parts. Yet it is nonsense to say that a judgeable-content is greater than a concept, and doubtful even to state that an object (number, city, etc.) is smaller than a judgeable-content.[25] Wholes are often conceived to be completely determined by enumeration of their parts. Hence, from identity of parts we often infer identity of wholes. This principle manifestly fails for judgeable-contents; here we must introduce, as it were, differentiation of isomers. For example, the judgment that the function $\sin x$ is everywhere continuous has the same constituents as the judgment that $\sin x$ is uniformly continuous, but the position of two quantifiers is inverted. Different judgeable-contents can be constructed out of a common stock of unjudgeable-contents. Typically subtraction of a part from a whole uniquely determines the remainder (cf. SR, 65). This principle too fails for judgeable-contents. Underdetermination of a function by its values for a proper subset of its arguments has the corollary that a concept is never uniquely determined by subtracting one of its arguments from a judgeable-content; indefinitely many concepts (functions)

---

25. Frege addressed a parallel objection to Wittgenstein with reference to the thesis that a fact is something composite (cf. Wittgenstein, *Philosophical Grammar*, p. 200f.)

take the object $\Phi(A)$ as value for A as argument.²⁶ Taken together, these observations about the part/whole relation render unintelligible the idea of incommensurable decompositions of judgeable-contents into alternative sets of parts. Nothing could be decomposed without remainder into two parts that were an object and a concept and also decomposed without remainder into two parts that were a third-level and a second-level concept! The conclusion seems unavoidable that to the extent that he reflected explicitly on the matter, Frege must have thought the 'decomposition' of judgeable-contents into 'parts' to be a metaphor, not a literal truth. The only coherent alternative is that he thought the analysis of a judgeable-content into function and argument to be a metaphor! That hypothesis is obviously absurd. On the other hand, he seldom focussed his attention on this matter, and, partly under the influence of traditional patterns of thinking, he seems to have betrayed a tendency to mistake the metaphor for a literal truth. His thought here was neither clear nor consistent.

Apart from this one issue, the principal contours of Frege's conception of unjudgeable-content are plain as well as plainly important for understanding all of his writings, early and late. Amid the welter of detail, four points stand out prominently. They are the main beacons for anybody wishing to navigate the unexplored waters of *Begriffsschrift* and to avoid the sunken reefs of the *Foundations*.

(i) What logic analyzes into function and argument are *judgeable-contents,* not *declarative sentences* expressing them. Frege's decomposition of judgeable-contents cannot be simply equated with modern semantical analysis of sentences.

(ii) What everywhere shapes Frege's logic are the basic ideas and norms of *function theory*. His logical analysis of judgeable-contents is an application of a much extended and generalized concept of a function. In particular, function theory provides all of the materials for his innovations in formal logic and his analysis of count statements. (The grammatical analysis of declarative sentences plays no role in this account.) The hierarchy of types in function theory supplies the framework of his logical theory.

(iii) The relation of *designation (Bezeichnung)* is the key to unjudgeable-contents. Only what is designated is relevant to determining the cogency of an argument in which a logically perspicuous formula plays a part. Any formula representing the judgeable-content expressed by a declarative sentence conforms directly with the Augustinian picture of language.

(iv) Symbols assigned unjudgeable-contents are typically systematically ambiguous in content. They can be variously viewed as having contents located at different levels in the hierarchy of function-types. This feature is the correlate of the unfettered functional abstraction which is Frege's primary tool in the logical analysis of judgeable-contents. The type of content assigned to a subsentential expression is not uniquely determined by its syntactic features.

Most of the central ideas of content-analysis *persist* in Frege's later thought

---

26. Frege overlooked this problem in arriving at relations apparently by subtracting objects from a single judgeable-content (FA, §70).

even after the bifurcation of judgeable-content into thought and truth-value. In particular, he retained the theses that names designate objects and that concept-words designate concepts; that numbers are objects; that concepts are species of functions; that functions are differentiated into types according to the levels and numbers of their arguments; that quantifiers stand for second-level concepts. He continued to hold that logic does not analyze declarative sentences (but rather thoughts), and he continued to acknowledge the legitimacy of radically different logical decompositions (of thoughts) and the systematic ambiguity of symbols in concept-script. Attention to his conception of unjudgeable-content makes clear the fact that Frege's mature theory, the *Basic Laws* and the attendant clarificatory articles, grew organically out of his earlier ideas. A proper grasp of the general contour lines of his account of content is necessary to establish the trajectory of his later reflections on logic and language. This is a valuable source of insights—and one that has hitherto gone untapped.

8
Contextualism

1. The problems of a principle

'Only in the context of a sentence has a word a meaning.' This dictum is a familiar aspect of Frege's theorizing. With slight stylistic variations, it is repeated four times in the *Foundations,* and at each occurrence it is highlighted. There is no doubt that what it expresses is central to his early thought. But there is doubt about what exactly it does express. Evidently the proposition it formulates is closely related to what Frege explicitly and recurrently avowed to be the leading novel idea of his logic (ACN, 94; PW, 15f., 253), viz. the primacy accorded to the judgment over the concept for the purpose of the logical analysis of inference. Challenging the traditional conception of logical analysis which accorded priority to the science of terms over the investigation of judgments and inferences, he declared that unjudgeable-contents originate from the decomposition of judgeable-contents. If the contextual dictum expresses a distinct thesis at all, then what it expresses must be a complementary idea, viz. that the assignment of judgeable-content to the *sentence* takes priority over the assignment of unjudgeable-contents to *sentence-constituents*. Unless this thesis is to collapse into the primacy of judgeable-content over unjudgeable-content, it must be taken to ascribe to the sentence itself (its constituents and structure) a pivotal role in the logical analysis of what it expresses. The decomposition of a judgeable-content must depend on the articulation of the sentence used in its formulation.

A cousin of this idea is a fundamental principle informing most contemporary work in philosophical logic. The validity of an argument depends on relations among the meanings of its constituent sentences, and the task of semantic analysis is to show how the meaning of an arbitrary sentence depends systematically on its constituents and its structure. According to this conception, the sentence rather than the word has the primary role in a philosophical theory of meaning. Meanings are ascribed to words solely in virtue of their making contributions to the meanings of sentences. Since the contextual dictum of the *Foundations* seems at first sight to formulate this fundamental semantic principle, there is a strong temptation to credit Frege with its discovery. The consequence of this reasoning is that the slogan 'Only in the context of a sentence has a word a meaning' must be considered to be 'the most important philosophical statement that Frege ever made'.[1]

This interpretation is deeply flawed. Our extensive examination of his theory of content provides ample reason for denying him the dubious honour of the title 'Father of Modern Philosophical Semantics'. He was concerned with a precise determination of the cogency of inferences, especially in mathematics, but in his view this turned essentially on relations among judgeable-contents (what is asserted) and not, save indirectly, on any relations among declarative *sentences*. The arguments previously surveyed for distinguishing content from the modern conception of meaning constitute a powerful case against taking Frege's contextual dictum to formulate the basic principle of modern philosophical semantics. This verdict can be reinforced by reference to the alternative decompositions of a single judgeable-content countenanced in the *Foundations*, e.g. decomposing the content expressed by the sentence 'Jupiter has four moons' into an identity (viz. that the number of Jupiter's moons is identical to the number four), and the canvassed possibility of carving up the content expressed by the sentence 'The line *a* is parallel to the line *b*' as an identity of directions (FA, §§57, 64). These instances show that decompositions of a judgeable-content are in no way restricted by, or grounded in, the articulations visible in the declarative sentence actually used to formulate this content. There is here no dependence of content-analysis on sentence-structure. On the contrary, Frege licensed decompositions of a judgeable-content that bear no intelligible relation to the articulation of a given sentence that expresses this content. This possibility is precisely what the modern conception of semantic analysis excludes. The conflict on this fundamental issue can be resolved only by declaring his judgments about content-identity to be blunders and his discussion of carving up judgments of one-one correspondence to yield identities between numbers to be unintelligible.

If the contextual dictum does not formulate the modern principle of the primacy of the sentence in semantic analysis, then it is urgent to establish exactly what idea it does express, and to clarify how, if at all, this differs from the primacy of judgeable-content in logical analysis of inference. This is a sizeable exegetical task. To start with, we must clarify the intended scope of the contextual dictum. The primary aim of the *Foundations* is the logical analysis of arithmetic. Since simple truths of arithmetic are typically expressed by the use of numerals in equations, the dictum must surely be intended to cover such expressions. Since numerical equations, on Frege's view, actually belong to concept-script, the dictum must apply to sentences (formulae) of concept-script. This thought is obviously reinforced by the fact that the programme of the *Foundations* is evidently built upon

---

1. Dummett, 'Nominalism' p. 38 in *Truth and Other Enigmas*.

the theory of conceptual-content propounded in *Begriffsschrift*, even though the exposition is informal and independent of concept-script.[2] This already presents a formidable array of problems associated with the application of the dictum to concept-script. How does the principle (or principles) expressed by the dictum cohere with the account of content-analysis which informs the construction of concept-script and Frege's system of logic?

First, the primacy of judgeable-content is associated with the principle that properties, relations, and concepts have no existence independently of judgeable-contents (PW, 17; PMC, 101). How does the contextual dictum mesh with this thesis? Does it give any reason to draw a basic (and sharp) distinction between names of objects and concept-words? Does the occurrence of the dictum in the *Foundations* cheek by jowl with the insistence on an absolute distinction between concept and object mark a substantial development in the doctrine of content presented in *Begriffsschrift*?

Second, it is natural to interpret the contextual dictum as an authorization to assign unjudgeable-contents to any expressions that make essential contributions to the judgeable-content expressed by a given formula in concept-script (cf. FA, §60). But knowledge that a formula '$\Phi(A)$' expresses a particular judgeable-content together with knowledge that the content of 'A' is a specific entity which is the argument of a function designated by 'Φ' does not suffice to fix the content of 'Φ' uniquely. How can the content of any expression viewed as a function-name be recovered from the decomposition of any judgeable-content of which it is a constituent?

Third, a corollary of the claim that an expression has a content only in the context of a sentence expressing a judgeable-content seems to be that any explanation or stipulation of the content of a sentence-constituent must refer explicity to some *sentence* in which it occurs, and must involve the derivation of its unjudgeable-content from the established content of this sentence. But how is this compatible with the explicit licence given (although never exploited) in *Begriffsschrift* §8 for explicit formal definitions which do not involve sentential paraphrase? And how does it square with the explicit definitions of numerals endorsed in the *Foundations* itself (FA, §§74ff.)? Do such definitions not ascribe contents to expressions in complete independence of any particular sentences in which they contribute to the formulation of judgeable-contents?

Putting concept-script aside, it is equally clear that Frege intended his contextual dictum to apply to declarative sentences and words of natural language, and invoked it in criticizing his adversaries' misguided analyses of conceptual-contents expressed in natural language. This in turn opens up a new range of problems demanding elucidation.

2. Indeed, the correspondence with Marty and Stumpf (PMC, 99ff., 171f.) suggests that Frege initially thought to execute the programme of the *Foundations* and the *Basic Laws* in a single volume, and that the absence of explicit reliance in the *Foundations* upon the formal apparatus of concept-script stemmed from his following Stumpf's advice 'to explain your line of thought first in ordinary language and then—perhaps separately on another occasion or in the very same book—in conceptual notation' (PMC, 172).

First, the contextual dictum is indubitably linked to the doctrine of the primacy of judgeable-content in logical analysis. The primacy of judgeable-content is indissociable from function/argument decomposition. But while the articulation of formulae in concept-script into function-names and argument-expressions is transparent, this is not true of sentences of natural language! What argument supports the claim that grammatical predicates in atomic sentences are function-names? By repeatedly invoking the allegedly absolute distinction between concept and object, a distinction manifestly to be interpreted in terms of function and argument, Frege applied his function-theoretic apparatus to natural language more explicitly, extensively, and boldly than in *Begriffsschrift*. But does he make out any cogent case for this? And what would remain of the idea formulated by his contextual dictum if we were to subtract from it any reference to function/argument analysis?

Second, the conception of judgeable-contents as abstract objects and the framework of function theory together make it unintelligible that there should be any limitations on the different logical types of function as the value of which for different arguments a given judgeable-content can be presented. This galaxy of possible decompositions is integral to the idea of function/argument decomposition. But how can this be married with any proposal that the structure of a given *sentence* formulating a particular judgeable-content should play an essential role in determining the logical decompositions of the content expressed? Would this not be as obviously absurd as the suggestion that the structure of a particular symbol '$x^2 \rceil_2$' designating the number 4 determines the range of possibilities for specifying this number as the value of a function for an argument? Is there any intelligible way to harmonize function-theoretic analysis or judgeable-contents with any conception of the primacy of the *sentence* in logic?

Third, any straightforward application of the contextual dictum to expressions in natural languages seems impossible. Does it imply that words have no meaning when not employed in sentences? But how does this mesh with the fact that we do employ and understand words not incorporated in sentences (e.g., the use of name-plates, greetings, number-plates, signposts)? If words have no meanings outside sentences, then it would seem that we could not explain their meanings except by showing how to derive them from the meanings of sentences in which they occur. But can we not ask what a word means independently of citing a sentential context? And do dictionaries not standardly give acceptable answers to such questions? Is the application of the contextual dictum to natural language not at best counterintuitive, at worst trivially false?

These puzzles by no means exhaust the exegetical problems presented by the use of the contextual dictum in the *Foundations*. Three further issues loom large.

First, careful inspection of the occurrences of the dictum in the book reveals that it does not express a single invariant thesis, that it is not used for only one purpose, and that some of its applications seem unsound. In one occurrence it appears to underpin the methodological principle 'never to ask for the meaning of a word in isolation' (FA, x). But by constructing explicit formal definitions of '0', '1', '2', ... (FA, §§74, 77), did Frege himself not violate this principle by giving the meanings of these numerals in isolation? Elsewhere, a crucial step in his analysis of numbers seems fallacious precisely at the point at which the contextual dictum is invoked:

> *Only* in a sentence have the words really a meaning. ... It is *enough* if the sentence taken as a whole has a sense; it is this that confers on its parts also their content. (FA, §60; our italics)

Is this transition not a blatant *non-sequitur?* Is a necessary condition for a word to have meaning or content not here transformed without explanation into a sufficient condition? No less puzzlingly, Frege seems to have employed the dictum to justify giving a contextual definition of 'the number which belongs to the concept F' (FA, §62). Yet he proceeded to declare this a failure (for seemingly scholastic reasons), ultimately replacing it with an explicit definition (FA, §68). Is the *raison d'etre* of the contextual dictum in the overall strategy of the *Foundations* only to license a fruitless detour in the reasoning? Should it not then be dismissed as fulfilling no essential role in the architectonic of the book?

Second, given that the *Foundations* is a crucial stage in the execution of Frege's logicist programme, and given that the contextual dictum is held to express a cardinal principle of the book, what precisely is its role in the logicist programme?

Third, it is remarkable that the contextual dictum is never explicitly restated in writings subsequent to the *Foundations*. It is not even alluded to in what Frege saw as the definitive synthesis of his thought, the *Basic Laws*. This generates an apparent dilemma in the interpretation of the whole corpus of his work. On the one hand, the disappearance of a principle so prominent and apparently fundamental to the *Foundations* must constitute a *prima facie* case for ascribing to him a radical change of view. This would be as dramatic a change as his dropping one of his other two remaining 'fundamental principles', viz. the distinction between concept and object, or between the logical and psychological (FA, x). On the other hand, he was perfectly explicit in acknowledging fundamental alterations to his framework of thought. Hence his failure openly to repudiate the principle expressed by the contextual dictum and his failure to criticize it constitute a strong *prima facie* case against ascribing to him such a change of view. This case is strengthened by the affinity between the idea formulated by the contextual dictum and the principle of the primacy of judgeable-content in logical analysis, since this second principle persisted (in the guise of the primacy of the thought (PW, 253)) in his mature system. Here we reach an impasse. Should we conclude that the contextual dictum expresses ideas that only appear to disappear from his thought? Are they indeed reasserted, but disguised under some new verbal clothing? Or is there an unacknowledged, perhaps even unrecognized, repudiation of what this dictum expressed in the *Foundations*? If so, what is the evidence for such discontinuity, and with what changes in his thinking was this correlated? If the principle is as fundamental as it seems to be in the *Foundations,* and if the *Basic Laws* builds on the results of the *Foundations,* then its repudiation should reverberate throughout his whole system. Is it plausible that he failed to notice or avow this? Or should we, absurdly, treat the whole dilemma as evidence that he attached no real importance to the contextual dictum even in the *Foundations?* A correct perspective upon the contextual dictum, the principles it expresses, and its later fate is clearly crucial for understanding the articulations of his thought, and essential for a proper grasp of its development.

Finally, Frege's dictum has proceeded to enjoy a life of its own, independent of his thought, especially of his doctrine of content and his concept-script. This historical fact calls for detailed examination of the interpretations put on his dictum by others in order to clarify the relations among these different conceptions and to pinpoint their resemblances and differences with the ideas which he used the dictum to formulate. Wittgenstein invoked the dictum in the *Tractatus* as expressing a thesis fundamental to the picture theory of meaning.[3] Later he cited the dictum in the course of a far-reaching criticism of the *Tractatus:*[4] he construed it then as expressing a central component of the radically distinct conception of meaning embodied in the slogan 'The meaning of a word is its use in the language'.[5] Some contemporary philosophers have also invoked Frege's dictum, though working in a direction antithetical to Wittgenstein's later conception of meaning, i.e. towards the construction of 'a theory of meaning for a natural language'; and even they have seen this Proteus in different forms. Is it plausible that any definite idea forming an integral part of Frege's doctrine of content should be found acceptable to a number of philosophers who can agree with each other on almost nothing? Can it be maintained that the contextual dictum in the *Tractatus* and the *Philosophical Investigations* expresses one and the same fundamental principle which these diametrically opposed works have in common?[6] Should we not rather conclude that a philosophical dictum has a meaning only in the context of a philosophical system?

Our aim in this chapter is not to clarify and evaluate the whole rake's progress of the contextual dictum. Rather, it is confined to disentangling the various ideas that cluster, more or less closely, around the dictum in Frege's writings prior to the 1890s, and to evaluating them as constituent parts of his doctrine of content. Subsequently we will trace the evolution of this cluster of ideas through his later thought.

2. The contextual dictum in the *Foundations*

Our investigation of Frege's contextual dictum, must start from a close examination of its occurrence in the *Foundations*. For only there did he state it, at least in a completely general form, and only there did he assign it an explicit role in

---

3. Wittgenstein, *Tractatus Logico-Philosophicus,* 3.3.
4. Wittgenstein, *Philosophical Investigations,* §49.
5. Ibid. §43.
6. For a detailed discussion of Wittgenstein's position, see Baker and Hacker, *Wittgenstein: Understanding and Meaning,* pp. 264ff.

argument. In the order of investigation the *Foundations* must be the starting point. But this methodological policy does not render irrelevant the examination of his earlier writings, nor does it rule out the possibility that the contextual dictum is firmly rooted in theses elaborated in *Begriffsschrift* and in other supporting elucidations of his concept-script. In the analytic, rational reconstruction of his thought, the *Foundations* need be neither the beginning nor the end of the story about the contextual dictum.

We now turn to examine the argumentative context of the four occurrences of the dictum in the *Foundations:*

(i) In its first occurrence, the dictum expresses one of the three fundamental principles of the book: 'Never to ask for the meaning of a word in isolation, but only in the context of a sentence' (FA, x). No *independent* rationale is given for this heuristic principle. It is, however, explicitly linked with the previously explained principle 'always to separate sharply the psychological from the logical'. For if we do ask for the meanings of words in isolation we are 'almost forced to take as the meaning of words mental pictures or acts of the individual mind'. Why this is so Frege did not explain here. He merely presented the maxim as a prophylaxis for a psychologistic disease.

(ii) In its second occurrence, the dictum recapitulates the heuristic principle and associates failure to heed it with the errors of psychologism.

> That we can form no idea of its content is . . . no reason for denying all meaning to a word. . . . We are indeed only imposed on by the opposite view because we will, when asking for the meaning of a word, consider it in isolation, which leads us to accept an idea as the meaning. Accordingly, any word for which we can find no corresponding mental picture appears to have no content. (FA, §60)

The only novelty here is the explicit rationale for the heuristic principle which explains its connection with antipsychologism. Ideas, Frege emphasized, are irrelevant to the validity of arguments:

> Time and time again we are led by our thought beyond the scope of our imagination, *without thereby forfeiting the support we need for our inferences.* (FA, 60; our italics)

This is cited as the reason for the claim that the ability to form an idea of its content is not a necessary condition for a word to have a meaning. By implication, the heuristic principle is grounded in the general principle (familiar from *Begriffsschrift*) that the meaning or *content* of a word is limited to its contribution to determination of the validity of inferences that may be drawn from the judgeable-contents in the expression of which it occurs. This general principle of content seems to support both the heuristic maxim and the antipsychologistic one equally directly; it constitutes the link Frege saw between the prohibition on asking for the meaning of a word in isolation and the demand to separate sharply the psychological from the logical.

However, the contextual dictum, in its second occurrence, has two further distinct roles in his chain of reasoning. First, it is used to state a thesis which we shall call 'the Restrictive Condition':

> ... we ought always to keep before our eyes a complete sentence. *Only* in a sentence have the words really a meaning. (FA, §60, our italics)

This Restrictive Condition is much more radical than the heuristic principle. The latter appears, in isolation, to be a methodological guideline which, if followed, will prevent certain kinds of error. The Restrictive Condition, however, denies that there is any such thing as a word's having a meaning or content except in the context of a sentence. Consequently the heuristic maxim obviously follows from the Restrictive Condition. If there is no such thing as a word's having a meaning except in the context of a sentence, then to ask for the meaning of a word outside the context of a sentence is not merely to run the risk of falling into error, it is already to have done so. At first sight, the Restrictive Condition seems absurdly strong. It is obvious that words have a meaning outside sentential contexts, that they have significant non-sentential uses and that their meanings can be explained outside such contexts, as is done in dictionaries. The absurdity is diminished, however, as soon as we relate this invocation of the contextual dictum to Frege's doctrine of content rather than to our ordinary intuitive use of 'meaning'. *Content* is limited to what is relevant to the validity of inferences. For, given that any chain of argument must be expressed by sentences, we can immediately derive from this general characterization of content the Restrictive Condition, namely that a word has a content (meaning) *only* in a sentence, since it is only in a sentence that it contributes to the determination of valid inferences.

The final role of the contextual dictum in this, its second, occurrence is to state a doctrine that we shall dub 'the Sufficiency Principle':

> That we can form no idea of its content is ... no reason for denying all meaning to a word. ... It is *enough* if the sentence taken as a whole has a sense, it is this that confers on its parts also their content. (FA, §60; our italics)

Here the dictum apparently affirms that the meanings of sentence-constituents are in principle derivative from the senses of sentences.[7] In the logical analysis of inference and inferential powers, the content of sentences (the vehicles of judg-

---

7. This interpretation coheres with the apparently close connection of the contextual dictum with the principle of the primacy of judgeable-content over unjudgeable-content, since that principle clearly implies that unjudgeable-contents are in principle derivative from judgeable-contents. On the other hand, the quoted passage may be given a weaker reading. According to this interpretation, it is not *essential* that content be ascribed to a word by reference to the content of a sentence in which it occurs; rather, it is *sufficient* for it to be assigned a content that it make a definite contribution to the judgeable-content expressed by a declarative sentence. This would render the Sufficiency Principle parallel to Frege's conception of the role of mental images in inference. He did not declare images to be uniformly worthless (cf. KS, 1ff.); instead, he claimed that their absence does not render inference impossible (FA, §60). Similarly, the weaker reading removes the apparent inconsistency between the Sufficiency Principle and the presence of (explicit) formal definitions in concept-script and in the logical analysis of arithmetic. It would be absurd to saddle the *Foundations* with a principle that excluded the possibility of giving definitions of names of objects in the form of identities! Nonetheless, the considerations supporting

ments) are logically prior to the contents of words. It is noteworthy that this Sufficiency Principle coheres smoothly with the preceding anti-psychologistic move. That we do not associate a word with an idea as its meaning or content does not mean that we 'thereby forfeit the support we need for our inferences'. In the sentence 'The number of visible stars is 0', we have no idea corresponding to the numeral '0' (FA, §58), but this does not deprive the sentence of meaning or content, nor rob it of a role in correct inferences or calculations, e.g. in licensing the conclusion that the number of visible stars is less than 10^{10} (FA, §59).

Three further features of the Sufficiency Principle are noteworthy. First, it is a principle of content-*analysis*. The possibility of making non-trivial applications of it presupposes that it is possible to apprehend the sense of a sentence prior to assigning meanings to its constituents. Otherwise no one could ever employ the sense of a sentence to determine the meaning of its constituents, and hence the sense of a sentence could never *confer* meanings on its parts. Second, the Sufficiency Principle is equivocal in one important respect. It might be interpreted to imply that each distinct sentence-constituent must have a distinct meaning conferred on it by the sense of the sentence in which it occurs. Alternatively, and more plausibly, the principle might be construed hypothetically: if a sentence-constituent has a meaning at all in the context of a given sentence, then its meaning there will be exhausted by its contribution to the sense or content of that particular sentence. This leaves open the possibility of sentence-constituents without meanings or content. Some might lack content in any sentential context, e.g. the copula (PW, 63f., 90f.). Others might possess content in one sentence, but lack it in another; e.g. the concept-word 'satellite of the earth' does not have a content in the sentence 'the satellite of the earth has no atmosphere' (since when prefixed with the definite article 'it ceases to count as a concept-word' (FA, §51)), and it is unclear what content, if any, Frege meant to assign to number-words used as adjectives (e.g. 'four' in the sentence 'Jupiter has four moons' (FA, §57)). Finally, he called attention to the wide range of applicability of the Sufficiency Principle. It would, he remarked, clarify many difficult concepts, e.g. that of the infinitesimal. Frege employed it himself to explain the content of imaginary numbers (FA, §104), infinity (FA, §84), and the notion of point at infinity in projective geometry (KS, 1). Moreover, 'its scope is not restricted to mathematics either' (FA, §60).

(iii) In its third occurrence, the contextual dictum is the cornerstone of his strategy for giving a constructive account of the meanings of number-words:

> How, then, are numbers to be given to us, if we cannot have any ideas or intuitions of them? It is only in the context of a sentence that words have any meaning. So our problem boils down to explaining the sense of a sentence in which a number-word occurs.... When we have ... acquired a means of arriving at a

---

the weaker reading of the Sufficiency Condition also seem to count against the doctrine of the primacy of judgeable-content, which Frege openly embraced, and therefore they do not constitute an overwhelming objection to ascribing to him the strong interpretation of the Sufficiency Condition.

determinate number and of recognizing it again as the same, we can assign it a number-word as its proper name. (FA, §62).

This passage prefaces Frege's examination (and subsequent rejection) of the claim that one-one correspondence provides an adequate content-analysis of numerals (FA, §§63–67). The contextual dictum is here invoked to justify something that looks very like what we now call 'contextual definition'.

Although the exact analysis of the argument is a matter of some uncertainty and controversy, there are three clearly distinguishable respects in which the contextual dictum is obviously involved. Two are already familiar, but the third has hitherto gone unmentioned. First, Frege's reasoning demands the Restrictive Condition. This is needed to justify limiting attention exclusively to *sentences* containing numerals when we attempt to specify the meanings of number-words. Second, the argument invokes the Sufficiency Principle too. This is required to justify treating the specification of the sense of a given sentence containing a number-word as constituting all that is necessary for assigning meaning to the number-word itself as it appears in that sentence. Only the conjunction of the Restrictive Condition and the Sufficiency Principle validates the reduction of the problem of determining the meanings of number-words to the problem of explaining the sense of a sentence in which a number-word occurs.

The third gloss on the contextual dictum is necessary to connect the conclusion of the argument with the Sufficiency Principle. The problem to be solved is to explain the sense of a sentence in which a number-word occurs. This is to be accomplished by picking out for each number-word a determinate object which can be assigned to it as its content. Since a number-word has independently been identified as a proper name, this procedure constitutes a schematic solution to the problem of assigning it a content only on the assumption that to specify which object a proper name stands for is sufficient for assigning a content to this name. This assumption is readily equated with the thesis that the content of a proper name *is* the object it designates. Frege's argument *here* does not commit him to this position; but it does commit him to a particular interpretation of the Sufficiency Principle as applied to proper names. The contribution of a proper name to the content of a sentence in which it occurs is exhausted by its standing for a determinate object. This thesis is not a trivial consequence of the abstract Sufficiency Principle, but a substantial and independent claim.

The third employment of the contextual dictum is remarkable in two respects. First, it explains why Frege contrasted the problem of explaining the meaning of number-words with the problem of explaining expressions that stand for entities of which we have ideas or intuitions (FA, §62, cf. §§104, 106). In the case of objects that are so given (e.g. a particular geometrical point on the circumferences of a specific circle given directly in perception (BS, §8)), we can assign meanings to their names simply by associating the proper names with the given objects (via our ideas or intuitions). The difficulty with proper names of numbers, however, is that numbers are not 'given' to us at all in perception or intuition. Hence, we cannot assign meaning or content to numerals *in this way* (cf. BLA, ii, §147). Second, the idea that correlating a name with an object suffices to assign this name a meaning naturally suggests that a name so defined has a meaning independently of its sentential occurrence. Frege, however, denied this consequence. Though he insisted that numbers are self-subsistent objects, he added

> The self-subsistence which I am claiming for number is not to be taken to mean that a number-word signifies something when removed from the context of a sentence.... (FA, §60)

He saw no conflict or inconsistency here.

(iv) In its final occurrence, the contextual dictum appears as a prominent part of his recapitulation of his earlier constructive account of the meaning of number-words:

> Number ... emerged as an object that can be recognized again, although not as a physical or even a merely spatial object, nor yet as one of which we can form a picture by means of our imagination. We next laid down the fundamental principle that we should not explain the meaning of a word in isolation, but only as it is used in the context of a sentence: only by adhering to this can we, as I believe, avoid a physical view of number without slipping into a psychological view of it.... The problem, therefore, was now this: to fix the sense of a numerical identity.... (FA, §106)

This passage contains no surprises. The contextual dictum is restated here in the form of the original heuristic maxim. But it must also be interpreted as encapsulating both the Restrictive Condition and the Sufficiency Principle if it is to authorize the reduction of the problem of explaining the meaning of number-words to the problem of fixing the sense of a numerical identity. Finally, it must be treated as incorporating Frege's conception of what it is to assign meaning to a proper name. For this gives the rationale for the claim that the Sufficiency Principle alone offers an escape from the dilemma of choosing between a physical and psychological view of number. Although assigning meaning to a number-word must take the form of correlating this proper name with a definite object, the possibility of such a correlation does not entail that the object is *given* to us. If this entailment did hold, then number-words would have meanings only if numbers were given to us in perception as external objects (the physical view) or in introspection as ideas, intuitions, or acts of the mind (the psychological view). Independently of this argument, Frege's claim would be left quite unsupported.

The upshot of this careful scrutiny of the *Foundations* is clear evidence for the claim that the contextual dictum as employed by Frege is a tool with a number of quite distinct functions. It is not so used that it invariably expresses a single thought. Instead, it often formulates more than one thought in a given occurrence as well as different thoughts in different contexts of argument. We have discerned at least three distinct thoughts expressed by the dictum: a heuristic maxim, the Restrictive Condition, and the Sufficiency Principle. In addition, it is associated with a particular conception of what it is to assign meaning to a proper name and with a general conception that ties the meaning of a word in a given sentence to the impact of this word on the validity of inferences incorporating the judgeable-content of that sentence. Frege evidently thought that all these ideas were intimately interwoven. Yet the *Foundations* fails to provide an explicit framework for relating them to each other. This suggests that a definitive account of those aspects of his thinking must extend well beyond the close study of the passages actually incorporating the contextual dictum. We must take a wider look around in search for the foundations of the *Foundations*.

3. Beneath the *Foundations*—*Begriffsschrift*

A direct approach to tracing the ancestry of the contextual principles of the *Foundations* would concentrate on a search for precursors of the contextual dictum in the text of *Begriffsschrift*. This immediately yields some promising results. There are two obvious echoes of the contextual dictum, though the precise interpretation of each of them raises considerable difficulties.

The first echo occurs at the very beginning of the main text. Frege divided the symbols of concept-script into two kinds, 'those which one can take to signify various things' *(unter denen man sich Verschiedenes vorstellen kann)* and 'those which have a completely fixed sense' *(die einen ganz bestimmten Sinn haben)* (BS, §1). In making this distinction, he generalized the distinction between variables and constants familiar from algebra and real analysis. Like variables in mathematics, the letters of concept-script represent something indeterminate *(unbestimmt)*, and they are principally used for the expression of generality. Nonetheless,

> no matter how indeterminate the meaning of a letter, we must insist that *throughout a given context* the letter retain the *meaning (Bedeutung)* once given to it. (BS, §1, our italics)

This latter observation is not a model for precision or clarity of exposition. What counts as a single context? And how is a symbol which is essentially indeterminate in meaning to be given a meaning which it may retain throughout a given context? Are there indeterminate meanings, parallel to completely fixed senses, which can be assigned to letters? Does the principle require that the same 'indeterminate meaning' be assigned to the same letter uniformly in a context and different 'indeterminate meanings' to different letters in a single context? Frege's practice in using letters in concept-script makes clear what the point is that he strove to articulate. His logical derivations employ free variables, e.g. representing judgeable-contents (BS, §15). Their cogency depends upon the understanding that any substitution of constants for free variables in an instantiation of a pattern of reasoning should be carried out uniformly throughout the entire derivation. So a single context must include a whole argument, *a fortiori*, any sentences (or formula) in which a letter occurs. And the 'retention of an indeterminate meaning' through a context might better be expressed as the requirement that replacement of a letter by a symbol with determinate meaning be carried out uniformly. Frege's explanation of the use of letters seems to have been muddied by the notion that variables have indeterminate meanings whereas constants have determinate meanings.

The fact that this first echo of the contextual dictum occurs in the context of a discussion of variables is of the utmost importance. It explains three separate matters that otherwise seem unfathomable. First, it suggests a peculiar appropriateness of the contextual dictum to sentences incorporating free variables. A constant has the same meaning in whatever formulae it may appear. What would be more natural than to contrast variables with constants by claiming that a variable has a meaning only in the context of a sentence (formula)? *Begriffsschrift* subscribes to this principle (infra, p. 207ff.), and it has a close cousin in Frege's later writings.[8] Second, linking the contextual dictum with variables accounts indirectly for the association between the doctrine of the primacy of judgeable-content and the thesis that properties (or concepts) and relations have no existence independently of judgeable-contents (PW, 17, PMC, 101). It explains the apparent asymmetry between individuals and concepts. The missing link here is the idea that a free variable occurs in an expression standing for a concept. The judgeable-content expressed by '$2^4 = 16$' may be decomposed in one way into a number (designated by the constant '2') and a concept (designated by the expression '$x^4 = 16$'); the sign for the concept '4th root of 16' is '$x^4 = 16$', and the sign for the concept 'logarithm of 16 to the base 2' is '$2^x = 16$' (PW, 16f.).[9] If a variable has a meaning only in the context of a sentence, then so too does any complex expression containing a free variable, and therefore it follows that the expression for a concept differs from names of individuals precisely in the respect of having a meaning only in the context of a sentence! This appears to be a reflection in the symbolism of the point summarized by claiming that concepts, unlike objects, do not exist on their own (PMC, 101). Finally any link between the contextual dictum and the explanation in *Begriffsschrift* of the constant/variable distinction would help to explain the absence of the dictum from Frege's later writings. He explicity repudiated his early explanation of variables; he castigated the error of speaking of 'terms whose meaning is indeterminate' or 'signs [which] have variable meanings', and he preferred instead to describe these signs as having no denotations (presumably *'Bedeutungen'*) at all (PMC, 181n.). To the extent that the contextual dictum has a rationale derived from his early reflection on variables, it too must be deemed to be defective. Moreover, his later account of variables generates a tension with one of his applications of the contextual dictum. It is obvious that free variables used to express generality do contribute to what a formula asserts (e.g. that '$m + n = n + m$' and '$m + n = m + n$' differ in content). Consequently, if content is conferred on sentence-constituents in virtue of their contribution to the expression of a judgeable-content (FA, §60), then free variables must be considered to have content *(Bedeutung)* in the context of a formula stating a generalization. This conforms to Frege's original view (BS, §1): variables have mean-

---

8. Letters in mathematics are claimed to be completely different in nature from numerals. 'They are not at all intended to designate [anything]; . . . rather, they are intended only to indicate so as to lend generality of content to the sentences in which they occur. Thus it is only in the context of a sentence that they have a certain task to fulfill, that they contribute to the expression of the thought. But outside this context they say nothing' (FG, 67). This principle seems to be the resultant of applying the distinction between reference *(Bedeutung)* and indication *(Andeutung)* to the putative principle that a variable has a meaning only in the context of a sentence.
9. This has an exact parallel in Russell's notation for propositional functions.

ings, albeit indeterminate ones, in particular contexts. But it conflicts with his later denial of sense and reference *(Bedeutung)* to free variables: their function is not to refer *(bedeuten)*, but to indicate *(andeuten)*. This doctrine eliminates the possibility of arguing that an expression that makes an essential contribution to the expression of a judgeable-content *eo ipso* is assigned an unjudgeable-content, although that argument is fundamental to the strategy of applying the contextual dictum to assign content to numerals (FA, §§60, 62). This awkwardness would be exacerbated by any reaffirmation of the contextual dictum after the emergence of the thesis that variables indicate but do not make reference.

The positive account of letters in concept-script (variables) which accompanies the first echo of the contextual dictum in *Begriffsschrift* has an importance for the interpretation of Frege's dictum. First, it explains why he linked the dictum with a criticism of the traditional doctrine that a judgment consists in a combination of two concepts independently given as the result of abstraction, in particular why he regarded the dictum as an effective weapon against psychologism. If asked to explain the judgment expressed by a mathematical formula containing free variables, a philosopher cannot proceed by specifying what concepts or ideas these variables stand for, since it is obvious that they stand for no definite ideas at all (they are signs *'unter denen man sich Verschiedenes vorstellen kann'* (BS, §1)). If it were claimed that the variables in '$m + n = n + m$', used to state that addition of signed integers is commutative, stood for the concept of a signed integer, then it would be obvious that this assignment of a concept to the variables would be correct only relative to this particular formula (and it would be opaque how to explain the difference in the roles of the two variables). In either case the strategy sanctioned by traditional logic and by philosophers who construct judgments out of pre-existing ideas correlated with words is frustrated. Generalizations expressed by formulae with free variables seem to administer a mortal blow to the synthetic conception of judgments. They can be used to demonstrate that not all words have meanings independently of the sentential context in which they occur. Of variables Frege thought it to be obviously true that they have meanings only in the context of particular sentences.

Second, his supposition that free variables do have (indeterminate) meanings or contents tends to support an assimilation of variables to ambiguous expressions.[10] The contrast between determinate and indeterminate meaning is often

---

10. This idea may now seem incredible, but this was not always so. Indeed, it was fundamental to Russell's explanation of the notion of a propositional function. In his view, 'a variable is ambiguous in its denotation' (Russell and Whitehead, *Principia Mathematics to *56*, p. 4). He then defined a propositional function ϕx as 'a statement containing a variable x and such that it becomes a proposition when x is given any fixed determined meaning', adding that ϕx is 'not a proposition, since owing to the ambiguity of x it really makes no assertion at all' (ibid. p. 14). He harped on this matter: a propositional function 'differs from a proposition solely by the fact that it is ambiguous: it contains a variable . . .' (ibid. p. 38). Hence 'the essential characteristic of a function is *ambiguity*. . . . ϕx *ambiguously denotes* ϕa, ϕb, ϕc, etc., . . . [i.e.] "ϕx" means one of the objects ϕa, ϕb, ϕc, etc., though not a definite one, but an undetermined one' (ibid. p. 39). Even the basic principle of the theory of types, viz. that a function cannot be its own argument, rests on the contention that 'a function . . . is a mere ambiguity awaiting determination'

invoked in discussions of lexical ambiguity. A word such as 'light' or 'bank', or a phrase such as 'on the wagon', may well be said to have no (definite) meaning on its own, although with respect to a particular utterance of such an expression in a sentence there will usually be no difficulty in assigning it a (definite) meaning in that context. Similarly, many prepositions seem ambiguous, displaying 'not so much a number of meanings as a body of meaning continuous in several directions',[11] but typically there is no problem in explaining what a preposition (or a phrase incorporating it) means in the context of a particular uttered sentence. In any such case, we might plausibly reply to a request for an explanation of the meaning of a word or phrase, 'It has no definite meaning in isolation; *only* in the context of a sentence has it a definite meaning'. Because Frege held that a free variable has indeterminate content, he might have been inclined to adopt these tactics and affirm 'Only in the context of a sentence (formula) has a variable a content'. And this might be subsumed under the generalization 'Only in the context of a sentence has a word a meaning'. By this subtle alchemy, a slogan applicable to ambiguous expressions is transformed into a disinfectant allegedly effective against incursions of psychology into logic.

The second obvious echo of the contextual dictum in *Begriffsschrift* occurs in a discussion of generality.

> If we compare the two sentences 'The number 20 can be represented as the sum of four squares' and 'Every positive integer can be represented as the sum of four squares', it seems to be possible to regard 'being presentable as the sum of four squares' as a function that in one case has the argument 'the number 20' and in the other 'every positive integer'. We see that this view is mistaken if we observe that 'the number 20' and 'every positive integer' are not concepts of the same rank. What is asserted of the number 20 cannot be asserted in the same sense of 'every positive integer'. . . . The expression 'every positive integer' does not, as does 'the number 20', by itself yield an independent idea *(selbständige Vorstellung)* but acquires a sense (Sinn) *only from the context of the sentence.* (BS, §9; our italics)

This passage, too, is not a model of precision or clarity. It confuses mention and use, apparently conflates concept and object (' . . . can be presented as . . .' and 'being presentable as . . .'), and uncritically employs the term 'idea' with all its potentialities for confusion. More importantly, the intended scope of the dictum is unclear; it can hardly express a wholly general thesis since the phrase 'the number 20' is implied to have a meaning independent of the context of sentences in which it occurs. What is asserted about the phrase 'every positive integer' is also opaque.

These mysteries can be unravelled. The key is to consider the representation of this pair of judgeable-contents in concept-script. The statement about the number

---

(ibid. p. 48). Frege eventually discerned some of the muddles in these ideas (PMC, 81ff., 180n.), even though he began from a very similar conception of variables.

11. This is the position adopted by F. Waismann, 'Language Strata', in A.G.N. Flew, *Logic and Language,* Second Series (Blackwell, Oxford, 1961), p. 16.

20 ascribes a definite property to a particular object, and hence in concept-script it will be represented by a configuration of constants similar to '⊢──── $2^4 = 16$' (PW, 18). The other statement is a generalization; since at this stage in *Begriffsschrift* the only available notation for expressing generality is free variables, this statement must be represented in concept-script in the form appropriate for expressing subordination of two definite concepts, i.e. in a formula similar to

$$\vdash\begin{array}{l} x^4 = 16 \\ x^2 = 4 \end{array}$$

(PW, 18). Consequently, the similarity in the grammatical form of the sentence 'The number 20 can be represented as the sum of four squares' and 'Every positive integer can be represented as the sum of four squares' masks a vital logical distinction. 'It seems that logicians have clung too much to the linguistic schema of subject and predicate, which surely contains what are logically quite different relations' (PMC, 100f.). In particular, because the same sentence form is used for both cases, they fail to distinguish 'the case of one concept being subordinate to another from that of a thing falling under a concept' (PW, 18). These cases are clearly distinct in representations in concept-script, and this proves that, if we follow the traditional practice of analyzing judgments as relations of concepts, the concepts corresponding to 'the number 20' and 'every positive integer' cannot be construed as concepts of the same rank of which a single property is predicated.

The connection of this passage with the constant/variable distinction in concept-script is directly supported by the continuation:

> For us, the different ways in which the same conceptual content can be considered as a function of this or that argument have no importance as long as function and argument are completely determinate. But if the argument becomes *indeterminate,* as in the judgment: 'Whatever arbitrary positive integer we take as argument for "being representable as the sum of four squares", the [resulting] sentence is always true.', then the distinction between function and argument acquires a *substantive (inhaltlich)* significance. (BS, §9).

The contrast between the determinate and the indeterminate is the contrast between constants and variables in concept-script (BS, §1). On pain of *non sequitur,* Frege must be taken here to have argued that the argument in the judgeable-content expressed by 'Every positive integer is the sum of four squares' is indeterminate, i.e. that it is expressed by a free variable in concept-script. Of this variable, then, it is true that it yields no independent image *('selbstandige Vorstellung')* since it belongs among signs that one can take to signify various things *('unter denen man sich Verschiedenes vorstellen kann')*. By contrast, the numeral '20' does have a completely fixed sense *('ein ganz bestimmte Sinn')* and therefore does not apparently acquire a sense *(Sinn)* only in the context of a sentence. Hence the context of the second echo of the contextual dictum itself echoes the context of the first echo in *Begriffsschrift*. This reveals that a comment seemingly targeted on the expression 'every positive integer' is in fact directed at the variable used in concept-script to express a generalization about positive integers. Frege tried to expose 'an illusion to which the use of [ordinary] language gives rise', but he carried out this project by making an observation about the phrase 'every positive integer' which manifestly fails to characterize it correctly but which clearly fits his conception of letters in concept-script!

This interpretation, however, raises the spectre of a radical asymmetry in the account of unjudgeable-contents. While variables and concept-words are manifestly assigned a content only in a sentential-context, did Frege here argue that some singular referring expressions ('proper names'), e.g. 'the number 20', have a content ('a completely fixed sense') independently of their occurrence in a sentence? And if so, what exactly did he mean by having a content independent of the sentential-context in which they occur?

On one construal, an expression has a content independent of its sentential-context if it has the same content in every sentence which expresses a judgeable-content, i.e. if it is free of content-ambiguity. But this feature cannot be employed to draw a general distinction between proper names and concept-words, since some referring expressions (e.g. 'I' and 'the present population of Berlin') are context-dependent for their content and some concept-words (e.g. 'is a prime number') are not. Moreover, Frege held that *every* name is ambiguous in content since its content in an identity-statement differs from its content in all other sentential-contexts (BS, §8). On pain of serious inconsistencies, absolute uniformity of content is not a characteristic of names at all.

On another construal, an expression has a content independent of the context of a sentence if it has content whether or not it constitutes part of a sentence, e.g. if it retains its content when it is used as a label, when it appears on a signpost or a nameplate, or when it is uttered as an expletive. But the possibility of such extra-sentential uses does not provide any ground for distinguishing names from concept-words. Nor does it offer any reason to assign *content* to any subsentential expression so used. For content is characterized *ab initio* as being 'only that part of judgments which affects ... possible inferences' (BS, §3). At best a *potential* content could be assigned to an expression independently of whether it occurs in a declarative sentence, but this potential content is actualized only in the context of a sentence used to formulate a judgment.[12]

On yet another construal, the contrast between expressions correlated with independent ideas and those having meaning only in the context of a sentence makes a metaphysical claim. It seems clear enough, from the fact that concepts are species of functions, that they lack self-subsistence or independent existence, having an ontological status parasitic on objects. For it would be absurd to conceive of a law of correlation of entities as existing 'independently' of the entities which it correlates. It seems natural to consider the category of objects to be prior to the category of concepts since it provides the materials for constructing the explanations of concepts of various levels. On the other hand, it is altogether opaque how this metaphysical thesis bears on the issue of an asymmetry of the

---

12. According to this conception, an expression would not have a content when it occurs as part of a sentence-question, an imperative, a WH-question, or an optative.

contents of proper names and concept-words in respect of dependence on sentential-context.[13] Of course, an object which is the content of an expression which occurs in a particular sentence and which is identified as a proper name possesses whatever self-subsistence it may have quite independently of its being the content of any name, but this truism carries no implications about the circumstances under which an expression is correctly held to have this self-subsistent object as its content. By the time of writing the *Foundations,* Frege was careful to *deny* that the self-subsistence of numbers implies that numerals (proper names of numbers) signify numbers outside the context of a sentence (FA, §60).

The crucial point that must govern any acceptable interpretation of the remark that 'the number 20' is linked with an independent idea is Frege's insistence that logical analysis of judgeable-contents is analytic, not synthetic. The judgment is not 'formed' by the composition of independently given terms, but rather logic starts from judgments and arrives at unjudgeable-contents by splitting up judgeable-contents (ACN, 94; PW, 17; PMC, 101). Although concept-formation is stressed in this account, it is unclear by what principles the derivative status apparently assigned to all unjudgeable-contents is to be *restricted* to concepts. It seems impossible to marry this strictly analytic conception of logical analysis with the idea of assigning unjudgeable-contents to any expressions independently of their occurrences in sentences expressing judgeable-contents. Moreover, the very possibility of assigning a definite content to some names independently of their sentential-contexts presupposes that some expressions in concept-script or in natural language can unequivocally be identified as names. But since any judgeable-content in concept-script decomposes into function and argument according to how we choose to regard it, we can analyze $\Phi(A)$ as the value of Φ for the argument A or as the value of A for the argument Φ (BS, §10); or $2^4 = 16$ as the value of $x^4 = 16$ for the argument 2, or as the value of $2^x = 16$ for the argument 4, or as the value of $2^4 = x$ for the argument 16 (PW, 16f.). Therefore, none of these constituent symbols in '$\Phi(A)$' or '$2^4 = 16$' can be classified as a proper name *tout court*. A similar observation holds for sentences of natural language (cf. BS, §9). Consequently, there seems no viable way sanctioned by *Begriffsschrift* to filter out of sentences any subset of expressions the content of which is not existence-dependent on the judgeable-contents of sentences in which they occur.

The upshot of investigating this second echo of the contextual dictum is the discovery of a clear doctrine, couched in obscure prose, about the content of real variables and expressions for concepts incorporating real variables, and an obscure hint at a thesis about the content of expressions 'yielding an independent idea'.

---

13. Or any other semantic issue. Frege may have drawn some such consequences. In particular, given his metalinguistic account of identities, the only ground that he might have had for denying the legitimacy of formulating formal definitions where concept-words flank the sign '≡' is that these expressions designate nothing that exists independently of objects and hence lack content in such occurrences (cf. PW, 17). Russell drew logical consequences from parallel metaphysical claims. From the premise that 'the ϕ in ϕx is not a separable and distinguishable entity, [but] it lives in propositions of the form ϕx, and cannot survive analysis,' he concluded that a function cannot be its own argument (Russell, *The Principles of Mathematics,* p. 88).

The central doctrine is not one which appears in the *Foundations*, although it coheres with the Sufficiency Condition. But its affinities with the guiding ideas of Frege's style of logical analysis encourages further exploration of *Begriffsschrift* in the hope of finding the ancestry and unifying source of the various contextual ideas of the *Foundations*. Having exhausted the strategy of frontal assault, we must now resort to an indirect approach.

Certain contextual principles seem to emerge naturally from the fundamental ingredients of the doctrine of conceptual content elaborated in *Begriffsschrift*. On any reckoning, the following five theses are central to the book: the identification of content as what is relevant to the cogency of inferences; the primacy of judgeable-content in logical analysis; the decomposition of every judgeable-content into function and argument; the denotational conception of unjudgeable-contents; and the analysis of identities as statements about symbols. Together these theses converge on a representation of judgeable-contents in concept-script which conforms to strong contextual principles. Only in the context of a formula expressing a judgment (a sentence of concept-script) has a symbol a content, since only there has it any relevance to the cogency of inference. Moreover, what content such a symbol has is determined by its contribution to the judgeable-content of the sentence in which it occurs, since the articulation of the symbols in a sentence in concept-script exactly matches the decomposition of a judgeable-content into its constituent unjudgeable-contents. The primacy of judgeable-content implies that, relative to concept-script, the content of subsentential expressions is to be derived from the content of sentences. Hence the fundamental ideas of content-analysis lead directly to the applicability of the Restrictive Condition and the Sufficiency Condition to symbols in concept-script.

Begriffsschrift conceives of concept-script as one language among others (special only in meeting the standards of logical perspicuity). Hence it draws no distinction in principle between sentences in concept-script and sentences in natural languages. Frege described everyday words and phrases in ways that really fit only symbols in concept-script (e.g. treating the phrase 'every positive integer' as if it were a variable (BS, §9)). He was not alive to what we would regard as major logical asymmetries between sentences in concept-script and everyday instruments of assertion. Consequently he found it completely natural to transfer contextual principles from concept-script to sentences of natural languages. The basic ideas of content-analysis thus lead, via consideration of symbols in concept-script, to the generalizations that a word has a content only in the context of a sentence and that what content it has is derivative from its contribution to the expression of a judgeable-content by this sentence. The trajectory of thought in *Begriffsschrift* is a wholly unrestricted application of the Restrictive Condition and the Sufficiency Principle.

Against this general background we might briefly re-examine *Begriffsschrift* in search of manifestations of the ideas explicitly associated with the contextual dictum in the *Foundations*. The latter Heuristic Maxim is never there formulated. Yet it could be conceived to be an invisible guideline throughout the argument whose effects are everywhere visible. The analysis of generality in terms of second-level functions presupposes forms of concept-formation (for first-level concepts) not explicable in terms of the logical addition or multiplication of simple concepts each of which is derived by abstraction from perceived or intuited properties or relations. Rather, Frege's quantification theory rests on the possibility of unrestricted functional abstraction that begins from judgeable-contents (and their representation in concept-script). His logic of generality is inaccessible to somebody who always asks for the content of an expression in isolation, and therefore it depends on conformity with the directive to seek the content of an expression of generality only in the context of a sentence used to formulate a judgeable-content.

The Restrictive Condition is also an integral part of the original doctrine of content, required by Frege's analysis of identities (BS, §8). In the context of identity-statements, names do not stand for their ordinary contents but rather for themselves, and therefore any sign that can flank the identity-sign is systematically ambiguous. What such a sign has as its content in a particular sentence can be determined only by reference to the (content of the) sentence in which it occurs. Hence, only in the context of a sentence has such a sign a content.[14]

The Restrictive Condition is directly implicated in another argument in *Begriffsschrift*. Frege considered the possible objection that the sorites paradox demonstrates the unacceptability of the logical thesis that every property F hereditary in an f-sequence must be a property of y if y follows x in the f-sequence and x has the property F. For, if F is the property of being a heap of beans and f is the procedure of decreasing a heap of beans by one bean, it appears to follow from this putative logical law that a single bean or even no bean at all would be a heap of beans! His rebuttal of this argument turns on the principle that an expression, even if its content is explained by correlating it with a perceptible object or property, has a content only if it occurs in a sentence which expresses a judgeable-content. Because the content of 'heap' is indeterminate, it has borderline cases: if 'a' names such an aggregate of beans, then 'Fa' does not express a judgeable-content (BS, §27). This reasoning excludes an obvious argument drawn from the synthetic conception of judgments, namely the argument that 'Fa' *must* express a judgeable-content because 'a' has a content (since it names a perceptible object and has a content in some other sentences) and 'F' has a content (since it stands for a concept in sentences predicating the property of being a heap of non-borderline aggregates of beans).[15] Blocking this argument presupposes a strong interpretation of the Restrictive Condition; an expression has a content only if it occurs in a declarative sentence which itself has a judgeable-content.

The Sufficiency Condition is indispensable for the decomposition of sentences expressing judgeable-contents into *function-names* and argument-expressions, i.e.

---

14. This version of the Restrictive Condition is not completely general. Some signs of concept-script cannot properly be used flanking the identity-sign; this restriction appears to hold for any function-name in isolation (i.e. a function-name not accompanied by the designation or indication of appropriate arguments (PW, 17)).
15. Frege later authorized such arguments by the assimilation of vagueness of concept-words to reference-failure for names. According to that view, either *every* formula of the form 'Fx' has a sense and a reference or else *none* has.

for assigning functions to certain expressions as their unjudgeable-contents. Together with the notion of a function, Frege took over from mathematics the common conception of what is required to assign significance to a function-name (what it is to 'define a function'). The proper procedure is simply to specify the value of the function for every admissible argument, i.e. to specify what entity is designated by the complex expression formed by completing the function-name with the name of any admissible argument. No additional manoeuvre is necessary; in particular, there is no question of having to correlate the defined function-names with any independently given entity. Applied to sentences expressing judgeable-contents, this conception licenses the possibility of assigning content to a subsentential expression independently of whether it can be directly correlated with any acknowledged property or relation (PW, 17). Specifying a law correlating entities of a particular type with judgeable-contents is sufficient to confer content on an expression provided it is treated as a function-name. Names of first-level concepts (e.g. 'logarithm', 'prime number') and of higher-order concepts (e.g. quantifiers, 'continuous function') can be assigned content by this principle. It is enough to grasp the content of a concept-word to know how it contributes to the judgeable-content of any significant sentence in which it occurs, i.e. to know what judgeable-content is expressed when its argument-place is filled by the name of an admissible argument.[16]

This central application of the Sufficiency Condition in *Begriffsschrift* has several noteworthy features. First, it is restricted to expressions identified as concept-words; it is not used to assign content to expressions taken to be names of objects. Second, it enshrines a partial liberation from psychologism; it recognizes modes of concept-formation not dependent on logical addition and multiplication of concepts each of which is abstracted from something given in perception or intuition. In allowing concept-formation to proceed from judgeable-contents, Frege may have considered functional abstraction to be a form of abstraction, but, if so, it is abstraction on materials not given at all in perception or intuition. Third, the Sufficiency Condition does not license assigning content to every expression used in formulating a judgeable-content. Concept-script contains various expressions without content, namely the judgment-stroke, the content-stroke, and bound variables. Sentences in natural languages have a larger variety of content-less expressions, e.g. the copula and pronouns serving the role of bound variables in concept-script. Finally, although a judgeable-content can typically be decomposed in many different ways and expressed in many different sentences, there is no guarantee that any particular function/argument decomposition can be matched to the expressions in a particular sentence formulating a given judgeable-content.

---

16. Note that this conception of defining functions does not imply that the content of a function-name can be recovered from the judgeable-content expressed by a single sentence in which the given function-name appears, but only from the totality of judgeable contents expressed by the totality of significant sentences in which it occurs. Hence *Begriffsschrift* sows the seed from which Frege later harvested the principle of completeness of definition for concept-words (BLA, ii. §56).

This holds even in concept-script. The pair of formulae '⊢——— $\Phi(x)$' and '⊢—$\overset{a}{}$— $\Phi(a)$' expresses the same content, but, whereas the second must be construed as a second-level function whose argument is the function Φ (BS, §11), the first must be so construed that 'Φ' names a function whose argument is indeterminate (BS, §9). Consequently, each of these formulae mirrors one articulation of the same judgeable-content, but neither of them contain symbols that may be assigned unjudgeable-content in accordance with the logical decomposition transparent in the other one. The Sufficiency Condition does not license apportioning unjudgeable-contents in an automatic way to subsentential expressions occurring in a particular formulation of a given judgeable-content.

This survey makes clear that the various ideas linked with the contextual dictum in the *Foundations* grow out of seeds planted in *Begriffsschrift*. What unifies these superficially diverse ideas is the doctrine of content. They are welded together by the primacy of judgeable-content and by the function/argument decomposition of judgeable-content in logical analysis. That these connections are less emphatic in the *Foundations* than earlier is the result of Frege's shift of attention from concept-words to number-words, which he held to be proper names of objects. The strategy introduced in *Begriffsschrift* to account for problematic concept-words (especially quantifiers) is now exploited to account for problematic names (viz. number-words). The *Foundations* also manifests an unequivocal application of the principles of content-analysis to assign content to sentences and subsentential expressions of natural language. The novelty in tactics, however, cannot mask the continuity in strategy.

4. Back to the *Foundations*

It is now evident that the ideas associated with the contextual dictum in the *Foundations* are not a random scatter of theses, but the most important members of a cluster of ideas bound together by Frege's doctrine of content. In the light of this substantial unity we shall refer to this cluster of ideas as 'contextualism'. This term, like 'idealism', 'psychologism', or 'verificationism', does not designate a sharply demarcated cluster of individually perspicuous ideas, but that does not derogate from its usefulness. A cluster can survive as an identifiable entity despite fuzzy boundaries, gradual gains or losses in its membership, and qualitative changes in some of its members. Just such continuities through change typify Frege's contextualism.

The contextual dictum is used in the *Foundations* to formulate a minimum of three distinct principles: the Heuristic Principle, the Restrictive Condition, and the Sufficiency Principle. It was further associated with a form of antipsychologism and employed as the key to the 'epistemological' problem of assigning content to names of objects not given in perception or intuition. These ideas together belong to the indisputable core of the contextualism of the *Foundations*. But they may not exhaust this core. Further proposals for membership have been tabled. The central idea is often alleged to be the insight that a sentence is 'what "has meaning" in the primary sense',[17] that 'the sentence is the unit of meaning',[18] or that only by the utterance of a sentence can a speaker make 'a move in the language-game'.[19] Alternatively the guiding idea is claimed to be that sentences take primacy over words in an account of what it is for expressions to have meanings.[20] More specifically, Frege purportedly subscribed to the principle that the meaning of any subsentential expression is to be understood as 'consisting in the contribution which it makes to determining the sense [i.e. the truth-conditions] of any sentence in which it may occur'.[21] The purpose of highlighting these doctrines in the *Foundations* is claimed to be the legitimation of the procedure of 'contextual definition', which he applied in defining number-words in terms of equinumerosity of concepts (FA, §§62ff.).[22] Alternatively, the purpose is to authorise the definition of the concept-word 'number' in terms of the equivalence relation of equinumerosity in the domain of concepts.[23]

What these proposals have in common is a desire to assign to Frege a major share in the development of modern philosophical semantics. In particular, some aspects of his contextualism are supposed most clearly to manifest both the modernity and the permanent value of his philosophical speculations. Our previous discussions of conceptual content should inculcate a general scepticism about this line of interpretation. But a more definitive verdict on the significance of his contextualism must rest on a more detailed analysis of the argument in the *Foundations* where the contextual dictum and associated contextualist ideas play an obviously pivotal role. Grand claims about the significance of his contextualism may depend as much on a misunderstanding of his argument as they do on charitable impulses.

The *Foundations* aims to establish two theses beyond doubt. First, count-statements (sentences of the form 'There are n F's') assert properties of concepts, not of objects or aggregates of objects; the content of such a sentence is therefore the value of a second-level concept for a first-level concept as argument. Second, number-words (both numerals and names of the form 'The number belonging to the concept F') are proper names;[24] the content of such an expression is therefore the object designated by it. The main constructive task in the *Foundations* is to yoke these two theses together into a unified account. The chief problem here is to explain how to conjure objects out of second-level concepts. This was a lasting preoccupation:

---

17. J. L. Austin, *Philosophical Papers* (Oxford, 1961), p. 24.
18. W. V. Quine, 'Two Dogmas of Empiricism', *Phil. Review*, Vol. 60, 1951; cf. Dummett, *Frege: Philosophy of Language*, p. 3.
19. Cf. Wittgenstein, *Philosophical Investigations*, §49.
20. Dummett, *Frege: Philosophy of Language*, p. 4; cf. p. 193f.
21. Ibid. p. 4 (glossed by reference to p. 5); cf. p. 194.
22. Ibid. p. 495ff.; also *Truth and Other Enigmas*, pp. 95, 108.
23. Cf. W. Hodges, *Logic* p. 187ff.; cf. Bell, p. 146ff.
24. At least in *some* occurrences (e.g. in '2 > 1') according to *one* way of viewing the judgeable-content expressed (since numerals can also be viewed as function-names in every such formula (cf. BS, §9)).

> Since a statement of number based on counting contains an assertion about a concept, in a logically perfect language a sentence used to make such a statement must contain two parts, first a sign for the concept about which the statement is made, and secondly a sign for a second-level concept. These second-level concepts form a series.... But still we do not have in them the numbers of arithmetic; we do not have objects, but concepts. How can we get from these concepts to the numbers of arithmetic in a way that cannot be faulted? (PW, 256f.)

Contextualism is meant to be the agent for effecting this conceptual alchemy and to whatever extent it is intertwined with the justification of Frege's definitions of number-words as extensions, it must be a *permanent* part of his thought even if later neither stressed nor even mentioned.

The central argument of the *Foundations* is deeply perplexing. The contextual dictum is invoked to vindicate already accepted definitions of number-words in terms of one-one correspondence—definitions which we would now regard as 'contextual definitions'. Frege then rejected this procedure for rather opaque reasons and substituted definitions of number-words in terms of extensions. Apparently contextualism is introduced only to license something immediately dismissed as a defective procedure. How can something allegedly so important play so pitiful a role?

This quandary suggests that we should make a fresh start.

> How, then, are numbers to be given to us, if we cannot have any ideas or intuitions of them? Since it is only in the context of a sentence that words have any meaning, our problem becomes this: to define the sense of a sentence in which a number-word occurs.... [W]e have already settled that number-words are to be understood as standing for self-subsistent objects. And that is enough to give us a class of sentences which must have a sense, namely those which express our recognition of a number as the same again. If we are to use the symbol a to signify an object, we must have a criterion for deciding in all cases whether b is the same as a.... In our present case, we have to define the sense of the sentence
>
> "the number which belongs to the concept F is the same as that which belongs to the concept G";
>
> i.e., we must reproduce the content of this sentence in other terms.... In doing this, we shall be giving a general criterion for the identity of numbers. When we have thus acquired a means of arriving at a determinate number and of recognizing it again as the same, we can assign it a number-word as its proper name. (FA, §62)

In this passage, the contextual dictum appears to introduce some major innovations. A search is launched for a sentence containing a number-word which can be used to confer a content on the number-word itself (cf. FA, §60). Frege then proposed making use of identity-statements, thereby suggesting that such sentences are logically on a par with other sentences incorporating number-words. (This departs from his previous analysis of identity-statements (BS, §8).) He then broached the novel general thesis that assignment of meaning to a proper name presupposes provision of a criterion of identity for the object named. (This departs from his earlier dismissal of the relevance of the way of determining an object to the content of a name (BS, §8), and it seems to adumbrate the distinction of sense and reference for names.) Finally, the proposed method for defining the name 'the number which belongs to the concept F' actually *exploits* the possibility that two sentences differing in grammatical form may express a single judgeable-content. Apparently Frege envisaged expressing the judgment that the concept F is equinumerous with the concept G by the sentence 'F's are in one-to-one correspondence with G's', then carving up this judgeable-content in a fresh way (cf. FA, §64), and finally apportioning out this content to the three logical constituents of the sentence 'The number which belongs to the concept F is identical with the number which belongs to the concept G'. (This departs from any precedent. *Begriffsschrift* did not clearly exploit the possibility of radically asymmetrical alternative analyses of judgeable-contents in this way.) Should we conclude that Frege here moved off in fundamentally new directions without giving any notification of a change of tack?

A less paradoxical account results from interpreting this crucial passage as looking backwards to *Begriffsschrift* rather than as prefiguring 'Sense and Reference'. In general contours the strategy pursued in the *Foundations* resembles a line of argument shaping Frege's earliest publication, his dissertation of 1873. This work opens with the formulation of a general problem: although the validity of the axioms of geometry is held to rest on the deliverances of intuition, the science of geometry contains imaginary structures whose properties not only cannot be traced to intuitions but also may even contradict our intuitions. As instances he cited infinitely distant structures (so called 'points, lines, planes at infinity' in projective geometry).

> Literally understood the expression 'infinitely distant point' is a *contradictio in adjecto;* for such a point would have to be the end point of a distance which has no end. The expression is therefore an improper one and stands for the fact that parallel lines stand to each other in projective relations just as do lines which intersect in the same point. 'Infinitely distant point' is therefore only another expression for what is common to all parallel lines, which we call direction in other contexts. Just as a line is determined by two points, so too is it given by a point and a direction. This is just one example of the general principle that generally, where projective relations are in question, a direction can take the place of a point. (KS, 1)

The justification for the use of the phrase 'infinitely distant point' in projective geometry is that this simplifies the formulation of geometrical laws by amalgamating principles that would otherwise require separate statements; it promotes *Übersehbarkeit*. Frege's procedure involves exploitation of contextualist ideas for the analysis of projective geometry, though no such ideas are explicitly stated. From the fact that the expression 'infinitely distant point' is apparently self-contradictory we cannot deduce that it has no legitimate place in geometry. Instead, we must investigate how it is used in the formulation of geometrical laws. Provided that it has a uniform and intelligible role, it has a content in the context of geometrical propositions and arguments. Inspection reveals that this condition is indeed met. Every sentence incorporating the phrase 'infinitely distant point' can be rephrased using the term 'direction' without thereby altering what is expressed by the sentence. The sentence 'Lines a and b intersect in an infinitely distant point' can, for example, be reformulated 'Lines a and b have the same direction' (or 'The direction of line a is the same as that of line b'). Frege here employed pairs of sentences identical in content for the very same purpose for which he later considered using the pair of sentences 'The number belonging to the concept F is identical with the number belonging to the concept G' and 'F's are in one-to-one correspondence with G's' in the *Foundations*. Equivalence in content justifies the ascription of content to one expression ('infinitely distant point') occurring in one sentence of the pair but not the other. The (known) content of a sentence is thus used to confer content on one of its constituents (cf. FA, §60). The application of the Sufficiency Principle shines through this earlier procedure, undimmed by the facts that the *definiendum* is an 'improper expression' and that it does not occur specifically in an identity-statement.

The second major aspect of the argumentative strategy of the *Foundations* makes its appearance at the opening of Frege's second publication, the dissertation of 1874. Although some specific concepts of magnitude (e.g. length, area, volume) are grounded in intuition, this is not true of all concepts of magnitude (e.g. angle), still less of the general concept of magnitude itself. ' ... [W]hat *magnitude* is in itself, what length and area have in common, evades intuition. . . . It is clear that so comprehensive and abstract a concept as that of magnitude cannot be an intuition' (KS, 50). How then can a general science of pure magnitude be developed independently of intuitions about particular species of magnitudes? The crucial matter is 'a determination of the concept of identity of magnitude *(Grossengleichheit)*':

> If we can judge in every case when objects agree in a property, then we clearly have a proper concept of the property. Hence could we specify under what circumstances identity of magnitude is realized, we would thereby determine the concept of magnitude. A particular magnitude, say length, is accordingly a property in which a set of things, independently of their ordering, can agree with a particular thing of the same kind. (KS, 51)

Frege then mentioned that it would take him too far afield to demonstrate 'how the content of arithmetic is contained in the established properties of magnitude, and how from this standpoint [of the general science] particular kinds of magnitude, such as natural number *(Anzahl)* and angle, can also be defined' (KS, 51). Instead he elaborated an account of how to apply magnitude to the analysis of iterated operations.

Three features of this introductory argument are noteworthy. First, the quarry is determination of a concept (here magnitude in general) which is not given in intuition (cf. FA, §62). Second, the key to the clarification of such a concept is the specification of identity-conditions (here the conditions for identity of magnitude). Third, Frege indicated that the same method of definition applies to particular concepts of magnitude not derived from intuition (in particular number and angle). The *Foundations* attempts to carry out an identical analysis of the concept of number *(Zahl)* by laying down the conditions of numerical identity (*'ein allgemeines Kennzeichen für die Gleichheit von Zahlen'* (FA, §62)). The aim of concept determination is crystal clear in the counterpart discussion of directions of lines:

> The judgment "line *a* is parallel to line *b*" ... can be taken as an identity. If we do this, we obtain *the concept of direction,* and say "the direction of line *a* is identical with the direction of line *b*".... We carve up the content in a way different from the original way, and this yields us *a new concept.* (FA, §64; our italics)

This is the procedure familiar to modern logicians as the introduction of a concept defined by reference to an equivalence relation. The only novelty of the *Foundations* is the repudiation of this definition of the concept of number because it is *logically* defective, whereas previously Frege had clearly regarded it with approval (KS, 51). There is no reason to equate this criticism with a blanket rejection of appeals to the Sufficiency Principle in ascribing content to sentence-components of a sentence with known judgeable-content. The 'logical doubts and difficulties' about defining numerical identity in terms of one-one correlation might well be specific, not wholly general.

The third ingredient here in the *Foundations* is the analysis of identity-statements developed in *Begriffsschrift*. Although the content of a name is typically the object named, in the context of an identity-statement its content is the symbol itself; or rather, the identity '⊢——— A = B' states that two names 'A' and 'B' designate a single object (BS, §8). If the content of 'A' in the identity 'A = B' were the object A, then the identity, if true, would not require a proof, but would exemplify the logical axiom 'A = A'; the fact that identities make assertions about symbols makes possible non-trivial proofs of identities in cases where two different names flanking the identity-sign are associated with different ways of determining an object (*Bestimmungsweisen* (BS, §8)). This account is actually repeated in the *Foundations* with the implication that it is generally applicable to numerical identities. In the sentence 'The number of Jupiter's moons is four' the word 'is' is claimed to have the sense of 'is identical with'.

> So that what we have is an identity, stating that the expression "the number of Jupiter's moons" designates the same object as the word "four". And identities are, of all forms of sentence, the most typical in arithmetic. (FA, §57)

This metalinguistic account of identity-statements is also implicated in the reasoning of the *Foundations* (even if it is not unwaveringly adhered to). It meshes with the citation of Leibniz's Law as a definition of identity (FA, §65). It is also required to explain away an apparent inconsistency. Frege explicity sought 'a general criterion for the identity of numbers'. Provided we had thus acquired 'a means of arriving at a determinate number and of recognizing it again as the same', we could assign it a number-word as its proper name (FA, §62). This argument seems to imply that a means of determining an object and of recognizing it again must be associated with any name that is properly assigned a meaning. This might suggest the thesis that different such means would correspond to differences in meanings of names. On the other hand, in advocating his own definitions, he made plain that it is a matter of indifference what means of determining an object and of recognizing it again is associated with the number-word 'zero' (FA, §74, cf. §§74n., 75, 53), and he later changed his preferred means of determination without explicit comment (BLA, §41). The explanation of this apparent oddity is straightforward, though unrecognized. The strategy of using an equivalence relation (one-one correspondence) to define the concept of number demands specifying the conditions of identity of numbers. But numerical identities in mathematics typically need the support of non-trivial proofs (cf. FA, §6). This is true in particular of names of numbers taking the form 'the number which belongs to the concept F', and therefore any definitions of number-words not making possible rigorous proofs of numerical *identities* is scandalously unsatisfactory (FA, §56). According to *Begriffsschrift*, such numerical identities express relations among names, and their proof turns on showing that the different ways of determining numbers associated with each name of a pair do in fact yield one and the same number (BS, §8). Hence numerical *identities* have definite contents (proof-conditions) only if each number-word is associated with a definite means of determination *(Bestimmungsweise)*, which may serve also as a means of recognizing something as the same again. On the other hand, means of determining an object have nothing to do with the content of sentences other than identities which incorporate a name with content (supra, p. 168f.). The reasoning here about the content of number-words treads precariously along the tightrope of the 'bifurcation in the content of names' brought about by identity-statements. It thus relies on the weak version of the Restrictive Condition, viz. that what content a name has depends on the sentential context in which it occurs.

From Frege's standpoint, the rationale of the putative definition of 'the number which belongs to the concept F' is clear and consistent. The justification for the definition is built on contextual ideas and actually appeals to the contextual dictum. It combines arguments prominent in his earlier writings. The heart of the procedure is the exploitation of the possibility of alternative decompositions of a single judgeable-content. A given content is carved up in two radically different ways; one decomposition is correlated with the constituents of one (and only one) of a pair of sentences expressing this content, the other decomposition with the constituents of the other (and only the other). Thus the content of the sentence 'F's are one-one correlated with G's' can allegedly be carved up either into a second-level relation between the concepts F and G or into a first-level relation (identity) between objects (numbers);[25] the second decomposition cannot be correlated with the constituents of the original sentence but only with the constituents of the

---

25. Provided that we ignore the official doctrine that identities state relations among *symbols*.

sentence 'The number belonging to the concept F is identical with the number belonging to the concept G', and conversely the first decomposition cannot be correlated with constituents of the latter sentence but only with those of the original one. This strategy could be viewed as a direct result of the Sufficiency Condition. That licenses conferring content on the name 'the number belonging to the concept F' because the sentence in which it occurs has a known content which decomposes into two objects and a first-level relation. Yet absence of any such name from the sentence 'F's are one-one correlated with G's' means that there can be no question of assigning these objects as unjudgeable-contents to the expressions in this sentence even though the two sentences express the same judgeable-content. Given an inclination to identify judgeable-content with sentence-meaning, we will undoubtedly find this proposal unintelligible. There can be no such thing as asymmetrical semantic analyses of a pair of sentences that have the same meaning. But from Frege's perspective, there is no problem even worthy of discussion. Any judgeable-content, being a Platonic object admissible as the value of a function, must in principle be capable of being the value of indefinitely many different functions of any logical type (i.e. having any number of arguments of any level in the hierarchy of concepts). Hence it is obvious that there are many different function/argument decompositions of a given judgeable-content which will have no structural resemblance whatever to a given decomposition (or to its linguistic expression). The possibility of asymmetrical decompositions of the content expressed by a given sentence is no more surprising than the possibility of specifying the number designated by '$x^2 \big]_2$' as the value of the second-level function $\frac{d}{dx}(\ )_{x=1}$ for the function x^4 as argument. Why should this way of presenting 4 as the value of a function for an argument be reflected in the structure of constituents of the designator '$x^2 \big]_2$'? In both arithmetic and logic what is decomposed into function and argument is a Platonic entity, and the limitless variety of function/argument decompositions guarantees that it is a mere accident if any particular decomposition is reflected in the structure (and constituents) of a particular expression designating the given entity. Consequently, the possibility of asymmetrical content analyses of sentences expressing a single judgeable-content is absolutely integral to Frege's conception of function/argument decomposition of judgeable-contents. He could not have denied this possibility without renouncing a fundamental principle of function theory in the course of generalizing the mathematical concept of a function to logic.

This whole investigation yields a powerful case for Frege's *accepting* the proposed definition of 'the number belonging to the concept F'. The general method of definition seems unexceptionable in his view, and it yields a result enjoying widespread acceptance among contemporary mathematicians (FA, §63). Indeed, not only did Schröder, Kossak, and Cantor accept this definition of 'number' (FA,

§63n), but also Frege himself had earlier done so (KS, 51). Finally, the definition is applicable not only to number-words in natural languages but also to number-words in mathematical concept-script. The Humean definitions have a formal analogue: viz. $(F(\rho) \breve{\rho} G(\rho)) = (N_r F_r = N_r G_r)$.[26] Hence they supply the materials for a formalization of arithmetic. How then can we account for his *rejecting* definitions that he apparently should have accepted? For, of course, he did reject the Humean definitions of number-words; these definitions raise 'at once certain logical doubts and difficulties' (FA, §63), and the subsequent argument sustains at least one of them (FA, §§63–7). 'Seeing that we cannot by these methods obtain . . . any satisfactory concept of number . . . , let us try another way' (FA, §68). With this preamble he introduced his own definition: 'the number which belongs to this concept F is the extension of the concept "equinumerous with the concept F"' (FA, §68). What justified this manoeuvre? Did he suddenly turn his back on contextualism? If so, why? Whatever the answer may be, it cannot be that abandonment of contextualism depends on the assimilation of sentences to names of truth-values![27]

We might speculate that the 'logical doubts and difficulties' which so impressed Frege were the very ones that now weigh with us. Two in particular would reduce his definitional procedure to incoherence. First, the sentence 'F's are one-one correlated with G's' makes an assertion arguably about concepts or classes, but certainly not about symbols; it belongs to the object language. Frege would agree. But, in his view, the identity 'The number of F's is the same as the number of G's' makes an assertion about symbols, not about numbers, concepts, or classes; it belongs to the meta-language.[28] This excludes without more ado the possibility that these two sentences should have the same judgeable-content. Devastating though we judge this objection to be, Frege did not produce it. In none of his early work did he face the consequences of his formalist account of identity, even for the cogency of formal inferences (cf. BS, §20f.). Even though he officially held the identity 'the number which belongs to the concept F is identical with the number which belongs to the concept G' to make an assertion about symbols, he did not regard this fact as invalidating the inference of this identity from the premise 'F's are one-one correlated with G's'! (FA, §73). A second charge of incoherence against the Humean definitions of number-words arises from the essential asymmetry in the content-analysis of the pair of formulae '$F(\rho)\breve{\rho}G(\rho)$' and '$N_r F_r = N_r G_r$' which putatively express the same judgeable-content. If what a sentence expresses (a proposition) can be assigned a structure only on the basis of the structure of the sentence expressing it, the very idea that a proposition might have a decomposition that could be assigned to only one of a pair of sentences each of which expresses this proposition in absurd. Hence the proposal to carve up what

---

26. Frege evidently explored building an analysis of arithmetic on this foundation (cf. N 47 in Scholz's catalogue on the Nachlass (printed in M. Schirn (ed.). *Studien zu Frege* I, p. 85ff.)).
27. Cf. Dummett, *Frege*, pp. 7, 196, 495.
28. There are, of course, indications of wavering about this analysis of identities; cf. especially FA, §67.

the sentence 'F's are one-one correlated with G's' expresses as a relation between objects would be unintelligible. But this too was not Frege's reason for repudiating the Humean equivalence. Far from advancing this objection, he thought that possible decompositions of a judgeable-content are in no way dependent on the structure of any particular sentence used to express this content.

The difficulties raised in our minds by Frege's exposition of the *prima facie* case for Humean definitions of number-words belong to the very framework of his thought in the *Foundations*. Neither the metalinguistic analysis of identities nor the essential independence of the analysis of judgeable-content from sentence-structure is something that he was in a position to call into question. This has the consequence that our reaction to the Humean definitions differs profoundly from his, with the attendant danger that we misconstrue both his general stance and his specific objections. Many modern logicians think that the strategy of using one-one correspondence to generate 'contextual definitions' of number-words is perfectly acceptable, and hence they embrace the Humean definitions canvassed by him. They accept this product, but would criticize his conception of the process of manufacturing it. Frege, however, rejected the product, viz. the so-called 'contextual definition' of number-words, but did not criticize the general programme now deemed unsatisfactory. His quarrel with the Humean definitions is very different from ours. Hence clarification of his criticism is needed to establish whether it manifests any turning away from contextualism. We need to establish what 'logical doubts and difficulties' he considered and which ones he sustained.

First, it might be objected that a special definition of numerical identity must be logically awry. The concept of numerical identity is determined by the general concept of identity and the concept of number; hence it should not be defined afresh. Composite symbols, each constituent of which has a determinate content, cannot properly be defined at all (BLA, §33; cf. PW, 210). Frege's reply turns the objection back on itself. The general concept of identity is indeed taken to be given; the Humean definition does not constitute a special definition of identity between numbers, but rather 'a means of arriving at that which is to be regarded as identical' (FA, §63). The concept of number is not something already given, as the objection presupposes, but something as yet to be determined. The first logical difficulty is thus deflected. Frege's reasoning clearly appeals to the Sufficiency Principle. The fact that the judgeable-content of the sentence 'F's are one-one correlated with G's' is identical with that of the sentence 'The number belonging to the concept F is identical with the number belonging to the concept G' confers content on the two names in the identity-statement provided the general concept of identity is given. The content of subsentential expressions can in appropriate cases be derived by subtracting the content of one subsentential expression from the established judgeable-content expressed by an entire sentence. Contextualism thus plays an essential role in defusing the first worry. Frege further rallied the faint-hearted by the observation that this apparently 'very odd kind of definition' is not altogether unheard of. He appended examples from geometry (FA, §64); others could have been added from number theory and analysis (FA, §§60n., 104, 109).

The second objection concerns the possibility of coming into conflict with the laws of identity, in particular with Leibniz's Law (FA, §65). This passage uses the definition of direction in terms of parallelism which matches the Humean definition of number. Suppose that the sentence 'the direction of line a is identical with the direction of line b' is stipulated to have the same content as 'line a is parallel to line b'. Conformity to Leibniz's Law requires that, in any sentence containing the name 'the direction of line a', the name 'the direction of line b' can be substituted without altering the judgeable-content expressed. How can this principle of substitution be proved from the hypothesis that line a is parallel to line b?[29] Frege's reply purports to show that this demand for a proof is inappropriate: '... [W]e are taken initially to know of nothing that can be asserted about the direction of a line, except the one thing, that it coincides with the direction of some other line' (FA, §65). The putative definition covers the content of all sentences of this form. And this, allegedly, is the only requirement to be met: 'The meaning of any other type of assertion about directions would first of all have to be defined, and in defining it we can make it a rule always to see that it must remain possible to substitute for the direction of any line the direction of any line parallel to it' (FA, §65). This argument is sophistical since its force is wholly dependent on the identification of concepts with functions. Every function taking as arguments the objects putatively designated by names of the form 'Da' must be constructed by stipulating values for these objects as arguments, and its value for any particular 'direction' Da will be independent of how this 'direction' is determined. Consequently, if every concept under which a direction falls is a function, the extensionality of functions will guarantee the validity of Leibniz's Law if directions are identified with the objects denoted by expressions of the form 'Da'.

The third objection is the fence that fells the putative definition of number-words (FA, §66). The counterpart of the Sufficiency Principle is a stringent criterion of adequacy for definitions (i.e. assignments of unjudgeable-contents to subsentential expressions): a definition is adequate only if it alone suffices to assign that content to an expression which is uniformly conferred on it by the judgeable-contents expressed by all the declarative sentences in which it may occur. In other words, a defined expression must make a uniform contribution to the judgeable-contents expressed by sentences in which it appears, and the definition must by itself stipulate what this contribution is. Consequently, a definition is inadequate if it does not suffice to settle the content of a sentence known to have a content and consisting, apart from the *definiendum,* only of expressions with known contents. By this standard, Frege argued, the Humean definitions are inadequate. They provide determinate content to identities only when the signs flanking the identity-sign are both of the form 'the number belonging to the concept F'. Only in this case can the definition settle whether two ways of determining a number yield the same result, i.e. whether two signs of the form 'the number belonging to

---

29. Frege explicitly raised the parallel question about number-words in a lost item from the *Nachlass* (N40 in Scholz's catalogue).

the concept F' have the same content; for only then is the question reduced, by the definition, to the question whether F's correspond one-one to G's. In any other case, the definitions provide no method for assigning content to identities. They do not themselves settle whether the judgeable-content expressed by 'Julius Caesar is the number of the moons of Jupiter' is to be affirmed or denied; they do not furnish a procedure for proving or disproving what this sentence expresses since the road back to one-one correspondence is blocked by the fact that the name 'Julius Caesar' is not of the form 'the number belonging to the concept F'. (The issue is not the falsity of this identity, which Frege took to be obvious, but rather the possibility of *proving* its falsity *from the definitions* of number-words!) In other words, Frege contended that the proposed criterion *(Kennzeichen)* of numerical identity does not constitute a genuine way of determining an object *(Bestimmungsweise)*. For, given any two names associated with definite distinct *Bestimmungsweisen*, the question whether the two ways of determination determine the same object can always be answered in principle, even if we cannot in fact decide the issue (cf. FA, §62). The paradigm here is the possibility of comparing the value of one function for one argument with the value of another function for another argument. The Humean criterion of numerical identity falls short of associating a definite *Bestimmungsweise* with the name 'the number belonging to the concept F', since if it did so every identity including a name of this form would have a definite content in virtue of the Humean definition alone.

This main objection is buttressed by two supplementary arguments (FA, §66). The first is a reminder that the putative definitions are supposed to define not only specific number-words but also the concept-word 'number' itself. This eliminates the possibility of appealing to the concept of a number to assign content to identities incorporating names other than names of numbers; viz. by the supplementary stipulation that any sentence of the form 'q = the number belonging to the concept F' expresses a content that is to be denied if q is not a number. What we are supposed to lack is the general concept of a number. The second argument notes the circularity that would result from trying to define the concept-word 'number' by the stipulation 'q is a number = there is a concept F whose number is q' since this presupposes that definite content is already assigned to all sentences of the form 'q = the number belonging to the concept F'. Finally, Frege noted the error of contending that q is a number only if it is introduced by the Humean definition, i.e. only if the name 'q' abbreviates an expression of the form 'the number belonging to the concept F' (FA, §67). This violates the condition that how an object is determined is not a *property* of the object determined. This condition stems from the principle that an object specified as the value of a function must be identifiable independently of its being a value of this particular function. (Indeed, if the expression 'N_rF_r' is construed as designating the value of the function $N_x\phi_x$ for the argument $F\xi$, then the demand that the number designated by 'N_rF_r' be capable of being designated otherwise than as a value of $N_x\phi_x$ is just an application of this general principle of function theory.)

The upshot of Frege's discussion is clear. The Humean definitions of number-words do not yield any concept of number with sharp limits to its application (FA,

§68); they do not associate with each number-word a definite way of determining an object and recognizing it again as the same (or different!); hence they do not suffice to endow all identities with judgeable-content. Their formal translations in concept-script $(F(\rho) \;\bar{\bar{\rho}}\; G(\rho) = (N_\tau F_\tau = N_\tau G_\tau))$ do not constitute an adequate foundation for the science of arithmetic. These deficiencies are remedied by a minor modification: 'the number belonging to the concept F = *the extension* of the concept "equinumerous with the concept F"' gives an adequate definition, at least provided that the concept of an extension of a concept is assumed to have sharp limits to its application (cf. FA, §68n.). Neither the objections to the Humean definitions nor Frege's own definitions of number-words betray any criticism of contextualism. On the contrary, components of contextualism play an explicit role in the defence against the first objection and in the exposition of the third objection. There is no querying of the legitimacy of analyzing the content expressed by a sentence into unjudgeable-contents not reflected at all in the structure of the given sentence. The point is rather that this does not guarantee the adequacy of the Humean definitions. The true light of contextualism shines undimmed through the tissue of argument in the *Foundations*.

In discussing the role of contextual ideas in the central constructive argument in the *Foundations* we have deliberately refrained from characterizing the Humean definition '(the number belonging to the concept F = the number belonging to the concept G) = (F's are one-one correlated with G's)' as a *contextual definition* of the expression 'the number belonging to the concept F'. We have done this in full knowledge that the *raison d'être* of Frege's contextualism is sometimes claimed to be vindicating the propriety of contextual definitions in philosophical semantics. That thesis seems both anachronistic and pregnant with possibilities for misconstruing fundamental aspects of the *Foundations*. For numerous reasons we have decided to pass by on the other side of the road.

First, there is some initial plausibility in treating the Humean definitions as contextual definitions of number-words, for these definitions do not define names of the form 'the number belonging to the concept F' in isolation, but rather provide a method for systematically paraphrasing numerical identities into equivalent sentences containing no number-words at all. Yet the supposition that Frege's contextualism is primarily designed to vindicate contextual definitions renders the strategy of the *Foundations* unintelligible. For he sustained one objection against the Humean definitions, and then replaced them by what are manifestly to be classed as *explicit definitions* in terms of the extensions of concepts (FA, §68). Why then did he go to the length of introducing a cluster of complicated doctrines to authorize a procedure which he forthwith repudiated in favour of a method of definition hallowed by unbroken tradition?

Second, according to the modern conception, contextual definitions are essentially reductive. They identify the meanings of a pair of sentences having asymmetrical structures. Since semantic analysis presupposes that the meaning of any sentence must reflect its (real) structure, the supposition that two sentences differing in structure have the same meaning demands that at most one of the two makes its real semantic structure manifest; the other has a real structure discrepant with its apparent construction out of its constituents. Hence, the point of a contextual definition of a sentence is to demonstrate that a proper analysis of its meaning does *not* conform to its apparent grammatical structure. Standard paradigms fit this pattern. The most celebrated, Russell's theory of definite descriptions, shows that definite descriptions, contrary to appearances, do not stand for objects; rather, they are 'incomplete symbols'. A sentence of the form 'The Φ-*er* Ψ-*s*' appears to attribute a property to an object, but the proposition expressed really has a totally different form which is perspicuous in the sentence-form '$(\exists x) (\Phi x \,\&\, (\forall y) (\Phi y \rightarrow y = x) \,\&\, \Psi x)$'. If these sentence-forms are identical in meaning, then the same semantic analysis must fit both, and therefore we have to choose between the thesis that the proposition expressed is logically subject/predicate in structure (matching the surface articulation of 'The Φ-*er* Ψs') and the thesis that it is logically existential in structure (matching the surface structure of the quantified formula). To assign *both* semantic structures to the proposition expressed by 'The Φ-*er* Ψ-*s*' would be incoherent. But the allegedly incoherent parallel in the case of number-words is precisely the proposal whose intelligibility Frege never questioned and whose adequacy he explored in detail! He did not reject the Humean definitions because they would reduce judgments apparently about numbers to judgments about concepts. On his view, equivalence in content of the sentences '$F(\rho) \stackrel{\smile}{\rho} G(\rho)$' and '$N_\tau F_\tau = N_\tau G_\tau$' would not call in question the claim that number-words are proper names. There is no need whatever to choose between the decomposition of this content as a relation between concepts and its decomposition as an identity between objects. Because 'contextual definitions' are essentially reductive, they are wholly unlike anything that Frege ever contemplated.

Third, the allegation that the adoption of an explicit form of definition of number-words in the *Foundations* rests on a general repudiation of the legitimacy of contextual definitions is also without foundation. His criticisms can be generalized only to sets of definitions that purport to stipulate both the content of a concept-word ('number', 'direction') and the contents of names of objects falling under the defined concept (number-words, direction-names). They have no force at all against apparently contextual definitions of concept-words (e.g. FA, §60n.). Even in the case of number-words, he offered his explicit definitions more in the spirit of a modification that a repudiation of the Humean definitions (that he himself had earlier accepted! (KS, 51)).

> In the same way with the definitions of fractions, complex numbers and the rest [especially infinite numbers (cf. FA, §84)], everything will in the end come down to the search for a judgment-content which can be transformed into an identity whose sides precisely are the new numbers. In other words, what we must do is fix the sense of a recognition-judgment for the case of these numbers. (FA, §104, cf. §109)

Contextual definitions are here treated as the beginning of wisdom, not as signposts on the way to perdition. Once again, logical doubts force us to modify this initial insight as before, and then 'the new numbers are given to us as extensions of concepts' (FA, §104). If contextual definitions had earlier been proved worthless, this argumentative detour in the *Foundations* would be inexplicable.

Fourth, it would be absurd to cite Frege's later tirades against piecemeal and 'implicit' definitions as evidence that he made a global repudiation of contextual definitions in the *Foundations*. Not only is any such argument from the future to the past *prima facie* suspect, but also these later criticisms are not even applicable to the *Foundations*. The paradigm for piecemeal definitions is the common mathematical practice of defining the values of a function for a limited range of arguments (e.g. the operation of addition over the integers) and then redefining the function for arguments from successively larger domains (e.g. the rationals, the reals, complex numbers). The *Foundations* tolerates such procedures (FA, §99ff.). Moreover, the Humean definitions of number-words openly aim at exhaustive specification of the contents of number-words in any possible occurrence, and therefore they escape any strictures of principle against piecemeal definitions. The archetype of objectionable implicit definition is provided by Hilbert's conception of Euclidean geometry; there the axioms are treated as generalizations that jointly constitute the definitions of the primitive terms ('line', 'point', 'lies on', etc.), so whatever systems of entities verify all of the axioms can be correctly characterized as lines, points, etc. Frege advanced two main arguments against this account. First, the axioms, if treated as definitions of the primitive concept-words, are defective because they do not provide any way of determining whether an arbitrary object (e.g. a pocket-watch) does or does not fall under one of these concepts (e.g. whether it is a point (FG, 18, 31, 63ff.)). This has a counterpart in one of the objections of the *Foundations*, viz. that the Humean definitions do not associate definite *Bestimmungsweisen* with number-words (FA, §§66f.). Second, interpreting the axioms of Euclidean geometry as generalizations (about *whatever* satisfies them) demands taking the primitive terms to be expressions for second-level concepts. On pain of incoherence, the axioms cannot simultaneously be treated as definitions of the first-level primitive concepts of geometry (FG, 19, 36, 68f.). This second argument from function-theoretic type-theory Frege thought to administer the *coup de grâce* to Hilbert's conception of axioms as implicit definitions, but it has no analogue in the *Foundations* and cannot be turned against the Humean definitions.

Finally, the very distinction between 'explicit' and 'contextual' definitions depends on a distinction that Frege deliberately refrained from drawing. In his eyes, every formal definition has the form of an identity-statement (BS, §8). It stipulates that two expressions are names of the same object. This characterization fits his so-called 'explicit' definitions of number-words: he would formulate them by the identity-scheme 'the number which belongs to the concept F = the extension of the concept "equinumerous with the concept F"' (FA, §68). In his view the characterization of formal definitions is equally fitting for his so-called 'contextual' definitions. These too are expressed by identities (or identity-schemata), though in this case the objects named happen to be judgeable-contents and not what we now consider to be proper objects (PW, 35f.). Frege would have formulated his provisional definitions of directions and numbers as identities between judgeable-contents: viz. '(the direction of a = the direction of b) = $a \parallel b$' (FA, §64) and '$(N_r F_r = N_r G_r) = F(\rho) \breve{\rho} G(\rho)$' (cf. FA, §63). In his view, all formal definitions of concept-words have the form of identities between judgeable-contents because they incorporate free variables, and such definitions constitute the vast bulk of the formal stipulations of *Begriffsschrift* and the *Basic Laws* alike. In particular, any definitions of second-level concepts would have this form; e.g. the explanations '$R(\xi,\zeta)$ is reflexive = $(\forall x) Rxx$'; and '$\Phi(f) = f(2)$'. Frege did not mark any logical distinction within the class of formal definitions. From a logical point of view the class is homogeneous, whether the names linked by the identity-sign designate judgeable-contents or objects other than judgeable-contents. The need for a differentiation between 'explicit definitions' and 'contextual definitions' arises only from adopting the modern conception that sentences and names belong to distinct semantical categories. Having from the outset assimilated judgeable-content to objects, Frege saw nothing to discuss. *A fortiori*, the idea of justifying 'contextual definitions' never crossed his mind.

In one way the harvest from this further consideration of the *Foundations* seems meagre. No fresh ideas emerge that deserve a place in the cluster of ideas tied to the contextual dictum. In a different way, however, the harvest is substantial. There is considerable clarification of the ideas indisputably part of the contextualism of the *Foundations*. In particular, it is now evident that there is nothing in Frege's thought parallel to the principle of the primacy of the *sentence* in semantic analysis. The possibilities of decomposing a given judgeable-content are not dependent on the structure of any sentence used to express this content. Therefore, apart from implying a prophylaxis against psychologism, the contextual dictum expresses the very same principle as the thesis of the primacy of judgeable-content in logical analysis. Frege assigned no *direct* role to sentences and sentence-structures in the doctrine of content. It is also now clear that he did not abandon any aspect of his contextualism in the course of the *Foundations*. On the contrary, he reaffirmed its core ideas in arguing to the need for his 'explicit' definitions of number-words as extensions of concepts. He apparently regarded the definitions of the *Foundations* as a refinement of the Humean ones. He also retained the idea of a close connection between the content of '*F*'s are one-one correlated with *G*'s' and that of 'the number belonging to the concept F = the number belonging to the concept G'. The two judgeable-contents are, however, not identical; the transition from the first to the second is a substantial inferential step (formalized in the *Basic Laws* in the notorious Axiom V). The persistence of the 'explicit definitions' of number-words and the retention of the conception of this basic step of inference suggest that the thought of the *Basic Laws* is no less grounded in contextualism than the reasoning of the *Foundations*. A proper account of contextual ideas in the *Foundations* supports the expectation that contextualism is a thread unifying Frege's earlier reflections with the mature system elaborated in the *Basic Laws*.

II
THE MATURE VISION: THE DOCTRINE OF FUNCTIONS AND SENSES

9
Function and Concept

1. Propaideutic

What Frege himself called a 'thorough-going development' of his logical views (BLA, p. ix f.) separated the original exposition of his logical system in *Begriffsschrift* and the outline of a formal proof of the analyticity of arithmetic in the *Foundations* from the revised and extended system of logic and the attempted formal execution of the logicist programme in the *Basic Laws*. This evident and important reorientation of the course of his thinking is taken to justify grouping together his earlier writings as manifesting the 'early phase' of his thought and grouping together the *Basic Laws* and the associated articles of the 1890s as the expression of the 'mature phase' of his philosophical reflections. The change fundamental to the 'thorough-going development' of his logic consisted in the splitting of judgeable-content into thought and truth-value, or, more generally, in splitting conceptual content into sense and reference. Since the sense/reference distinction persisted as a basic feature of his thinking for the rest of his life, the 'mature phase' of his thought might be deemed to encompass *all* of his later writings, even if further subdivisions may usefully be drawn within this corpus. At any rate, the primary and incontestable writings of his mature phase are the two volumes of the *Basic Laws* and the celebrated articles 'Function and Concept', 'On Concept and Object', and 'On Sense and Reference'.

These mature writings form an integrated structure. The formal proof of the *Basic Laws* resembles the vault of a cathedral, the informal exposition and the contemporaneous articles (and also the *Foundations*) serve as supporting members. Though these have their own interest and contribute directly to the beauty of the edifice, there can be no doubt that their *raison d'être* is their structural role. Only the vindication of the thesis that arithmetic is a branch of logic would ensure that Frege's logical system was, as it were, an adequate manifestation of the glory of God.

Like a cathedral, this awesome intellectual creation invites different kinds of analysis. One is a clarification of the principles of engineering manifested in the construction—a delineation of the lines of force and a specification of the structural role of each distinguishable element. Does the sense/reference distinction constitute the deepest foundations of the *Basic Laws?* Or is it a buttress erected to prop up something else which was already shakily in place somewhere in the superstructure? Would removing the doctrine that sentences are proper names of truth-values purify the building of a noxious decorative feature? Or would it bring Frege's edifice down in ruins? To what extent does the *Basic Laws* derive support from the *Foundations?* And conversely, to what extent does it damage the stability of the *Foundations?* The naked eye of reason is powerless to settle these questions of technology. Only painstaking analysis of the elements of the structure and their interactions will reveal the secret of supporting a vault of stone on a wall of stained glass.

A second kind of clarification is historical. Our understanding and appreciation of a complex piece of architecture is greatly enhanced by knowledge of its past and the stages of its evolution. The present state of a cathedral may become more fully intelligible from the discovery that the choir was rebuilt atop an earlier crypt, that Gothic decoration was carved into heavy Norman piers, or that a discontinuity in the vaulting of the nave was the product of a project of rebuilding only partly executed. Frege's mature system of logic calls for parallel clarification. On the foundation of *Begriffsschrift* he erected a nearly complete structure, a book laying out a formal derivation of basic truths of arithmetic and exhibiting them as analytic truths (PMC, 99f.; BLA, p. ix). What parts, if any, of this building were incorporated in the fabric of the *Basic Laws?* Should we view the *Basic Laws* as something built on wholly new foundations? Or rather as a superstructure rebuilt on the original foundations? Or even as the superposition of new decoration on constant structural members? We know also that the *Foundations* elaborated a blue-print for a formal proof of the analyticity of arithmetic. But, since this preceded the 'thorough-going development' of Frege's logical views, we cannot assume that the *Basic Laws* was built according to that plan. How does the completed edifice relate to the earlier design? To what extent does the *Foundations* illuminate the 'logic of the building'? Frege's mature thought, like a typical English cathedral, visibly has a history of decisive importance for understanding its structure. Careful investigations in this dimension are necessary to explain mysterious juxtapositions of elements and apparent stylistic incongruities.

These two kinds of architectural analysis are obviously not independent of each other. Absurdity results from producing an historical rationale for a feature whose presence is necessitated by principles of technology, and conversely from manufacturing a scientific explanation of a feature whose existence is a relic of some past structure now mostly demolished. The interdependence of the two means that historical ignorance can have catastrophic consequences on the appreciation of a building. There are many exact parallels in the interpretation of the writings of Frege's mature phase. Indeed, general knowledge of his work is limited to his mature theory, and many examinations of his ideas, sometimes deliberately, sometimes without premeditation, simply begin from the three famous articles of the 1890s. The result is a caricature of every aspect of his thought. If the fundamental insight underlying his invention of quantification theory were an appreciation of how sentences of natural languages expressing generality are constructed in a sequence of stages from atomic sentences, then was the formalization of the logic of generality in *Begriffsschrift* openly constructed without any foundations at all and did it hang in the air for a dozen years until Frege built some rickety supports in 'Function and Concept'? Did sheer ignorance of the possibilities of distinguishing connotation from denotation, and the content of a concept from its extension, explain the failure to incorporate something along the lines of the later sense/reference distinction into the original exposition of his formal system? The *Foundations* appears to clarify the logical significance of number-words, both in mathematical symbolism and in sentences of natural languages, and Frege seemed to judge that it gives adequate philosophical support to the purely logical definitions of names of natural numbers in the *Basic Laws*. Yet this must seem a mystery if the sense/reference distinction is the key to his logical analysis of language. The arguments of the *Foundations* rest on the undifferentiated notion of content. Their cogency, it seems, must depend on the fact that the *Foundations* was already groping towards the sense/reference distinction—or even on the fact that the contextual dictum and the demand for criteria of identity for objects embody substantial progress towards the mature system, so that the *Foundations* must be acknowledged to be not 'fully intelligible' without recourse to the sense/reference distinction. Even Frege's mature thought appears incongruous. The *Basic Laws* is clearly intended to be the definitive exposition of his views on logic and arithmetic. But from its exposition of his revised logical system, nobody would be able to reconstruct the contents of 'On Sense and Reference' or 'On Concept and Object', and the remarks on sentences or phrases in natural languages in 'Function and Concept'. To the extent that Frege held the semantic analysis of language to provide the foundations of logic, and to the extent that a clear conception of the sense/reference distinction is the key to semantics, his own presentation of the logical system of the *Basic Laws* must be adjudged radically defective—hardly worthy of what he meant to be his *chef d'œuvre*. The wisdom of hindsight does nothing to illuminate the earlier stages in the evolution of a cathedral, while ignorance of them may distort our perception even of what is present before our very eyes.

The preceding analysis of the leading ideas in the early phase of Frege's philosophical reflections on logic should make possible a deeper and better informed appreciation of his mature writings. It would be ill-advised to conclude that he meant to demolish the foundations of his earlier work in absence of weighty evidence for such a radical reconstruction of his thinking. And it would be rash to argue that his ideas moved off in a fundamentally new direction if he did not avow any such intention and if his mature thought can be interpreted as a smooth continuation of evident and openly acknowledged aspects of his earlier work. By adhering to these principles, we shall discover that the remarkably linear evolution of Frege's thinking is even more linear than commonly thought.

In so far as this thesis is established, the distinction of his work into an early phase and a mature one seems inapposite. But in any case we might well cavil at the label 'mature' for the writings of the 1890s. At the very least it is question-begging to characterize *Begriffsschrift* and the *Foundations* as 'not yet mature', or 'immature', as if they were mere juvenilia! For has *Begriffsschrift* not a reasonable claim to being hailed as 'the most important single work ever written in logic'?[1] And can a commentator who excludes the *Foundations* from Frege's mature writings consistently describe it as the single most perfect work of philosophy ever written, conceived 'at the very height of his powers',[2] and formulated without the least trace of the 'scholasticism' disfiguring the *Basic Laws* and the associated articles?[3] Though disturbing, such oddities are not the most damaging aspect of the label 'mature'. We are prone to infer that the mature form of Frege's thoughts is somehow already implicit in his early writings, as if his previous ideas were intrinsically incomplete and had a propensity to develop in a particular direction in order to realize themselves fully (as it were, by entelechy). But surely the exposition of logic in the *Basic Laws* is no more foreshadowed in *Begriffsschrift* than the present structure of Canterbury cathedral was in the original twelfth-century building. As a prophylactic, one might liken the maturation of Frege's system to the lignification of a tree: shoots originally supple and adaptable thicken and harden into branches whose ultimate configurations show the effects of disease, parasites, and pressures of the environment as well as the genetic endowment of the organism. We should at least consider whether the so-called 'mature' writings do not manifest the decline of Frege's thinking, its deformation by external criticism or internal weakness, or even its decay. In any case, we should be on guard against allowing the widespread and convenient distinction (which we shall continue to use) between his early and his mature works to settle the issue of when his logical reflections reached their acme.

2. The problem of genre-identification

The foregoing preamble about methodology establishes no more than the framework for clarifying Frege's thought. It leaves the entire picture to be sketched in. But it does indicate the importance of a detailed comparison of the 'mature doctrines' with their predecessors in the early writings, and it calls attention to the value of ascertaining Frege's own intentions in introducing alterations into his logical system. The question of how to view his mature theory arises at the outset

---

1. Quoted from the introductory comments to the translation of *Begriffsschrift*, in J. van Heijenoort (ed.), *From Frege to Godel: A Source Book in Mathematical Logic 1879–1931* (Cambridge, Mass., 1967), p. 1.
2. Dummett, *The Interpretation of Frege's Philosophy*, p. 21.
3. Dummett, *Frege: Philosophy of Language*, p. 643.

in the enterprise of interpreting his writings. This question cannot be answered by reference to the details of his theory; rather, how correctly to interpret details often turns on a prior decision about a proper general orientation towards his texts. We must take a wider look around, trying to obtain a synoptic view (or *Übersicht*) of his philosophical reflections.

In executing this programme, two principles should be taken as guides. First, at least *ceteris paribus,* we should presume that the order of composition of his various texts reflects the order of development of his ideas, and hence too the order of explanatory priority among them. Therefore, e.g., we should expect that the basic innovations in 'Function and Concept' are prior, both temporally and logically, to the elaboration of the sense/reference distinction in 'On Sense and Reference'. Second, in establishing Frege's intentions, we should assign paramount importance to his own declarations. Therefore, e.g., his statement 'Concept and relation are the foundation stones upon which I erect my structure' (BLA, §0) constitutes a strong *prima facie* case that he saw the sense/reference distinction as an adjunct of his redefinition of concepts and relations in the *Basic Laws,* not as the ultimate philosophical foundation of his logical system, in spite of the fact that the hallmark of the transformation of the earlier system into the later one is the bifurcation of conceptual content into sense and reference. Together these two principles suggest that we should be better advised to seek to ground our basic orientation towards the mature writings in 'Function and Concept' than in the more familiar text of 'On Sense and Reference'.

Gazed upon from afar, 'Function and Concept' has a number of features that stand out sharply above the welter of detail. Some are conspicuous from any angle; others become prominent only from the vantage-point of Frege's earlier ideas or from the usual outlook on his later writings.

(i) The generalization of the notion of a function underlying the symbolism of concept-script is prefaced by an explanation of the mathematical concept of a function which is to serve as the basis for generalization. This procedure contrasts with the exposition in *Begriffsschrift,* which takes an understanding of the mathematical concept entirely for granted. No doubt Frege reasoned that the incomprehension of that book, especially among its critics, rested on the fact that the term 'function' was meaningless to non-mathematicians and even misunderstood by most contemporary mathematicians, and hence he tried to remedy matters by various elucidations of the unanalyzable concept of a function.

(ii) The mathematical concept of a function is generalized by lifting all restrictions on what entities a function may take as its arguments and values and also on what operations may play a role in the specifications of functions. On the basis of this very abstract and general concept of a function Frege erected a formal calculus, drawing attention to truth-values as potential values for functions and singling out for special consideration functions of various logical types whose values are always truth-values. In particular, he suggested that open equations and inequations can be regarded as standing for such functions. The striking features of this account are that the two truth-values play a pivotal role from the outset and that the resulting formal calculus is elaborated prior to any specification of its intended interpretation or of its theoretical utility. This procedure too contrasts with the exposition of concept-script in *Begriffsschrift*, where well-formed formulae were explained from the beginning as standing for the traditionally recognized objects of logical investigation (rechristened by Frege 'judgeable-contents'). In 'Function and Concept', and in the *Basic Laws*, a subsequent argument is needed to connect the initial formal calculus with logic.

(iii) The original version of concept-script is extended by the addition of new primitive symbols (especially the operator '$\text{\'{e}}\Phi(\epsilon)$' mapping concepts onto courses-of-values) and new axioms (especially the axiom licensing inferences from generalizations of identities to the identity of courses-of-values). Moreover, truth-values take over the role previously assigned to judgeable-contents in the definitions of what concepts and relations are. These alterations to *Begriffsschrift* are presented without any explicit rationale. The revised version is expounded quite dogmatically, as if there were nothing serious to discuss in spite of its major differences from the original one (presumed to be familiar).

(iv) The sense/reference distinction plays only the most minimal role in the exposition. It appears only in the form of the distinction between what a mathematical formula designates (its truth-value) and what it expresses (a thought), and even then the sole use of the distinction is apparently to rebut a criticism of Frege's account of the truth of certain identities where equations or inequations flank the identity-sign. On the other hand, that certain formulae expressing mathematical judgments can be construed as standing for truth-values is indispensable to the actual exposition of the formal calculus. Frege was indeed forced into treating sentences as proper names. But the reason was not the emergence of the sense/reference distinction. Rather it was his commitment to employing function theory in the logical analysis of inferences. It is the sense/reference distinction, not the True and False, which seems to be only loosely integrated into his mature system of logic.[4]

(v) The identification of concepts and relations as functions whose values are uniformly truth-values is intertwined with a doctrine about the logical analysis of atomic singular statements in natural language (e.g. 'Socrates is wise' or 'Caesar conquered Gaul'). The grammatical articulation of such a declarative sentence into subject and predicate is claimed to correspond to the logical articulation of its truth-value into argument and function—the argument being the object named by the grammatical subject, the function being what is designated by the grammatical predicate. Since a sentence of this kind had standardly been taken to assert that an object falls under a concept, and since the assertion that the object named by the subject falls under the concept presented by the predicate is true if and only if the function singled out by Frege takes the True as its value for the named object as argument, he concluded that 'we may say at once: a concept is a function whose value is always a truth-value' (FC, 30). This argument apparently

---

4. If he had cut §32 from the text of the *Basic Laws* (and the four sentences mentioning sense from §2), would we have immediately noticed a grave lacuna in his exposition of his formal system?

presupposes a *strict* correspondence between logical and grammatical analysis, at least in the case of singular atomic statements.[5] Although grammatical considerations played a significant role in his analysis of arithmetical statements in the *Foundations,* it is a novelty to find them directly connected with the exposition of his logical system and a shock to be informed that nobody who lays down logical rules at all can avoid basing them on grammatical categories and linguistic distinctions (CO, 45).

These five features should strike any perceptive reader of 'Function and Concept'. Significantly, too, the first four are equally characteristic of the initial exposition of the logical system in the *Basic Laws,* whereas the fifth is absent (there being no discussion of grammatical distinctions or the structures of declarative sentences). What light is thereby thrown on the nature of Frege's mature thought?

One strategy of interpretation is to discern the emergence of an interest in how language contrives to express thoughts and a dawning of the realization that the logical analysis of inference must proceed by means of the semantic analysis of declarative sentences expressing judgments. This approach emphasizes the sense/reference distinction, assigning it a more important role than it appears to have. It explains this distinction as originating in reflection on the logical significance of singular referring expressions. Frege allegedly discovered compelling reasons for holding that logicians must ascribe senses to proper names in addition to reference, e.g. because the mere association of a name with the referent does not explain in what understanding a name consists. At the same time it supposedly dawned on him that the real structure of a sentence corresponds exactly to the structure of the thought expressed by it. These fresh discoveries provide the tacit rationale for the reconstruction of the logical system originally presented in *Begriffsschrift.* But in carrying out this project, Frege was unfortunately duped

---

5. Frege vacillated on this point. Sometimes he made it clear that he accepted this principle. He argued that the grammatical subject of such a sentence (a 'proper name') cannot stand for a concept (even in the case where it is a phrase of the form 'the concept Φ'!), and that the predicate (a 'concept-word') cannot stand for an object. On this basis, he concluded that a concept is simply 'the reference of a grammatical predicate' (CO, 43n.). On the other hand this doctrine of isomorphism of logic and grammar, even though limited to singular atomic statements, contradicts prominent earlier ideas whose persistence is presupposed. It conflicts with the general licence for functional abstraction: to view any expression (even one grammatically certified as a proper name) in a declarative sentence as invariant while other expressions are regarded as being replaceable is to view this expression as a *function-*name (designating a function of an appropriate logical type). Unless this account is revoked, it excludes the very possibility of marrying *Auffassung-*relative function/argument analysis in logic with the absolute syntactic classification of the constituents of declarative sentences. Moreover, even a doctrine of limited isomorphism sits precariously in a logical system expressly designed to break the tyranny of grammar over logic and to replace the distinction of subject/predicate by the distinction of function/argument. Of course, Frege did not reintroduce the traditional error of conflating generalizations with singular judgments, but he did commit himself to the doctrine that all singular atomic declarative sentences must be analyzable, *inter alia,* into the proper name of an object (which may be the Sun, 2, red, or the direction of line A) and a function-name. Hence he treated expressions with wholly different uses as possessing the same logical form.

by a flimsy analogy between sentences and proper names, perhaps too by a passion for metaphysical simplicity; consequently he wrongly took truth-values to be objects and erroneously assimilated sentences to proper names, thereby partly spoiling his guiding insight that sentences play the primary role in a theory of meaning. Nonetheless, on this view, the emergence of concern with the semantic analysis of declarative sentences, and derivatively of their constituent expressions, is the key to the mature phase of Frege's logic. An understanding of his earlier conception of logic reveals that such an interest in language must be reckoned a real novelty. Neither external influences nor the trajectory of his earlier thought accounts for its emergence. Of course, if we consider the grounding of logic in semantics to be a deep truth, we might accept its acknowledgement by Frege as a revelation, no more and no less miraculous than any other epoch-making discoveries in philosophy or science. Nevertheless, this interpretation is viable only if one recognizes a dramatic kink in the evolution of his thought. *Ceteris paribus,* this hypothesis should be avoided.

Most of the current interpretations of Frege's mature theory develop some version of this general approach. He is viewed as having given birth to the semantic conception of logical validity, i.e. to the truth-conditions conception of meaning that has dominated philosophical logic in the twentieth century. Yet a dollop of scepticism at the outset of investigating his mature system seems entirely justified by our survey of the obvious general features of his exposition of logic. The prevalent view divorces his reconstruction of the logical system of *Begriffsschrift* from any motivation arising out of the programme of reducing arithmetic to logic. It prevents our taking the *Basic Laws* as the definitive exposition of his conception of logic. It makes his avowal of the centrality of the notion of a concept to his thinking misleading, or even dishonest. It ignores the stress that he placed on the functional character of concepts. It distorts the role of the sense/reference distinction in his exposition of concept-script by treating what is central (that sentences stand for truth-values) as something dubious and peripheral, while taking what is peripheral (that proper names have senses in addition to reference) to be crucial. It imputes to Frege a muddled perception of the primacy of sentences in constructing a theory of meaning. It convicts him of error in supposing that the semantic roles of concept-words and sentences can be explained by correlating entities with expressions. And it compels us to conclude that nowhere in his writings is there a definitive exposition of his philosophy of language, even though this should be seen as the foundation of his formal logic and his philosophy of mathematics.[6] We would do well to ponder upon the fact that these apparent deficiencies in his mature writings might vanish if we could substantiate that we had misidentified the genre to which his thinking belonged.

What alternative strategy can we imagine to develop a framework for interpreting his mature theory? The most obvious view would be to regard Frege as applying abstract function theory to the traditionally recognized subject-matter of logic (viz. concepts, judgments, and inferences) for the purpose of executing a

---

6. Dummett, *The Interpretation of Frege's Philosophy,* p. 19.

proof of the reduction of arithmetic to logic. We shall show that this interpretation is indeed correct. But as a preamble, we must look more closely at what he declared to be the centrepiece of his mature system of logic, his analysis of concepts and relations. In doing so, we shall respect both of our basic methodological principles at once: we shall heed his declarations of intent, and we shall begin from the earliest formulation of his mature views.

3. Starting afresh

The *Basic Laws* elaborates a system of formal logic expressed in concept-script for the declared purpose of vindicating the thesis that arithmetic is a branch of logic. Frege noted that other contemporary investigations into the foundations of arithmetic employed the terms 'set' and 'correspondence', but he complained that 'for the most part there is lacking any deeper insight into what they are really intended to mean' (BLA, §0). This he thought to obscure the epistemological nature of the truths of arithmetic and to doom foundational enquiries to confusion. Since both notions (under the labels 'the extension of a concept' and 'the relation of equinumerosity') were incorporated into his own theory (FA, §68), he himself had to secure an impeccable grasp of each of them to execute the formal proofs needed to substantiate the basic thesis of the *Foundations*. In his view, he had to clarify what concepts and relations are, because these notions provide the *points d'appui* for his entire logical system. He made this the primary task of the opening sections of the *Basic Laws*.

The same purpose informs 'Function and Concept', which he intended to serve as an introduction to the mature system of logic formally expounded in the *Basic Laws*. He confessed his faith that 'arithmetic is a further development of logic', and then declared 'I base upon [this opinion] the requirement that the symbolic language of arithmetic [i.e. real analysis (cf. FC, 40f.)] must be expanded into a logical symbolism' (FC, 30). The overriding aim of the article is to explain how this 'expansion' is to be accomplished, and the means adopted is to clarify the terms 'concept', 'extension of a concept', and 'relation'. It is no surprise to find a close correspondence between this article and the exposition of the *Basic Laws*.

At one level Frege held the task of clarification to be trivial. He considered all three expressions to be definable; he thought it possible to analyze each of the corresponding concepts into more primitive ones. Indeed, a concept is a function of one argument, a relation one of two or more arguments, whose value is always a truth-value, and the extension of a concept is the course-of-values of such a function of one argument (FC, 30; BLA, §3). Though rigorously correct and scientific, these definitions are worthless for effecting a sound grasp of the defined concepts among his readers. In Frege's view they had three shortcomings. First, they presuppose an understanding of what functions and courses-of-values are, i.e. a grasp of certain basic mathematical concepts. Second, they assume a knowledge of what truth-values are and of how to extend the mathematical concept of a function to encompass these logical objects as admissible arguments and values. And finally, they take for granted sufficient insight into the logical notions of concepts, relations, and extensions of concepts to recognize that the stated definitions really do constitute correct analyses of the notions already deployed by logicians. In the conviction that none of these conditions is satisfied, Frege addressed himself to each issue in turn. This accounts for the major articulations in the discussion of concepts presented in 'Function and Concept'.

The first part of the account clarifies the mathematical concept of a function, i.e. 'the original meaning of the word "function" as used in mathematics'. A function is not a mathematical formula containing a variable (e.g. '$2x^3 + x$'); this would conflate a sign with what is designated by a sign. Rather, the expression for a function is that part of complex designations for its values which are not (viewed as being) specifications of its arguments;[7] alternatively, it is that part of a formula containing a variable which is over and above the variable. Such a function-expression *designates* or *refers to* a function. Although in mathematics functions are indicated by such expressions as '$2x^3 + x$', a more perspicuous notation might be one with gaps for argument names '$2(\ )^3 + (\ )$' or mere placeholders '$2\xi^3 + \xi$'. Such a notation makes clear the fact that the variable (in '$2x^3 + x$') is not part of the expression for the function, that a function-name is unsaturated or in need of completion. To this in turn corresponds the fact that the designated function itself is incomplete or unsaturated. Though the argument 'goes together with the function to make up a complete whole', it is 'not to be counted as a part of the *function*, but serves to complete the function'. This unsaturatedness of functions is the respect in which they 'differ fundamentally from numbers'. Functions, in fact, are *laws of correlation* of numbers (arguments) with numbers (values), which can typically be represented in graphs that are pictures of correlations of numbers. In Frege's view, no part of this elucidation constitutes a *definition* of what a function (in mathematics) is: the concept of a function is allegedly unanalyzable, and hence the most that can be achieved is a compilation of hints that will bring about a proper grasp of it (FC, 21f., 24f.; BLA, §1).

Frege supplemented this elucidation of 'function' with an elucidation of an associated unanalyzable concept, the concept of a course-of-values (of a function). Two functions that have the same value for each argument are said to have identical courses-of-values. The course-of-values of a function is claimed to be an object, a saturated or complete entity. Hence to state that two functions have identical courses-of-values is to make a genuine assertion of identity. A function, being unsaturated, is distinct from its correlated course-of-values, even though the mapping of functions into courses-of-values is presumed to be one-to-one. This mapping spans the logical divide between unsaturated and saturated entities. Correspondingly, the inference from the fact that two functions take the same value for each argument to the claim that their courses-of-values are identical is taken to be a substantial step of inference requiring a special axiom for its justification (FC, 26; BLA, §§3, 10, 20).

---

7. Obviously, any complex designator of a number may be viewed in various ways as presenting the value of a function for an argument; e.g. '$2 \cdot 3$' can be regarded as designating the value of $2x$ for the argument 3 or the value of $3x$ for the argument 2.

The second part of Frege's account concerns the extension of the mathematical conception of a function. Mathematicians had extended the range of operations used to construct functions by adding the operation of proceeding to a limit and by admitting numerical correlations expressed in words. They had also extended the field of admissible arguments and values of functions to include complex numbers as well as functions themselves. They had not, however, employed all the signs of analysis in the formation of function-names. Frege lifted this restriction by adding to the usual signs (e.g. '+', '−') such signs as '>', '<', '=', so that expressions such as '$\xi^2 = 4$' or '$\xi > 2$' count as function-names. (By implication, any complex well-formed expression in the symbolism of mathematics can be viewed as the designation of the value of a function for certain arguments.) This extension of the concept of a function makes sense only if some entities are specified which are fit to serve as the *values* of the functions allegedly designated by such expressions as '$\xi^2 > 2$'. Otherwise there would be no possibility of discerning here any expressions for laws of correlations of entities. Frege solved the problem by discovering and introducing two *truth-values*. '"$2^2 = 4$" stands for the True [and "$2^2 = 1$" stands for the False] precisely as "2^2" stands for the number 4.' Like the number 4, the True and the False are objects, and therefore the function named by '$\xi^2 > 2$' is on a par with the numerical function x^2 (i.e. both are of first-level). Essential to viewing open equations and inequations as function-names is the extension of the domain of admissible values of functions to include the True and the False. *Pari passu*, the range of admissible arguments should be similarly widened. To avoid any arbitrary restrictions, it is advisable to admit 'objects in general' as possible arguments and values of functions. By this step-by-step extension we ultimately arrive at a completely general concept of a function. Correspondingly, we also attain a completely general concept of a course-of-values (of a function). Principles of inference applicable to reasoning about mathematical functions and courses-of-values will be extended over this wider domain of functions. In particular, the inference from the fact that two functions take the same value for each argument to the conclusion that their courses-of-values are identical is a substantial step of inference requiring the support of a special axiom. (FC, 28ff., 38f., 41; BLA, §9f.).

Simply on the basis of the supposition that the True and the False are objects admissible as arguments and values of first-level functions, we may proceed to construct a formal calculus, an extension of mathematical function theory that includes theorems about functions whose values or whose arguments are truth-values. Using solely function theoretic notation, we would evidently obtain a calculus that would be correctly described as an expansion of the symbolic language of complex analysis ('arithmetic'). On this apparently austere and schematic basis, it is possible to introduce the primitive symbols of Frege's concept-script as function-names, to justify the truth of the formulae selected to be axioms in his formal calculus, and to validate his primitive rules of inference. Such an axiomatic calculus is presented in the *Basic Laws*. Certain primitive symbols are introduced as names of specified first-level functions (the horizontal, the condition- and negation-strokes, identity), and further primitive symbols are introduced as names of specified second-level functions (the universal quantifier, the operator 'ὲΦ(ε)' mapping concepts onto their extensions, and the operator '\ξ'). Axioms are constructed which govern these primitive functions, and theorems are derived. The result is an axiomatized calculus which is a branch of abstract function theory.

Frege's calculus can be viewed as a formal system developed independently of its intended interpretation. Whether it has any coherent interpretation at all, in particular whether it can be applied to the analysis of inferences, is a question independent of the elaboration of the calculus by the derivation of theorems from the axioms. The calculus is simply an autonomous extension of function theory, though its interest turns on the possibility of its application. But the fact that it is a branch of function theory puts a strong constraint on any admissible interpretation. In specifying any function, and hence in explaining the significance of any function-name, one must *lay down* what value the defined function takes for every admissible argument. Since the True and the False are *ex hypothesei* admissible arguments and values, one must specify for which arguments a given function takes the True as its value, for which arguments the False, as well as what values it takes for the True and the False as arguments. This has the consequence that it is an *intrinsic* feature of any given function whether or not it takes the True (or the False) as its value for any particular argument. In other words, the function and argument alone determine the truth-value which is the value of a function for an argument. To abrogate this principle would be to cease to view Frege's calculus as he did, namely as an extension of *function* theory.[8]

---

8. The fact that the value of a function is always determined *a priori* by the function and its argument raises an obstacle to applying Frege's calculus to the analysis of inferences incorporating empirical judgments. The very idea that the relation of an object's falling under a concept is always an internal relation strikes us as ludicrous. Consequently we are inclined, without any warrant, to foist on Frege the thesis that the truth-value of a thought is determined by a concept and its argument *together with a specification of the (relevant) facts*. But, in so arguing, we overlook the fact that this interpretation completely undermines an essential ingredient of his function/argument analysis of truth-values. In his view, it is unintelligible to hold that the value of a well-defined function for a given argument may depend on observation or experiment. This would be tantamount to rupturing the internal relation between a function and its course-of-values. Of course, we might instead contend that a concept should be construed as a more complex function which contains parameters whose values must be specified by observation, thereby restoring the internal relation between a function and its arguments. But this manoeuvre would undermine Frege's case for representing the judgment expressed by 'A falls under the concept Φ' by a unary function-expression 'Φ(A)' in concept-script; instead, we would be driven to the absurd conclusion that he had no idea how to represent any empirical statement in concept-script!

It might be objected that our contention that Frege conceived of an object's falling under a concept uniformly as an internal relation is a *reductio ad absurdum* of our *interpretation* of Frege's system. Absurd though his idea seems now, it is an ineluctable consequence of his conception that concepts are literally *functions*. This implication of his notion of a function has not passed entirely unnoticed, though its consequences have not been followed up. It is expressed in Dummett's worry that Frege's account of concepts threatens to make all atomic statements (not merely identities) cognitively trivial (*Truth and Other Enigmas*, p. 133) It is also directly implied by Church's understanding of the principle that the reference (truth-value) of a sentence is a function of the references of its parts (*Introduction to Mathematical Logic*, p. 8f.).

Even as an uninterpreted calculus, the system has two important oddities. First, by treating identity as a function-name (for a first-level function of two arguments whose value is always a truth-value) it violates the syntax or formation-rules of function theory. Expressions of the form 'A = B' or '$\Phi(\Gamma) = \Psi(\Delta)$' may be substituted in place of any expression designating an object, and therefore such formulae as '(A = B) = Γ', '(A = B) = ($\Gamma = \Delta$)', and '$\Phi(A = \Psi(\Gamma))$' must be considered well-formed (cf. FC, 29), contrary to the standard syntax of function-theoretic notation. Second, the formal system contains no simple designations of either truth-value. Consequently, specification of the values of a function whose values are always truth values must diverge from the canonical pattern in complex analysis. There the standard form exploits the canonical designations of numbers, e.g. using the symbol '4' (not the symbol '$\left(\sin \frac{\pi}{4}\right)^{-4}$') to specify the value of x^2 for the argument 2. But in Frege's extended calculus the only method for specifying a truth-value is to employ a complex symbol which presents the True (or the False) as a value of a function for an argument. Hence, to stipulate that '$\Phi(A)$' designates the True, he would have to use the identity '$\Phi(A) = \Phi(A)$' (which would be circular) or some identity of the form '$\Phi(A) = \Psi(B)$'. But given the standard understanding of identities, a stipulation of the latter form states only that two functions take the same value for certain arguments without stating what the value is; it could be properly used to make an assertion by somebody who was ignorant of what the value of $\Psi(B)$ was. Therefore, in the absence of simple designations for the True and the False, it seems impossible to frame in Frege's calculus definitive stipulations of the values of functions whose values are truth-values. We cannot, in conformity with the stipulation '$x^2 \Big]_2 = 4$', *say* that the value of Φ for A as argument is the True. To accomplish this feat we would first have to extend his concept-script. This fact about his calculus sowed some of the seeds of deep rooted confusions about truth (infra, p. 347ff.).

Of course, Frege never presented his calculus as a wholly uninterpreted formal system. He provided a limited interpretation in introducing the True and the False. These objects are what certain mathematical formulae (equations and inequations) stand for. His only explanation of what the True is consists in stating that '$2^2 = 4$' and '$2 > 1$' stand for the True just as '2^2' stands for the number 4 (FC, 28f.). His calculus, on his view, rests on the fundamental insight that equations and inequations can be construed as names of abstract objects. The calculus would be an empty symbolic game in the absence of the two truth-values. Moreover, the application of function theory to the analysis of equations and inequations would be checkmated. It is solely because '$2^2 = 1$' stands for the False that we can articulate this equation into a function-expression '$x^2 = 1$' and an argument-expression '2' (by appeal to the license for functional abstraction). And given Frege's presumption that mathematical symbolism is part of his concept-script, this doctrine about equations and inequations is indispensable to his claim that function/argument decomposition is the universal norm of representation for complex symbols in concept-script. Although his calculus in the *Basic Laws* is elaborated against the background of this very restricted interpretation, the isolation of the True and the False as what equations and inequations designate is indispensable to his enterprise (FC, 28f.; BLA, §2).

Though backed by this restricted interpretation, the calculus of functions among whose values and arguments occur the truth-values scarcely looks like a system of formal logic at all. The third part of 'Function and Concept' fills in this lacuna. It indicates how to interpret concept-script as a 'logical symbolism'. Frege sketched this account with a few bold strokes of the pen, filling in little detail and offering little argument. He started from mathematical symbolism. The fact that the function x^2 takes the value 1 for the argument -1 he expressed by the statement that the function $x^2 = 1$ takes the True as its value for the argument -1, whereas in traditional logic the same fact would be phrased as the statement that the number -1 falls under the concept of being a square root of 1. Consequently, he would state that the function $x^2 = 1$ takes the True as its value for a number n as argument if and only if a traditional logician would agree that n falls under the concept: square root of 1.

> We thus see how closely what is called a concept in logic is connected with what we call a function. Indeed, we may say at once: a concept is a function whose value is always a truth-value. (FC, 30)

Moreover, two functions whose values are always truth-values have the same course-of-values if and only if they take the same value for each argument, i.e. if and only if any object which falls under one concept also falls under the other. But this is the condition for identity of the extensions of two concepts. Hence the extension of a concept *is* the course-of-values of a function whose value is always a truth-value, and consequently is an object (FC, 30f.; BLA, §3). On the basis of these observations we come to appreciate that Frege's calculus really symbolizes relations among concepts (and between objects and concepts) despite the fact that it does not immediately appear to do so from his explanations of its symbols. It is indeed a logical symbolism.

Frege turned to reinforce this argument by observations about non-mathematical symbolism. This constitutes a fourth part of the argument of 'Function and Concept' (and the only one for which there is no parallel in the *Basic Laws*).

The linguistic counterparts of equations and inequations are declarative sentences. Like an equation, a declarative sentence expresses a thought and (typically) stands for a truth-value. Thus 'Caesar conquered Gaul' stands for the True just as '2^2' stands for 4. Any declarative sentence, like any well-formed compound mathematical symbol, can be imagined as split into a complete part and an unsaturated part, the former standing for an object, the latter for a function. Thus, e.g., the sentence 'Caesar conquered Gaul' splits into the two parts 'Caesar' and 'conquered Gaul'. Since no sentence of the form 'x conquered Gaul' has a value other than a truth-value, the predicate 'conquered Gaul' must stand for a concept, the value of which for Caesar as argument is the True. Accordingly, a concept is what a grammatical predicate stands for (i.e. its reference (*Bedeutung*)). Of course, the recognition of such a function presupposes that objects without restriction must be admitted as the arguments and values of functions, e.g. persons, cities, nations, truth-values, courses-of-values. Furthermore, 'scientific rigour' demands that every concept be defined for every object as argument, e.g. that 'ξ is a prime' be assigned a truth-value for the Sun as argument. A predicate whose definition does not conform to this canon of completeness of definition does not stand for a concept (FC, 30ff.). Later all of these points are packaged economically in the dicta that a concept is predicative and that it is the reference of a grammatical predicate (CO, 43).

On the basis of these swift moves one can, it seems, bring the whole panoply of function theory to bear upon the analysis of everyday thought by merely following the lead of the grammatical structures of declarative sentences. Since objects without restriction must be admitted as the values (and arguments) of functions, any complex designation of an object can be decomposed into a function-name and an argument-expression (by regarding some of the constituent expressions as open to replacement by others); e.g. 'the capital of the German Empire' designates Berlin as the value of a function (designated by 'the capital of ξ') for the German Empire as argument (FC, 31f.). The complexity of singular referring expressions thus mirrors a logical 'decomposition' of its reference. Sentence-connectives appearing in mathematical proofs become equally transparent. Sentences are names of truth-values, and hence precisely defined functions whose arguments and values are truth-values can be recognized as what is designated by such sentence-connectives as 'both ... and ...', 'neither ... nor ...', 'it is not the case that ...', etc. (FC, 33ff.). The composition of compound sentences mirrors the presentation of a truth-value as the value of a truth-function. The stratification of functions into types or levels holds the key to the analysis of sentences expressing generalizations. Precisely defined second-level functions whose values are always truth-values are designated by certain sentence-forming operators on predicates, e.g. 'all ... are ...', 'there are ...', 'there are exactly n ...', etc., and hence a sentence expressing a generalization perspicuously represents a truth-value as the value of a second-level concept (FC, 35ff.), and so on. The resources of function theory can be applied to reveal that the structure of declarative sentences and of complex sentence-components reveals the logical articulations of what is designated—if not perfectly, at least *far* more closely than traditional logicians or the author of *Begriffsschrift* had ever suspected. 'Function and Concept' seems to blossom into a vision that the business of logic is closely entwined with the analysis of language.

The impression that Frege was captivated by this idea is readily strengthened by reflection on the associated texts of 'On Concept and Object' and 'On Sense and Reference'. Against Kerry, Frege argued that logical rules must be based on linguistic distinctions, and in particular he stressed that the reference of any phrase of the form 'the concept Φ' is an object, not a concept. This committed him to the thesis that 'The concept *man* is not empty' makes a statement about an object (CO, 47), and apparently to the thesis that sentences of the form 'The concept Φ is subordinate to the concept Ψ' or 'The object Γ falls under the concept of a Φ' make assertions about binary relations among objects (cf. CO, 47f., 51). These comments seem difficult to marry with the suggestions that the same statements are made by sentences of different forms, e.g. 'Something is a man', 'All Φ's are Ψ's', and 'Γ is a Φ', which do make assertions about concepts. Here the grammatical structures of declarative sentences seem to be thrust upon the logical articulations of truth-values without any independent motivation from function theory and in spite of the impossibility of giving a clear explanation of what objects are designated by expressions of the form 'the concept Φ'. Rigorous adherence to grammar generates a mystery in place of Frege's earlier apparent adherence to the familiar idea that 'The concept Φ is subordinate to the concept Ψ' is simply an idiomatic variant of 'All Φ's are Ψ's' (cf. FA, §47). His later official doctrine requires that 'The concept Φ is subordinate to the concept Ψ' does *not* make an assertion about concepts, and this seems incompatible with its expressing the same thought as the sentence 'All Φ's are Ψ's' which does express a relation among a pair of concepts. That he embraced apparent paradoxes and impaled himself on these thorny problems attests to a willingness to be guided by grammar in the logical analysis of truth-values, at least where grammar seemed to lead in directions in which he wished to go.

'On Concept and Object' also adumbrates the idea that the *thought* expressed by a sentence mirrors the grammatical structure of the sentence (CO, 54f.). Just as a singular atomic sentence splits into a saturated and an unsaturated expression, so does the thought expressed decompose into a corresponding saturated and an unsaturated thought-component. This rudimentary and limited idea of sentence/thought isomorphism is developed no further. Much later (1914) this dormant seed germinated into the doctrine that a sentence is a model of the thought which it expresses, so that a novel thought may be apprehended by constructing it out of thought-building blocks corresponding to the significant parts of declarative sentences (cf. PW, 225, 243). Here again are indications of a willingness to be led by grammar.

Finally, 'On Sense and Reference' seems to exploit clues of grammar to open up new vistas to logical analysis. By employing the notion of sense, Frege gave an account of reports in indirect speech in conformity with the surface sentence structure (viz. a noun-clause which appears as the direct object of a transitive verb), and he followed up what he saw to be this notable success with a sketch of how to analyze what is expressed by a wide range of sentences containing varieties of subordinate clauses. These activities seem to manifest his autonomous interest in aspects of language altogether remote from his concerns in formal logic and the philosophy of mathematics. Once again they seem too to show his willingness to model logic on grammar.

It would be foolish to deny that Frege took any interest in expressions of natural languages. And it would be equally absurd to claim that he thought logicians to be wholly unjustified in following clues from grammar in analyzing judgments and inferences. On the other hand, the reasoning of 'Function and Concept' and the text of the *Basic Laws* give no grounds for asserting that he advanced beyond the commonplace idea of a rough correspondence between logic and grammar to any conception that the true business of logicians is a science of language (semantics). And this modern notion conflicts with his Platonism about senses and truth-values as well as with the correlative idea that neither thoughts nor truth-values have unique analyses (into functions and arguments). The hypothesis that he intended to lay the foundations of logical semantics is implausible. Nonetheless, we might judge that he had inadvertently done so. Could we pursue the tentative suggestions that he made, thereby erecting this mighty science? How solid are the foundations that we might dig out of 'Function and Concept', 'On Concept and Object', and 'On Sense and Reference'?

4. Logical doubts and difficulties

Frege's explanations about concepts and relations resemble the patter of the proficient conjurer. While we have attended to the flow of his words, what magic he has worked! Out of thin air he has produced a pair of hitherto unknown objects, and with prestidigitations too rapid for the mind to grasp he has clothed familiar linguistic entities in the shimmering garb of a hitherto unheard of extension of function theory. The spectator is left with a feeling of vertigo, a wave of excitement or a pang of anxiety as he feels the solid ground of conventional wisdom start to slip from beneath his feet.

Before we succumb to the mesmerizing quality of Frege's performance, we might be well advised to jolt our benumbed critical faculties and to observe more carefully what is before us. Of solid argument there is little; but precipitate haste to enter into the intricacies of a highly abstract calculus is evident everywhere. His work seems pre-eminently the product of a mathematician's mind. We might suspect that in logic

> ... The mathematician ... is only the ingenious technician, the constructor, as it were, who ... builds up his theory like a technical work of art. As the practical mechanic constructs machines without needing to have ultimate insight into the essence of nature and its laws, so the mathematician constructs theories ... without ultimate insight into the essence of theory in general, and that of the concepts and laws which are its conditions.[9]

Instead of accepting Frege's apparatus and manipulating it mechanically to see what it might produce, we should investigate the *philosophical* underpinnings of his system of logic. A promising point to begin is where the conjurer's trickery seems most transparent.

Truth-values as objects: It is to what we assert (judgeable-contents, thoughts) that we ascribe the properties of truth and falsity. How then can we recognize two novel objects, the True and the False? What are they? Of course, Frege did not think it possible to give proper definitions of 'the True' and 'the False'. According to his fullest elucidation, the value of the function $\xi^2 = 4$ for the argument 2 is the True if and only if '$2^2 = 4$' expresses a true thought, i.e., '$2^2 = 4$' stands for the True if and only if the thought expressed by '$2^2 = 4$' is true (BLA, §2). Alternatively '$2^2 = 4$' stands for True if and only if the sentence '$2^2 = 4$' is true

---

9. E. Husserl, *Logical Investigations*, tr. J. N. Findley, Vol. I, p. 244f.

(cf. SR, 63; PW, 233). Similarly, '$3^2 = 4$' stands for the False if and only if the thought expressed by '$3^2 = 4$' is false. Consequently, he concluded, 'These two objects are recognized, if only implicitly, by everybody who judges something to be true' (SR, 63). By parity of reasoning, there must be a whole galaxy of objects (the Necessary, the *A priori,* the Analytic, the Paradoxical), hitherto unsuspected, but implicitly recognized by every logician who judges some truth to be necessary, *a priori,* etc.! This vista of Platonic objects seems uninviting, and the argument adduced is without force. Still worse, how is the idea that true judgments *stand for* the True to be married with the idea that truth is a *property* of what is asserted? The preferred elucidation of what the True and the False are, on pain of vacuity, presupposes that these two ideas are compatible and that truth and falsity are identifiable properties of thoughts (or sentences). At the same time, the two ideas seem plainly incompatible (cf. SR, 64; PW, 233). Frege himself was tortured by the Sisyphean task of keeping them in harmony (infra, p. 347ff.).

Sentences, equations and inequations as names: Though it is primarily the objects of possible assertion that are true or false, declarative sentences or mathematical equations can also be derivatively so characterized (PW, 129). Hence they must stand for the True and the False (FC, 28; BLA, §2). From a logical point of view they are *names* of truth-values (BLA, §26). Quite literally, '$2^2 = 4$' stands for the True just as '2^2' stands for 4 (FC, 28f; BLA, §2). Sentences must differ from proper names of persons only as persons' names differ from numerals or names of ships. How can this alleged homogeneity be reconciled with the sharp differentiation commonly and rightly recognized to hold between sentences and names? And if sentences can be shown to be designations of entities, what proves that the designated entities are the two truth-values rather than, say, facts? Atop these sins of omission are piled sins of commission. If sentences and equations are names of objects, logically on a par with names of persons or numbers, then it should be legitimate to substitute these expressions for numerals or persons' names, thereby obtaining something expressing a thought. Hence, if '$2^4 = 4^2$', is well-formed, so must be '$4 = (2 > 1)$' and '$(4 = 2^2) = (2 > 1)$', as Frege supposed (FC, 29). By parity of reasoning, '$(4 = 2^2) > (2 = 1)$', '$4 = 2^2$ conquered Gaul', and 'The planetary orbits are ellipses conquered Gaul' must also be well-formed. Yet all of these are gibberish. Is Frege's fiat strong enough to push back the bounds of nonsense?

Proper names and objects: Excluding from consideration entire declarative sentences, let us scrutinize what Frege called 'proper names'. He advanced certain syntactic criteria for distinguishing 'proper names' from other expressions ('concept-words'). In particular, the definite article with a singular noun-phrase constitutes a proper name, and a noun lacking a plural is to be classified as a proper name, whereas the indefinite article with a noun-phrase and plural noun phrases without any article are distinguished as concept-words (FA, §§51, 66, 68; CO, 45). These criteria are obviously parochial, being applicable only to languages with definite and indefinite articles. They are also not watertight; Frege acknowledged exceptions even in German (CO, 45, 50). What his criteria roughly isolate is the syntactic category of singular referring expressions. For everyday purposes his account seems quite satisfactory, but can it really serve as the foundation for a rigorous logical analysis of declarative sentences? Does every uncontestable singular referring expression stand for an object? What about 'the whereabouts of the Prime Minister', 'the lack of purchasing power', 'the ability to ride a bicycle', 'the way in which Borg serves'? And what about 'the function x^2', 'the concept of being a square root of -1'? Is there not a risk that 'serious objects' will be overwhelmed by a host of 'frivolous and spurious ones'?[10] Conversely, does every expression satisfying none of Frege's criteria for being a proper name *ipso facto* fail to designate an object? What about personal pronouns (e.g. 'I', 'he') and demonstrative pronouns (e.g., 'this' and 'that')? And what about plural definite descriptions (e.g. 'the Houses of Parliament', 'the Galapagos Islands')? Further problems arise in respect of expressions that only partly satisfy the syntactic criteria for proper names. How are declarative sentences to be analyzed whose subjects are names of stuffs (e.g. 'iron', 'butter'), other mass-nouns (e.g. 'light', 'shade', 'rain'), and pseudo-mass-nouns (e.g. 'cutlery', 'money', 'machinery')? In the face of these difficulties, can we glean from Frege's account of proper names any precise rules of correspondence between the syntactic articulations of declarative sentences and the articulations of the corresponding formulae in concept-script? Is the syntactic classification of an expression an infallible marker of its semantic role? The generalization that an expression is a proper name if and only if it stands for an object could be interpreted to state a truism; otherwise it is evidently false. In either case, syntax does not provide the key to the metaphysical category of objects.

Worries about the indeterminacy of the boundaries of Frege's concepts of proper names and objects are matched with puzzlement about the heterogeneity of objects and about the significance of grouping entities together into this super-category. Apparently we must count among objects such diverse things as minds, physical objects, numbers, events, facts, states of affairs, dispositions, properties, ideas, and thoughts. What can be informatively stated about entities of one category (e.g. when and where an event takes place) often cannot be intelligibly said about other 'objects' (e.g. facts do not take place, and have neither spatial nor temporal locations). This makes it doubtful whether anything significant can be said about objects in general. Frege said remarkably little—in effect nothing—

---

10. Cf. Dummett, *Frege: Philosophy of Language*, p. 70ff. Frege had no such qualms; Wittgenstein reports him as countenancing the simultaneous occurrence of a court case and a lunar eclipse as an object. Dummett, however, endeavours to sort the wheat from the chaff by stipulating that only expressions associated with a criterion of identity are genuine proper names of objects. This will allegedly include number words, colour-names, names of directions, lengths, weights, etc., but exclude designations of modes of action, names of tactile qualities, etc. It is, however, unclear why, if 'the direction in which Borg served' is a serious object, 'the way in which he served' is a frivolous one. Certainly Frege never claimed that *every* proper name must have associated with it a criterion of identity. Second, what he understood by 'a criterion of identity' differs *toto caelo* from Dummett's notion. Finally, if a criterion of identity is a characteristic mark in virtue of which we can identify something as the same again, then surely there are criteria of identity for, e.g., distinctive ways of hitting a backhand in tennis, putting in golf, opening at chess.

apart from the elucidation that objects are whatever are not functions (FC, 32; BLA, §2). That manoeuvre merely replaces one insoluble problem with another. The apparent vacuity of his concept of an object is not the worst of its defects. Objects constitute a super-category whose unity transcends ordinary category-differentiations. We are invited to look upon category-restrictions as irrelevant to logic and to adopt a more enlightened, more abstract point of view according to which what can be said of one object can be intelligibly asserted of any object. As if it were narrow-mindedness that prevented our making sense of fitting events into holes! Or of locating ideas or facts in space! Category-distinctions correspond to limitations on what combinations of words make sense. Even if it made sense to abolish them by fiat, it would be unprofitable. For the cash-value of recognizing an expression as a proper name would shrink almost to nothing. Practically no consequences would follow from such classification. A perfect vacuum lies at the end of this continued ascent on the wings of abstraction.

Concepts as functions: The sole argument that concepts are functions whose values are truth-values turns on the claim that the sentence 'The number 2 falls under the concept: square root of 4' is equivalent to the sentence 'The True is the value of the function $\xi^2 = 4$ for the number 2 as argument' (FC, 30).[11] This pattern of argument must be fallacious by Frege's own lights since he denied that 'The courses-of-value of $\Phi(\xi)$ and $\Psi(\xi)$ are identical' makes the same assertion as '$\Phi(\xi)$ and $\Psi(\xi)$ have the same value for each argument' in spite of the fact the two sentences are true in exactly the same circumstances! With equal impropriety we could infer from the true premise that 'It is a fact that snow is white' is equivalent to 'The thought that snow is white is true' the false conclusion that facts are true thoughts (infra, p. 344f.). Is the 'proof' that concepts are functions not worthless? Moreover, is its conclusion not simply false? The terminology appropriate for discussing functions is inapplicable to concepts. Concepts cannot be said to have arguments or values; they do not map or transform objects into truth-values. Some of Frege's descriptions of 'concepts' are unintelligible as characterizations of concepts. Others are incorrect in crucial ways. According to his account of concepts as functions, the relation between an object and a concept under which it falls (or under which it does not fall) is always an *internal* relation. For it is an intrinsic feature of any function that it takes a particular value (e.g. the True) for a given argument. Hence his conception of concepts conflicts with the conception standard among logicians according to which an empirical singular judgment states that an *external* relation holds between an object and a concept. This common conception admits the possibility that an invariant concept might have different extensions in different possible worlds; it allows that different concepts may as a matter of fact have identical extensions; and it makes no requirement that a

---

11. The *Basic Laws* offers an even less direct argument. Since the courses-of-values of functions whose values are always truth-values are identical only if the extensions of certain corresponding concepts are considered to be the same, 'it seems appropriate to call directly a *concept* a function whose value is always a truth-value' (BLA, §2).

concept be so specified that whether or not an arbitrary object falls under it is laid down in advance. Frege's conception differs in all three respects. The extension of a concept is one of its intrinsic features,[12] and hence he had to embrace the ludicrous conclusion that the *concept* of being red would have differed from what it is had my copy of the *Tractatus* been bound in navy blue rather than red! Equally oddly, he took co-extensionality to be the criterion of concept-identity and he laid down the apparently bizarre requirement of completeness of definition of concepts,[13] as if no genuine concept could have borderline cases of correct application (e.g. as if it were illegitimate to speak of the concept of a heap or the concept of being bald). Frege's functional conception of concepts not only excludes too much, but it is also too liberal. Any function whose value is always a truth-value is a concept on his view, and hence he took 'a round square' to signifiy an empty concept (as does 'a unicorn') although there is no such thing as the concept of a round square.

Frege's radically mistaken conception of a concept becomes most apparent from examining his account of completeness of definition (BLA, ii, §56) and his function-theoretic conception of what that consists in. His account is disastrous on two fronts. On the one hand, a predicate whose definition does not conform to the canon of completeness of definition for functions is alleged not to stand for a concept at all. Vagueness (thus understood!) is assimilated to reference-failure for singular referring expressions, and every declarative sentence in which such an 'ill-defined' predicate occurs is held to lack a truth-value. On this view it must surely be doubtful whether *any* sentence in English can be used to make an assertion! Moreover, most of the concepts which we conceive ourselves to possess will transpire to be still covered with 'irrelevant accretions which veil [them] from the eyes of the mind' (FA, vii), and most of the expressions which we do conceive as concept-words will turn out not to stand for concepts at all. While this requirement intolerably restricts the concept of a concept, Frege's conception of stipulating reference for a function-name results in absurd expansion of the concept of a concept. Suppose, e.g., we take the names in the London telephone directory: 'A_1', 'A_2', ... 'A_n' and define a function $\Phi(\xi)$ as follows:

$\Phi(A_1)$ = the True iff A_1 owns a Jaguar

$\Phi(A_2)$ = the True iff A_2 lives in Richmond

$\Phi(A_3)$ = the True iff A_3 has two sons

$\Phi(A_4)$ = the True iff A_4 attended a concert on January 1, 1981

---

12. This is an obvious corollary of defining the extension of a concept as the course-of-values of a function, since any alteration to its course-of-values is inseparable from an alteration in the function itself.
13. This legislation seems an unmotivated blunder to anyone who holds that the relation of a concept to its extension is external. But it has a transparent rationale when this relation is viewed as being always internal. For then to admit the possibility that the extension of a concept is intrinsically indeterminate would be to deny that a concept is uniquely and fully characterized by its extension.

$\Phi(A_n)$ = the True iff A_n has been divorced more than twice
For all other x, $\Phi(x)$ = the True iff x is a number
or x is an inhabitant of Oxford.
Otherwise $\Phi(x)$ = the False.

According to Frege, '$\Phi(\xi)$' stands for a concept. But what concept? There is no alternative to rehearsing the whole unsurveyable explanation just sketched, and if the original explanation left room (as it did) for the question 'What concept?', then so too will the second edition.[14] Should we conclude that one does not, save indirectly and *per accidens,* define a concept by stipulating a truth-value as the value of a function for every object as argument? Or is it rather[15] that conceptscript pushes back the bounds of sense?

Concepts are objects of understanding, not what we speak *about* in typical assertions; they more closely resemble the senses of concept-words rather than the references. The very linguistic expressions which Frege took as signifying concepts are, from our point of view, deviant. We do not speak of 'the concept *law*' let alone of the concept 'ξ is a law', but of the concept of law. The concepts of justice, truth and beauty are of perennial interest, but the 'concepts' *just,* is *true,* and ξ *is beautiful* are not new stars in the philosophical firmament, but new aberrations in the philosophical imagination. Not only did Frege speak incoherently of concepts, but he also banned as incoherent ways of speaking of concepts that are perfectly intelligible. First, he disallowed the practice of speaking of concepts as identical or different (since identity and difference are relations between objects, not concepts), whereas we discuss with perfect propriety the resemblances and differences between his concept of sense and the everyday concept of meaning. Second, he denied that such a sentence as 'The concept of understanding is widely misunderstood by philosophers' can be used to make an assertion about a concept, whereas we can and do use it for exactly this purpose. Likewise he considered remarks such as 'The concept of mind is a difficult concept to survey', or such truisms as 'The concept of a number is a concept' to be nonsense. His argument is that an expression of the form 'The concept of A' is a proper name standing for an object, not for a concept, and that 'ξ is a concept' stands for a second-level concept, since concepts are first-level functions which cannot be *arguments* of functions lower than second-level. Hence 'The concept of A is a concept' violates type restrictions, an antinomy Frege expressed in his notorious claim that the concept *horse* is not a concept (CO, 46; cf. PW, 119ff., 177, 193; PMC, 136). It is

---

14. The contention that this stipulation introduces a concept distorts the notion of a concept in just the way that the notion of a property is distorted by the philosophical thesis that the number 2, the Eiffel Tower, and the last movement of the Jupiter Symphony share a common property, viz. the 'disjunctive property' of being even or being built of steel girders or having rondo form. (Cf. Wittgenstein, *Philosophical Investigations,* §67.)
15. The issue is not merely theoretical, since questions such as 'What concept is designated by Frege's horizontal?' produce just this quandary (infra, p. 341).

allegedly a deep defect of natural language that one cannot say of a concept or function that it is a concept or function without immediately distorting the thought one wishes to express, and that one cannot coherently state which concept a particular predicate stands for.[16] But it is no more a defect of language that 'the concept A' cannot coherently refer to a Fregean 'concept' than it is a defect that no expression can coherently refer to the square root of a fact or the velocity of 2^2. Concepts, on the other hand, can be referred to by such a Fregean 'proper name', and the sentence 'The concept of a horse is a concept' states a platitude. The very idea that there is even an apparent antinomy itself needs examination. In arguing that the concept *horse* [sic!] is not a concept, Frege presupposed a connection between type-restrictions in function-theory and expressions in language. The real issue is what right he had to apply to definite descriptions and predicates any restrictions grounded in the differentiation of mathematical functions into types according to the levels of their arguments.

In view of these gross discrepancies, how can Frege's definition of concepts as functions be accepted as an analysis of our concept of a concept? Far from clarifying the real nature of concepts, the imposition of the terminology, categories and norms of function theory produces fundamental distortions and incoherences. What Frege called 'concepts' are not to be found in heaven or earth, but only dreamt of in his philosophy.[17]

Predicates as function-names: Starting from the claim that a concept is the reference of the grammatical predicate in a singular atomic statement, Frege apparently moved towards the conclusion that an expression stands for a concept if and only if it is a predicate. This is suggested by his insistence that concepts are predicative in nature and by his explanation of a concept as the reference of a grammatical predicate (CO, 43n.). Here too it seems that syntax is meant to provide the key to metaphysics. What then did Frege understand by the term 'pred-

16. Dummett (*Frege: Philosophy of Language*, p. 212ff.) and Geach ('Critical Notice of *Frege: Philosophy of Language*', *Mind* (1976): 438) credit Frege with finding a solution to his antinomy (cf. PW, 119ff.). An expression of the form 'what the concept-word "is A" stands for' stands for precisely what 'is A' is held to stand for. Hence, although the concept *horse* is not a concept, and although 'the reference of "is a horse"' does not name what 'is a horse' names, still, what 'is a horse' stands for is what Black Beauty is.

We might accept that 'ξ is a horse' stands for what Black Beauty is, *if* we accepted that 'ξ is a horse' is a *concept-word,* that it *stands for* anything at all, that what it stands for is an unsaturated entity which maps Black Beauty onto an object called 'the True', and so on, and on. To chortle over this 'solution' as a triumphant vindication of Frege against the Philistines (as Geach puts it) is like crowing over a successful tonsillectomy performed on a patient with acute appendicitis.

17. One might claim that Frege's discovery was not that concepts are functions, but that *properties* are, since 'A falls under the concept ϕ' = 'A has the property ϕ' = 'The True is the value of ϕ for the argument A' (cf. FC, 30; BS, §10; CO, 51). This too is futile, since the same fallacious argument must be invoked to support it as is employed to show concepts to be functions. Moreover the conclusion is false. The terminology of function theory is inapplicable to properties, the identity- and existence-conditions for Frege's concepts do not fit properties, and his characterizations of concepts (e.g. that they are uniformly imperceptible and mind-independent) clash with features of properties.

icate'? He openly vacillated. Even in the case of singular atomic sentences there is an unclarity. Sometimes his arguments presuppose that the copula is not part of the predicate (e.g. CO, 43f; PW, 90ff.), as if he accepted the traditional parsing of 'Socrates is wise' into subject plus copula plus predicate. At other times, he implied that the deletion of a proper name from a simple mathematical formula or a singular atomic declarative sentence yields a predicate or predicative expression (e.g. FC, 31; CO, 50);[18] on this view, the copula (if it occurs) would be part of the predicate, and even the grammatical subject of the sentence might be part of the 'predicate' (e.g. 'Caesar' would be so if we were to parse the sentence 'Caesar conquered Gaul' into the proper name 'Gaul' and the unsaturated expression 'Caesar conquered x'). Frege's use of 'predicate' becomes even more idiosyncratic in respect of more complex declarative sentences. In the sentence 'All mammals are red-blooded', the traditional identification of 'mammals' (or 'all mammals') as the grammatical subject is not incompatible, on his view, with the predicative nature of the concept of a mammal (which is the reference of this expression), nor is the agreement of 'all' (in gender, e.g., in Latin) any obstacle to affirming that 'all' logically belongs with the predicate (CO, 47f.). Would any grammarian identify as a predicate the expression derived by deleting a proper name from any declarative sentence, however complex it might be? And would anybody infer from the fact that 'All mammals are red-blooded' is synonymous with 'If anything is a mammal, it is red-blooded' that 'mammals' in the first sentence functions grammatically as a predicate? Was there available to Frege any account of grammar that even begins to make plausible the thesis that an expression designates a concept if and only if it is a grammatical predicate? It would even be mistaken to suppose that we could now fill this lacuna in conformity with his practice. He

18. His equivocation is neither accidental nor trivial. For purposes of emphasizing the incomplete, unsaturated, or predicative nature of expressions which allegedly stand for concepts, he opted for explaining concept-words as what remains of a singular atomic sentence from which a proper name has been deleted, e.g. 'Jumbo is a pink elephant' yields as the 'predicate' or 'concept-word' 'ξ is a pink elephant'. But to show that 'exists' is a second-level concept-word, taking a first-level concept-word as argument-expression, the copula must be dropped. For 'A pink elephant does not exist' cannot be seen, without *ad hoc* assumptions about 'deletion operations in depth-grammar', as obtained by filling in the gap-holder in 'ξ is a pink elephant' with the second-level concept-word 'exists' and the negation-sign. Nor will switching to a paraphrase, e.g. 'There is no pink elephant', avail, since here too the 'first-level predicate' is represented by the words 'pink elephant', not 'is a pink elephant'. In neither case does ordinary grammar suggest that the existential generalization is derived from an incomplete expression obtained from a singular atomic sentence by deletion of a proper name. Nor will further paraphrasing have any point, since the structure of an equivalent paraphrase can throw no light on the structure of the original sentence which does not fit Frege's analysis.

Even worse difficulties occur with second-level generalization. For while we may pass to 'Jumbo is something (or other)' or 'There is something which Jumbo is', if it is 'ξ is a pink elephant' that is really the predicate, then the generalization must apply to it, not merely to a part of it. But can we make any sense of 'There is something which Jumbo something' let alone of 'There is something which Jumbo somethings'? (Cf. B. Rundle, *Grammar in Philosophy*, pp. 31ff., 110ff (Clarendon Press, Oxford, 1979).

vehemently denied that a proper name can be a (complete) grammatical predicate. But his account of functional-abstraction committed him to the possibility of viewing any expression (even one certified by syntax to be a proper name) as a function-name. No refinement of syntax can deliver into our hands the key to his notion of a concept and the correlative notion of a predicate.[19]

Even if we waived these difficulties and assume that the deletion from any singular declarative sentence of one or more 'proper names' yields a 'predicate', we would have to concede that he offers no argument whatever[20] for supposing that such an expression (a 'predicate') stands for or designates any entity at all. Is this (traditional) idea not absurd? We usually distinguish the role of referring from that of predicating, allocating the former to 'names' (proper names, demonstratives, definite descriptions, etc.) and the latter to 'predicates'. To describe predicates as referring to concepts seems to obscure this fundamental distinction. If the role of predicates is to stand for concepts, that of names to stand for objects, and that of sentence-connectives and quantifiers to stand for yet other kinds of concepts, then to characterize an expression as 'standing for something' is to say nothing about what its meaning is. Is it not either vacuous or false to assert that predicates stand for concepts? How does the sophistication of what predicates are supposed to stand for (functions) compensate for the gross crudity involved in thinking that their logical significance should be explained solely in terms of what they stand for?

Even if we concede that predicates stand for entities commonly called concepts, how would this license the conclusion that they are *function*-names? An indirect route would rest on the uncompelling argument that concepts are functions. But an alternative direct route seems available. Is it not apparent that what can be asserted by the use of paradigmatic function-names in mathematical formulae can also be asserted by the predicates of corresponding declarative sentences? For example, the formula '$4 = 2^2$' (or better: '$4 = x^2 \big]_2$') says exactly the same thing as the sentence 'four is the square of two', and the two-place grammatical predicate 'ξ is the square of ζ' here discharges the role of the function-name 'x^2' in the equation. This is not enough to prove that every predicate is a function-name, but does it not conclusively show that *some* are? This appearance is deceptive. By Frege's own lights, the argument rests on conflating two usages of the word 'is' (cf. CO, 43). For, in the predicate 'ξ is the square of ζ', 'is' is employed to express identity. Hence the same predicate could be expressed by 'ξ is identical with the square of ζ', and now it is obvious that the function-name 'x^2' corresponds not to the whole predicate, but only to a proper part of it (the phrase 'the square of ζ'),

---

19. He himself hinted at this negative thesis, especially in affirming that 'the grammatical categories of subject and predicate can have no significance for logic' (PW, 141n.) and in warning logicians 'not to attach too much weight to linguistic distinctions', especially the subject/predicate structure of sentences (PW, 143).
20. Unless PW, 192f., can be dignified by the name of 'argument'.

while the remainder of the predicate is expressed by the sign of identity in the formula '$4 = x^2 \Big]_2$' (cf. CO, 43f.). To resuscitate this argument would require treating the mathematical signs '=', '>' '<' as admissible components of function-names, contrary to standard practice. But even this would not be enough. What would authorize the inference from the fact that a declarative sentence can be used to make the same assertion as a mathematical formula decomposing into a numeral and a function-name to the conclusion that the predicate is therefore a function-name? Does the fact that the True is the value of the function $2^2 = \xi$ for the argument 4 show that the predicate of the true sentence '4 is the square of 2' stands for this function? Could expressions with different roles not be combined together into a number of sentences that made the same assertion? Or is it known already that any name for a truth-value (expression of a thought) must contain at least one function-name? How could this generalization be established? Would it not presuppose the truth of the claim that predicates are function-names?

Not only does the argument that predicates are function-names seem flimsy, but also the conclusion appears false, or even unintelligible. Function-names have two prominent features difficult to reconcile with predicates of everyday declarative sentences. First, every function has a definite number of argument-places, and therefore any correct notation must ensure that the name of an n-argument function appears always with exactly n argument-places, neither more nor less. Predicates, by contrast, display variable polyadicity.[21] The verb 'surrender', e.g., seems to have a different number of argument-places in the sentences 'I surrender', 'I surrender my gun', 'I surrender this gun to the police', 'I surrendered the gun immediately to the police', etc. Does each verb in English have a canonical valence? Are sentences where a given verb appears to have different valences derived from implicit canonical sentences by ellipsis?[22] Second, Frege demanded of every acceptable function-name that it be defined for every appropriate argument-expression, at least for 'scientific rigour'; in particular, every first-level function-name must be defined for every non-empty proper name as the argument-expression. This condition is not met by predicates, even those employed in the most rigorous scientific reasoning. Nor could it be met. Category restrictions abound in natural language and limit what combinations of names with predicates make sense. In what way would the abolition of these restrictions promote the rigour of our reasoning? Would their abolition leave the significance of predicates unaltered? Does it even make sense to suppose that grammatical *predicates* in English are subject to no combinatorial restrictions apart from those imposed by Frege's type-theory? This pair of observations alone suggests that it is far from plausible to maintain that predicates are function-names.

---

21. Cf. A.J.P. Kenny, *Action, Emotion and Will,* p. 156ff. (Routledge and Kegan Paul, London, 1963).
22. Davidsonian quantification over events does not resolve this problem, but rather, by 'transforming a concept into an object', as Frege would put it, recognizes its insuperability.

Would a weaker thesis perhaps serve Frege's purpose, viz. that predicates may be viewed as function-names from a logical point of view? Though less liable to disproof, this seems scarcely intelligible. According to his conception of a function, this thesis presupposes the dubious doctrine that sentences may be viewed as proper names of objects. Moreover, it distorts the notion of a function-name beyond recognition. Function-names are explained by exemplification in the symbolism of complex analysis. Mathematicians learn to pick out certain symbols as function-names in contrast to other symbols which have different roles; e.g. in the formula '$\sqrt{2} = \sin \frac{\pi}{4}$', '$e^{i\pi} = -1$' or '$\log_e \pi > 1$', the symbols '$\sin \xi$', '$e^{\xi}$', and '$\log_e \xi$' would be counted as function-names, but not '>', '=', or 'i'. What then are we asked to do when directed to imagine that '>' or '=' are function names? Would it not make just as much (or as little!) sense to tell somebody to treat all these symbols as names of objects? If not a general licence to extend the range of well-formed formulae to encompass such gibberish as '$(\pi > 2) > 1$' and '$e^{(\sqrt{2} = \sin \pi/2)} = 1$', what is the content of the instruction to extend the range of function-names to include '>' and '='? Or to encompass predicates of natural language?

From our point of view, what Frege needed is a licence to translate mathematical formulae and declarative sentences into the notation of function-theory, to represent what is asserted by such formulae as '$\Phi(A)$' or '$\Psi(A,B)$'. Should we not see the claim that predicates are (or can be viewed as) function-names to constitute this authority? There is no reason to think that Frege felt the need for any such licence, still less to interpret his claim as satisfying this need. Even were this not so, the formulation would be misleading. What right have we to draw any conclusions from the possibility of such paraphrase about the role of predicates in declarative sentences? If the sentence 'Lotze was a better logician than Frege thought' were translated into French, the comparative clause would require a (pleonastic) '*ne*'; should we therefore infer that a negation governs the verb 'thought' in the English sentence? Would this claim and Frege's putative principle not have to stand or fall together?

Truth-values, first- and second-level functions are the tools for constructing the first-order predicate calculus to modern specifications. The slogans 'Predicates are function-names' and 'Concepts are functions' are part of the contemporary logician's catechism, recited in elementary logic textbooks for the improvement of the mind and taken for granted as fundamental and incontestable truths in a multitude of learned discourses. It is hardly surprising, then, that some modern logicians are disposed to look with favour upon Frege's exposition of his logical system. But it is absolutely astounding that they should suppose that he shared their modern understanding of the building-blocks of logic (e.g. that his understanding of what a function is exactly matches theirs) and the modern interpretations of these allegedly basic truths (e.g. that classifying an expression as a predicate involves identifying the depth structure of a sentence). The acme of the miraculous, however, is thinking that he put modern concepts and doctrines on firm foundations. A careful reading of his writings drives one to precisely the opposite conclusion. To the extent that modern ideas rest on the same foundations as Frege's, they are supported neither by cogent argument nor by compelling insight. Any attempt to build on his writings any weighty philosophical principles is doomed to sink into a quagmire of sophisticated nonsense.

There seems to be no obvious respect in which everything has become simpler and sharper through the introduction of truth-values. On the contrary, this innovation raises serious doubts about the significance of his formal system which Frege's remarks about language exacerbate. If his aim was to rebuild the notions of concepts and relations on foundations more solid than those of *Begriffsschrift*, it seems evident that he did not succeed. Yet he had notable success in stealing the critics' fire. He launched himself in a new ship and beckoned to logicians on the shore to join him for a voyage on the deep. The glittering superfices of the vessel and the fascinating mathematical gadgetry in the engine room diverted their attention from the leaks, and as time has passed more and more have thronged to sign on as ship's company. But alert to the questionable soundness of the principles of construction of this vessel, would we not be more prudent to remain on *terra cognita,* leaving the exploration of truth-values and functions whose values are truth-values to those who are both bolder and blinder?

5. Logic and language

Frege's attempt to forge a link between his function-theoretic logical system and the grammatical structures of declarative sentences of natural languages must be judged to be radically defective. It manifests the same schizophrenic strategy conspicuous in most nineteenth century logic. He tried to combine the policy of following the lead of grammar with castigating the grammar of natural language for being misleading. On the one hand, the traditional notions of subject and predicate should be banished from logic, to be replaced by the notions of concept and object, function and argument. On the other hand, no proper name designates anything other than an object, and a concept can be represented only by a grammatical predicate. In cases of conflict Frege was inclined to reformulate principles of grammar, thereby distorting the meaning of 'proper name' and 'predicate' in the interests of perspicuity! One might with justice characterize his distinction of concept and object as a *sublimation* of the subject/predicate distinction,[23] concealing as many *different* logical forms as the grammatically rooted distinction it was meant to overthrow. His putative logical analysis of declarative sentences merely reflects his commitment to using the categories of function theory as a *norm of representation* of judgment and inference.

What conclusion should we draw from Frege's failure to establish any sound principles relating the articulation of a thought in the symbolism of concept-script to the structure of a declarative sentence expressing this thought? We might well suppose the consequence to be a total catastrophe, the collapse of his mature sys-

---

23. Cf. Wittgenstein, *Philosophical Grammar*, pp. 205, 265ff; F. Waismann, *The Principle of Linguistic Philosophy*, p. 227f. (Macmillan, London, 1965).

tem of logic and hence too the destruction of his demonstration of the logicist thesis. This verdict seems ineluctable provided we pass judgment on his work from the 'enlightened' modern point of view! We are apt to subscribe to the thesis that the validity of an inference depends on the semantic analysis of the sentences used to formulate it. We then consider that Frege's observations about the grammatical structures of declarative sentences are intended to vindicate the claim that function/argument decomposition is the key to semantics. Finally, we conclude that his employment of function theory in the logical analysis of inferences is justified solely by the conjunction of these two premises. According to this view, the only way to make sense of a function-theoretic logical system is to view it as grounded in a function-theoretic analysis of sentences of natural languages. Therefore, the only way to make sense of Frege's concept-script is to accept the premise that his remarks connecting function/argument decomposition with grammatical categories were intended to illuminate the workings of language. On this view, the claim that predicates are function-names was meant by him to give an insight into the true nature of the meanings of predicates which would serve as the foundation for function-theoretic semantics and logic.

However plausible this reasoning may seem, it involves a grotesque distortion of Frege's conception of his whole enterprise. The basis of his function/argument analysis in logic is a radical generalization of the concept of a function and the recognition of certain Platonic objects (first judgeable-contents, then truth-values). His guiding idea is that any object can be presented as the value of a function for an argument. Consequently he thought it to be perfectly intelligible to develop a logical system exploiting abstract function theory independently of any doctrine about how declarative sentences stand for the objects of logical investigation, and *Begriffsschrift* contains just such an exposition of logic. Frege's Platonism about judgeable-contents and truth-values secures the possibility of a purely abstract *a priori* science of logic parallel to arithmetic (the *a priori* science of numbers). The widely recognized claim that the only human access to the proper objects of logic is through declarative sentences no more converts logic into a science of symbolism than the parallel recognition that numerals are the only available designators of numbers transforms arithmetic into the science of number-words. If it could be proved that declarative sentences did perspicuously reflect the structures of the thoughts expressed, this would add to the attractions of Frege's function/argument schema of logical analysis. But nothing within his formal system hangs on the doctrine that formulae in concept-script mirror the grammatical structures of declarative sentences. Nor, of course, does the execution of his demonstration of the logicist thesis depend in any way on this doctrine. The contention that everything would be brought down in ruins if this doctrine crumbled away is mistaken because his mature system of thought rests on independent foundations provided by Platonic objects.

To think that the weaknesses of Frege's observations about grammar threaten the stability of his edifice is to misidentify the purpose of his 'analysis of language' and to misinterpret what he saw to be its role in his philosophical reflections about logic. To make this verdict convincing, however, we must offer a positive account of what his interest in language was and what motivated him to rest philosophical arguments on statements about the grammatical structures of declarative sentences.

If we assume provisionally that he had clear motives for construing formulae in concept-script and declarative sentences as names of truth-values, then one rationale for his concern with language has already been indicated. Frege had to vindicate the claim that his formulae expressed in the symbolism of function-theory constitute a formalization of logic.[24] His reasoning turned on the acceptance of the traditional conception that the subject-matter of logic is inferences, judgments, and concepts, and that inferences must be analyzed by breaking down the constituent judgments into their component concepts. In addition, he exploited the standard conception of a rough and ready correspondence of logic with grammar, i.e. the idea that grammar is a fallible guide to logical form. By appealing to such traditional doctrines as that a concept is a possible predicate, he sought to establish that his formal system is *logic* presented in the guise of function theory (rather than in the guise of abstract algebra or the theory of point-sets).

Frege's case falls into two parts. One is an argument to show that formulae which he characterized as names of truth-values are expressions of judgments. The reasoning is immediate. To produce the formula '$2^2 = 4$' with the intention of making an assertion is to intend it to be understood that the thought expressed by '$2^2 = 4$' is true. By appeal to his elucidation of what the True is, this in turn is to intend it to be understood that '$2^2 = 4$' designates the True. Moreover, since the thought expressed by '$2^2 = 4$' is none other than the thought that $2^2 = 4$, and since the thought that $2^2 = 4$ is true if and only if $2^2 = 4$, to intend it to be understood that the thought expressed by '$2^2 = 4$' is true is simply to assert that $2^2 = 4$. Therefore, to construe the utterance '$2^2 = 4$' as intended to designate the True is to construe it as the assertion or judgment that $2^2 = 4$. This apparently proves that a formula (or sentence) viewed as a name of the True expresses just the same judgment that it is ordinarily taken to express, and that in turn demonstrates that the well-formed formulae of concept-script do express judgments. The crucial step in this argument is the thesis that the thought that p is true is the same thought as the thought that p. It is noteworthy that this idea surfaced in Frege's thinking only after he began to view sentences as names of truth-values. In any case, this 'redundancy theory' of truth is essential to establish even a *prima facie* case that the formulae of concept-script are expressions of judgments. Unless 'predicating [truth] is always included in predicating anything whatever' (PW, 129), how could names of the True be shown to be expressions of judgments?

The second part of the case that concept-script is a logical symbolism is explicitly presented in 'Function and Concept'. This is the brisk (and fallacious) argument that a concept is a function whose value is always a truth-value since to assert that A falls under the concept Φ is equivalent to stating that the value of the function $\Phi(x)$ for the argument A is the True. Frege reinforced this reasoning

---

24. Note that this problem did not arise in respect of the symbolism of *Begriffsschrift*, since there complete well-formed formulae stood for judgeable-contents!

by accepting the standard opinion of contemporary logicians that the subject/predicate structure of singular atomic sentences perspicuously represents the logical articulation of the judgment expressed. In decomposing into subject and predicate, such a sentence yields a pair of expressions one complete in itself and one in need of completion which match in this respect the pair of symbols, an argument-name and a function-name, which occur in the representation of the same judgment in concept-script. It seems natural (though fallacious) to conclude that the subject and predicate of the declarative sentence must have the same roles as the argument-expression and function-name in the formula in concept-script. From this doctrine about symbols it is further natural to conclude that the designation of the predicate (viz. a concept) must be what is designated by the function-name (viz. a function whose value is always a truth-value). It apparently seemed to Frege that traditional wisdom about the correspondence of predicates with concepts could be marshalled in support of the claim that his function/argument articulations of judgments carry on the logician's proper business of analyzing judgments into concepts (and objects). In this respect his interest in grammar was indirect; he sought primarily to clarify the true nature of concepts, not to develop a semantic theory about the role of predicates in natural languages. But, since predicates were the traditional handle for laying hold of concepts, how could he make plausible his analysis of concepts independently of any appeal to grammatical categories?

One rationale for Frege's observations about grammar is the need to show his concept-script to be a fit instrument for the formalization of *logic*. A second rationale is equally apparent. Although he thought the concept of a concept to be analyzable, the proper definition of 'concept' (viz. 'function whose value is always a truth-value') itself introduces an indefinable term, 'function', not widely understood correctly. No doubt the proper procedure would be a short course in mathematics, from which the unenlightened would grasp the mathematical concept of a function. Alas, most would rebel at this initiation into the conceptual foundations of the *Basic Laws,* complaining *'Mathematica sunt, non leguntur'*. Hence Frege sought a short-cut to mastery of the concept of a function and the functional character of concepts. Syntax seemed to him to provide what is required.

The proper expression for a function in mathematics is either a formula containing gaps for the insertion of argument-names (e.g. '$2(\ )^3 + (\ )$' (FC, 24)) or one containing variables construed as mere place-holders for argument-expressions (BLA, §1). Either representation makes perspicuous crucial features of (first-level) functions: (i) the expression does not stand for a number; (ii) it combines immediately with a numeral to form a complex designator of a number; and (iii) filling its argument-place with another such function-name never yields a designator of a number. Hence mathematical notation makes transparent the type-difference between numbers and functions. It does so because the variable (or gap) in a function-name 'renders recognizable the particular type of need for completion that constitutes the specific essence of the function designated' (BLA, §1). The nature of what is symbolized shines through features of the symbol.

Frege decided to capitalize on what he perceived to be a parallel in the symbolism of natural languages and to elucidate the functional character of concepts directly by appeal to features of symbols standardly taken to stand for concepts. Here again he exploited the doctrine that the predicate of a singular atomic sentence stands for a concept. In his view, such a sentence can be split in accord with its subject/predicate structure into a part complete in itself (a proper name of an object) and an expression needing to be completed (a predicate designating a concept). This 'unsaturatedness' of the predicate can be made quite perspicuous by using a notation containing a gap or a variable holding a place open for a proper name. Thus the sentence 'Caesar conquered Gaul' can be split into two parts, the proper name 'Caesar' and the expression 'conquered Gaul' or 'x conquered Gaul' (cf. FC, 31). Such visibly unsaturated expressions, constructed on the model of function-names in mathematical symbolism, Frege rather oddly called 'predicates', and the incompleteness of these expressions he described as their 'predicative' character. On this understanding, the unsaturatedness of a predicate is claimed directly to mirror the nature of what it symbolizes (viz. a concept), and the distinction between proper names and predicates is held to reflect the essential difference between objects and concepts. Construed as an argument that concepts are really functions, this reasoning would be transparently question-begging. But as an elucidation of Frege's notion of the functional character of concepts, his observations might have some utility. From his preferred symbolism, the unsaturatedness of concepts shines through the unsaturatedness of the predicates which symbolize them. This makes perspicuous the difference in logical type between concepts and objects as well as burying traditional qualms about the 'unity of the proposition'. But if concepts are conceived to be what the predicates isolated in traditional grammar stand for, then the degree to which the 'predicative nature' of concepts is transparent in natural language is proportional to the degree of analogy (if any) between these expressions and the visibly unsaturated expressions that Frege called 'predicates'. Whatever the success of his elucidation of the functional character of concepts may be judged to be, his intentions are clear enough. His purpose in stressing that concepts are predicative was not to illuminate the semantic role of predicates, but to clarify what he considered an essential feature of concepts.

Even if we prized his elucidation of concepts by appeal to syntax,[25] we would have to concede that it conveys only a partial and provisional understanding of the conception that concepts are functions. In particular, it leaves type-differentiations among concepts completely opaque. It does nothing to make clear that each concept, being a function, must have a determinate and invariant number of

---

25. It is noteworthy that Russell followed the same strategy in his early elucidation of the concept of a propositional function. Having remarked that '[a]lthough a grammatical distinction cannot be uncritically assumed to correspond to a genuine philosophical difference, yet the one is *prima facie* evidence of the other, and may often be most usefully employed as a source of discovery', he went on to explain the distinction between *concepts* and *things* by reference to the subject/predicate decomposition of singular declarative sentences and to explain what a propositional-function is in terms of replacing a term of such a proposition by a free variable (Russell, *Principles of Mathematics*, pp. 42ff., 83ff.).

argument-places. And it leaves the notion of higher-level concepts shrouded in mystery; for, apparently, the name of a second-level concept must contain a gap for an argument-expression, and so must the name of a first-level concept, and therefore the combination of the two, being visibly an incomplete expression, seemingly *cannot* stand for an entity complete in itself (a truth-value)! The surface features of predicates may provide useful hints about some aspects of the functional nature of concepts to persons who lack understanding of the mathematical concept of a function, but at best the indications of syntax must be supplemented by further elucidations drawing on different considerations. Presumably Frege hoped that, having mounted the ladder of syntax to a preliminary grasp of what a concept is, the initiate could kick it away and proceed upwards to an apprehension of the other aspects of concepts essential to his construction of quantification theory.

There are clear indications in his writings of motivations for his reflections on the grammatical articulations of declarative sentences which do not require the ascription to him of the alleged insight that the logical analysis of inferences is inseparable from the semantic analysis of the expressions of judgments. His comments on grammar no more constitute a solid foundation for the philosophy of language than do Mill's in *A System of Logic,* and he did not consider them in any way central to the construction of his logical system. For the most part he treated grammar as no better than a fallible guide to the analysis of thoughts. To the limited extent that he went clearly beyond this position he fell into scholastic nonsense, such as the claim that 'The concept Φ is subordinate to the concept Ψ' expresses a relation between a pair of altogether mysterious objects. But of course even the rough generalizations that he presented as the beginning of wisdom are themselves nonsense of a less obvious sort. We can make sense of his attaching importance to the claims that predicates stand for functions and that concepts are essentially predicative even if we cannot make sense of these claims that he and many of his successors have found so important.

6. Putting the *Basic Laws* on proper *Foundations*

The task of identifying Frege's motivation in the mature phase of his thought and hence the task of specifying the genre to which his mature writings should be seen to belong collapses into the double-headed enterprise left over from our initial survey of the general features of 'Function and Concept'. First, we must account for the paucity of argumentative support provided for the basic ideas of his mature logical system. And second, we must explain why his overriding innovation was the thesis that expressions of thoughts (equations, inequations, and declarative sentences) are names of the True and the False from a logical point of view. It is reasonable to expect that the answers to both questions depend on an accurate description of the direction from which he approached his fundamental problem, the clarification of the notions of concepts and relations. If we could reconstruct exactly how he saw matters, we might find that his angle of vision prevented him from seeing that there was anything to discuss where we discern things that are highly problematic. We must now explore whether we can discover any interpretation of his mature thought that integrates at least the great bulk of it smoothly in his grand design to prove the analyticity of arithmetic.

Neither 'Function and Concept' nor the *Basic Laws* start from scratch. Both build on what is laid down in *Begriffsschrift* and the *Foundations*. 'Function and Concept' was intended for an audience (supposedly) familiar with Frege's earlier system of logic and ready to accept its general contours; the lecture sets out to explain how this earlier work is now to be modified (FC, 21). Likewise the *Basic Laws* explicitly incorporates the main results of the *Foundations* (BLA, p. ix, §0); the core of this project is presenting a formal and rigorously established solution to the problem of how to map certain second-level concepts (viz. the referents of numerically definite quantifiers) onto objects (viz. the natural numbers) in conformity with the constructive analysis of the *Foundations* (cf. PW, 256f.). Frege assumed that he had already justified a number of fundamental principles, and therefore he omitted a re-exposition of the supporting arguments. Our previous examination of his doctrine of conceptual content makes clear that there is a massive overlap of ideas between his earlier logical system and the one presented in the *Basic Laws*. It is particularly noteworthy that many of these permanent ideas are just those for which we now think that he should have provided argumentative support in his later writings. Let us briefly review the main points of continuity which bear on his mature account of concepts:

(i) Sentences (also equations and inequations) stand for objects. Previously, any expression of a judgeable-content designated (*bezeichnet*) a judgeable-content (BS, §2); now, any expression of a thought stands for (*bedeutet*) a truth-value. In both cases, sentences are 'proper names'.

(ii) Concepts are functions. Previously, concepts were functions whose values were always judgeable-contents; now, concepts are functions whose values are always truth-values. *Begriffsschrift* rested on the principle that the concept/object decomposition of a judgeable-content is an analysis into function and argument. In both cases, concept-words are function-names.

(iii) Logical constants are function-names (concept-words). Previously, propositional connectives stood for functions whose argument and values were judgeable-contents, while quantifiers paradigmatically stood for functions mapping first-level concepts onto judgeable-contents. Now the explanations are altered only by the substitution everywhere of 'truth-values' for 'judgeable-contents' (FC, 30ff.; BLA, §5ff.).

(iv) The distinction of objects from concepts is fundamental in logic and absolute (FA, x; BLA, §2; FC, 32; CO, 43ff.).

(v) The notion of the extension of a concept is essential for the possibility of effecting a transition from concepts to objects (PMC, 140f.; cf. PW, 257). Extensions of concepts are objects (FA, §68n.; BLA, §§2f.), and they appear to stand in one-to-one correspondence with concepts. Hence hope of finding a logical operation to transform concepts into objects is parasitic on the notion of the extension of a concept. Numbers in particular are identified as the extensions of certain concepts (FA, §68; BLA, §41f.).

(vi) Specification of functions by analytical formulae containing free variables is the paradigm of formal definitions of concept-words in concept-script (BLA, §33ff.; PW, 228ff.). Function theory thus provides a standard of adequacy or a norm of representation for explanations of concept-words, whether these explanations take the form of elucidations of primitive symbols in concept-script (e.g. BLA, §5), formal definitions of abbreviations in concept-script (e.g. BS, §24; BLA, §34), or analyses of concept-words already in use (e.g. PW, 23ff.; cf. 210f.).

(vii) The unlimited possibilities for functional-abstraction license a limitless variety of modes of formation of complex and fruitful concepts. Even for formulae in concept-script, decomposition into function and argument is typically a matter of how we choose to view the thought expressed. If '——— $\Phi(A)$' expresses the thought that 2 is prime, then we can take '$\Phi(\xi)$' to designate the concept of being prime and 'A' to name the number 2, or we may view the concept of being a prime as the argument of a second-level concept (BLA, §22; cf. BS, §10), or we can take the horizontal as a second-level function that transforms any first-level function of one argument into a concept (BLA, §5) and take '——— $\Phi(A)$' to stand for the value of this function applied to the concept of being a prime when A is taken as argument. (All of the symbols of a formula in concept-script are systematically ambiguous!)

(viii) Every formula of concept-script fits the Augustinian picture of language. Previously, every symbol in the expression for a judgeable-content had a content (except for the content-stroke and the judgment-stroke). Now, with the exception of free variables and the judgment-stroke, every symbol in the expression of a thought has a sense, and provided that scientific rigour is achieved, it will also stand for something (have a reference).

This survey of points carried over from *Begriffsschrift* and the *Foundations* to the *Basic Laws* makes clear that Frege would not have thought it worthwhile to discuss whether or not sentences stand for objects and whether or not concepts are functions. These are fixed points in his framework of thinking. From his perspective, the serious open questions are *what* objects sentences stand for and what the arguments and values of concepts are. Of course, it remains to explain why he had become dissatisfied with his earlier answers to these questions, i.e. why he felt it necessary to reopen these questions; and also to explain why the introduction of truth-values provided him with what he now thought to be satisfactory answers. From our stand point, however, his proceedings are deeply unsatisfactory. Frege merely tinkered with comparative details within a pre-established system, while we feel an urge to call the entire enterprise into question and to demand a cogent justification for the *Grundgedanken* of his logical investigations. To follow his reasoning will at best make intelligible his reaching the conclusions he did, but may do nothing to make the conclusions he reached intelligible. Nonetheless, there is value in inspecting his reasoning carefully because we will thereby be forced to reconsider what differentiates our own conception of logic from the confusions evident in his and ultimately to reopen the question about the fundamental justification for our own discrimination between sense and nonsense.

What needs clarification now is why Frege felt the need for modifications, both formal and philosophical, to the logical system of *Begriffsschrift* and what constraints he imposed on what would constitute acceptable solutions to the problems that bothered him. The programme of the *Basic Laws* is the presentation of complete formal proofs in concept-script of fundamental truths of arithmetic from which the whole body of arithmetical truths may be deduced. Frege aimed to vindicate the view of arithmetic which he had outlined informally in the *Foundations*, viz. to prove that arithmetic is a branch of logic from the definition of numbers as the extensions of certain concepts. Does the logical system elaborated in *Begriffsschrift* suffice for this purpose? A negative verdict is glaringly obvious. There is no notation for extensions of concepts in this original system. *Begriffsschrift* lacks an operator which maps concepts onto extensions, and hence it affords no means for determining relations among extensions of concepts from relations among concepts themselves. Without supplementation, therefore, it does not provide the possibility of converting the proof-sketch of the *Foundations* into the required complete proofs in concept-script. The vital advance in the formal system of the *Basic Laws* is the filling of this lacuna: the introduction of the notion of the extension of a concept and the novel notation, a smooth breathing, for designating extensions of concepts (BLA, p. ix f.). If the expression '$\Phi(\xi)$' stands for a concept, then the complex symbol '$\grave{\epsilon}\Phi(\epsilon)$' names the extension of this concept. The smooth-breathing can be viewed as a variable-indexed variable-binding operator standing for a function that maps each concept onto its extension. The presence of this symbol among the primitive symbols of concept-script is the main formal innovation differentiating the *Basic Laws* from *Begriffsschrift*.

The introduction of this operator into the symbolism of concept-script is a necessary condition for a formal vindication of the central thesis of the *Foundations*, but by no means a sufficient condition. The designated operation: $\Phi(\xi) \rightarrow \grave{\epsilon}\Phi(\epsilon)$ is subject to one overriding constraint. It must be a *logical* operation. If numbers are to be defined as the extensions of concepts, then it is not sufficient for demonstrating that numbers can be defined in pure logic that the concepts whose extensions they are are purely logical.[26] In addition, the extension of these concepts must be determined by logical considerations alone. Since numbers are defined as sets of equinumerous concepts, this condition can be met only if the transition from a concept to its extension is effected in every instance by a logical operation. What this means is somewhat obscure, but at the least it must require that the extension of the concept $\Phi(\xi)$ can be calculated *a priori* from the concept $\Phi(\xi)$, i.e. that the value of $\grave{\epsilon}\Phi(\epsilon)$ depends only on the argument $\Phi(\xi)$ and not in addition on matters of fact. This conception is contrary to the standard logicians' view of the relation between concepts and their extensions. In the 1890s (as now too) the extension of a concept was commonly held to depend, in typical cases, on

---

26. Although the concept designated by '$x \neq x$' is purely logical, and so too the second-level relation of equinumerosity of pairs of concepts, the issue of which concepts are equinumerous with the concept $x \neq x$ seems to turn on matters of fact (e.g. the identification of the concept of a unicorn as such a concept). If the identity of a set depends on what members it has, this reasoning precludes taking the set of concepts equinumerous with the concept $x \neq x$ as a *logical* object.

facts of experience; which objects, and how many objects, fall under the concept of being an inhabitant of Berlin at midnight on 1st January 1893 can be determined only by observation and not from close inspection of the concept. For this reason, no doubt, Kerry called the extension of a concept its *'empirische Umfang'*.[27] This background set Frege an important task. The successful execution of his logicist programme demanded that he exhibit the operation: $\Phi(\xi) \rightarrow \acute{\epsilon}\Phi(\epsilon)$ as a purely logical one. In his view he had to prove that 'by means of our *logical faculties* we lay hold on the extension of a concept by starting out from the concept' (PW, 181, our italics.) This task is of course philosophical and stands apart from the formal proofs of the *Basic Laws*. His general account of concepts and relations is directed at discharging this task, not at all at justifying his entire scheme of logical analysis *de novo*. He was absolutely right to discern a need for philosophical argument on the narrower front, even if insensitive to a parallel need on the wider one.

It is important to note that the informal explanations accompanying the formal system in *Begriffsschrift* do not supply the materials necessary for constructing the appropriate philosophical backing for introducing into the *Basic Laws* the operation: $\Phi(\xi) \rightarrow \acute{\epsilon}\Phi(\epsilon)$. A straightforward argument leads to this conclusion. Frege did not give an explanation of the expression 'the extension of a concept'; he took it over from traditional logic, assuming its meaning to be altogether familiar and unproblematic (cf. FA, §68n.). If pressed, he would probably have produced some standard explanation; that A belongs to the extension of the concept Φ means that A falls under the concept Φ, or that A has the property Φ, or that it is true that $\Phi(A)$. Alternatively, he might have declared that the extension of Φ is the set of all those objects of which Φ is true. The crucial point is that the notion of the extension of a concept is internally related to the notion of truth (and falsity). But truth plays no direct role in the framework of *Begriffsschrift*. From a logical point of view, all that matters is judgeable-content, and any judgeable-content can be apprehended independently of knowing its truth-value. Hence determining the truth or falsity of a given judgeable-content falls outside the purview of logic. This has repercussions for concepts. Since a concept is a function, it is completely characterized by stipulating its value for every admissible argument, and since its values are always judgeable-contents, this characterization is wholly independent of establishing the truth or falsity of the particular judgeable-contents which are determined as its values for the various arguments. According to *Begriffsschrift*, therefore, the transition from a concept to its extension cannot be a logical operation. The determination of the extension of a concept from the concept itself is always indirect (FA, §53) and typically depends on considerations external to logic. Consequently the programme adumbrated in the *Foundations* demands a modification and extension of the informal explanations of *Begriffsschrift* as well as changes within the formal system.

The philosophical value of the system of proofs in the *Basic Laws* rested on the solution of a philosophical problem for which Frege's earlier logical reflections

27. B. Kerry, 'Ueber Anschauung and ihre psychische Verarbeitung, IV', *Vierteljahrsschrift für Wissenschaftliche Philosophie*, 11 (1887): 264.

could not supply the answer. Though not yielding the solution, *Begriffsschrift* did mould his thought. It supplied a number of obvious constraints on what he could reasonably count as a solution, and thereby it steered his investigations of possible options in certain directions.

(i) The notion of the extension of a concept is a datum. Hence the only method to secure that the operation: $\Phi(\xi) \rightarrow \acute{\epsilon}\Phi(\epsilon)$ is purely logical is to modify the notion of a concept given in *Begriffsschrift*.

(ii) The notion of an extension is internally connected with the concept of truth via the explanation that the extension of $\Phi(\xi)$ consists of all and only those objects x of which it is true that $\Phi(x)$. Therefore whether or not a given object falls under $\Phi(\xi)$ must be built into the identity of the concept $\Phi(\xi)$ if the operation $\Phi(\xi) \rightarrow \acute{\epsilon}\Phi(\epsilon)$ is to be guaranteed to be independent of matters of fact.

(iii) Since concepts are functions, since the range of objects that fall under a concept is something given, and since the identity of a function is wholly determined by correlating with each argument the value of the function for this argument, modification to the notion of a concept presented in *Begriffsschrift* must take the form of some fundamental change either in the nature of objects or in what objects are admissible as values of the members of that subclass of functions which count as concepts.

(iv) Although it must be built into the identity of the concept Φ whether any given object does or does not fall under Φ, it is a fundamental datum that knowing what is asserted by a sentence or formula '$\Phi(A)$' is independent of knowing whether what is asserted is true, i.e. whether A does fall under Φ (PW, 7). Hence knowing what is asserted by '$\Phi(A)$' must be independent of knowing the extension of the concept Φ and, *pari passu,* independent of knowing what concept 'Φ' stands for.

It is one thing to enumerate these constraints, quite another to satisfy them all simultaneously. Frege attempted to do so by his account of the general notions of concept and relation, together with the supplementary distinction of sense and reference. But can each of these constraints be *justified* and is it *possible* to satisfy them *all together?* Frege has sown the seeds of a kind of philosophical schizophrenia. Both the interest and the intelligibility of his thesis that count-statements ascribe properties to concepts presupposes that we have a prior grasp of what concepts are, but the proof that arithmetic is a branch of logic requires there to be an internal, *a priori* connection between concepts and their extensions which is at odds on a fundamental point with our conception of a concept. According to the first of these strands of thought, he would see himself as analyzing the notion of a concept which we already possess, whereas, according to the second, he must view his enterprise as one of replacing our defective conception of a concept by something more adequate, more precise and explanatory. We might well expect to find Frege vacillating here, conceiving of himself now as clarifying what we already understand, now as discovering the hidden depths in what we mistakenly suppose ourselves to understand fully, and now as proposing to replace our defective conception with something better.

7. New logical machinery

The successful completion of the philosophical programme outlined in the *Foundations* rested on demonstrating a logical connection between concepts and their extensions, and Frege thought this in turn to depend on establishing an internal relation between the objects taken as the values of concepts and the truth or falsity of what is asserted in saying that objects fall under concepts. He had no option but to revise the philosophical basis of his concept-script as he had expounded it in *Begriffsschrift*. He had to argue either that the truth or falsity of a judgeable-content is logically related to the identity of the content itself or that objects other than judgeable-contents are the values of concepts. If both moves were blocked, the reduction of arithmetic to logic would be checkmated.

The less radical of the two options seems superficially to be a modification to the conception of judgeable-content. Why not reason that it is a blunder to divorce judgeable-contents from truth and falsity? Frege had evidently assumed that every judgeable-content is necessarily either true or false (BS, §27). This establishes a connection (the principle of bivalence) relating what is asserted to truth and falsity. Could this connection not be tightened up so that the truth or falsity of every judgeable-content could be determined independently of experience? This would require either that the property of truth must be self-evident from the apprehension of a judgeable-content which is in fact true[28] or that the truth of a content can be calculated *a priori* from some other property common to all and only those judgeable-contents which are actually true. Neither of these avenues was open to Frege. The first conflicts with his principle that it is possible to grasp a judgeable-content (e.g. a putative theorem in geometry) without knowing whether or not it is true (e.g. in ignorance of whether it has a proof). This idea *seems* immune to challenge. The second avenue also turns out to be a *cul de sac*. At a minimum it is required that there be some property whose possession by a judgeable-content provides an *a priori* justification for holding this content to be true. Presumably, the sentence expressing the fact that a given content has this property expresses a judgeable-content, distinct from the given content. This generates a vicious infinite regress in justifying the calculation of the truth of a content from its possession of some other property. Consequently, it makes no sense to suppose that the truth (or falsity) of a judgeable-content can always be determined *a priori* from the intrinsic features of this content. If truth is correctly viewed as a property of judgeable-contents, it must in general be an extrinsic property. That conclusion frustrates the hope of establishing any logical relation between concepts and their extensions.

The less radical option being foreclosed, only the more radical one remains. Objects other than judgeable-contents must be introduced to play the role of the value of concepts (viewed as functions), and unlike judgeable-contents, these fresh objects must have an intrinsic connection with truth and falsity. A few waves of the magician's wand and there spring full-blown from Frege's top hat two novel

---

28. As Descartes thought in proposing clarity and distinctness as a criterion of truth.

objects: the True and the False! This act of creative genius is accompanied by some reassuring patter: 'The thought that 2 is prime designates the True' says the same thing as 'It is true that 2 is prime', while 'The thought that 4 is prime designates the False' is equivalent with 'It is false that 4 is prime' (cf. BLA, §2; FC, 28f.).

If we succeed in swallowing the True and the False without choking, we find that our logical hunger is appeased. This pair of objects yields a solution to Frege's philosophical quandary which conforms with his constraints. The notion of the extension of a concept remains inviolate. Indeed, since concepts are still viewed as functions, this notion can even be analyzed by identifying the extension of a concept with the course-of-values of this concept (BLA, §3; FC, 30f.). Modification to the earlier notion of a concept has taken the form of tampering with the objects that constitute values of concepts for arguments. The possibility is secured of grasping what is asserted without knowing whether this is true or false. This possibility is now represented as the possibility of knowing what thought is expressed by a sentence without knowing which truth-value it designates—a special case of the general thesis that knowing the reference of an expression is not an immediate consequence of knowing its sense. Finally, the operation: $\Phi(\xi) \rightarrow \acute{\epsilon}\Phi(\epsilon)$ is manifestly a purely logical transformation of concepts into objects, since the identity of a function is completely determined by its course-of-values (the correlation with each argument of the value of the function for this argument) and the extension of a concept $\Phi(\xi)$ is uniquely picked out as those arguments for which $\Phi(\xi)$ takes the True as its value.

No wonder Frege boasted 'how much simpler and sharper everything becomes by the introduction of truth-values' (BLA, p. x)! On this innovation alone rests the justification for the claim that the formal derivations of the *Basic Laws* constitute a proof of the philosophical thesis that arithmetical laws reduce to laws of logic. The True and the False bear the weight of the *Foundations*. Frege did not advance the view that basic conceptions and principles in logic should be supported inductively by showing that they license the derivation of truths of arithmetic. But the requirements of his logicist programme no doubt aided him in seeing 'the true nature' of concepts. Indeed, he suggested that the point (*Zweck*) of construing concepts as functions is to make possible the reduction of arithmetic to laws of logic (FC, 30). A philosopher of a more candid pragmatic bent might have argued that the formal derivations of the *Basic Laws* by themselves justify 'postulating' the True and the False, perhaps adding the gloss that 'there is probably no better way to decide what is methodologically permissible in science than by investigating what successful science requires.'[29]

This defence need not silence us. We might press objections on two fronts. First, we should challenge the claim that Frege has made minimal conceptual modifications in introducing the two truth-values for his purposes. Rather, he has generated patent nonsense and enthroned it in place of sound understanding. It is

---

29. J. A. Fodor, 'The Mind-Body Problem', *Scientific American* Vol. 224, No. 1 (Jan. 1981), p. 132.

absurd to deny that truth is a property of assertions (infra, p. 347ff.), to assert that concepts are functions or to affirm that sentences stand for any objects whatever. To build the science of logic on these contentions would yield a body of propositions as full of interest as the science of the pitches of colours or the investigation of the causal interactions among the natural numbers. In logic, as Wittgenstein remarked, *all* differences are *big* differences. The second line of objection would focus on the conception of philosophy as the elaboration and testing of hypotheses. The issue of inductive support arises only in respect of suppositions that make sense. Therefore, somebody who points to the simplicity and power of Frege's logical system as an argument in favour of the existence of the True and the False has begged the central question, viz. whether any of Frege's contentions about these two objects make any sense at all. There is no question here of inductive support any more than there is a question of confirming hypotheses about the conditions under which the opponent of a patience player wins games of patience. On this view, philosophy concerns the clarification of what is in some sense familiar, not the discovery of hitherto unsuspected facts or entities.

Whatever our ultimate verdict on this clash of *Weltanschauungen* may be, we must be clear what Frege's angle of vision was. This emerges plainly from the background to his introduction of the True and the False. The reduction of arithmetic to logic was his central concern in logic; a logical connection of concepts with their extensions was indispensable to his plan for executing the logicist programme; and treating the True and the False as the values of concepts was the only possible means for forging the necessary link of concepts with their extensions. The two truth-values constitute the drive-shaft of his logical machinery, not some ornamental trim concealing the real works. His talk of the True and the False is not a *façon de parler;* rather, in his view, the existence of these *objects* is the *sine qua non* of exhibiting concepts as *functions*. To treat comments about these putative objects as figures of speech would be to reduce the whole of his logical system to the status of an elaborate metaphor. Frege's hopes for the salvation of his logicist programme rest entirely on the True and the False. Shorn of these two objects, his logical system would have appeared to him as flat, stale, and unprofitable as Christianity without the Passion and the Resurrection.

8. Spin-off from the new technology

The introduction of the True and the False into Frege's logical system acted as a catalyst to bring about a profound transformation that left no substantial part of the exposition of *Begriffsschrift* intact. Admission of courses-of-values into concept-script was a change 'extremely important in principle' (BLA, p. x). The two truth-values, in serving as the values of concepts for objects as arguments, required the support of the sense/reference distinction, and this implied a thorough-going modification of the earlier system of logic. Recognition of the True and the False has repercussions that were propagated throughout Frege's mature thought. Certain positive repercussions were most obvious to him, and these he highlighted in the *Basic Laws*. But there was as well a set of fresh problems, only one of which he explicitly faced and clearly resolved.

The first gain is the possibility of remedying an inelegant asymmetry in the informal explanations of *Begriffsschrift*. Though judgeable-contents were there treated as objects, they were distinguished from a logical point of view from all other objects. This was apparent even in the formation-rules for concept-script (BS, §2). But more importantly, judgeable-contents alone seemed to be exempted from the principle that the content of a proper name designating an object is simply the object designated, since the representation of a judgeable-content as the value of a particular function for a given argument is logically significant. Function/argument decomposition of judgeable-contents was the primary innovation of *Begriffsschrift*. The introduction of truth-values as what sentences designate sharpens the perception of this asymmetry and at the same time makes it appear less defensible. In analyzing inferences, a logician must distinguish between different sentences (complex expressions) standing for the True. Their logical significance is manifestly not exhausted by a specification of what object they designate; rather, it must extend to an indication of *how* this object is determined *as* the value of a function (concept) for an argument. But why not treat all proper names of objects in a parallel manner? Frege suggested that we should admit any object without restriction as a potential argument or value of a function. If an object is specified as the value of a function for an argument, then what the corresponding complex proper name stands for is this object; but the name also indicates how this object is determined as the value of a function for an argument, and this is logically significant. Indeed, recognizing this fact remedies one of the defects of *Begriffsschrift*, i.e. the inability to give a coherent account of why a proof may be required to justify substitution of one name for another in the formulation of a true judgment (supra, p. 168f.). In the *Basic Laws* every complex name of an object, whether or not this is a truth-value, has a double role: it stands for the object, and it indicates how this object is determined as the value of a function. Concept-script itself contains two primitive functions whose values may be objects other than truth-values: the description-operator '\$\backslash\xi$' and the course-of-values operator '$\acute{\epsilon}\Phi(\epsilon)$'. Expressions formed from these operators represent objects as determined in particular ways. Thus functions whose values are numbers or extensions of concepts have a legitimate place in concept-script,[30] even if concepts have a certain pride of place among the totality of functions.

A second gain is a greater clarity in the informal explanations of the basic logical operators: the negation-stroke, the condition-stroke, and the universal quantifier. *Begriffsschrift* in effect elucidated each of these symbols as a concept-word, but without the explicit claim that any of them stood for a concept (function). This left a degree of murkiness in these explanations. The account of the condition-stroke, e.g. specifies whether

---

30. Numerical functions and operations had earlier appeared in mathematical applications of concept-script (e.g. PW, 23ff.).

'⊢ B'
  ⊥ A

is to be asserted in terms of whether the judgeable-contents A and B are separately to be asserted (BS, §5). But how does knowledge of whether what is expressed by

'⊢ B'
  ⊥ A

is to be asserted constitute knowledge of what is asserted in uttering

'⊢ B'
  ⊥ A

with assertoric force? Is it then clear what judgeable-content (i.e. what value) is assigned to the function designated by the condition-stroke for any pair of judgeable-contents as arguments? The same problem beset the explanations of the other two primitive operators of *Begriffsschrift*. The introduction of the True and the False seems at first sight to disperse this fog. The condition-stroke, e.g. is now explicitly defined as a function-name:

'⊢ B'
  ⊥ A

has the value the True if 'A' stands for the False or 'B' stands for the True, and otherwise has the value the False (BLA, §12).[31] Similarly, '—$\overset{a}{\frown}$— Φ (a)' is explained as taking the True as its value if and only if 'Φ(A)' stands for the True whatever object is named by 'A' (BLA, §8). Consequently the two truth-values enabled Frege to give explanations of his primitive logical operators that are isomorphic with modern semantic definitions.[32] He could specify with perfect clarity what each molecular formula in concept-script stands for in terms of what its constituent symbols stand for.

A third apparent gain is the sharper characterization of 'the nature of the function as distinguished from the object', with the consequent clarification of 'the distinction between first- and second-level functions' (BLA, p. x). In retrospect the elucidations of *Begriffsschrift* appeared defective. Lack of clarity about the precise nature of the value of a concept for an argument infected with obscurity the elucidation of the notion of a concept and the distinction of concepts according to the type-hierarchy of function-theory. The two objects, the True and the False, remove all such doubts and difficulties. A concept, if defined as a function whose value is always a truth-value, is obviously and literally a function, exactly comparable with 'x^2' or 'sin x'. The distinction of functions according to the level of their arguments is directly and perspicuously applicable to concepts; the symbol

---

31. The explanation of the horizontal guarantees that this definition is complete, since '——— A' stands for a truth-value whatever object 'A' may stand for (BLA, §5).
32. To suppose that they are identical with these modern explanations is deeply mistaken (supra, p. 114ff.).

'——$\overset{a}{\frown}$—— Φ (a)' contains the name of a second-level function exactly comparable with second-level functions in analysis $\left(\text{e.g. } \dfrac{d}{dx} f(x)_{x=0}\right)$.

A fourth gain is a sharp analysis of the concept of an extension of a concept and the consequent resolution of controversy and perplexity about the real nature of extensions. Frege accused his contemporaries of making a fundamental mistake in building logical analysis of inferences on the concept of a set, aggregate, or system, as if this concept were primitive. The results of this misguided strategy were distinctive problems and confusions: an inability to give a coherent explanation of the empty set, an assimilation of empty concepts to names lacking referents, a tendency to confuse class-membership with class-inclusion, and an apparent asymmetry in the nature of finite and infinite sets (BLA, §0; SVAL, 86ff.). The remedy is to acknowledge that 'the extension of a concept or class is not the primary thing' in logic (PW, 184), that 'the concept ... takes logical precedence over its extension' (SVAL, 106). The concept of a concept is prior to the concept of a class, and classes are introduced into concept-script only by means of the primitive logical operation mapping each function $\Phi(\xi)$ onto its corresponding course-of-values $\grave{\epsilon}\Phi(\epsilon)$.

Did Frege reap all those benefits at no cost? Not even he was so sanguine. The price of overall simplification was the bifuraction of conceptual content into sense and reference. But he paid that price cheerfully, fancying that he derived further independent benefits. Of other hidden costs he manifested no awareness.

Introducing the True and the False as what sentences (including equations) stand for raises obvious objections. Frege considered three, it seems. First, the manoeuvre licenses taking certain identity-statements (e.g. '$(2^2 = 4) = (2 > 1)$') to express true judgments despite the fact that what is asserted by the two constituent sentences is different (FC, 29). Second, many sentences contain sentences as proper parts, and the substitution for a constituent sentence of another sentence of the same truth-value often fails to preserve the truth-value of the whole (SR, 65ff.). Third, simple identity-statements may require proof (e.g. '2 + 2 = 4' (cf. FA, §6)), although their content may seem no different from that of another identity not requiring proof (e.g. '4 = 4') (cf. SR, 56f.). Two of these objections can be side-stepped by the independently required distinction between what a sentence expresses and what it stands for (FC, 29), i.e. its sense (a thought) and its reference (a truth-value).[33] The other objection elicits the plausible response that a subordinate clause need not stand for a truth-value (cf. SR, 65ff.). But the distinction of sense from reference allows the transformation of this negative doctrine into an exciting positive thesis: in all cases where a clause does not stand for a truth-value or for an ordinary object or property (as in relative clauses), what it stands for is (at least in part) the thought ordinarily expressed by this sentence. This yields a persuasive analysis of indirect speech; an indirect statement stands

---

33. This involves a revision of the assessment of the truth of certain well-formed formulae in concept-script. In *Begriffsschrift*, the judgment expressed by '$(2^2 \equiv 4) \equiv (2 > 1)$' would be false, whereas the thought expressed in the *Basic Laws* by '$(2^2 = 4) = (2 > 1)$' is true.

for the thought expressed by the sentence in direct speech. It seems transparent that if I report 'Frege asserted that sense is always distinct from reference', then I describe a relation as holding between Frege and a thought, viz. the thought expressed by the sentence 'Sense is always distinct from reference'. A generalization of this argument constitutes an elegant account of *oratio obliqua*. Frege thus inspected his new creation, the separation of sense from reference, and saw that it was good: 'Only in this way can indirect discourse be correctly understood' (BLA, p. x). A deft manoeuvre transformed an objection into a brilliant argument in favour of his own position. No wonder Frege harped on his analysis of *oratio obliqua* in spite of his having no direct use for it in the main stream of his work![34]

Other fundamental problems raised by the revised analysis of concepts and relations had no direct reflection in formulae of concept-script. Frege gave them little, if any, explicit attention, and he did nothing to rebut obvious objections, still less anything to turn them to his credit. Astute excavations in carefully chosen spots will bring them to light.

---

34. 'In my *Begriffsschrift* I did not yet introduce indirect speech because I had as yet no occasion to do so' (PMC, 149). Does the analysis of indirect speech in concept-script play a notable role in the *Basic Laws?*!

10
Sense and Reference: An Area Survey

1. The rationale

The introduction of truth-values and the elucidation of concepts as functions whose values are truth-values was *the* innovation of the 1890s enabling Frege to complete the logicist programme. For the purpose of constructing logical systems, truth-values replaced judgeable-contents as values of functions for arguments. However, judgeable-content had two further roles, over and above its role as the value of functions for certain arguments, namely as the object of assertion (i.e. what is or can be asserted by uttering a declarative sentence in an appropriate context) and as what is proved or is proof-relevant. Truth-values, however, cannot fulfil these two additional roles of content. Although when we assert something we in effect say that it is true, *what* we assert is not a truth-value. *A fortiori* what distinct proofs prove to be true is not thereby identical, and the random substitution of one true judgment for another in a proof will typically impair the cogency of the argument. In short, the truth-value assigned to a sentence does not exhaust its logical significance, so truth-values cannot take over the entire burden carried by judgeable-contents in *Begriffsschrift*.

What is asserted by using a declarative sentence is an entity distinct from a truth-value. Frege rechristened this, calling it a 'thought' (in a non-psychological sense). What was formerly conceived as a judgeable-content must in effect be split up (horizontally, as it were) into two distinct entities, and the various roles it fulfilled in his early logical system must be distributed between the two. A sentence stands for or designates a truth-value, it expresses a thought. The truth-value is the *reference* of the sentence, the thought expressed is its *sense*. The truth-value fulfils one role of judgeable-content: it is the value of concepts *(functions)*. The thought fulfils the other two roles—the object of assertion, and whatever is relevant to the cogency of inference (as distinct from tone and colouring).

Although Frege said that he split judgeable-content into thought and truth-value (sense and reference), it would be more accurate to say that he split the *roles* of judgeable-content into two and allocated them between thought and truth-value, for truth and falsity were not originally parts of judgeable-content, but externally related to it. This manoeuvre prefaced a 'thorough-going development' (BLA, p. x) of his logic.

Frege ascribed to thoughts all of those features of judgeable-contents not directly connected with function/argument decomposition in logic:

(i) Thoughts are not psychological objects, but rather Platonic entities which exist independently of cogitation or language. They are *what* is judged, asserted or entertained, but their existence is independent of being judged, asserted or entertained. They are what is true or false. Since truth is objective, that which is true, viz. the thought, allegedly must likewise be objective.

(ii) Consequently thoughts stand in contrast to ideas, and fulfil the same role in Frege's antipsychologist polemics after 1890 as judgeable-content did in the *Foundations*. Thoughts are existence- and identity-independent of a bearer (thinker). They are public and shareable. Hence they fulfil the required role of an object of *proof,* making scientific knowledge possible.

(iii) A thought is not identical with the meaning of a sentence, nor with part of its meaning (e.g. its meaning minus its tone). For only declarative sentences express thoughts.[1] Yet, of course, non-declarative sentences have meanings.

(iv) Sense is distinguished from tone or colouring. Substitution of 'nag' for 'steed', of 'but' for 'and', may affect the colouring of the sentence (supposedly generating in the hearer different mental images) but it will leave the thought expressed intact. What belongs to the thought is restricted to what is relevant to inference.

(v) The thought expressed by a sentence is independent of the force annexed to the sentence, i.e. the thought is invariant no matter whether the sentence is used to make an assertion or only embedded in the antecedent of an asserted conditional (or merely uttered in fiction or drama, or used to formulate a formal definition).

(vi) The thought is connected with understanding only *qua* content of a possible assertion or judgment, not as meaning. To grasp a thought which is asserted is not to grasp the meaning of a type-sentence (though it may presuppose that) but rather to grasp what is asserted. Indeed thoughts, being existence-independent of language, can in principle be grasped (though not by us 'mere humans') quite independently of language (PW, 269).

(vii) Thoughts are *objects*. They can occur both as arguments and as values of first-level functions (and Frege indeed exploited both possibilities). Designations of thoughts, e.g. 'the thought that p', are proper names of these objects.

The genealogy of the thought seems clear enough, and its pedigree impeccable. The other branch of this family-tree is truth-values. These abstract objects for

---

1. 'The Thought' modifies this position: sentence-questions also express thoughts (cf. PW, 7, 137f.).

which declarative sentences stand (and which are the references of thoughts) inherit the remaining roles of the earlier conception of judgeable-content.

Splitting the judgeable-content into sense and reference was essential for logicism. But it doubtless seemed to resolve or clarify a host of further issues.

A first possible advantage is that the sense/reference distinction, by ensuring a public object of communication, agreement and disagreement, in conformity with the requirements of antipsychologism, enables us (in principle) to avoid the embarrassment of recognizing private (subjective) objects as unjudgeable-contents. For one might claim that as long as sense is objective the subjectivity of reference is unobjectionable. It is, however, far from obvious whether Frege clearly grasped this (cf. p. 53ff.).

Second, the sense/reference distinction makes it possible to distinguish sharply between expressing a thought and designating a thought, and consequently permitted a novel (though arguably incorrect)[2] analysis of indirect discourse. This would have enabled Frege to eliminate the conflation, in *Begriffsschrift,* of the content expressed by 'A' and 'the circumstance that A', since he could now claim that 'A' expresses what 'the circumstance that A' designates. He did not, however, explicitly acknowledge and rectify this error.[3] He multiplied confusion by insisting that a sentence which expresses a thought still designates *something,* namely a truth-value. Furthermore, shades of the old error persist, viz. that the dependent-clause 'that p' expresses the same thought, unasserted, as is expressed by the assertion of 'p' (PW, 251), even though the dependent-clause 'that p' is said to denote what 'p' expresses. Finally, by distinguishing between expressing and designating a thought *by means of the machinery of sense and reference,* Frege arrived at the unhappy conclusion that a person's beliefs are *about* quite different objects than his utterances, since if I believe that Fa, the subject of my belief is the sense of 'a', whereas if I utter 'Fa', the subject of my utterance is a. Our beliefs, as it were, are insulated from what we talk about by a layer of sense; which is surely nonsense.

Third, the distinction makes it possible to recognize differences between names with the same content, and hence to acknowledge the need for proof to substantiate interchange of coreferential names. This is an immediate consequence of taking the mode of presentation of a content (sense) as inference-relevant. This contrasts with *Begriffsschrift* where the *Bestimmungsweise* was held to be irrelevant to content.

Fourth, the sense/reference distinction made it possible to resolve the antinomy of identity (i.e. render the need for *proof* of an identity intelligible) without

---

2. See A. N. Prior, *Objects of Thought,* p. 48ff.
3. He even hovered on the brink of his earlier conflation of what 'A' expresses with what 'A' designates. For what 'A' designates is a truth-value. But he explained the truth-value of 'A' to be 'the circumstance that it is true or false' (SR, 63). Combining these claims, we conclude that what 'A' designates is what 'the circumstance that "A" is true' designates, and this is a mere hair's breadth from the claim that 'A' and 'the circumstance that A' designate the same thing.

recourse to the counterintuitive apparatus of *Begriffsschrift*. There the antinomy was resolved by attributing a systematic ambiguity to all names, holding that they stand for themselves in identity-contexts and making identities statements about signs. But the advantage of not construing all names as systematically ambiguous is immediately thrown away in the analysis of indirect speech, and Frege's criticism of the earlier account of identity does not turn on the decisive point that it conflated mention and use, but rather on a bogus requirement about the conditions for real knowledge (SR, 57).

Fifth, it made it possible to budget explicitly for complex (sub-sentential) proper names in laws of pure logic. This in turn permitted a *general* explanation of number words, since Frege could now treat $F(\xi)$ as an argument in his explanation of the complex proper name 'the number of *Fs*' as the extension of the concept 'equinumerous with $F(\xi)$'.[4]

Sixth, it allowed Frege to recognize a novel possibility, viz. that names without content (that do not 'stand for anything') may nevertheless contribute to the judgeable-content expressed (i.e. the thought). They have sense, but lack reference. This led to the revamping of the earlier criticism of formalists (FA, §96). Their error is now identified not as the illegitimate transition from concept to object without proof of uniqueness, but the admission of proper names without proof of reference. This in turn might have opened up the possibility of elucidating the cogency of arguments as independent of the actual truth of premises and conclusions, but Frege firmly repudiated any such idea.

Finally, distinguishing sense from reference enabled him to avoid the circularity in the specification of concepts which characterized the theory of content (supra p. 150ff.). It is now possible to give an independent specification of the value of a concept (function) for an argument (i.e. the value can be characterized as a truth-value, independently of being the value of a given concept for a certain argument). Yet, curiously enough, the same muddle is immediately reintroduced at the level of *thoughts*, conceived as values of sense-functions (infra p. 328ff.).

It is striking, and noteworthy, how little Frege capitalized on the potential powers of the sense/reference distinction to resolve difficulties that we perceive in his earlier system of logic. Gains evident to us went unnoticed by him, even when he canvassed philosophical support for making the distinction. His apparent concern was only indirectly philosophical; he aimed primarily to remedy technical weaknesses in the system of *Begriffsschrift* which thwarted his completion of the formal programme outlined in the *Foundations*. Splitting judgeable-content into thought and truth-value was less the fruit of philosophical insight than the product of an inexorable demand on concept-script. Though this brought mostly unharvested philosophical gains, it also provided a rich nutrient for the growth of novel confusions and fresh mysteries.

---

4. In the *Basic Laws* he slightly modified the definition given in the *Foundations*, substituting a double course-of-values in place of the set of concepts having the property of being equinumerous with $F(\xi)$.

2. Introducing the sense/reference distinction

Commentators typically focus upon 'On Sense and Reference' as the *locus classicus* of Frege's distinction. Hence the distinction is generally conceived to be motivated wholly by the problem of cognitively non-trivial identities. This is misleading, for it inclines one to think of proper names as the focal point for his application of the distinction, and hence to view its application to sentences and concept-words as derivative. Consequently, when we find various aspects of its application to sentences philosophically objectionable (e.g. that sentences have truth-values as their references), we all too readily think that this 'extension' can be pruned without harming the fundamental conception underlying the distinction.

In fact, less than a third of 'On Sense and Reference' concerns proper names, while the rest focuses upon the central thesis that sentences have truth-values as their reference. More important yet is the fact that 'Function and Concept' introduces the distinction between sense and reference first for the special case of equations and inequations, then for sentences in general, and then finally, but only implicitly for proper names. These moves are all executed in the course of the application of functional analysis to equations, 'proving' that sentences stand for truth-values. Indeed, Frege here admitted that his manoeuvre may appear 'arbitrary and artificial' and referred to his forthcoming essay 'On Sense and Reference' as providing *supplementary* support for his contention that sentences express thoughts which are their sense and stand for truth-values which are their reference (FC, 29n.). This suggests that it is misleading to view the sense/reference distinction as introduced solely or even primarily to resolve a special problem about identity-statements.[5] We should rather examine 'On Sense and Reference' in conjunction with 'Function and Concept' if we are to understand the rationale for Frege's manoeuvres.

Consequently, we should look for two different, and differently motivated, ways of introducing the sense/reference distinction. The first commences with sentences, distinguishes the thought as the sense of the sentence and the truth-value as its referent, and then applies the distinction to parts of sentences. This route is crucial, and tied to the further development of the function-theoretic logical system of *Begriffsschrift*. It is, however, artificial, and appears problematic. The second mode of introduction commences from sub-sentential proper names, distinguishing the sense of a proper name from the object which is its reference, and then extends the distinction to sentences and concept-words. This might be thought the more natural and intuitive way of building up a case for a generalized distinction, since it resembles familiar styles of analysing singular referring expressions. One might indeed say that the first route is the proper scientific one, the second a more popular or rhetorical and, one might hope, persuasive one. The

---

5. Scrutiny of Frege's discussions of sense and reference in BLA and the *Nachlass* confirms the suggestion that it is misleading to accord any *special* priority to introduction of the distinction for proper names rather than for sentences.

SENSE AND REFERENCE: AN AREA SURVEY

former mode of introduction is presented in 'Function and Concept' and most of the *Nachlass* essays, the latter in 'On Sense and Reference'. They are complementary, and we shall sketch both.

The first route

The pivotal point in this introduction of the sense/reference distinction is the extension of function theory to encompass '=' and '>' as function-names (FC, 28; BLA, §2). This in itself was no innovation, since both identity (BS, §21) and equality (CN, p. 205) had explicitly been treated as functions in the early work, and inequality, like every two term relation (BS, §9) was implicitly treated as a binary function. The innovation of the 1890s, relative to *Begriffsschrift*, was to treat these as functions whose values are truth-values, rather than judgeable-contents. Common to this conception *and the earlier one* is construing sentences (expressions appropriate for making assertions e.g. '2 = 1 + 1', '3 > 2') as names of the values of functions for certain arguments.

Yet although '$x^2 = 4$' and '$x > 1$' both have the value *True* for the argument 2, they do not *say* the same thing. '$2^2 = 4$' and '$2 > 1$' both 'stand for' the same thing (just as '2^2' and '$2 + 2$'), but they are not logically equipollent. They make different assertions, require different proofs, and their intersubstitution in a proof need not preserve cogency of argument. Hence they are now said to express different thoughts. Consequently the object of assertion, previously labelled 'judgeable-content', is now called 'a thought', the sense of a sentence. A sentence is no longer conceived as standing for a judgeable-content. It stands for a truth-value, and it *expresses* a thought.

Once truth-values are 'recognized' as objects (as values of functions for arguments), the way is open to treating statements in general as expressing thoughts which stand for truth-values, and to split up the expression for any statement into an argument-expression and function-name, the first (in the simplest case) being a proper name, the second a predicative-expression (a concept-word). For we can regard a sentence naming a truth-value as decomposing on analysis into a function-name which maps an object on to a truth-value, and a proper name designating the object. Concept-words thus immediately emerge as function-names (FC, 30f.). The crucial innovation here is the introduction of *truth-values* (to serve as values of concepts for arguments). The essential pay-off is the possibility of connecting the concept with its extension via the two logical objects, the True and the False.

Frege's manoeuvres thus far leave him with four entities related to a simple assertible expression, (i) a thought expressed, (ii) a truth-value designated, (iii) a function, (iv) an argument. The thought refers to the truth-value, and the function has that truth-value as its value for the argument. The problem this immediately generates is how to relate the thought to the function/argument structure that determines its truth-value (reference).

One possible move would be to conceive of the thought (like its ancestor, the judgeable-content) as decomposing upon analysis into the very same function (concept) and argument (the object designated by the proper name in the sentence expressing the thought). This Frege now refused to do. His explicit argument focused exclusively upon the constituent proper name. A proper name stands for an object, but that object cannot be part of a thought; e.g. in 'Mont Blanc is 4000 metres high', 'Mont Blanc' stands for the mountain, but a concrete object cannot be part of a thought (PW, 187), and a part of a thought cannot be in France (cf. PMC, 128). So the argument of the function cannot be a constituent of the thought expressed. Second, a proper name may not designate anything (e.g. 'Odysseus'), but though this deprives the thought of a truth-value, it does not affect what is expressed by the sentence containing the proper name (PW, 191). Since the function in question lacks any argument (and hence value) but the thought is unimpaired, the thought-constituent cannot be the argument of the relevant function. Third, the thought that $a = b$ may differ from the thought that $a = a$. But if the object designated by a proper name were part of the thought, these thoughts would not differ. Finally, the thought that Fa may differ from the thought that Fb even if $a = b$ (PW, 192; FC, 29).

Frege took these considerations to prove that a proper name has, apart from its referent (what it stands for) a sense. The sense of a proper name (its mode of presentation) *is* part of the thought. Its referent, which is not part of the thought, is the argument of the function whose value is the reference of the thought. Given that the thought splits up into constituents, one of which is the sense of the proper name, Frege took it for granted that the remainder of the sentence has as its sense the remainder of the thought. The thought (in the simplest case) is thus conceived as consisting of the sense of a proper name and the sense of the predicative expression or concept-word. The latter, in Frege's published works, remains unargued for and unclarified. At best this leaves a lacuna to be filled by describing the relations among the thought-constituents and between them and the thought of which they are constituents; at worst, it generates incompatible requirements on the concept of sense.

The second route

The second, familiar, method of introducing the sense/reference distinction is via proper names in identity-contexts. The problem harks back to *Begriffsschrift*. '$a = a$' and '$a = b$' may differ in 'cognitive value'; the former is trivially analytic, the latter may well be the expression of new knowledge (as in a novel proof), and is sometimes synthetic. The very possibility of arithmetic's being a science depends on the cognitive non-triviality of identities, since, in Frege's mature view (BLA, p. ix), equations are identities.

Why was Frege dissatisfied with his *Begriffsschrift* solution, viz. that '$a = b$' says that 'a' and 'b' are coreferential? His objection cannot be a general criticism of the systematic ambiguity introduced by that analysis (after all, he retains *that* feature of signs in his analysis of oblique contexts). Rather, his argument turns on the point that if identities were *about signs* they would not express 'proper knowledge', since the connection between an object and its name (its 'mode of designation') is arbitrary (SR, 56f.). This argument is opaque. One might claim that the connection between the great French playwright and the names 'Molière' and 'Jean-Baptiste Poquelin' is indeed arbitrary.[6] But presumably 'Molière is the same man as Jean-Baptiste Poquelin' expresses proper knowledge, and if so, is that knowledge *really* any different from what is expressed by '"Molière" and "Jean-Baptiste Poquelin" name one and the same man'?

We must, Frege implied, distinguish the mode of designation from the *mode of presentation.* Only if a difference in the former involves a difference in the latter is proper knowledge expressed in an identity statement. Hence, with respect to any singular referring expression we must distinguish what it stands for from its mode of presenting what it stands for. Where 'a', 'b', and 'c' designate the three bisectors of the angles of a triangle, 'the intersection of a and b' and 'the intersection of b and c' are different ways of designating one and the same point which correspond to different modes of presenting this point. So too, the names 'the Morning Star' and 'the Evening Star' indicate different modes of presenting the planet Venus. A difference in the mode of designating an object is a difference in the signs which stand for it. 'Four', '*vier*', '*quatre*', 'iv', 'IV' are all different modes of designating the number 4 but are not different ways of presenting it. The mode of presenting an object is not meant to be a sign, but something associated with one, namely a particular way of determining its referent, a 'route' to the referent. Different modes of determining a referent, e.g. '2^2', '$2 + 2$', '$\sqrt{16}$', 'serve to illuminate only a single aspect of the referent' (SR, 58), they 'lead to it from different directions' (PMC, 152).[7]

The notion of a mode of presentation is clearly heir to the *Begriffsschrift* conception of *Bestimmungsweise,* the mode of determination of a content (BS, §8).[8] There Frege had declared 'the way of determining a content' to be irrelevant to the content itself. Identity statements are only an apparent exception, for in them the sign is its own content. The identity accordingly states that two different signs have the same content in virtue of being associated with different ways of determining a single object. Now, having split content into sense and reference, sense *is* logically relevant in all contexts. It is the sense of the sign, not the sign itself, which presents or determines a definite object as its reference. A sign has a reference only via the sense which is associated with it: 'The regular connection between a sign, its sense, and its reference is of such a kind that to the sign there corresponds a definite sense and to that in turn a definite reference' (SR, 58).

6. Although one might also insist that the connection between 'Molière' and Molière is internal, since '"*N.N*" names *N.N*.' is a grammatical truth.
7. It is striking that Lotze had already claimed that the value of a non-trivial (arithmetical) identity lies in the fact that '*different paths* had led to the *same goal*', that the non-trivial identity is 'concealed under the different forms in which the two [references] were originally *presented*', that in it 'one and the same self-identical value [is] *presented under different forms*', and that in this very possibility is to be found 'the motive force of all fruitful reasoning in mathematical science' (H. Lotze, *Logic,* III-v-§352 (our italics)).
8. Also to 'the way in which [an] object . . . is introduced' or 'given' discussed in the *Foundations* (FA, §67).

Sense, *ab initio*, is treated as an abstract entity mediating between a sign and its designation. It has all the roles of a sign, e.g. combinatorial possibilities (syntactic properties), reference (semantic properties), equivalences (definability), although it is not a sign, is not 'perceptible to the senses'. It is, one might say, the *soul* of a sign.

How does incorporating this reified mode of presentation of what is 'given' by a symbol into the logically relevant features of a symbol resolve any problems? In particular, how does it resolve the difficulties about identities for which it was ostensibly designed? Frege's aim is to demonstrate that identities can express *proper* (i.e. non-linguistic) knowledge. What then shows the cognitive non-triviality of an identity? Not introspection. Rather, in mathematics, cognitive non-triviality is equivalent to need for proof, i.e. absence of self-evidence. Generalizing, an identity is cognitively non-trivial if and only if a proof is necessary to establish its truth conclusively. It is in virtue of this that such identities license non-trivial inference, i.e. constitute substantial premises (cf. FA, §67). With the introduction of the concept of sense, cognitive non-triviality of an identity is equivalent to a difference in the senses of the two coreferential expressions linked by the identity-sign (SR, 56f.). Absence of self-evident coreferentiality betokens difference of sense (PMC, 152) and the need for proof. It is this which gives substance to a non-trivial assertion of identity. The concept of sense is thus connected *ab initio* with the idea of *proof*, a connection commonly overlooked.

On the assumption that the notion of mode of presentation is perspicuous, Frege proceeded immediately to extend his distinction from proper names to sentences. In 'Function and Concept' the claim that sentences stand for truth-values is taken to be established as soon as equality and inequality are treated as functions to truth-values. There functional analysis wears the trousers. 'On Sense and Reference' presents a novel, specious, argument. A declarative sentence 'contains' a thought, i.e. what is asserted by its assertoric use. The thought cannot be the reference of the sentence since substitution of a coreferential constituent name in the sentence cannot affect the reference, but does affect what is asserted (i.e. the thought). Therefore, the thought is the sense.

This assumes that what is asserted is either a sense or a reference, without substantiating that it need be either. It also assumes that the reference of a sentence, if it has one, is determined by the references of its parts, an assumption relying on the (here) unargued premise that concepts are functions. Given that all we have been told of sense is that it is a mode of presenting a reference, we have no licence to assume that a *sentence* has a sense, i.e. expresses a mode of presenting anything at all, still less to assume that *what one asserts* is a mode of presenting anything. Frege quietly relied on assimilating the object of assertion with the idea of a way of presenting an object, an assimilation which is supposed to be part of the conclusion of this very argument, not one of its premises.

Having claimed that a sentence has a thought as its sense, he went on to inquire whether it has a referent. Given that sense has been explained as a way of presenting a referent, this ought to be a closed question. But he treated it as an open one. The references of parts of a sentence are irrelevant to its sense (SR, 63).[9] We concern ourselves with the reference of a constituent proper name only when we are concerned with the truth-value of the sentence. So the truth-value of the sentence *is* its reference. To be sure, we may 'concern ourselves' with the references of constituent proper names in a sentence only when we are concerned with whether it is true or not, but it in no way follows that the sentence has a reference at all, only that it may be true or false.

Finally, Frege explained what the True and the False are:

> By the truth-value of a sentence I understand the circumstance that it is true or false. There are no further truth-values. For brevity I call the one the True, the other the False. Every declarative sentence concerned with the reference of its words is therefore to be regarded as a proper name, and its reference, if it has one, is either the True or the False. These two objects are recognized, if only implicitly, by everybody who judges something to be true. (SR, 63)

Yet this is surely sophistry of a high order. Assertoric sentences may be true or false, but the *circumstance* that a given sentence, e.g. 'Grass is green', is true is not identical with its 'truth-value', for the circumstance that grass is green is *toto caelo* different from the circumstance that the sky is blue. If the truth-value of a sentence were identical with the circumstance of its being true or false, there would be not two truth-values, but as many truth-values as there are different circumstances (facts) making sentences true. Nor does the fact that a sentence is true or false license viewing it as a proper name of a strange object denominated 'a truth-value'.

It is difficult not to conclude that 'On Sense and Reference' misfires completely in attempting to render the 'artificial' moves made in 'Function and Concept' more natural and intuitive. Russell justifiably complained 'I cannot bring myself to believe that the true or the false is the reference of a sentence in the same sense as, e.g., a certain person is the reference of the name Julius Caesar' (PMC, 150f.). The reason for the weakness is simple. Whatever intuitive plausibility and traditional acceptability may be associated with distinguishing sense from reference in the case of proper names, the 'extension' of the distinction to sentences is and must be wholly dependent upon the function-theoretic innovations of 'Function and Concept'. The flimsy arguments of 'On Sense and Reference' add nothing by way of intuitive plausibility or naturalness to the functional analysis of sentences adumbrated in the previous essay. Whatever 'artificiality' attaches to that cannot be rubbed off by the 'alternative route' by which the sense/reference distinction is introduced in the later article.

---

9. With this claim too, made in this context, we might quarrel. For if the thought is what is (or may be) asserted, and if assertion involves typically predicating something of something, then if no referent exists, surely, by Frege's lights, no assertion is (or can be) made, for 'it is of the reference of the name that the predicate is affirmed or denied. Whoever does not admit the name has reference can neither apply nor withhold the predicate' (SR, 62).

Conclusion

The residue of this survey leaves us with five important points:

(i) The sense/reference distinction is not introduced in order to resolve a problem in the theory of meaning or of understanding. In neither method of introduction is understanding even mentioned, and our exposition has by-passed that notion altogether. Sense is an object of understanding only in the way in which a statement is, i.e. it is what is understood when we understand or grasp what is asserted by an utterance (or part of one). Consequently one common interpretation of 'On Sense and Reference' is surely wrong, viz. that Frege was addressing the problem of how it is possible to *understand* an identity-statement without knowing its truth-value (where it is assumed that one understands, knows the meaning of, the constituent proper names). But he was not concerned with understanding or knowing meanings, only with the question of why a *proof* (or confirmatory empirical evidence) of identity is necessary (or even possible). This he answered in terms of senses of proper names, conceived as different ways of determining or presenting a referent. Cognitive non-triviality is a mark of the need for, and possibility of, a proof (PW, 224). Where the question of proof does not arise, agreement over sense of sub-sentential constituents, indeed assignment of any sense to them, is, it seems, altogether unimportant. In the famous footnote on which Frege's so-called 'theory of proper names' rests, he declared that in the case of ordinary proper names different people may assign different senses to the same name (where, by implication, a proper name is individuated by its bearer!). However, 'so long as the reference remains the same, such variations of sense may be tolerated, although they are to be avoided *in the theoretical structure of a demonstrative science*' (SR, 58n., our italics). In a letter to Peano the same point is made even more emphatically:

> [O]ur vernacular languages are also not made for *conducting proofs*. And it is precisely the defects that spring from this that have been my main reason for setting up a conceptual notation. The task of our vernacular languages is essentially fulfilled if people engaged in communication with one another connect the same thought, or approximately the same thought, with the same sentence. For this it is not at all necessary that the individual words should have a sense and reference of their own, provided only that the whole sentence has a sense. *Where inferences are to be drawn* the case is different. (PMC, 115; our italics)

It is further noteworthy that this common misinterpretation is required to support the mistaken thesis that according to Frege the sense of a sentence is its truth-conditions (as Wittgenstein was later to argue in the *Tractatus*). Explaining how it is possible to understand a sentence without knowing its truth-value is indeed a central concern of the truth-conditions account of meaning, but it was not Frege's concern to address this problem even in the case of identities.[10]

---

10. Given his account of concepts, Frege should have found the problem just as pressing for atomic sentences as for identities. But, *mirabile dictu,* he turned his back on this wider issue! (Cf. Dummett, 'Frege's Distinction between Sense and Reference' in *Truth and Other Enigmas*, p. 133). The explanation is simple—this was not his concern!

(ii) Though senses fulfil part of the earlier publicity requirement on contents (as against psychologism), this is not because of general publicity requirements on meaning and Fregean intimations of Wittgenstein's argument against a private language, but because mathematics is a *science*. It establishes shared, public knowledge. What is asserted in a theorem, the thought expressed, must be public so that proofs may give common knowledge (BLA, ii, §58n.; PW, 165, 215; PMC, 126).

(iii) It is mistaken to think that Frege was concerned, in introducing his distinction, with so-called 'contingent identities'. It is true that the identity of the Morning Star with the Evening Star is one of his favoured examples, but it is quite evident that the heart of his preoccupation lies with arithmetical identities. His concern is to demonstrate the nature of the possibility of *cognitively non-trivial identities,* lest the analyticity of arithmetic collapse into its triviality. Indeed, since he moves without hiatus from discussing the cognitive non-triviality of '5 = 2 + 3' to that of geographical or astronomical statements of identity (PW, 224), we may conclude that in this respect he saw no difference between them. There is no evidence that he thought that *a posteriori* identities were contingent.

(iv) Frege's assimilation of sentences to proper names was not a *consequence* of introducing his distinction between sense and reference, nor was it first *introduced* by the weak argument of 'On Sense and Reference'. It was present in his work in *Begriffsschrift,* a prerequisite of his extension of function theory to logic. The innovation of 'On Sense and Reference' and 'Function and Concept' in this respect was to view the sentence as a proper name of a *truth-value* rather than of a content. If there is error here, it is fundamental; not an uncharacteristic peccadillo, but the 'original sin' of Frege's logic.

(v) It is noteworthy that in his published works he did not apply the sense/reference distinction to concept-words. Rather, he simply took it for granted. Concepts, he argued, are functions from arguments to truth-values. Concept-words stand for, refer to, these functions, which, though intimately related to extensions, are yet distinct from them, since functions are unsaturated and extensions are objects. The fundamental logical relation is that of an object's falling under a concept. But whether a concept-word stands for a concept is a distinct question from whether anything falls under it. In all of this there is no mention of what the sense of a concept-word is, of what the 'mode of presentation' of a concept consists. We may, perhaps, take the absence of any explicit account to be a result of his supposing that there was no need to state the obvious. Following the footsteps of his contemporaries, it seems, Frege identified the sense of a concept-word with its intension, a feature manifest in the fact that he thought that its sense is given by specification of its *Merkmale*.

The harvest from 'On Sense and Reference' is singularly meagre. Its arguments for the distinction are far from perspicuous and its support for generalizing it are unconvincing as well as incomplete. In particular, on its own it leaves us with little more than an intuitive grasp of what a mode of presentation is. Furthermore, the detachment (and reification) of the 'mode of presentation' from the sign and its mode of designation should leave us deeply suspicious. It is, after all, an instance of a common, misguided philosophical syndrome:

> ... we say "Surely two sentences of different languages can have the same sense"; and we argue, "therefore the sense is not the same as the sentence", and ask the question "What is the sense?" And we make of 'it' a shadowy being, one of the many which we create when we wish to give meaning to substantives to which no material objects correspond.[11]

3. Formal principles of sense and reference

We are still a long way from grasping the detailed nature and articulations of Frege's sense/reference distinction. To complete our area survey we must adumbrate some general formal principles relating expressions, senses and references. These will not tell us what senses are, nor how it is that they 'determine' references, let alone how they are 'associated' with expressions. But they will make clear the theoretical commitments of this novel notion. It remains to be seen whether anything can be conceived as satisfying these demands.

(i) Different expressions may have the same sense, and different senses may determine the same reference (SR, 58).

(ii) Identity of sense implies identity of reference, and conversely, difference of reference is incompatible with identity of sense. (This is a consequence of conceiving of functions as single-valued correlations.)

(iii) Since multiple senses may correspond to the same reference, identity of reference does not imply identity of sense (FC, 29; SR, 62). There is no 'route back' from reference to sense, from the value of a function for an argument to the argument and function *as* the value of which it is presented. (A function is underdetermined by a proper part of its course of values.)

(iv) Sense and reference are always distinct for any token expression.[12]

(v) Every symbol that has a reference must have a sense. There is no such thing as *direct* designation of entities in the realm of reference, i.e. no possibility of short-circuiting the flow from symbol to reference *via sense*. (This is an important unnoticed modification to the conception of symbols in Frege's theory of content. It also makes clear that the sense/reference distinction modifies, but does not repudiate, the Augustinian picture of language.)

(vi) An expression may have a sense but lack a reference. This is an innovation relative to content-analysis, since there an expression which stood for nothing trivially had no content. Nonetheless, there were certain conditions to be met for an expression to have content; in particular, a definite description has a content only if the corresponding concept has a unit class as its extension (FA, §74n.). The theory of sense builds some of these requirements into the senses of expressions,

---

11. Wittgenstein, *The Blue Book,* p. 36 (Blackwell, Oxford, 1958).
12. Restriction to token-expressions is required by Frege's analysis of oblique contexts. There different tokens of the same expression may so occur that the sense of one is the reference of the other, e.g. 'A believes that he, A, believes the whole of the Nicene Creed.'

thus opening up the possibility of inference from the known sense of an expression to the conclusion that it has (or lacks) a reference.

Since Frege held lack of reference to exclude the possibility that an array of sentences formulates a cogent argument, he thought it essential to establish the soundness of the derivations carried out in the *Basic Laws* by *proving* that every well-formed formula in concept-script has a reference. Consequently, each symbol in concept-script must be so explained that its sense supplies a logical guarantee that it has reference. A demonstrative science must, in Frege's view, employ symbolism that removes any scope for sense without reference.

In natural language, however, many expressions occur with sense but no reference. Proper names with sense but no reference are exemplified by names in fiction (SR, 62f.; BLA, ii, §64) and by such complex names as 'the least rapidly converging series' (SR, 58). (These are very different. Even if we could associate with each name in fiction a sense, could we consider it to be a mode of presentation of a *possible* reference? Is it mere chance that Sherlock Holmes does not exist? By contrast such a name as 'the least-rapidly converging series' cannot be viewed as a mode of presenting a *possible* object, for the object it purports to designate *necessarily* does not exist.) Concept-words with sense, but no reference (i.e. which do not stand for concepts) are allegedly exemplified by predicates with indeterminate extensions (PW, 122; PMC, 115).

(vii) The reference of a complex expression is a function of the references of its constitutents.[13] As a direct consequence, a declarative sentence (or a corresponding formula in concept-script) has a truth-value if and only each of its constituent symbols has a reference; it lacks a truth-value if and only some constituent has no reference at all.

Three corollaries are important. First, Frege held that the reference of any logically complex expression is quite literally a *function* of the references of its logically significant components. All complexity is analyzed in terms of decomposition into function and argument (PW, 254). According to Frege's conception of a function, this imposes a stringent condition on the connection of the truth-value of the judgment expressed by a declarative sentence and the references of its constituents.

Second, this principle committed Frege to a bizarre doctrine about fiction. Accepting the widespread idea that sentences in works of fiction do not make true (or false) assertions, he drew the conclusion that their lack of truth-values must depend on reference-failure of their constituent proper names. This absurdly implies that historical novels abound in true and false assertions about historical figures and that fictional sentences not containing any proper names must be true or false. No less absurdly, it threatens the distinction (in acting a play) between

---

13. This principle of interchangeability is commonly recognized to be one of the foundation-stones of 'Frege's semantics'. It is given prominence in many works, e.g. Carnap, *Meaning and Necessity*, pp. 96f and 122ff., Church, *Introduction to Mathematical Logic*, p. 86, and D. Kaplan, *Foundations of Intensional Logic* (unpublished Ph.D. thesis, University of California, 1964), p. 8.

a character's making a true or false assertion and an actor's making no assertion at all, as well as the distinction between characters and imaginary (or fictional) characters in fiction. (This body of doctrine is also unnecessary, since Frege could have denied that fictional sentences express thoughts, simply declaring them to be irrelevant to the construction of proofs and the study of the laws of valid inference (cf. PW, 129f.).)

Third, the principle of reference-dependence generates a strange account of the 'vagueness' of concept-words. On the assumption that whether or not a symbol has a reference is an absolute, not a context-relative matter, a vague predicate may well seem to deprive every sentence in which it occurs of a truth-value (PW, 122). If the predicate is to be faulted when application of a vague predicate to a borderline case yields a sentence without a truth-value, then it follows that every sentence incorporating this predicate lacks a reference provided that, ambiguity apart, there is no such thing as a symbol's having a reference in one sentence but lacking reference in another.[14] Logic must shun vagueness like the plague. The positive correlate is the demand for completeness of definition of concept-words, viz. that every concept-word be so defined that the sentence resulting from filling its argument-place with *any* name of an object has a truth-value (BLA, ii, §56; cf. §62; FC, 33).

(viii) The sense of a complex expression is a function of the senses of its parts.[15] As an immediate consequence, every complex expression constructed in conformity with logical type-theory has a sense if and only if each of its constituents does; it lacks a sense if and only if at least one constituent lacks sense.[16]

This principle is problematic in several respects. First, it is unclear whether the dependence of the sense of a complex expression on the senses of its parts is uniformly open to function/argument analysis or whether it should be differently conceived (e.g. on the model of the whole-part relation). Second, direct application of the principle to natural language or even to mathematical symbolism generates an unpalatable dilemma: either few (if any) expressions actually used in communication have senses at all or else many complex expressions, though apparently nonsensical, really do have senses (e.g. 'the sun plus one is dead'). Third, the possibility of sense without reference reveals a puzzling asymmetry. A well-formed complex expression may lack a reference although each of its constituents has a sense. Indeed, this constitutes Frege's mature objection to formalism in mathematics. But the possibility of sense without reference is opaque if the sense of an expression is a mode of *presenting* its reference. Moreover, the proposal to abolish the possibility in the case of well-formed expressions in concept-script by stipulating references arbitrarily wherever they would otherwise be

---

14. This is an important departure from the theory of content in *Begriffsschrift*. It virtually necessitates the (commonly unnoticed) doctrine that the sense of expression determines its reference independently of matters of fact.
15. This second principle of interchangeability is also commonly recognized as a second pillar of 'Frege's philosophy of language'.
16. Wittgenstein highlighted this implication (*Tractatus Logico-Philosophicus*, 5.4733).

absent seems doubly unintelligible, as if it made no difference to the identity of the sense of an expression *which* reference it presented in such a case! Although the principles of sense-dependence and reference-dependence might be expected to run strictly parallel to one another, they seem to diverge without good cause. (The discrepancy is required primarily for Frege's criticism of formalism, and that could have been redrafted as the thesis that the formalists misguidedly do less than is required for securing references for the constituents of complex expressions (cf. BLA, §29).)

4. Frege's later contextualism

Having introduced two fresh protagonists in the second act of the drama of Frege's thought, we must review the surviving *dramatis personae*. For while some of the key figures of the theory of content will not be seen again, others reappear in fresh guise, and are not always readily identifiable in their new costume. In particular we must trace the fate of the contextual ideas which played so prominent a role in the first act.

The dictum that a word has a meaning or content only in the context of a sentence expressed one of the pivotal ideas of the *Foundations:* the primacy of judgeable-content over unjudgeable-content in logical analysis. Under this umbrella shelter a number of distinct but related principles fundamental to the doctrine of content. (i) The Heuristic Maxim with its antipsychologistic implications; (ii) the Sufficiency Principle; (iii) the Restrictive Condition; and, (iv) the legitimacy of alternative decomposition of judgeable-contents. The *Basic Laws* represents the culmination of Frege's researches into the foundations of mathematics. Yet, amazingly, the contextual dictum is not repeated there, nor in any subsequent writings.[17] If its non-appearance is to be explained by reference to his *repudiation* of these various principles, this must indeed herald changes so deep as to undermine the very Foundations upon which the Basic Laws are built.[18] How extraordinary then that an author who described in detail the evolution of his mature views should forget to mention this!

---

17. Although there are obvious echoes of it in the polemical debates with Hilbert and Korselt (FG, 8, 53, 67, 71).
18. Dummett (*Frege: Philosophy of Language,* pp. 196, 645; *Truth and Other Enigmas,* pp. 95, 108) insists that with Frege's explicit assimilation of sentences to proper names in the 1890s he was forced to relinquish the principle that only in a sentence do words have a meaning, thereby abandoning his most profound insights concerning (i) the role of sentences in the performance of speech-acts, (ii) the primacy of the sentence in the theoretical explanation of meaning, (iii) the order of understanding (i.e. that we understand a sentence by 'calculating' its sense from the known senses of its constituents and their mode of combination), (iv) the legitimacy of contextual definition. But Frege assimilated sentences to proper names (designations of values of functions for arguments) *ab initio;* he never associated the contextual dictum with a theory of speech acts; he was never concerned with meaning but only with judgment; he gave a generative account of understanding only in 1914, by which time, according to Dummett, he had long abandoned contextualism; he never countenanced contextual definitions as we conceive them, viz. reductively, and never had qualms about flanking the identity-sign with sentences; and he never abandoned the principle of the primacy of the judgment in logical analysis.

Rather than propose so implausible and hasty a conclusion, we should do better to search for the descendants of the cluster of contextual principles in the *Basic Laws*. The disappearance of the *dictum* may have a trivial explanation. With the splitting of judgeable-content into sense and reference, any contextual principle of content likewise bifurcates. Each new such principle must be examined separately for cogency. It is hardly to be expected that *all* the resultant principles will be acceptable. Moreover, there is no single dictum on which one can cogently hang those which are.

The *Basic Laws* introduces the sense/reference distinction by splitting up the *judgeable*-content conveyed by declarative *sentences,* which are now conceived as expressing thoughts and standing for truth-values. Hence the primacy of judgeable-content in analysis immediately splits up into two theses, one for thoughts, the other for truth-values. In the theory of sense, thoughts enjoy pride of place. Frege's most important general statement about sense (BLA, §32) explains the sense of subsentential constituents as consisting in their contribution to the sense of the sentence in which they occur (cf. PW, 102, 231; PMC, 79). Just as he insisted in 1880 that unlike his predecessors he began, in his theory of content, from judgments rather than from concepts (PW, 16), so he emphasized towards the end of his life that

> What is distinctive about my conception of logic is that I begin by giving pride of place to the content of the word 'true', and then immediately go on to introduce a thought as that to which the question 'Is it true?' is in principle applicable. So I do not begin with concepts and put them together to form a thought or judgment; I come by the parts of a thought by analyzing the thought. This marks off my concept-script from the similar inventions of Leibniz and his successors.... (PW, 253)

On this matter of principle Frege never wavered.[19] The rationale for the primacy of the thought is inherited from the primacy of judgeable-content in the earlier theory. Logic, the science of the laws of truth, investigates the conditions under which one thought (the conclusion of a proof or inference) is true in terms of the truth of other related thoughts. The analysis of thoughts into constituents is necessary only for purposes of determining relations among the truth-values of thoughts. Hence analysis of thought-components is strictly subordinate to and derivative from the analysis of thoughts.

In logical analysis the principle of the primacy of the thought is correlative with

---

19. Note that when given an opportunity to revise or correct earlier errors, he not only did not retract his adherence to the primacy of the judgment, but added the remark: 'To found the "calculus of judgments" on the "calculus of concepts" (which is properly a "calculus of classes") is to reverse the correct order of things; for classes are something derived, and can only be obtained from concepts (in my sense) ... [C]alculation with classes must be founded on the calculation with concepts. And the calculation with concepts is itself founded on the calculation with truth-values (which is better than saying "calculus of judgments")' (PMC, 192 n.71). Given that this 1910 discussion was a survey of his philosophy from *Begriffsschrift* onwards, this restatement of the principle proves beyond doubt that Frege never abandoned it.

the principle of the primacy of the truth-values (PW, 122). The primacy of judgeable-content in logic was associated with the idea that concept-formation in logic proceeds from the (inference-relative) decomposition of judgments (PW, 16). This conception, duly modified, is retained. Concepts are explained as the references, not senses, of concept-words. The thought presents a truth-value *as* the value of a function for an argument. The concept *is* the function (which is the reference of part of the thought) which has that truth-value as its value. Similarly, the general explanation of what it is for proper names to have references presupposes truth-values and an antecedent explanation of what concepts are (BLA, §29). In Frege's theory construction, truth-values enjoy a priority over concepts: 'the calculation with concepts is itself founded on the calculation with truth-values' (PMC, 192 n.71). Their primacy is equally evident in the identification of particular concepts, since the *esse* of a concept is its correlating each object with the True or the False. Frege's liberal conception of concept-formation proceeds from his admitting any such correlation as specifying a concept. The truth-values are *primi inter pares* among objects. In his view, 'logic is not concerned with how thoughts, regardless of truth-value, follow from thoughts. . . . [T]he step from thought to truth-value—more generally, the step from sense to reference—has to be taken. . . . [T]he laws of logic are first and foremost laws in the realm of references' (PW, 122).

The direct heirs of the principle of the primacy of judgeable-content in Frege's early work are a pair of principles conspicuous in his mature system: the primacy of the thought in the theory of sense and the primacy of truth-values in the theory of reference. This important form of continuity through change motivates a search for descendants of the various contextual ideas identified in the *Foundations*.

(1) *The Restrictive Condition:* Applied to sense, this states that only in the context of a sentence has a word a sense. Frege did not unequivocally adhere to this principle. On the one hand, he held that many expressions are ambiguous in sense, i.e. that the same type-expression is assigned different senses in different utterances. Any such expression has a definite sense only in the context of a particular sentence. This qualification certainly applies to indexical expressions, and it would apply to every expression whatever if indirect sense is uniformly distinct from direct sense, since any expression can occur in a report of an assertion in indirect speech. This argument from the non-uniformity of the senses of tokens of a single type-expression can be reinforced by arguments of principle. Unless a sentence expresses a thought it can have no role in inference; therefore, since sense is restricted to inference-relevance, words have senses only when used in sentences that express thoughts. Furthermore, since sense determines reference and since a word has reference only in the context of a sentence expressing a thought (BLA, ii, §97), a word cannot consistently be assigned a sense outside the context of a sentence. Finally, the sense of a word is explained as its contribution to the thought expressed by a sentence in which it occurs, and this eliminates the possibility of its having a sense except in the context of a sentence expressing a thought. On the other hand, Frege was now inclined to accept a synthetic conception of thoughts. He argued that any sentence constructed in conformity with

type-restrictions out of components known to have senses in other sentential contexts *must* itself express a thought; i.e. a well-formed sentence lacking sense guarantees that at least one of its constituents has not been assigned a definite sense (cf. BLA, ii. §56ff.). He gravitated towards a compositional conception of sense according to which a thought is put together out of thought-building blocks. This part/whole analysis of thoughts seems to demand that the senses of sentence constituents are given independently of their occurrence in any particular sentential context.

Applied to reference, the Restrictive Condition soldiers on. Frege's official principle is not that an expression has a reference only in the context of a sentence that has a reference (truth-value), because some expressions in a sentence may have both sense and reference while others have sense but no reference (thereby depriving the sentence itself of a reference). Rather, his principle is that an expression has a reference only in the context of a sentence expressing a thought (BLA, ii, §97). What reference an expression has depends on the context of the sentence in which it occurs, since in any oblique context its indirect reference is its customary sense. His analysis of indirect speech introduces a bifurcation in the reference of every expression and hence a dependence of reference on sentential context which is formally parallel with the consequences of the analysis of identity in *Begriffsschrift*.

(2) *The Sufficiency Principle:* Applied to sense, this states that to fix the sense of an expression it is sufficient to determine how it contributes to the sense expressed by a sentence in which it occurs. Frege was ambivalent about this principle. On the one hand, he obviously thought it possible to apprehend a thought independently of any particular sentence expressing it, *a fortiori* independently of assigning senses to the constituents of a sentence known to express it (PMC, 115). Specific arguments for assigning sense to particular sentence-constituents actually exploit this possibility: e.g., the proper name 'Mont Blanc' must have a sense because the sentence 'Mont Blanc is 4000 metres high' expresses a thought (PW, 191f.). In all such cases, the thought expressed by a sentence is treated as conferring sense on subsentential expressions. On the other hand, free variables used in formulae of concept-script to express generality are denied sense (and reference) although they do make significant contributions to the thoughts expressed; their role is not to designate *(bedeuten)*, but rather to indicate *(andeuten)* something (BLA, §§1, 8, 17, 25; PW, 162; FG, 23n., 53, 67). Moreover, if a thought expressed by a sentence is itself the value of a function which is the sense of the predicate, then the principle that a function is not uniquely determined by specifying its value for one single admissible argument would block the derivation of the sense of a predicate from the sense of a sentence in which it occurs. If, however, a thought is related to the sense of subsentential expressions exclusively as a whole to its parts, there is little merit in describing the sense of a sentence as conferring sense on its parts, and Frege did in late writings drift into a synthetic or compositional conception of the thought expressed by a sentence as built up out of the senses of the subsentential expressions. At best these points call for qualifications to the principle that the sense of a sentence confers sense on its

parts, at worst for a blanket repudiation of the Sufficiency Principle applied to sense.

Applied to reference, the Sufficiency Principle states that the truth-value of a sentence suffices to determine the reference of its constituent expressions. This is obviously unacceptable. Frege's analysis of the use of free variables to express generality blocks the adoption of the principle that every logically significant expression in the formulation of a true judgment must have a reference. Moreover, since the truth-value of a sentence is subject to function/argument decomposition, i.e. since it is presented as the value of a function for an argument, the principle that a function is not uniquely determined by specifying its value for one argument (and that an argument is not uniquely determined in general by specifying the value that some particular function takes for this argument) guarantees that there is no direct route from the truth-value of a sentence to the reference of each of its constituents. Indeed, it is even impossible to 'calculate' the reference of a constituent expression from specifications of the reference of the whole sentence and of its remaining constituents. On the other hand, the Sufficiency Principle might take the form that the sense of a sentence suffices to confer reference on its constituents. This would be at least as plausible as the application of the Sufficiency Principle to sense. But it might seem even better supported: if the sense of the formula '$\Phi(A)$' is that a truth-value is presented as the value of the function Φ for the argument A, then the reference of each of the expressions in '$\Phi(A)$' can be recovered immediately from the specification of the sense of '$\Phi(A)$'! In view of the possibility of sense without reference, of alternative analyses of a single thought, and of the occurrence of complex designations of objects and functions, the principle that the reference of sentence-parts must be transparent in a specification of the sense of a sentence cannot be accepted.

There is a weaker version of the Sufficiency Principle that does apply to both sense and reference: viz. it is sufficient to assign sense (or reference) to a subsentential expression to determine its contribution to the sense expressed (or reference designated) by *every* well-formed sentence in which it may occur. This interpretation of the principle simply transfers to the extended application of function theory in logic the fundamental idea of what it is to specify a well-defined function. Understood in this way, the Sufficiency Principle provides the driving force of Frege's criticism of psychologism, and it gives a perspicuous rationale for the requirement of completeness of definition of concepts (BLA, ii, §56).

(3) *The Heuristic Maxim:* Although Frege's opposition to psychologism continued unabated, he did not reiterate the advice never to ask for the meaning of a word in isolation, but only in the context of a proposition. Lack of an associated idea does not debar an expression from having sense or reference, and the cogency of inferences is independent of associated mental images. Nonetheless, Frege saw no point in repeating the Maxim. Its purpose in the theory of content was to stress that not all unjudgeable-contents must be either actual or subjective, i.e. to make room for Platonic objects and concepts (functions). This point goes without saying for senses, since *all* senses are conceived as being objective and non-actual. It is applicable, however, to reference, since the references of some expressions are

material objects and the references of some others are mental entities. But instead of stating the Heuristic Maxim for reference, Frege formulated a positive general account of what it is to assign reference to an expression (BLA, §29). This makes it clear that a function-name or number-word may have a reference even though its reference is not something 'given' in perception or intuition. Any reformulation of the Heuristic Maxim would be redundant.

(4) *The legitimacy of alternative logical decomposition:* The most conspicuous implication of the principle of the primacy of judgeable-content in logic is the licence for producing radically dissimilar decompositions of a single judgeable-content. This licence is fundamental to Frege's mature system. Applied to the reference of a sentence, it is immediate. The True and the False can each be presented in innumerable different ways as the values of functions for arguments. But he also applied this licence to the thought expressed by a sentence. The sense of '$\Phi(2)$' (e.g. the thought that 2 is a prime) can be decomposed into the sense of a proper name combined with the sense of a first-level concept-word or into the sense of a second-level concept-word conjoined with the sense of a first-level concept-word (BLA, §22).[20] The unlimited possibilities of functional abstraction at the level of reference must be matched by unlimited possibilities of decomposing a thought in irreducibly different ways, unless a single sense can determine references at different levels of the hierarchy of objects and concepts.[21] To cancel the licence for alternative decompositions of thoughts would be to revoke the authorization of free functional abstraction which is essential to Frege's logic of generality.

The currents of contextualism still flow strongly through the *Basic Laws*, though not exactly in their earlier channels. This continuity is to be expected. The headwaters of the various contextual ideas are the fundamental function-theoretic inspiration of Frege's whole thought. His logic calls for decomposition of thoughts into functions and arguments in order to determine the cogency of inference. In the explanation of the concept of a function as well as in the definition of individual function-names, independently identifiable values are presupposed. The values (and arguments) of functions occupy, as it were, a station superior to functions themselves. Once judgeable-contents, or later truth-values and thoughts, are acknowledged to be the values of the functions basic to logic, then their primacy in logic is secured, and so too is the legitimacy of representing them in different ways as the values of functions for arguments. To relinquish the primacy of thoughts and truth-values from the *Basic Laws* would be to undermine the foun-

---

20. The sense of a proper name and the sense of a second-level concept-word must differ radically from each other (cf. CO, 54)! Frege also apparently authorized the decomposition of the thought expressed by 'There is at least one square root of 4' into the sense of a proper name and the sense of a first-level concept-word since he noted that this sentence expresses the same thought as the sentence 'The concept *square root of 4* is realized' which presents the True as the value of a concept for the object named by 'the concept square root of 4' as argument (CO, 49).
21. Indeed, if the sense of a concept-word is itself a function taking thoughts as values, then the possibility of generating alternative decompositions of a thought is guaranteed by the applicability of functional abstraction at the level of sense.

dations of Frege's mature system. Equally, to adopt the principle that the structure of a declarative *sentence* determines the structure of the thought expressed by it would be to move off in a novel direction. He himself, late in life, apparently made a very tentative exploration in this direction, suggesting that a thought is built up out of components which are the senses of the subsentential expressions used in its formulation (infra, p. 325ff.). He seems to have been unaware of the deep inconsistency between this conception and the idea of function/argument analysis of thoughts which is essential to his mature theory. What led ultimately to an obvious instability in the structure of his thinking were certain scarcely visible flaws in the supporting members of his account of sense. These will now be put under a magnifying glass. It will then be clear that the House that Frege Built must be condemned as unfit for occupation.

11
Sense and Reference: Digging Down

1. Mode of presentation

Frege gave only minimal general clarification of the notion of sense, no doubt considering this concept to be primitive and unanalyzable. He simply explained that the sense of an expression is the mode of presentation of the entity designated. He took it for granted that the notion of a mode of presentation is self-explanatory, or perhaps that it would be made crystal clear from his discussions of particular cases of identity and difference of sense.

His faith in this procedure has proved to be misplaced. Agreement does not reign about how to explicate the concept of a mode of presenting an entity. In the case of sentences and predicates (concept-words), we have difficulty in rendering intelligible the idea that these expressions present any designated entities at all. Even the case of proper names is problematic. His initial explanatory example makes use of definite descriptions (e.g. 'the point of intersection of lines *a* and *b*') which are said to designate points and at the same time to indicate modes of presentation. He must have supposed it to be transparent what mode of presentation such a description indicates and whether two such descriptions indicate the same or different modes of presentation (SR, 57). At least some definite descriptions must, as it were, have their senses written on their faces. But what about other kinds of singular referring expressions: personal pronouns (BLA, p. xvi f.; T, 24f.), proper names of persons (SR, 58n.), or of planets or of mountains (PMC, 80; PW, 224f.)? What about demonstratives, numerals, names of kinds of stuffs, class-names, etc.? Does it make sense to suppose that every singular referring expression indicates a mode of presenting an object? Does it do so openly or covertly? If covertly, is this because it is equivalent to a definite description that openly indicates the relevant mode of presentation? Did Frege intend to convey that every proper name of a person is an abbreviation for some definite description (cf. SR, 58n.)? Or is it sufficient that there be associated with every singular referring expression some way of recognizing an object as its referent? How are such ways of recognizing objects to be individuated? In the case of a proper name of a person, can any way of recognizing the bearer survive his demise? And does the fact that someone typically 'recognizes' certain persons (e.g. his wife and his children) in a multitude of ways (by the sound of their voice, their appearance, their characteristic gait, etc.) imply that he associates with their names a multitude of different senses?

Before concluding that 'mode of presentation' is unintelligible and hence that the concept of sense is an impenetrable mystery, we might review the rationale for introducing the sense/reference distinction. Three matters stand out. First, truth-values are the values of concepts for objects as arguments; hence, a complete formula (sentence) in concept-script such as '——— Φ(A)' or '———Ψ(A, B)' stands for the True or the False. Second, specifying the truth-value of a sentence does not exhaust its logical significance; the cogency of arguments turns not only on the truth-values of constituent sentences, but also on what is asserted by them, i.e. what thoughts they express. Third, formulae in concept-script expressing thoughts are standardly decomposed, as in *Begriffsschrift*, into function-names and argument-expressions; they have such forms as '——— Φ(A)', '——— Ψ(A, B)', etc. These three observations combine together in a natural way. What thought is expressed by a formula such as '——— Φ(A)' must be transparent in the notation of concept-script (otherwise the symbolism would have no point!). But grasping what thought is expressed cannot be a matter of knowing the truth-value of '——— Φ(A)', since different formulae expressing different thoughts may have the same truth-value as '——— Φ(A)'. The only feature not yet exploited which is transparent in the formula '——— Φ(A)' is that a truth-value is designated *as* the value of the function (concept) Φ for the object A as argument. This is something quite different from designating the same truth-value as the value of the same function for a different argument or as the value of some other function for some argument. Moreover, the formula '——— Φ(A)' could plausibly be described as indicating how a truth-value is designated as the value of a function for an argument, and what is thus indicated could intelligibly be characterized as a way of presenting or representing an object, viz. *as* the value of Φ for A as argument. Provided that Frege did not call into question the representation of judgments in concept-script by such formulae as '⊢——— Φ(A)', '⊢——— Ψ(A, B)', etc., then nothing could be more natural than to characterize such symbols as indicating ways of presenting truth-values. They explicitly designate truth-values as the values of functions for arguments, and this aspect of these symbols is of crucial importance in logic. Since Frege wished sense and reference jointly to exhaust the logical significance of an expression (*a fortiori*, expressions in concept-script), and since how a truth-value is determined as the value of a function for an argument is not the reference of a sentence in concept-script, it must belong to its sense; indeed, it must exhaust the sense of such a symbol unless concept-script fails to capture everything of logical significance. Consequently, a formula such as '——— Φ(A)' has two aspects: it stands for a truth-value (its reference), and it indicates how this object is determined as the value of a function for an argument (its sense, i.e. its mode of presenting a truth-value).

This interpretation treats the phrase 'mode of presentation' literally, not metaphorically. It is to be understood in terms of function theory. This background renders the expression immediately intelligible and gives scope for fairly general application. The basic idea is simple enough. One can calculate the value of a function for given arguments, and one can represent entities *as* the values of functions for given arguments. The name of any function whose value for a given argument is a particular entity can be incorporated into a complex expression specifying (naming) this entity. Thus, e.g. '$x^2]_2$', '$x+2]_2$', '$x^x]_2$', '$4 \cos x]_0$' all specify the number 4, each 'indicating' a different method of 'arriving at 4'. We might conceive of '$4 \cos x]_0$' as associating with 4 a particular route that leads to it, or a particular manner in which 4 'presents itself' to us. So too for numerical functions (as well as numbers): '$\frac{d}{dx}\left(\frac{x^3}{3}\right)$', '$\int 2x\,dx$', '$(x+1)^2 - 2x - 1$', all represent the function x^2 (over real numbers). Here too identity and difference of these modes of presenting this function turn on the identity or difference of methods of calculating what function is specified, not on superficial linguistic forms or arbitrary notational devices. Obviously the calculations required by '$x^2]_2$' and '$4 \cos x]_0$' have nothing in common. Hence, while '2 = 2' is uninformative, '$2x^3 + x]_1 = x^2 - 1]_2$' is informative, since it signifies that two different routes (methods of calculation) lead to the same value.

Frege's idea was to capitalize on this aspect of expressions in function theory. Every sentence in concept-script stands for the True or the False. But in addition, it represents a truth-value as the value of a function for an argument. Hence it indicates how a truth-value is presented (a mode of presentation of a truth-value) just as the expression '$x^2]_2$' indicates how 4 is determined as the value of a function for an argument. The fact that formulae in concept-script have the structure of formulae in function theory makes possible a direct transfer of the conception of a mode of presentation to the logical analysis of judgments. Much of Frege's conception of sense is immediately clarified by this interpretation. In particular, it makes perspicuous that every sentence in concept-script expresses a sense by indicating how a truth-value is determined.

To capitalize fully on the power of this strictly function-theoretic account of sense, some initial clarification seems needed. The first is the rationale for treating the sense of a sentence in concept-script (a thought) as an object. Though a sentence may indicate a mode of presenting an object, is it not altogether bizarre to conceive of 'how a truth-value is presented as the value of a function for an argument' as the proper name of an object? A bit of linguistic alchemy suppresses this worry. The sense of the formula '———— $\Phi(A)$' could well be specified as '(the thought) that the True is the value of the concept Φ for A as argument', and this, Frege agreed, is just another formulation of 'the thought that A falls under the concept Φ' or 'that A has the property Φ' (FC, 30). These noun clauses seem much more plausible candidates for proper names of objects.

A second argument shows how to extend the function-theoretic account of sense to expressions outside concept-script. Frege contended that a simple declarative sentence designates a truth-value and expresses a thought; also that it can be analyzed into a function-name (predicate) and argument-expression (proper name) (FC, 31). Accordingly a declarative sentence, by its grammatical structure, indicates how a truth-value is determined as the value of a function (concept) for an object as argument. Similarly, he articulated a complex singular referring expression ('the capital of the German Empire') into a function-name ('the capital of x') and an argument-expression ('the German Empire'); although he failed to introduce the term 'sense' here, he must have thought that the phrase, as well as designating Berlin, indicated how Berlin is presented as the value of a function for an argument (FC, 31f.), and it is tempting to suppose that he would have offered a parallel decomposition into function-name and argument-expression of any complex singular referring expression. The upshot is that Frege thought the idea of a mode of presentation derived from function theory to be applicable to every complex expression which is, or is part of, a declarative sentence of a natural language. What seems bizarre to us is central to his vision, and it accounts for the fact that he viewed the phrase 'mode of presentation' as an illuminating general explanation of his concept of sense.

This elucidation of Frege's conception does not account for all of his remarks about sense, but it does make a great deal fall into place.

(i) The central idea, that a sense is a mode of presentation of an entity, stems from the same source (function-theory) that informed *Begriffsschrift*. The concept of sense is not an alien import, but a further refinement within a fixed frame of reference.

(ii) It is transparent that many senses may correspond to a single reference. Any entity that can be identified as the value of some function for some argument can be picked out as the value of indefinitely many functions. This basic norm of function theory will hold equally for numbers, truth-values, persons, concepts, and relations.

(iii) This, in turn, clarifies the remark that the sense of a name 'serves to illuminate only a single aspect of the reference. . . . Comprehensive knowledge of the reference would require us to be able to say immediately whether any given sense belonged to it' (SR, 58).[1] Frege apparently thought that comprehensive knowledge of an object would consist in the ability to state all of its properties and all

---

1. He seems tacitly to have abandoned his earlier belief that the various ways of viewing an object

of its relations with other objects. But if any sense belonging to an object consists in a way of determining this object as the value of a function whose arguments are objects or first-level concepts, then this comprehensive knowledge must settle whether the given object is identical with the value of Φ (e.g. 'the capital of x') for the argument A (e.g. 'the German Empire'). Since there are infinitely many ways of presenting any object as the value of a function for an argument, it is evident that 'To such knowledge we never attain' (SR, 58)!

(iv) It is apparent now both *that* sense determines reference and *how* it does so. A properly defined function can be said to determine its value for each admissible argument. Therefore, by indicating a way of determining an entity as the value of a function for an argument, a complex expression makes manifest that the entity designated is determined as the value of some function for some argument and also how it is so determined (viz. in the way indicated).

(v) It is obvious that identities are typically not cognitively trivial. Generally a proof will be needed that two different functions yield the same value for a pair of different objects as arguments. (This is not a reflection of intellectual weakness, but a norm of constructing a rigorous demonstrative science.)

(vi) The possibility for alternative symbolizations of functions and arguments immediately differentiates the mode of presentation of an entity from the mode of designation of an entity (by a sign). A single mode of presentation may correspond to different signs (e.g. '$x^2 \rceil_2$', '$f(2)$', '2^2').

(vii) Frege's verdicts about identity and difference of sense for particular sentences, mathematical as well as non-mathematical, have a transparent rationale. It is because '2 + 2' indicates *one* mode of presenting the number 4 (viz. as the value of the addition operator for the argument-pair $\langle 2, 2 \rangle$) and '2^2' indicates *another* mode of presenting 4 (viz. as the value of the function x^2 for the argument 2) that these proper names differ in sense. Similarly, since '$2^2 = 4$' presents the True as the value (say) of $\xi^2 = 4$ for the argument 2, and since '2 > 1' presents the True as the value of the different concept $\xi > 1$ for the argument 2, it follows that these two sentences differ in sense. The argument in both cases is direct, appealing to nothing more than an intuitive conception of difference of functions (here, the extensional criterion of function-identity). Converse arguments to identity of sense are equally direct. Why does '2 > 1' have a unique sense although it can be viewed now as indicating the True as the value of the concept $\xi > 1$ for the argument 2, now as indicating the True as the value of the different concept $2 > \xi$ for the argument 1? Frege's answer must be that both these ways of viewing the True can be subsumed under viewing the True as the value of the relation (binary function) $\xi > \zeta$ for the pair of arguments $\langle 2, 1 \rangle$. It is self-evident that

---

do not constitute specifications of its objective properties and relations. What he here called 'aspects' *(Seiten)* and argued to be objective (SR, 60) he elsewhere (FC, 23) called 'ways of regarding' *(Auffassungen)*.

the value of $\xi > \zeta$ for any argument pair $\langle n, 1 \rangle$ is identical with the value of $\xi > 1$ for the argument n. Hence it is apparent without calculation that $\xi > 1 \big]_2$ has the same value as $2 > \xi \big]_1$. In effect, both ways of viewing what '$2 > 1$' indicates amount to presenting the True as the value of a single function for a pair of arguments, viz. as $\xi > \zeta \big]_{2,1}$. Here we need appeal only to an intuitive conception of identity of functions (the intensional criterion of function-identity).[2]

(viii) It is obvious that what is expressed by an identity is cognitively trivial if the signs flanking the identity-sign have the same sense. Since each sign designates the same object as a value of the same function for the same argument(s), no calculation or proof is needed to establish that the identity is true.[3]

(ix) Transparency of identity of sense is a natural corollary too. Two expressions have the same sense only if they indicate an entity as the value of the same function for the same argument. Depending on the criteria for identity and difference of functions and arguments, identity of sense might be conceived as sometimes requiring proof or alternatively as uniformly self-evident. The latter position would result from a strong intensional criterion of function-identity: viz. two function-names stand for the same function only if they are correlated with the same procedure for calculating values from arguments. According to this criterion, two expressions have the same sense only if the expressions designating the indicated functions themselves indicate these functions as determined in the same way. A model test-case is the pair of expressions '$\frac{d}{dx}(x^3) \big]_1$' and '$3x^2 \big]_1$'. Since $\frac{d}{dx}(x^3) = 3x^2$, these do indicate a number in two ways as the value of a single function for the argument 1 according to the extensional criterion of function-identity. But, since '$\frac{d}{dx}(x^3)$' and '$3x^2$' indicate this function as determined in different ways (as the value of second-level functions), the expressions '$\frac{d}{dx}(x^3) \big]_1$' and '$3x^2 \big]_1$' can be claimed to indicate different ways of presenting 3 as the value of functions for the argument 1. This verdict seems to conform to Frege's conception of sense, and it leads to the conclusion that two expressions having the same sense must self-evidently have the same sense (cf. PW, 210f.).

(x) An apparent puzzle about formal definitions in concept-script is removed. A formal definition stipulates a sense for a novel simple sign apparently by equating its sense with that of a complex expression correctly formed out of familiar

---

2. Church elucidates the identity of senses of concept-words by reference to this criterion of function-identity (*Introduction to Mathematical Logic*, p. 16).
3. It is assumed that these signs have references.

signs. But Frege described this procedure as stating that the *definiendum* shall have the same reference[4] *(gleichbedeutend)* as the *definiens* (BLA, §§27, 33). Does a formal definition then *say* that two expressions have the same reference and somehow *show* that they have the same sense?[5] How are formal definitions distinct from statements of coreferentiality that do not carry the implication of identity of sense, e.g. '(—$\stackrel{a}{\frown}$— $\Phi(\mathfrak{a}) = \Psi(\mathfrak{a})) = (\dot{\epsilon}\Phi(\epsilon) = \dot{\alpha}\Psi(\alpha))$' (cf. BLA §10 n.16)? Or does a formal definition assert that two senses are identical? How then does it 'go over directly' into an identity stating that these expressions have the same reference (BLA, §27)? This thicket of problems can easily be avoided. Since the *definiens* in a formal definition is always a complex expression, it must, in Frege's view, present a reference as the value of a function for an argument; it indicates a sense as belonging to the reference designated by the *definiens* and assigned to the *definiendum* by fiat. Any other complex expression designating the same reference would also indicate a sense belonging to this reference, and it could be used with the newly introduced *definiendum* to formulate a true (non-trivial) identity statement. What distinguishes the formal definition is not then its *form* (an identity), i.e. not its doing something over and above specifying the reference of the *definiendum* (which the non-trivial identity does as well). Rather, it is the *role* of this particular specification of a reference (as opposed to other possible ones) in respect of constructing logical proofs:

> In mathematics, what is called a definition is usually the stipulation of the reference of a word or sign. A definition differs from all other mathematical propositions in that it contains a word or sign which hitherto has had no reference, but which now acquires one through it. . . . Once a word has been given a reference by means of a definition, we may form self-evident propositions from this definition, which may then be used in constructing proofs in the same way in which we use principles [i.e. sentences expressing axioms]. (FG, 23)

A formal definition is therefore distinct from any other specification of the reference of an expression in that it alone licenses the substitution of an expression (the *definiens*) anywhere in a proof in place of the *definiendum* without the need for a supplementary proof of coreferentiality. It is a *canonical* presentation of the reference of an expression *as* the value of a function for an argument. That it alone among true identities determines the sense of the *definiendum* is another description of its distinctive role.

(xi) One reason is clear why Frege found it appealing to speak of the sense of a part of a sentence (or formula) as part of the sense of the whole sentence, at least in respect of complex subsentential expressions. The model here is a multiply complex name of a number, e.g. 'log (sin πe^x) $\Big]_2$ '. The sense of the whole expression is the mode of presenting a number as the value of a function for an argu-

---

4. Mistakenly puzzled by this account, Furth argues himself into mistranslating *'gleichbedeutend'* (BLA, §27n.).
5. Cf. Dummett: *Truth and Other Enigmas*, p. 105; *Frege: Philosophy of Language*, p. 227.

ment. Calculating the value of this function decomposes into a series of calculations, each of which takes the result of the preceding calculation as its basis. First we work out the value of e^2, then we take this number as the argument of the function sin $\pi(\xi)$ and calculate the corresponding value, and finally we do the same with this result in respect of the function log ξ. It is natural then to redescribe these nested calculations by transferring the relation of part to whole from the calculation-procedures themselves which are indicated by the symbols to the senses of the nested symbols (i.e. what they indicate).

(xii) The twin ideas that sense is a feature of every (logically) complex expression and that the sense of any expression is to be identified with how an entity is determined as the value of a function for an argument mesh together to yield one of Frege's explicit principles of logical analysis: as far as logic is concerned, complexity is always a matter of 'completing something that is in need of supplementation' (PW, 254), i.e. uniformly a matter of exhibiting an entity as the value of a function for an argument. This principle demands that the logical significance of any complex expression be exhausted by its correlation with a way of determining its reference as the value of a function, and this would be impossible unless the sense of a complex expression were identified with the way of determining its reference as the value of a function for an argument! No alternative elucidation of the concept of sense would yield this consequence.

Explaining the notion of a 'mode of presentation' in the framework of abstract function theory immediately clarifies many otherwise opaque aspects of Frege's thought about sense. It links up apparently disparate ideas, and it goes far in explaining why he thought what he did. But it does even more. It pinpoints important problems and reveals overall weaknesses in his conception of sense, and it highlights further independent sources of some of his fundamental contentions. The elucidation of sense as mode of presentation is as significant in respect of what it fails to accomplish as in respect of what it succeeds in doing.

First, this explanation of sense leaves an apparent lacuna: it provides no rationale for assigning sense to simple expressions, whether proper-names or concept-words. That is an absolute gap in Frege's discussion of sense. Although he insisted that every expression, simple or complex, has a sense, he was content to rest with this dogmatic assertion, giving no explanation of how to determine what sense a simple expression might have. He did not elaborate a theory of (simple) proper names. An aside on this subject (SR, 58n.) suggests that different speakers may attach different senses to the name 'Aristotle' which are specified by different definite descriptions (cf. T, 24f.). This leaves us in the dark about whether every name is equivalent in sense (for each speaker) to some definite description; and the comment is also unsatisfactory in that the descriptions supplied themselves contain proper names, so that the question of what their senses are threatens to lead to an infinite regress. Similar problems arise for simple concept-words which cannot be analyzed by means of *Merkmal*-definitions. What, we wonder, is the sense of the predicate '(is) red'? We look in vain for any answer from Frege.

The complaint that simple expressions expose a lacuna in his account of sense is too weak. If a mode of presenting an object is the way indicated by an expression for determining this object as the value of a function for a given argument, then there is no such thing as a mode of presentation associated with a simple name. To the naked eye of the mathematician,[6] the complex expression '$x^2 \rfloor_2$', does present 4 as the value of x^2 for the argument 2, but the numeral '4' does not present 4 in any way at all! Similarly, the function-name 'log (sin x)' does present a function as the value of the (second-level) composition-function for the argument-pair $\langle \log x, \sin x \rangle$, but the function-name 'sin x' does not present a function as the value of any (second-level) function. The real question is not how Frege would have proceeded in answering questions about the senses of simple expressions, but how to make sense of this notion. Russell rightly complained: 'I see the difference between sense and reference only in the case of complexes whose reference is an object, e.g., the values of ordinary mathematical functions like $\xi + 1$, ξ^2, etc.' (PMC, 169). No wonder! Frege's rationale for talking of the senses of simple expressions has nothing to do with the idea of a mode of presenting an object. It has a wholly independent source: the ideas that the reference of an expression cannot be part of its sense (e.g. that the number 3 is not part of the thought that $3 + 5 > 7$) and that a thought must decompose into parts.

Recognizing this defect in Frege's conception of sense may forestall a wild goose chase. Commentators have often attempted to give an epistemological or even psychological interpretation to the phrase 'mode of presentation of an object'. The sense of a proper name is then conceived to be a preferred way to ascertain with certainty whether a given object is its referent, or perhaps a description of the combination of characteristics that somebody actually exploits in recognizing the bearer of the name, or even just a recognitional capacity to pick out the bearer.[7] There are a few faint hints of such a conception in Frege's writings. The clearest is the thesis that the sense of 'I' in soliloquy is the particular, primitive way in which the speaker's self is presented to himself and to nobody else (T, 25f.). But this account leads into a tangle of questions to which Frege indicated no answers. It produces a host of mysteries and confusions. How are ways of recognizing objects to be individuated? How can they survive the destruction of the objects to be recognized? How can they be communicated? How can they be constituents of Platonic entities such as thoughts? It is even dubious whether the questions to be answered make sense. Is there in every case any such thing as a

---

6. Through Frege's logical telescope, of course, '4' can be seen to be analyzable as the value of a function for an argument. This does not blunt the objection, though it would call for a different example.
7. Dummett speaks variously of knowing 'under what conditions some other terms will stand for the same object', of knowing 'some particular *means* by which [the association of a name with its reference] is effected', of connecting 'the name with a particular way of identifying an object as the referent of the name', of commanding 'some means of identifying' the referent, and of 'fixing definite senses' for the purpose of establishing a 'clear means ... to achieve a resolution' of conflicts about truth-values of disputed statements, and of having 'a capacity for recognizing an object ... when presented with it' (*Frege: Philosophy of Language*, pp. 73, 93, 95, 99, 104f.; *Truth and Other Enigmas*, p. 129).

procedure for making certain that a given object is the referent of a name? (And what counts as being 'given' an object?) Does one always (or ever?) recognize an object as the reference of a name some*how?* These issues seem utterly alien to Frege's thinking, and he certainly contributed nothing to the discussion of such epistemological and psychological questions, brushing them aside with apparent disdain (cf. PW, 145). His behaviour here is justifiable to the extent that he had an altogether independent way of explaining the phrase 'mode of presentation of an object', as he in fact had. His conception of sense as the mode of presentation of an entity extends only over what is required to make intelligible the notion of something's being designated as the value of a function for an argument, and hence it does not trespass at all into the territory of epistemology or psychology.

A second limitation of Frege's account of sense also becomes explicable from the background of function theory. On the one hand, he introduced the sense of a name as the mode of presentation of an object. On the other hand, he emphasized that a name may have a sense although it lacks a reference. How, we wonder, can a mode of presenting an object fail to present an object? Are we to imagine (or postulate!) free-floating modes of presentation that may or may not present things? This seems incoherent. Why did Frege not address the problem? No doubt he saw matters differently. The same perplexity arises in function theory in respect of so-called 'singularities'. The function x^2 determines a number as value for every real number as argument. By contrast, the function $\frac{1}{x}$ determines a number as value for every real number as argument *except* 0; it is not defined for the argument 0—the number 0 is correlated with a singularity. We might say that the (nature of) the function $\frac{1}{x}$ determines that it lacks a value for the argument 0. Should we then deny that the complex expression formed by filling the argument-place of a function-name with the name of a real number always indicates a way of presenting a number as the value of a function? Should we distinguish thus between the expression '$\left.\frac{1}{x}\right]_0$' and '$\left.\frac{1}{x}\right]_2$'? This might seem to punish the expression '$\left.\frac{1}{x}\right]_0$' twice for a single offence. It would also seem to war against the inclination to say that the function $\frac{1}{x}$ itself *is* a way of relating one real number (a value of this function) to another real number (the corresponding argument). Indeed, the way in which ½ is related to 2 is the same as the way in which ⅓ is related to 3, ¼ to 4, etc.; hence, we might conclude that nothing is related to 0 in the way that ½ is related to 2. This is separated by only a hair's breadth from the contention that a way of presenting an object as the value of a function may sometimes present no object! The root idea that a mode of presentation is a way of determining something as the value of a function leads by a plausible (if not cogent) chain of reasoning to assimilating the notion of a name's having a sense without a reference to the notion of a concept's having null extension. It is one of

Frege's central contentions that this last idea is wholly unproblematic. What is crucial to his conception of mathematics as a genuine science is that it always be possible to determine whether or not a given entity is the value of a given function for a given argument. In his view, this carried the corollaries that it be possible to prove that a given entity is not the value of a given function for any argument and that a given function may have no entity as its value for a given argument (i.e. that the concept of being a value of this function for this argument is empty!). If the locution 'mode of presentation of an object' makes this seem problematic, we can fall back on the underlying general principles of function theory.

Three fundamental weaknesses at the root of Frege's conception of sense conceived in function-theoretic terms can now be brought to the surface.

First, the legitimacy of functional abstraction seems to license ascribing indefinitely many distinct senses to every complex expression. If, e.g., the formula '——— $\Phi(A)$' has a sense in virtue of its indicating how a truth-value is determined, viz. as the value of the concept Φ for the argument A, then it must be assigned another sense in virtue of the open possibility of viewing it as indicating how the same truth-value is determined as the value of a second-level concept whose argument is the first-level concept Φ. And so on, *ad infinitum,* by the abstraction of concepts of ever higher level. By generalizing this reasoning we seem driven to the conclusion that every complex expression must have as many distinct senses as there are possible ways of viewing it as designating an entity as the value of functions for arguments. In respect of sense, it must be considered *systematically* ambiguous. This conclusion is not palatable. It contradicts Frege's evident assumption of the uniqueness of the sense of a typical sentence in concept-script or natural language, and it renders incoherent his countenancing alternative logical decompositions of thoughts, i.e. his contention that the same thought can be correlated with the different modes of presenting an entity which are generated by functional abstraction (BLA, §21ff.). The function-theoretic conception of sense together with the boundless possibilities of functional abstraction lead directly to the paradoxical conclusion that there is no such thing as *the* sense of any complex expression.

Can Frege be rescued from these dire straits? Salvation would require an argument that functional abstraction never generates *genuinely* different ways of presenting an entity as the value of a function for an argument. And it would require a supplementary demonstration that difference in level of functional abstraction does not affect the cogency of arguments (and therefore is not necessary in specifying the senses of expressions). This second point might readily be conceded. Although the elucidation that '——— $\Phi(A)$' presents a truth-value as the value of a second-level function whose argument is Φ may be important for securing acceptance of the logical law $\vdash$——— $\Phi(A) \to (\exists \phi)\phi(A)$ (cf. BLA, §25), the truth of this general law is independent of this elucidation (just as the axioms of geometry are not logical consequences of the informal explanations of the primitive terms). The pair of logical laws

$$\vdash\!\!\!\text{———}\, \Phi(A) \to (\exists \phi)\phi(A)$$
$$\vdash\!\!\!\text{———}\, \Phi(A) \to (\exists x)\Phi(x)$$

are equally fundamental and immediate; the very same formula '$\Phi(A)$' expressing a single thought occurs in both, even if we are inclined to view the indicated concept and argument now at one pair of levels, now at another. This reasoning suggests an approach to the fundamental issue. Why not argue that the free availability of functional abstraction is precisely what precludes taking it to introduce essentially different ways of viewing an entity as the value of a function? The possibility of viewing '——— $\Phi(A)$' as indicating how a truth-value is determined as the value of Φ for the argument A is inseparable from the possibility of viewing it as indicating how the same object is determined as the value of the second-level function $\mathfrak{F}$ for the argument Φ (where, on our modern view, $\mathfrak{F}$ is specified by the schema: $\mathfrak{F}(\phi) = \phi(A)$). These are not real alternatives. Each carries with it the other. It makes no sense to deny that '——— $\Phi(A)$' indicates a truth-value as the value of a second-level concept, as the value of a third-level concept, etc.; nor does it make sense to describe '——— $\Phi(A)$' as pre-eminently indicating one of these ways of viewing a truth-value rather than some others. More positively, the crucial fact is the *self-evidence* of the identity of the values, for the appropriate arguments, of the various functions of different level abstracted from the thought expressed by '——— $\Phi(A)$'. There is therefore no more reason to infer multiplicity of senses from the possibility of higher-level functional abstraction than to conclude from the different possibilities of same-level functional abstraction that what '2 > 1' expresses is not a unique thought. In each case an apparent difference between modes of presentation of an entity is denied to be a real difference on the grounds that two different routes self-evidently lead to the same goal.

Although this defence has some plausibility and even charm, it lands Frege in further difficulties. First, it completely undermines the intuitive clarity of the function-theoretic explanation of 'mode of presentation'. Our original instructions were to assign different senses to two complex expressions if they designated the same entity as the value of different functions for different arguments. The most dramatic possible contrast between functions is difference in logical type (difference in level, difference in number of arguments). But now, *mirabile dictu,* we are informed that we should overlook such differences in certain cases and assign a single sense to an expression in spite of its indicating different ways of determining an entity as the value of different functions. What is now left of the notion of a mode of presentation of an entity? The second difficulty concerns the constituents of a complex expression. The argument that functional abstraction does not introduce multiple senses for a complex expression concedes the point that the component expressions are systematically ambiguous in respect of *reference*. The expression '——— $\Phi(A)$' indicates that a truth-value is the value of a first-level concept Φ for the object A as argument, but equally that it is the value of a second-level concept for the concept Φ as argument, etc. How does this leave room to deny that the reference of 'A' in '——— $\Phi(A)$' can be conceived to be a second-level concept with just as much right as its being conceived to be an object? This suggests that the defence of the uniqueness of the sense of complex expressions is of little value. The cost of that victory is systematic ambiguity in the reference of component expressions. And, unless sense is denied to simple expressions, there seems no way to prevent a further outbreak of systematic ambiguity of sense in the course of the logical analysis of thoughts.

A second fundamental commitment of Frege's explanation of the concept of sense is that the route from sense to reference must be independent of matters of fact. The sense of an expression presents its reference as the value of a function for an argument. But it is of the essence of a *function* that it be possible to *calculate* its value for any given argument.[8] Hence the function indicated by an expression must, as it were, determine by itself what the reference of the expression is, i.e. what value this function takes for the argument also indicated by the expression.[9] Anybody who knows what function and what argument are indicated

---

8. It is natural to declare that its indicating a method for calculating its values for suitable arguments is an essential characteristic of a mathematical function, or similarly that the standard specification of a function is an analytical formula that actually lays down a uniform computation procedure for arriving at a value for any substitution of names for the variables of the formula. These ideas then seem to be threatened by the extension of the concept of a function within nineteenth century mathematics, especially by Dirichlet's characteristic function on the rationals (which take the value 1 for a real number if and only if it is rational, otherwise taking the value 0). This calls for a refinement of the notion of calculation involved in the primitive mathematical conception of a function. Here we might differentiate between a psychological and a logical connection between calculations and functions. The psychological connection depends on the strengths and weaknesses of the human intellect. It seems that we resort to calculation only when we lack direct awareness of the answer to the question 'What is the value of the functions ... for the argument ...?'. There is no difference of principle between saying what the value of x^2 is for the argument 1 (where we need make no calculation) and saying what the value of log sin x is for the argument $\frac{2\pi}{17}$ (where we would have to carry out a complex computation). The concept of a function would not be affected by increases (or decreases) in the abilities of mathematicians to dispense with computations. What is crucial is not that it is necessary for us to calculate every value of whatever can be correctly called a function, but rather that it is possible to calculate such values if we find it necessary to do so because we cannot immediately see what they are. This raises the logical point. Mathematicians must specify functions whose arguments are entities belonging to infinite or unsurveyable totalities. Consequently, there is no such thing as defining such a function by a list which gives its value for each admissible argument. The definition must indicate some laws or rules, appeal to which would in principle settle whether any particular assignment of a value to the function for a particular argument is correct. Liberalization of this concept of a function in mathematics, e.g. by extending it to Dirichlet's characteristic function on the rationals, leaves the logical requirement intact (at least on the assumption that any inability to decide whether a given number is rational or irrational merely reflects a human intellectual deficiency). The fundamental issue is the contrast between calculation and experiment. The connection between the argument of a well-defined function and its value is considered to hold *a priori*, to obtain independently of observation and experience. For this reason, the resolution of any dispute must be reached by following the directive 'Calculemus'. Mathematicians, in extending this concept of a function, did not until relatively recently abrogate the principle that any function determines its value *a priori* for any admissible argument.

9. Church apparently incorporates this conception into his exposition of a logical semantics explicitly modelled on Frege's ideas. In his system, the reference of a name is a function of its sense, i.e. there is a function f such that

    reference of $N = f$ (sense of N)

    for all names N which have a reference. Like Frege (WF, 112), he excludes as senseless the

by a complex expression is already in possession of all the *knowledge* required in principle for carrying out a calculation of the reference. Frege could not abandon this idea without surrendering the vital thesis that the transition from a concept to its extension is a purely logical operation. Yet the thesis that the sense indicated by an expression determines its reference independently of matters of fact threatens to make nonsense of the very idea of applying the concept of sense in the analysis of statements of empirical knowledge. The only scope left for the intervention of facts in the determination of the truth-values of an assertion is in the specification of the *sense* of the asserted sentence. Once we know what function (concept) and what argument are indicated, then we can resort to calculation. Hence, it must be presumed that we are ignorant of the sense of the sentence if we must in principle make recourse to matters of fact. But this is altogether mysterious. How can we search for the sense of an expression? By what are we to be guided? What questions are there here which descriptions of facts might answer? How can we tell whether the sense has successfully been unearthed? These queries do not have any air of profundity; rather, they seem absurd. The point of raising them is to call attention to an important objection of principle to extending the concept of a mode of presentation of an object outside mathematical function theory. For declarative sentences making empirical statements there is no such thing as a thought, grasp of which enables one in principle to *calculate* its truth-value. *Pari passu*, we recognize no such entity as a concept, grasp of which allows the *a priori* calculation of the extension of any concept whose extension depends on matters of fact. This points towards a fundamental asymmetry glossed over in Frege's system. He paraphrased the formula '⊢——— $\Phi(A)$' as the judgment that A has the property Φ (BS, §10). But whether '——— $\Phi(A)$' designates the True can be calculated from specification of the concept Φ and the object A, whereas whether an object has a property Φ cannot be calculated at all if 'A' is a simple name and Φ is a property the possession of which by an object is settled by observation or experiment. His paraphrases in terms of 'property' disguise these fun-

---

statement that one thing is a function of another (e.g. that velocity in free fall is a function of the time) unless it is possible to answer the question 'What function is it?'. The answer to this (e.g. '$v = gt$') must *stipulate* a value for each admissible argument, thereby laying down a correlation each element of which is independent of experience. (A. Church, *Introduction to Mathematical Logic, Vol. I* (Princeton, 1956), p. 9, 19.) Other commentators commonly deny that the relation between sense and reference is invariably an internal relation, building interpretations of 'Frege's semantics' on this basic misconception. Kaplan makes the point explicit in his presentation of the 'Fregean picture' of singular terms. The sense of a singular term is 'determined by the conventions or rules of the language', while the reference of this term 'is, in general, an empirical relation, [viz.] the individual who falls under the concept (i.e. who, uniquely, has the qualities)' (*Demonstratives*, p. 4). According to this picture, the reference of an indexical expression is determined by an invariant sense together with information about the context of its utterance, whereas Frege held that the *sense* of a context-dependent expression depended on the context of its utterance. Kaplan also presupposes that a context-independent definite description (i.e. one tailored for an eternal sentence) would have an invariant sense in different possible worlds so that it determines the actual reference only in virtue of certain matters of fact. This presumption is nearly universal among expositors of Frege, but it has no direct warrant in his texts.

damental asymmetries, and they serve to mask the deeply paradoxical consequences of his notions of a concept and a relation.

A third defect of Frege's account of sense is equally radical. Since an expression indicates its sense by presenting its reference as the value of a function for an argument, the notion of sense is applicable only to *complex and analyzable expressions,* and furthermore, only to such as contain at least one *function-name.* For only if an expression contains a function-name can it be described as indicating how its reference is determined as the value of some function. Frege suggested that every declarative sentence contains at least one function-name (typically the grammatical predicate), and he hinted that every complex proper name can be construed as consisting of a function-name with suitable argument-expressions (FC, 31f.). Presumably he held it to be obvious that every (logically significant) complex expression could be articulated into a function-name and argument-expression(s).[10] In his view, then, this latter restriction is vacuous. But this is precisely the point that we cannot now concede. The application of the sense/reference distinction extends only as far as the frontiers of the domain of *complex expressions containing function-names.* There is reason to doubt whether this domain includes any sentences in any natural language, and no reason whatever to maintain that it includes every declarative sentence in every natural language. The intelligibility of the explanation that senses are modes of presentation of entities seems to be inversely proportional to the range of its intelligible applications.

2. Poisoned springs

A simple expression can be ascribed a sense only if it is logically equivalent to (immediately replaceable in a proof by) some complex expression. Otherwise there is no such thing as its indicating how an entity is presented as the value of a function. Yet, notoriously, Frege ascribed senses to simple unanalyzable expressions, and indeed he distinguished their senses in every case from their references. The rationale for this manoeuvre must have a source wholly independent of the connection of the 'mode of presentation' of an entity with function theory. To illuminate this matter we must search for other springs from which he took deep draughts.

Light may come from reflecting on preconceptions that we no longer share. Certain ideas which Frege inherited so influenced his thinking that they supported contentions quite independent of, even inconsistent with, the function-theoretic background to his conception of sense. From our point of view, three ideas that he took for granted were of crucial importance:

(i) *Uninhibited Platonism:* Like other nineteenth-century Platonists, he had no philosophical qualms about abstract objects. Hence, if a complex expression can properly be said to indicate *a way* of presenting an entity as the value of a function or *how* an entity is determined as the value of a function, then *ipso facto* what it

---

10. This would account for his assumption that standard *Merkmal*-definitions give senses to defined concept-words.

indicates is an object. There is on this view no gap between acknowledging this role of expressions and asserting the existence of senses of expressions, therefore no room for 'postulating senses'.

(ii) *Intension/extension:* Philosophical tradition sanctioned distinguishing a pair of logically significant features of typical words, variously called '*compréhension*' and '*étendu*', 'connotation' and 'denotation', 'intension' and 'extension'. Two general theses accompanied these distinctions. First, the second member of each pair was assigned to any expression only indirectly, i.e. as a consequence of assigning the first; e.g. the connotation of the word 'man' settles that it is applicable to Socrates. Second, the two members of each pair were taken typically to belong to different ontological strata; connotations and intensions of concept-words were conceived to be either mental entities or abstract entities, hence on a different level of being from the perceptible world. Propositions or judgments were considered to be put together out of ideas or concepts, not out of concrete objects and perceptible properties. The distinction between intension and extension corresponds roughly with a major category-distinction, although which one was a matter of dispute. Frege was aware of this background and eager to correct it in one respect: in his view, philosophers placed concepts into the wrong pigeon-hole, treating them as category-homogeneous with intensions, whereas in fact they should be assigned to the same level as extensions (PW, 123). His concern here seems parasitic on his general acceptance of the idea of some metaphysical gulf between intensions and extensions.

(iii) *Complexity of judgments:* Inferences were universally held to consist in transitions from certain judgments to other ones, and their correctness was thought to depend on patterns according to which these entities were built up from simple entities (concepts, ideas, terms of judgments). It is solely in virtue of the complexity of judgments that some inferences are correct, others incorrect. The very possibility of formulating laws of logic presupposes that judgments may be decomposed into simple constituent parts. (Modern logic too accounts for logical relations in terms of the complexity of certain entities, but it ascribes structures and parts to sentences themselves rather than to statements, propositions, etc. Consequently, it is an unshakeable article of faith that sentences can be so analyzed in logic that their logical relations are determined by their structures and their constituents.)

The juxtaposition of these three ideas in Frege's thinking yields as an obvious consequence the principle that every expression,[11] whether simple or complex, must have a sense as well as a reference if it is a logically significant part of an expression that expresses a thought. Thoughts must be complex, since assertions stand in logical relations to one another. But thoughts are not category-homogeneous with concrete objects. Therefore, concrete objects cannot be parts of thoughts; the two belong to different metaphysical levels. Finally, since thoughts are complex, there must be thought-constituents related to logically significant sentence-parts as thoughts are related to sentences. These are the senses of sen-

---

11. Barring exceptions such as variables and the assertion-sign.

tence-constituents, and they lie on a different metaphysical level from the references of sentence-constituents. This chain of reasoning sweeps us along to the conclusion that all significant sentence-constituents, whether complex or unanalyzable, *must* have senses, though we must concede, of course, that we have no idea what the sense of an unanalyzable expression might be.

The details of this reasoning are highly questionable. But in outline it has considerable momentum, and it leads to a conception of the sense/reference distinction that has the charm of simplicity. It suggests a bifurcation of conceptual content, as it were horizontally, into a pair of metaphysical strata. This picture captivated Frege immediately upon his introducing the sense/reference distinction. He contrasted the 'level of thoughts' with the 'level of reference', and he argued that combining senses into a thought cannot affect an 'advance' from one level to the other, from a thought to its truth-value (SR, 64). From these seeds there grew a gradually increasing tendency to contrast 'the realm of truth' with 'the realm of fiction' (PW, 130, 191, cf. 232), and ultimately 'the realm of reference' with 'the realm of sense' (CT, 541). Frege presented the separation and parallelism of entities in these different strata by means of a diagram (PMC, 63):

sentence	proper name	concept-word
↓	↓	↓
sense of the sentence (thought)	sense of the proper name	sense of the concept-word
↓	↓	↓
reference of the sentence (truth-value)	reference of the proper name (object)	reference of the concept-word (concept)

Here a complex network of relations seems to generate isomorphic structures on three different levels (language, like sense, standing over and against 'the world' (reference)). The complexity of the thought expressed by a logically perspicuous singular atomic sentence mirrors the structure of the sentence itself.[12] The sentence seems to express this thought in virtue of the fact that its constituents (subject and predicate) express or indicate counterpart constituents in the realm of sense. The sentence is a model of the thought, for its sense (a thought) is presented as a construction out of senses corresponding to the expressions out of which the sentence itself is constructed. Similarly, both the sentence and the thought mirror relations in the realm of reference. If the sentence is true, then it designates the True in virtue of the fact that the object designated by the proper name falls under the concept designated by the concept-word; similarly, the thought designates the True because the object determined by the sense of the name falls under the concept determined by the sense of the concept-word. What guarantees this triple-layered isomorphism is the idea that sentence-constituents refer to objects and concepts only *via* entities in the realm of sense. That any signs refer to entities at all is depicted as if it were parasitic on the fact that senses are the real agents of

---

12. Disregarding complications stemming from the fact that structurally distinct sentences may express the same thought, and that a thought may be decomposed in radically different ways.

reference. The best we mere mortals can do is, as it were, to hitch words onto senses and then rely on the grace of God to make an advance to the level of reference.

Frege's thinking manifests a deep commitment to a trichotomy of reality into language, the realm of sense, and the realm of reference. In respect of distinguishing language from 'the world', his conception was quite orthodox. Though fundamentally defective, this schema continued to be influential in logical atomism and logical positivism, and it still survives despite Wittgenstein's powerful criticisms.[13] What is less straightforward is the separation of the realm of sense from the realm of reference. Though this too has reputable ancestry, it is unclear how to make this bifurcation mesh with the explanation of sense as a mode of presentation of an entity and how to give the picture of stratification any coherent application, particularly within the parameters of Frege's own thought. His picture is clear, and it is clearly important for him, but scrutiny of its immediate implications suggests that it is misleading and distorting.

There are four major reasons for concluding that the picture does not harmonize with the function-theoretic explanation of sense as 'mode of presentation':

(i) The reification of senses in accordance with this questionable picture presupposes that a unique sense is standardly correlated with each type-expression.[14] Given that senses are not mind-dependent (SR, 59f.), Frege entertained no general qualms about the uniqueness of senses regularly connected with signs (cf. SR, 58). He felt no impropriety in employing the expression 'the sense of the expression "A"' (SR, 59); he held that a formal definition fixes the sense of the defined simple expression, and that an analysis of a simple expression already in use might have the same sense as the analyzed expression (PW, 210f.). There is, however, a *prima facie* conflict between these obvious points and the conception of sense as mode of presentation. For the limitless possibilities of functional abstraction guarantee that any complex expression that indicates one way in which an entity is designated as the value of some function for some argument also indicates other ways in which the same entity is designated as the value of other functions for other arguments. The only means for avoiding outright contradiction with the assumption of uniqueness of sense must concede that the constituent symbols of a given complex expression are themselves systematically ambiguous in reference (supra, p. 311f.). But now, if senses are assigned to these constituents in conformity with the principle that a sign refers to something only via its sense, and if difference in reference presupposes difference in sense, then these constituent symbols must have as many distinct senses as they have distinct references, i.e. each one must have indefinitely many. Therefore, it is doubtful whether there is any such thing as *the* mode of presentation of a truth-value which is indicated by a

---

13. E.g., the celebrated debate between Austin and Strawson about whether facts are part of language or part of the world.
14. There will, of course, be many kinds of exception: demonstratives and indexicals, ordinary proper names ('Tom', 'Dick', and 'Harry'), and lexically ambiguous concept-words call for obvious qualifications, as do oblique contexts.

sentence, and it is certain that there is typically no such thing as *the* sense of any sentence-component. In particular, the constituent symbols of a formula of concept-script are systematically ambiguous in respect both of sense and of reference.

(ii) The conception of senses as thought-constituents mediating between signs and their references ascribes senses to logically significant expressions irrespective of whether they are simple or complex and of whether they are primitive or analyzable. Frege held that there must be indefinables in any language or system of signs; that these terms must have senses since they contribute to the expression of thoughts; and that the denial of sense to indefinables would deprive of sense any complex expression in which they occurred (PW, 191). Without the least inkling of what the senses of these expressions might be, we know that there *must* be such entities in the realm of sense. This claim is an exemplary instance of philosophical dogmatism. It is also in head-on collision with the explanation of sense as mode of presentation. For, in the case of indefinables, there is no such thing as the indicated way in which an entity is determined as the value of a function for an argument. Frege's concept of sense is constructed out of two incompatible ingredients.

(iii) That a given function takes a certain value for a particular argument is an intrinsic feature of its being that function (not another one); a function determines its values, as it were, by itself. Consequently, the conception of sense as a mode of presentation of an entity as the value of a function forges an internal connection between the sense of an expression and its reference. This is threatened by the reification of senses. The very idea of logical relations between distinct existents is obscure; there is an inclination to suppose that all relations among them must be external or extrinsic. Frege succumbed to this temptation at least intermittently. In admitting the possibility that an expression might have a sense without having a reference, he apparently[15] conceded that the connection might be contingent: the phrase 'the heavenly body most distant from the earth' certainly has a sense, but it may well lack any reference (SR, 58). *A fortiori,* the reference cannot always be *calculated* merely from the sense of an expression. Although this brings the concept of sense into line with the traditional conception of connotation or intension in this respect, it threatens to undermine the thesis that the relation of a concept to its extension is purely logical. If, for example, '────── Φ(A)' designates a truth-value as the value of the concept Φ for the argument A, and if the identity of the concept Φ alone settles whether A falls under Φ or not, then the truth-value of '────── Φ(A)' must in principle be calculable from the sense which it indicates. The same conclusion must hold for any complex expression designating something as the value of a function for an argument. Platonism about senses and the demands of logicism pull in opposite directions.

(iv) Reification of senses generates a mystery about the relation of the senses of the parts of a sentence (proper names, concept-words) to the sense of the whole sentence. In the case of sentence-parts that are themselves complex, a readily comprehensible figure of speech might justify speaking of their senses as being

---

15. But not if the modality of the *possibility* of reference-failure is merely epistemic.

parts of the sense of the whole sentence. There would, however, be no warrant for treating this as a literal description of a relation between abstract entities (e.g. as if the sense of 'log (sin x)' were *smaller than* the sense of 'log (sin (πe^x))'!). Nor would speaking here of parts and whole carry the implication of homogeneity of *parts:* a sentence may contain two complex expressions, one of which indicates a mode of presentation of a concept as the value of a function, the other of which indicates a mode of presentation of an object as the value of a function (just as 'log (sin x)' designates a function, 'πe^x' a number). Platonism about senses threatens to transform a harmless figure of speech into a misleading metaphysical principle governing entities in the realm of sense. We slip easily into thinking of the sense of any sentence-part as part of the sense of the whole sentence, and then we are apt to make use of arguments presupposing that this principle discloses the quasi-chemical composition of thoughts out of homogeneous constituents. Apart from the puzzles and obscurities thereby generated, there is a fundamental mystery about the senses of simple constituents of sentences. Suppose for ease of exposition that there are indefinable proper names, and that '——— $\Phi(A)$' designates a truth-value by means of a simple concept-word '$\Phi(\xi)$' and a simple proper name 'A'. Now '——— $\Phi(A)$' indicates, i.e. presents as its sense, how a truth-value is determined, viz. as the value of $\Phi(\xi)$ for the argument A. What do '$\Phi(\xi)$' and 'A' indicate? Unless this question is rejected as nonsensical (and sense is denied to simple expressions), then the requirement that the sense of a symbol be transparent in concept-script carries the implication that '$\Phi(\xi)$' and 'A' must indicate the concept $\Phi(\xi)$ and the object A if '——— $\Phi(A)$' is to indicate that a truth-value is determined as the value of $\Phi(\xi)$ for the argument A! Sense would not be transparent if anything else were indicated by either '$\Phi(\xi)$' or 'A'. But the supposition that '$\Phi(\xi)$' indicates the concept $\Phi(\xi)$ and 'A' indicates the object A contradicts the principle that sense is always distinct from reference for any token-expression. Consequently, if objects and concepts are banned as possible thought-constituents, it is unclear that any other entities satisfy the constraints placed on the senses of simple expressions by the explanation that senses of complex expressions are the mode of presenting entities as the values of functions. This leaves both the nature of thought-constituents and the relations among them shrouded in mystery and subject to inconsistent demands.

Conflict with the explanation of sense as mode of presentation, though of fundamental importance, is not the only major defect in Frege's reification of senses and his separation of the realm of sense from the realm of reference. Other deep problems arise from his Platonism, some integral to his thinking, others more apparent to us than to him.

First, the hypostatization of senses together with the dogmatic principle that every significant expression has a sense legitimates, indeed presupposes, the idea that the senses correlated with expressions of natural language exist independently of an apprehension of them and hence await discovery. Presumably Frege thought himself to have revealed for the first time the senses of the numerals '0', '1', '2', ... ! But it is wholly obscure what method is appropriate to the search for senses, what the criteria of success are, or even whether there is any such thing as unknown entities that play roles in the logical assessment of inferences. Unless these matters can be clarified or dismissed as irrelevant, we may be forced to deny senses to simple expressions, to which Frege dogmatically assigned senses.

Second, regarding senses as entities generates perplexity about what constitutes a proper specification of the sense of any expression. Questions of the form 'What is N?' standardly elicit an answer incorporating some expression which *designates* whatever is named by 'N', on the model of answers to the question 'What is the Eiffel Tower?' or 'Who was Aristotle?'. Given that 'the sense of "A"' names an object, this suggests that specifications of sense should be identity statements exemplifying the pattern 'The sense of "A" is (identical with) the sense of "B"'. But in fact Frege never explained or defined any expression in this way even when he claimed to be establishing its sense. Had he always done so, it would be unclear how his explanations would allow us, as it were, to break into the circle of sense. In his actual practice, the sense of an expression is always specified by an identity statement which *uses*, but does not mention, the expression to be explained. In the simplest cases the *definiendum* is stipulated as designating the same object as the *definiens*, and by implication the two expressions are taken to indicate the same mode of presentation of this object. Whether or not an identity statement carries this implication is something external to it, not a matter of its form but of how it is employed (especially in connection with proofs). This gives rise to a feeling of disquiet or bewilderment. It seems as if senses are ineffable, that we cannot *say* what the sense of an expression is, but merely gesture in that direction. It even seems unclear how to formulate a proper *answer* to the question 'What is the sense of "A"?'.[16] How, we may wonder, does a formal definition succeed in singling out one abstract entity (the sense of the *definiendum*) by means of other entities (the reference of the *definiens* and the *definiendum*)? Platonism about senses fosters questions that are both urgent and intractable.

Third, it must be intelligible that there can be multiple independent ways of identifying any genuine entity. Platonism about senses is intelligible only to the extent that it meets this requirement, in particular with respect to any specific thought. Two methods for doing so might be proposed, but there are serious doubts whether either is coherent within the framework of Frege's system. The first method would exploit the role of thoughts as objects of assertion. In a given context, there are invariably numerous distinct indirect statements that will be acceptable as correct reports of what a certain speaker said (asserted). On Frege's view, each of these indirect statements will itself designate the sense of the sentence the assertoric utterance of which is reported. Consequently, there will always be distinct ways of specifying the sense of any declarative sentence. The conclusiveness of this reasoning is open to doubt. It is unclear whether the require-

---

16. Cf. Dummett, 'What Is a Theory of Meaning?', pp. 97ff., in S. Guttenplan, ed., *Mind and Language* (Clarendon Press, Oxford, 1975).

ment of independent specifiability of what is *expressed* by a sentence is met by producing a variety of expressions which *designate* what is expressed by it. Should we perhaps not demand alternative independent ways of *expressing* what is expressed by any given sentence? Frege's account apparently allows us to fulfil this requirement too. For an indirect statement (e.g. 'that grass is green') is claimed to designate what the enclosed sentence ('grass is green') expresses, and therefore we can extract from any set of reports in indirect speech of what a given speaker asserted a set of sentences identical in sense. But this contention is incompatible with the demand that sentences identical in sense be equipollent in proofs. The availability of alternative correct reports in indirect speech makes (justifiable and context-relative) presuppositions about the co-referentiality of expressions: in a rigorous proof, their substitutions might require supplementary proofs. Here fresh difficulties break out. If we must strike out from the set of indirect statements all those containing sentences not meeting this equipollence condition, it is not clear which one to take as privileged and it is doubtful whether the resulting set would not typically consist of a single member only! On the other hand, if we insist that multiple indirect statements are correct reports of what is asserted and yet acknowledge that identity of sense requires equipollence in proofs, we would be driven to conclude that every utterance of an assertoric sentence expresses as many distinct thoughts as there are non-equipollent correct reports in indirect speech! In neither case do we arrive at multiple independent specifications of a single thought.

The second method for attempting to satisfy this requirement exploits the idea that singling out an entity as the sole thing having certain attributes can be viewed in every case as a matter of determining that entity as the value of a function for appropriate arguments. Consequently, independent ways of identifying an entity must, on Frege's view, correspond to different ways of determining it as the value of a function for an argument. Can this condition be satisfied for a thought? Two possibilities must be considered. The first is that every thought must itself be indicated as the value of a function for an argument. This case will be deferred (infra, p. 323ff.). The second is that any thought, though not necessarily presented as the value of a function, can be so presented in various different ways. Consider the thought expressed by '——— $\Phi(A)$'. Could we infer that the same thought is expressed by the double negation of this formula, viz. '—┬┬— $\Phi(A)$'? Frege seems to have accepted this thesis (CT, 548, 553ff.; cf. N, 132ff.) and also to have supposed that such formulae expressed a thought as the value of a function (whose arguments are thoughts). But the idea fits poorly with the conception that formulae expressing the same thought must be interchangeable in proofs without further ado. For, in substituting '—┬┬— $\Phi(A)$' in place of '——— $\Phi(A)$', any given proof must be altered in respect of one application of the basic logical law ├─┬┬— B = ——— B (BS, Preface).[17] Frege's thinking here was subject to irresolvable tensions. Accepting that logical operators are genuine function-names would drive him into holding that '—┬┬— $\Phi(A)$' presents a truth-value differently

---

17. For simplicity, appeal here is made to one of the logical axioms of *Begriffsschrift*.

from '——— Φ(A)' and hence into denying identity of sense, while denying that they are genuine function-names would blow a hole in the programme of function/argument analysis of truth-values. It is doubtful whether logical operators can be exploited to yield different ways of expressing a thought as the value of different functions, and hence this method too provides no clear way of meeting the requirement of multiple independent ways of identifying senses necessary to back Frege's Platonism about senses.

Fourth, it is highly problematic to conceive of senses as essential intermediaries between symbols and what they stand for. Mystery envelops the very idea that one entity (a sense) can 'determine' another entity (a reference). Confusion is piled on mystery by attributing to senses both the properties of signs (e.g. strictly speaking thoughts, not sentences, designate the True and the False (SR, 64f.; PW, 129)) and properties incompatible with signs (e.g., thoughts are imperceptible abstract objects). Still worse, an impenetrable veil of *Sinn* seems to fall between symbols and what is symbolized. The thought, or sense of a sentence, becomes a shadowy go-between connecting a sentence with its truth-value. It has the status of a curious entity towards which we may stand in various relations apparently independent of, and in competition with, our relations to truth-values or the states of affairs described by sentences.[18] When someone asserts truly that grass is green, what he asserts is not simply what is the case, viz. that *grass* is *green,* but rather something else, the thought *that* grass is green, which stands in an obscure relation (of 'designating the True') to what is the case.[19] Symbols, it seems, cannot make direct contact with what is symbolized. The ethereal machinery of senses determines what symbols refer to, independently of what language-users take their references to be. If Frege did first discover the sense of the numeral '0', then it must have rested on repeated acts of grace that anybody had ever previously succeeded in referring to nought by this symbol! Reflection on the role of senses as intermediaries between signs and their references, together with the inscrutability of the mechanics of the determination of references by senses should no doubt instil in us all 'a thorough sense of the omnipresence ... of that *Almighty Spirit* ... [upon whom] we have a most absolute and immediate dependence'[20] in employing symbols in thinking and communicating!

3. Thought-constituents: intolerable tensions

A third ingredient in Frege's conception of sense has remained so far unexplored. He held that the cogency of an inference depends on the relations among its constituent asserted thoughts. In common with traditional logic, he found this to be intelligible only on the assumption that thoughts are complex entities. Logical

---

18. A. N. Prior, *Objects of Thought,* p. 51ff.
19. This account does have the advantage of showing 'how it is possible' to make false assertions or to hold false beliefs. For that is no more (and no less) mysterious than making true assertions or holding true beliefs.
20. Berkeley, *Principles,* §155.

relations among thoughts can arise solely from their sharing some thought-constituents. The presumed complexity of thoughts is apparent in his separating off the sense of a name from its reference: he reasoned that if the reference of 'Mont Blanc' is not part of the thought that Mont Blanc is more than 4000 metres high then something else (the sense of 'Mont Blanc') must be. His picture of the strata of sense and of reference depicts this complexity too. By splitting judgeable-content 'horizontally', Frege populated the realm of sense with counterparts of each of the entities in the realm of reference. Thoughts correspond to truth-values, while the senses of names and of concept-words correspond to objects and concepts. 'The analysis of the sentence corresponds to an analysis of the thought, and this in turn to something in the realm of reference, and I should like to call this a primitive logical fact' (PMC, 142).

Hitherto we have focused on how these senses of expressions are supposed to be related to the expressions themselves and to their references. But it is equally problematic how they are related to each other. How do the constituents of a thought combine together to constitute a thought?

At first blush, the priority in logical analysis accorded to judgeable-contents carries over into the priority of the thought over the senses of subsentential expressions. This primacy of the thought is essential to making sense of the possibility of alternative decompositions of a single thought—a procedure much exploited in the *Basic Laws*. The natural corollary of this conception would be a transfer of the function/argument decomposition of judgeable-contents to the analysis of thoughts. This is borne out by the observation that 'Where logic is concerned, . . . every combination of parts results from completing something that is in need of supplementation' (PW, 254). In other words,[21] logical complexity is uniformly a matter of decomposition into function and argument.

The analysis of thoughts into function and argument emerges clearly from the description of the thought-constituents expressed by the proper name and concept-word in a simple atomic sentence. Just as the references of a proper name and a concept-word are respectively a complete, self-subsistent or saturated entity (an object) and a predicative, incomplete, or unsaturated one (a concept), so too their senses are symmetrically characterized as being respectively 'complete' or 'saturated' and 'predicative', 'incomplete', or 'unsaturated' (CO, 54; PW, 119n., 191ff., 254f.). In the same way, the logical analysis of a generalization turns on the possibility of isolating within a thought unsaturated parts corresponding to predicative expressions (PW, 189ff.).

The characteristic of incompleteness or unsaturatedness, in Frege's view, is the hallmark of functions. Consequently, to describe the sense of a concept-word or predicate as unsaturated is to classify this entity as a function, and to analyze a thought into a complete part and an unsaturated one is to impose on it a decomposition into function and argument. The construction of sentences in concept-

---

21. Note Frege's gloss: 'The functions of Analysis correspond to parts of thoughts that are thus in need of supplementation' (PW, 255)!

script out of function-names and argument-expressions accurately depicts the structure of the thought expressed.

The transfer of the distinction between saturated and unsaturated entities from the level of reference to the level of sense involves the transfer of ideas that this metaphorical distinction was supposed to highlight. First, it so explicated the type-heterogeneity between concepts and objects that the function-theoretic type-restrictions on well-formed sentences have a transparent rationale. Second, it made clear that the fundamental logical relation between concept and object is not some *tertium quid*.

> [T]he unsaturatedness of the concept brings it about that the object, in effecting the saturation, engages immediately with the concept, without any need of special cement. Object and concept are fundamentally made for each other, and in subsumption we have their fundamental union. (PW, 178)

Third, the fact that concepts are unsaturated entities makes perspicuous that language mirrors ontological relations in one important respect: the predicative nature of concept-words, the gappiness of predicates, reflects the unsaturatedness of the designated concepts, and therefore the subject/predicate articulation of a simple atomic sentence corresponds to the object/concept 'decomposition' of the corresponding truth-value. Frege now reduplicated all of these basic contentions in respect of the *senses* of subsentential expressions. Type-restrictions can be justified at the level of sense: '[N]ot all the parts of a thought can be complete; at least one must be "unsaturated", or predicative; otherwise they would not hold together' (CO, 54). The unsaturatedness of thought-components solves the traditional and persistent problem of explaining 'the unity of the proposition', i.e. of accounting for the fact that the parts of a judgment coalesce instead of remaining a mere scattered assemblage of entities. No metaphysical glue is necessary, nor need the humble copula be endowed with miracle-working powers. A thought is literally the value of a function (one thought-constituent) for an argument (another thought-constituent) just as its truth-value is the value of a function (a concept) for an argument (e.g. an object). Wonder at the unity of a thought is as out of place as surprise that 2^2 *is* 4! Finally, the predicative nature of concept-words or predicates now is a visible reflection of the unsaturatedness of the corresponding senses as well as of the designated concepts. The subject/predicate structure of a simple atomic sentence is isomorphic with the structure of the thought expressed as well as with the decomposition of the designated truth-value. In general, a declarative sentence can be regarded as a mapping or model of the thought expressed (PW, 255; cf. 207, 225, 243, 259; N, 123; CT, 537ff.), though only sentences in concept-script are guaranteed to manifest thought-structures exactly.

The idea of functions whose values are thoughts is also integral to Frege's analysis of logical connectives. The phrase 'the negation of A' designates a thought provided that 'A' designates a thought, and this thought is presented as the value of the function 'the negation of ξ' (N, 131ff.). On the assumption that 'the negation of the thought that $2 + 3 > 4$' designates what '⊢─┬─$2 + 3 > 4$' expresses, we seem bound to conclude that the connective '―┬― ξ' has as its sense the function which is the reference of 'the negation of ξ', i.e. a function whose argument and value are thoughts. Hence the thought expressed by '―┬― $2 + 3 > 4$' is presented as the value of a 'sense-function' for the thought expressed by '$2 + 3 > 4$' as argument! It thus emerges that function/argument analysis in logic must be carried out twice, once at the level of reference, once at the level of sense (N, 132). Moreover, its application to thoughts exactly mirrors its original application to judgeable-contents. There the judgeable-content expressed by an atomic sentence was the value of a function (a concept, i.e. the content of the predicate) for an argument (an object, i.e. the content of the name), whereas here the thought is taken to be the value of a function (the sense of the concept-word) for an argument (the sense of the name). The function/argument decomposition of thoughts is the direct heir of the earlier application of function theory to judgeable-contents (cf. PW, 253). It alone makes at all intelligible the important idea that a thought may be decomposed in indefinitely many ways into incommensurable thought-components which correspond to the concepts of different levels that may be extracted by functional abstraction from a single 'decomposition' of a truth-value. Different sense-functions for different arguments may apparently take a single thought as their value.

Side by side with this conception of function/argument decomposition of the thought Frege held the principle that the thought is a *whole* consisting of *parts* which are the senses of the subsentential constituents of the sentence expressing it. On this view, the sense of an atomic sentence is a composite object whose constituent ingredients are the sense of a concept-word and the sense of a proper name. This conception informs the general explanation in the *Basic Laws* of the sense of a sentence-constituent as its contribution to the expression of the thought expressed by the sentence. 'If a name is part of the name of a truth-value, then the sense of the former name is *part* of the thought expressed by the latter name' (BLA, §32, our italics; cf. PW, 231; CO, 54). Frege made frequent use of the phrase 'part of a thought'. In particular, he reiterated the thesis 'As the proper name is part of the sentence, so its sense is part of the thought' (PW, 191), as well as the more general principle 'As a sentence is generally a complex sign, so the thought expressed by it is complex too: in fact it is put together in such a way that parts of the thought correspond to parts of the sentence' (PW, 207; cf. 225, 243). It is 'the whole-part relation' that holds between a thought and the senses of subsentential expressions, and further subdivision of these senses yields parts of parts of thoughts (PW, 255). The contention that a thought expressed by a sentence is literally *composed* of the senses of the subsentential expressions is the explicit point of his late sketch of a generative theory of understanding (infra, p. 381ff.). It is by fitting together thought-building-blocks *(Gedankenbausteinen)* that humans allegedly have the capacity to express and grasp thoughts never previously expressed or entertained (PW, 225; cf. CT, 538).

Taking *au pied de la lettre* the claim that the senses of sentence-constituents are *parts* of the sense of a sentence leads to immediate conflict with other main aspects of Frege's conception of sense. Logic requires that thoughts be decomposed into function and argument, but neither the argument nor the function itself are typically parts of the value of a given function for an argument. Moreover, part/whole analysis of thoughts would destroy isomorphism between the correlates of sentences in the realm of sense and in the realm of reference. For the reference of a part of a sentence is claimed *not* to be part of the reference of the sentence;[22] and more generally, the reference of the parts of any complex expression are not typically parts of the reference of the whole, e.g. Copenhagen is the capital of Denmark, but Denmark is not part of Copenhagen! These considerations suggest that we should perhaps take the apparent part/whole analysis of thoughts with a large pinch of salt.

A case for discounting the importance of the terminology of 'part' and 'whole' in Frege's discussion of sense can easily be constructed. These terms are members of a family of expressions including 'analysis', 'articulation', 'decomposition', 'splitting up', 'composed of', 'consist in', 'constituents', 'components', etc., some of which apparently must appear in any intelligible exposition of a system of logic. Should they be so construed that it is self-contradictory to speak of 'function/argument *analysis* or *decomposition*' of a thought or judgment? Is it incoherent to *split* or *carve up* a judgeable-content into a relation and a pair of objects (FA, §64)? In drawing these queer conclusions, would we not be behaving 'like savages, primitive people, who hear the expressions of civilized men [and] put a false interpretation on them'?[23] The terminology of 'part' and 'whole' would seem as antiseptic as the commonplace phrase 'to analyze a thought or proposition'. It was part of the patter of logicians in the nineteenth century (e.g. Bolzano, Lotze), even among those who emphasized the very diverse nature of the components of judgments. Perhaps the most obvious trap to fall into is the assumption that all the parts of a given whole must be homogeneous. But Frege clearly avoided that pitfall. In dissecting thoughts, he harped on the fundamental differences between the 'parts': in particular, he stressed the distinction between the unsaturated parts (sense-functions) and saturated parts (sense-objects). He even called attention to the fact that such terms as 'made up of', 'consist of', 'component', and 'part' when applied to thoughts might lead to our looking at matters in the wrong way:

> ... the thought that contradicts another thought appears as made up of that thought and negation ... If we choose to speak of parts in this connection, all the same these parts are not mutually independent in the way that we are else-

---

22. Originally he had 'transferred the relation between the parts and the whole of the sentence to its reference, by calling the reference of a word part of the reference of the sentence, if the word itself is part of the sentence'. But he added that he thereby 'used the word "part" in a special sense.... This way of speaking can certainly be attacked ... because the word "part" is already used in another sense of bodies' (SR, 65). Hence, although Mont Blanc is 'part' of the True (because 'Mont Blanc is more than 4000 metres high' denotes the True), its relation to the True is not that of a chair-leg to a chair. But this point was precisely the reason later advanced for denying that the reference of constituents of a complex expression are parts of the reference of the whole. Therefore constancy in his view of the matter is marked by a change in his terminology. He evidently decided that the special sense given to 'part' was misleading.
23. Wittgenstein, *Philosophical Investigations*, §194.

where used to finding when we have parts of a whole.... The two components, if we choose to employ this expression, are quite different in kind and contribute quite differently towards the formation of the whole. (N, 131f.)

Ultimately in reiterating the thesis that we can distinguish parts in the thought corresponding to the parts of a sentence, he added an explicit caveat:

To be sure, we really talk *figuratively* when we transfer the relation of whole and part to thoughts; yet the *analogy* is so ready to hand and so generally valid that we are hardly ever bothered by the hitches which occur from time to time. (CT, 537, our italics)

In view of all of these remarks, would it not be petty-minded to accuse Frege of contradicting his own basic theses about sense in adopting a part/whole analysis of thoughts?

To absolve Frege from the charge of inconsistency would require a demonstration that his discussions of the parts of thoughts never carried commitments incompatible with function/argument analysis of thoughts. This cannot be produced. First, part of his case[24] for attributing a sense (distinct from its reference) to every proper name rests on the claim that a concrete object (Mont Blanc, Jupiter, the sun) cannot be a *part* of a thought. This thesis is patently absurd if a thought is conceived to be the value of a function. Otherwise we must infer that no concrete object is the reference of a proper name since none is a part of a truth-value! Second, the requirement that a thought may occur as part of a 'compound thought' depends on the claim that compound thoughts are literally put together out of thoughts. For if a compound thought is conceived to be the value of a function whose arguments are thoughts, the fact that a sentence may be a constituent of a molecular sentence does not prove that it expresses a thought (unasserted) which is part of, or occurs in, the thought asserted by the molecular sentence. Third, it is characteristic of the part/whole relation that the whole and all but one of its parts suffice uniquely to determine the remainder (cf. SR, 65). But no parallel holds for function/argument decomposition; in particular, the principle that a function is underdetermined by a proper part of its course-of-values guarantees that no function is uniquely determined by the fact that some entity is the value of some unknown function for a given entity as argument. Frege correctly drew this consequence in respect of the references of sentence-constituents (PMC, 128f.). But he also presupposed the recoverability of the sense of a part of a sentence from knowledge of the sense of the whole and the senses of its remaining parts. In fact, he arrived at the senses of predicates by appeal to this principle: *the* unsaturated part of a thought is obtained as the remainder of subtracting the sense of a proper name from the sense of the complete sentence (PW, 190f.). He seemingly embraced the more general principle that the sense of a sentence

---

24. The other part being the need to explain the cognitive non-triviality of cognitively non-trivial identities; but *this* leads to ascription of senses to proper names only on the assumption that the puzzle has to be resolved by postulating entities, an assumption which can and should be challenged!.

uniquely determines the sense of each of its constituents, since he characterized the sense of each constituent as the contribution it makes to the thought expressed by the whole (BLA, §32; PW, 231). Finally, Frege's late idea of isomorphism requires a literal interpretation of the thesis that thoughts are composed out of parts. This is apparent both from his terminology (thoughts are 'built up' out of 'thought-building-blocks' (PW, 225)) and from his unreflective commitment to part/whole decomposition of *sentences*.[25] There are deeper reasons too. If understanding sentences is to be explained in terms of constructing their senses from the familiar senses of their parts, then the sense of a sentence must be considered to be transparent from the senses of its parts. But this would not in general be so if the sense of a sentence were assigned function/argument structure. The value of a function is typically the product of a calculation, often of a calculation that may be complex and lengthy. Hence, it would be implausible to assert that understanding a sentence is a matter simply of grasping the senses of its parts (even together with their structure) if thoughts are decomposed into functions and arguments. This would be as absurd as to claim that a person understands a coded message as soon as he is handed the key for decoding it.

Unequivocally Frege committed himself to two different patterns of analysis of thoughts: part/whole and function/argument. It is amazing that he never noticed the incompatibility between these conceptions of the complexity of thoughts. Despite their fundamental inconsistency, arguments based on one occur in his writings cheek by jowl with arguments based on the other (e.g. PW, 254f.). His thinking here is deeply flawed: neither conception alone is either coherent or sufficient for his purposes, and the pair are irreconcilable. Had he noted this, he would have had to cast the whole concept of sense back into the melting pot.

The incoherence of the function/argument analysis of thoughts is the direct descendant of the basic problem about function/argument analysis of judgeable-contents (supra, p. 150ff.). The intelligibility of specifying a thought as the value of a function for an argument presupposes the possibility of identifying this thought in other independent ways, as the value of other functions for other arguments. It is clear that the possibility of expressing the same thought in different symbolisms does not satisfy this requirement. It is problematic whether any possibility of alternative analysis of thoughts does so. The simplest case of alternative decomposition arises from relational judgments. The truth value designated by '⎯⎯⎯ $\Psi(A, B)$' can be decomposed in at least three ways: viz. it is designated alike by '⎯⎯⎯ $\Psi(\xi, \zeta)\big]_{A,B}$', '⎯⎯⎯ $\Psi(\xi, B)\big]_{A}$' and '⎯⎯⎯ $\Psi(A, \zeta)\big]_{B}$'. Each of these expressions indicates the same thought, and on the assumption that references can be attached to signs only via isomorphic entities in the realm of sense, these expressions indicate seemingly distinct ways of determining the same thought as the value of a sense-function. But now the same argument invoked to

---

25. Frege never construed *sentences* as the values of functions (so-called 'linguistic functions' (supra p. 172 n.7)).

show that the three expressions share the same sense (viz. that the values of the designated functions for the designated arguments are self-evidently the same as the value of '$\underline{}\Psi(\xi, \zeta)\Big]_{A,B}$') proves the corresponding sense-functions *self-evidently* to have the same value. That claim relative to reference was taken to show that a truth-value is not really determined as the value of different functions for different arguments. By parity of reasoning, it now must establish that the thought is not really expressed as being the value of different sense-functions. Exactly parallel reasoning about the possibility of functional abstraction proves that in that case too what would appear to be different specifications of a thought as the value of different sense-functions must really collapse together. Otherwise, self-evident identity of the values of functions of different type would not suffice to establish uniqueness of sense in the face of the explanation of sense as mode of determining an entity as the value of a function. Therefore alternative analyses of a single thought secure the possibility of specifying it as the value of different functions only by undermining the case for the uniqueness of the thought to be analyzed! Even if the requirement could be met and a given thought could be indicated as the value of genuinely distinct functions by virtue of being expressed by different formulae in concept-script, then these formulae must indicate the same thought *differently!* This calls for a differentiation in the notion of 'indication' parallel to the original differentiation of 'indication' from 'designation', and hence for ascribing meta-senses to expressions in addition to senses. For if it makes sense to suppose that a thought could be presented as the value of independent functions, then there is no conceivable reason to claim that it can only be presented as the value of a function for an argument if it is *self-evident* that this function for this argument takes the given thought as its value. The same force that generates senses out of function/argument analysis at the level of reference also drives one inexorably upwards from the application of function/argument analysis at the level of sense.[26] It is doubtful, then, whether function/argument decomposition of thoughts is viable by itself, and it is certain that this conception could not alone deliver the whole of Frege's account of sense.

Abandoning function/argument analysis and trying to manage solely with the decomposition of thoughts into parts is equally problematic. Formulae in concept-script (and even sentences of natural language) are claimed to consist of function-names and argument-expressions. It would be altogether opaque how they represent logical analyses of *thoughts* unless each thought decomposes into function and argument in parallel with the structure of the formulae expressing it. How else can we begin to make sense of the 'primitive logical fact' of a correspondence between the analysis of a sentence at the level of sense and at the level of refer-

26. Some commentators hold Frege to be committed to an unending hierarchy of abstract entities by his notion of indirect sense (cf. PMC, 154); e.g. R. Carnap, *Meaning and Necessity*, §30, enlarged edition (University of Chicago Press, Chicago, 1956), and A. Church, 'A Formulation of the Logic of Sense and Denotation', in *Structure, Method and Meaning, Essays in honour of H. Sheffer*, (New York, 1951).

ence? And without this isomorphism in the analysis of thoughts and of truth-values, how could we explain Frege's fundamental principle that unsaturated thought-components correspond to concepts, saturated ones to objects? To deny that thoughts decompose into functions and arguments threatens to undermine the thesis that the business of logic is to analyze thoughts; it reduces the characterization of thought-components to childish babbling; and it makes part of the structure of formulae in concept-script irrelevant to the thoughts expressed. Furthermore, literal part/whole analysis of thoughts does not marry smoothly with the conception of the sense of a sentence as a mode of presentation of a truth-value. Can we ever make any *literal* sense of the claim that *how* an entity is presented as the value of a function consists of *parts?* And what force has the argument that the logical treatment of generality gives rise to 'the need ... *to analyze a thought into parts*' (PW, 187; cf. 201f.)? The basis of Frege's novel account of generality is the introduction of second-level concepts, i.e. his viewing truth-values as the values of second-level concepts for first-level concepts as arguments, and this has no obvious (or even intelligible) connection with splitting thoughts into parts![27] Finally, part/whole analysis is vested with implications difficult to reconcile with his fundamental ideas. The parts of a given whole under a single analysis must be roughly homogeneous, like the components of a motor. Frege, however, held saturated and unsaturated thought-components to be wholly unlike in their combinatorial powers. The thesis that they are nonetheless all literally *parts* of thoughts makes as little sense as the claim that a brick and the property of being red are alike parts of the fact that a particular brick is red. Part/whole analysis, without supplementation, generates a problem in differentiating isomers. How can two wholes differ which are constructed out of exactly the same parts? This problem had baffled traditional logicians seeking to analyze statements involving multiple generality. Frege's solution consisted in replacing part/whole analysis of judgments by function/argument decomposition of judgeable-contents. To resuscitate part/whole analysis of thoughts would be a retrograde step, reintroducing a puzzle at the level of sense to complement a more satisfactory solution at the level of reference. Platonist superphysics would be needed to 'explain' how the thought expressed by '$\Phi(A, B)$' may differ from that expressed by '$\Phi(B, A)$': the sense of 'A' and the sense of 'B' do not occupy the same 'positions' in the two different thoughts (cf. CT, 548).[28] Even this account may leave us in the lurch in the crucial task of differentiating the different

---

27. The notion of representing *how* a truth-value is determined as the value of a function itself as the value of a function is no more perspicuous. But it might have appeared unproblematic to Frege if he approached the analysis of thoughts from the presumption that references must be assigned to sentence-parts only via (isomorphic!) thought-constituents.
28. Frege adumbrated this account in the analysis of compound thoughts, perhaps as part of a misguided attempt to explain his logic without recourse to the technical apparatus of functions. Russell went through similar contortions, assigning different 'senses' to the relation R in the propositions aRb and bRa, although the 'sense' is puzzlingly neither a constituent nor part of the form of a proposition (B. Russell, *Philosophical Essays*, p. 183ff. (Longmans, Green, London, 1910)).

thoughts expressed by the formulae '$(\forall x)(\exists y)Rxy$' and '$(\exists y)(\forall x)Rxy$'![29] To tie Frege down to part/whole analysis alone would be to create mysteries and difficulties at the very core of his logical system.

Neither function/argument nor part/whole analysis of thoughts is coherent in isolation, and neither conception of the complexity of thoughts would by itself suffice for Frege's purposes. Both are clearly necessary. But the two are patently incompatible. The conflict between them cannot be alleviated since it stems from the fundamental point that neither the argument nor the function itself are typically *parts* of the value of a function for an argument. Built in the dockyards of function theory, Frege's logical system slides down the ways straight onto the rocks. 'The sense of a sentence' flags the wreck.

Is this account not incredible? Surely no more so than the fact that Descartes' theory of motion unwittingly committed him to the impossibility of motion,[30] that Berkeley's metaphysics inadvertently limited him to Tea for Two (himself and God), that Kant, in his eagerness to secure the freedom and autonomy of the moral self, banished it from the empirical world in which alone it might have any moral decisions to make, any moral values to respect, any moral self-knowledge to achieve. A secure niche in the philosophers' pantheon is secured no less by great illusions than by great insights, since finally insight may spring from disillusion. Frege's incoherent concept of sense no more provides secure foundations for contemporary philosophical investigations into language than his logicism provides foundations for arithmetic.

Ironically, what wreaked havoc with Frege's conception of sense was perhaps a dim perception of the truth. On the one hand, the possibility of exhibiting logical relations among thoughts seemed to depend on their having genuine complexity. On the other hand, both the argument and the function itself seemed to be swallowed up in the thought if the thought is determined as the value of a function for an argument. Instead of addressing the issue and becoming aware of the inconsistent demands made on the concept of the sense of a sentence, Frege let himself be pulled now in one direction, now in the other. Like Hume, he might have declared 'there are two principles which I cannot render consistent, nor is it in my power to renounce either of them.[31]

---

29. Dummett extricates Frege from this impasse by crediting him with the discovery of the guiding idea of modern transformational generative grammars. 'Frege's insight consisted in considering the sentence as being constructed in stages, corresponding to the different signs of generality occurring in it' (*Frege: Philosophy of Language*, p. 10). This allegedly enabled him so to specify the senses (truth-conditions) of sentences involving multiple generality that 'isomeric' thoughts could be clearly differentiated. 'Once we know the constructional history of a sentence involving multiple generality, we can from these simple rules determine the truth-conditions of the sentence, provided only that we know the truth-conditions of every sentence containing proper names in the places where the signs of generality stand' (op. cit. p. 11). This proposal attributes to Frege at least three major ideas that he never entertained in order to solve a problem that the function/argument decomposition of judgeable-contents (and later of thoughts) prevented him from even noticing.
30. Cf. A.J.P. Kenny, *Descartes*, p. 215 (Random House, New York, 1968).
31. Hume, *A Treatise of Human Nature*, Appendix.

This inconsistency between two divergent principles in the theory of sense is a surface manifestation of a far deeper and more destructive incompatibility in Frege's mature theory. Here the two grand principles which he could neither render consistent nor renounce are the function-theoretic analysis of judgment and the sense/reference distinction.[32] It is upon the application of function-theory to logic that his system rests, and it is this which is the engine of his logicist aspirations. Introduction of truth-values as the values of concepts for arguments was an essential step in connecting concepts with their extensions by a logical operation. This in turn forced the sense/reference distinction upon him, and made it imperative smoothly to coordinate the new theory of sense with the modified functional analysis at the level of reference. But, as everything in the foregoing investigation indicates, this is precisely what Frege was unable to do.

---

32. It is perhaps no coincidence that of the two most eminent followers of Frege, Dummett and Geach, one has written hundreds of pages on sense and reference, yet given negligible attention to the foundations of his function-theoretic analysis, while the other has emphasized his function/argument analysis of sentences, but dismissed his reflections on sense and reference as a sketchy and obscure theory which relates merely to a puzzle about *oratio obliqua* clauses, and hardly appears in much of Frege's work (see P. T. Geach, 'Frege,' pp. 162, in *Three Philosophers* (Blackwell, Oxford, 1967)). Had they focused sharply on *both* warring principles they might not have mistaken the brilliance of a falling star for an apotheosis.

12
Cracks in the Structure: Assertion, Truth and Thoughts

1. A Fregean philosophical grammar?

Frege's theory of sense is sketchy, riven with contradictions and fraught with inner tension between irreconcilable principles. The application of his functional analysis to declarative sentences of natural languages would require the support of principles which are metaphysically jejune, question-begging and philosophically ill-founded. Such flawed foundations are hardly encouraging grounds upon which to erect ambitious philosophical edifices. Yet some twentieth century philosophers, particularly of late, have found in his writings a primary inspiration for what they take to be *the* central task of modern philosophy, viz. the elaboration of a theory of meaning, the construction of a logical grammar.

Everything in the foregoing investigation suggests that such an idea was not only remote from *his* thought but actually inconsistent with the primary thrust of central elements in his theory. From the standard modern view-point, logic cannot be divorced from the semantic analysis of language. Hence logicians have as their explicit purpose the investigations of one aspect of the workings of language. They do not shy away from invoking and applying sophisticated mathematical tools for this purpose (e.g. infinite sequences of objects) in spite of the fact that the relevant mathematical concepts are not widely understood.[1] Some of what is clear about language may be rendered obscure. Nonetheless, mathematical techniques are the engine for progress in the construction of a theory or philosophy of language. Frege denied the underlying premise: on his view, logic is the science of certain concepts and objects (thoughts, truth-values, the concept of negation, etc.). These Platonic entities exist independently of language, and their relations, like those

---

1. On the contrary, the *recherché* or arcane aspects of their explanations seem to have a strong appeal, transforming banality (and even nonsense) into 'scientific' principles commanding widespread respect.

333

among numbers or mathematical functions, could in principle (though not by us) be studied directly, without having to cast side-long glances at expressions. The appropriate tool for their analysis is the concept of a function. Frege thought, however, that a pure logic expounded solely in terms of function theory applied to Platonic entities would be unintelligible to most philosophers, even to the formal logicians of his day. To make himself understood, to secure maximum impact for his ideas, he decided to try to *explain* his system without recourse to function theory in certain less technical writings (especially in the *Foundations,* 'On Concept and Object', 'Compound Thoughts', and some of his abortive attempts to write a book on logic). By explaining what he saw to be conspicuous features of language, he hoped to convey the notion of a function to the mathematically unenlightened and thereby secure, by an indirect approach, a proper understanding of his logic. Consequently, in relating the articulations of sentences to formulae in function theory, Frege was carrying out a programme diametrically opposed to modern logical semantics. He used features of language to indicate important principles of function theory, whereas we now employ mathematical tools with the purpose of arriving at a new understanding of how language works. Modern philosophers are prone to misconstrue his attempted explication to non-mathematicians of the rudiments of function theory as the first steps in the construction of a theory of meaning. We are inclined to borrow the garb of function theory to make our ruminations about language look impressive and scientific, whereas he made observations about expressions in order to render intelligible his application of function theory to the objects of thought.[2]

This divergence of purpose is of colossal importance. It should colour the interpretation of all of Frege's specific remarks about language. Nevertheless, even if he never dreamt of elaborating a 'philosophy of language', we might discern in his writings the tools necessary for carrying out this enterprise and a blue-print of how to proceed. The proposal that we might build a new edifice from carefully selected raw materials that he supplied is worth examining in its own right. Disregarding the deep flaws in his conception of generalising function theory, averting our eyes from the crumbling *Gedankenbausteine* of his theory of sense and reference, and ignoring his conviction that syntactic distinctions are but fallible

---

2. Russell followed a similar strategy. He appealed to features of language to elucidate the allegedly indefinable notion of a propositional function. In the sentence 'Socrates is a man', we may regard the name 'Socrates' as replaceable by others ('Plato', 'Aristotle', 'the number 2', etc.). In this way, the proposition expressed by 'Socrates is a man' generates a class of propositions differing from the original one in a single constituent. Russell then introduced the expression 'x is a man', which he described as ambiguously denoting a proposition in this class (just as 'x^2' ambiguously denotes a value of the function $y = x^2$). This excursus into language he then employed to illuminate the essential features of propositional-functions. They are themselves ambiguities in need of determination (Russell and Whitehead, *Principia Mathematica,* pp. 40, 47f.); they cannot take propositional-functions (of the same type) as arguments (ibid. p. 47ff.); and they are not 'separate and distinguishable' entities, but live only in propositions of certain forms and 'cannot survive analysis' (Russell, *Principles of Mathematics,* p. 88). Independently of whether the conclusions are correct, or even intelligible, the strategy of using features of language to explain the fundamental ideas of abstract function theory is transparent.

guides to the logical significance of expressions and little guide at all to the variety of possibilities for analyzing thoughts and truth-values, we might consider whether we can build a worthy Fregean monument by focusing on formal principles of sense/reference relations.

At the very outset of this investigation, we should remind ourselves of one hard fact. It makes no more sense to debate *Frege's* philosophical grammar than it does to discuss Coleridge's conclusion to 'Kubla Khan'. The most that we can do is to speculate about possibilities and to invent the necessary details. At best Frege's remarks about extending sense/reference and function/argument analysis to ordinary language are programmatic.

This point is manifest in his remarks on the categories of subsentential expressions.

(i) Proper names: the 'account' of the sense of ordinary proper names is useless as an explanation of the varied nature and use of such expressions in natural language (e.g. personal names, place names, names of fictional characters and places, of real persons in historical novels, of works of art, of days of the week or holidays, of God or gods).

(ii) Indexicals: The relation between their rule-governed use and their (variable) sense is never discussed. If a personal pronoun (e.g. 'I') has, on a given occasion of its use, a sense only insofar as it is *combined* with a person and occasion, is the speaker who is, according to Frege, the referent, also part of the symbol expressing that sense? It is unclear whether a speaker can be part of a symbol the other part of which he utters and the whole of which refers to himself. On the anaphoric or cataphoric discourse-reference of 'this' and 'that', on their use as noun or noun-phrase determiners, or as adjective intensifiers, Frege said nothing. Indeed the larger part of the subject of indexical referring expressions is left virtually unexamined.

(iii) On the subjects of definite and indefinite reference by means of compound descriptive phrases, Frege's analysis was sketchy at best. His remarks on the definite article are fragmentary. Comments on the referential uses of the indefinite article are virtually non-existent. A similar lacuna is evident when it comes to the detailed scrutiny of the subtle differences and similarities in the uses of the various applicatives of natural language which do not at all match the Fregean quantifiers.

(iv) Concept-words: Frege told us nothing of the sense of concept-words not analysable into *Merkmale* nor definable by explicit definition. Are there simple senses to be associated with, say, perceptual predicates? How are these to be explained? What counts as grasping or explaining them? How are we to explain internal relations between determinates of determinables and the inference rules that correspond to them? If there are vague predicates in natural language, do they really deprive *every* sentence in which they occur of any truth-value? Are we then wrong to attribute a truth-value to such sentences (or thoughts expressed by them)? How are we to budget for the common variable polyadicity of predicates while retaining the functional analysis of concept-words? Frege gave us no account whatever of modality in its various forms, of the adjectival occurrence of such expressions as number words or colour words. We are left altogether in the dark in the matter of the grammar of sortal count-nouns as opposed to non-sortal ones, or of mass nouns as opposed to count-nouns, or of attributive as opposed to predicative adjectives, or of verbs in contrast with 'adverbial verbs', or of psychological verbs displaying 1st/3rd person asymmetry. In all these cases, and many others, special philosophical difficulties of familiar kinds are attendant on analysis. Frege was silent on all these matters.

Nor is it only with respect to sub-sentential expressions that his account is of so little help to the aspiring theorist of meaning. His remarks on the sense and reference of sentences and on their possible structures are similarly of very restricted utility for purposes of constructing a philosophical grammar.

(i) The notion of the sense of an arithmetical sentence is explained in terms of proof-conditions. How this analysis should be extended to sentences of ordinary language is left to the imagination. Although it is common to attribute to Frege a truth-conditional account of sentence-sense, this has neither textual justification nor philosophical coherence (infra p. 373ff.) But even if it were justifiable, it would leave opaque the analysis of the sense of normative (deontic) and evaluative sentences in general, of ethical sentences in particular, as well as aesthetic and religious ones.

(ii) Frege had little to say about non-declarative sentences. If we are to speak of a philosophical grammar constructed on Frege's principles, we must discover an account of non-assertoric utterances consonant with them. Yet all there is to be found in his writings is the contention that sentence-questions express the same thought as corresponding declaratives. He did indeed claim that imperatives, optatives and WH-questions have a sense, but do not express thoughts. It is therefore obscure what their sense (or meaning) is, for if 'Shut the door!', 'Who is it?', 'If only it would rain', etc., do not indicate ways of presenting truth-values, what are they names of? If they are not names of anything, how are they to be decomposed into function and argument? Are concept-words to be construed as ambiguous, designating functions to truth-values in declarative sentences, and different functions, or not functions at all, in non-declarative sentences?

(iii) Colouring is a ragbag of features of words and sentences to which Frege allocated anything characterizing language that is irrelevant to cogency of inference. The little he said about colouring is demonstrably unsatisfactory. The differences between 'nag', 'steed', and 'horse' are not of the same kind as those between 'and' and 'but'. Moreover 'and' and 'but' are not, as Frege supposed, always intersubstitutable *salva veritate,* since in some contexts replacement of one by the other generates nonsense, e.g.

(1) He wanted to go because it was raining *and* because he had no umbrella.
(2) I came *and* I saw *and* I wept.
(3) He left in order to catch the bus *and* in order to accompany John.
(4) Tell me what is the matter *and* I will help you.

Frege was not really concerned with colouring. Yet if there is to be a global Fregean account of language, the boundary lines between sense and colouring must be redrawn, and the different types of linguistic phenomena wrongly lumped together must be carefully distinguished.

Frege did not produce anything remotely like a philosophical grammar. Nor is there any evidence that the very idea of one ever crossed his mind. To elaborate a *Fregean* philosophical grammar demands prodigious inventiveness, for what must be done is to examine a wide variety of *possibilities* consistent with the central tenets of Frege's system, which appear to have fruitful application to language. Such an endeavour must operate within fairly narrow constraints if it is to be a further development of *his* ideas. Four such fundamental constraints are noteworthy—and daunting.

(i) *Philosophy of mathematics:* Since the whole purpose of Frege's account is to show that mathematical propositions can be both analytic and non-trivial, any putative development of his theory that undermines this conclusion can hardly be deemed an extension of his own. A Fregean philosophical grammar or theory of sense for natural language must conform to the requirement that mathematical statements have a sense, that arithmetical equations are typically non-trivial, that definitions may be 'creative' (FA, §88) in so far as they may contain unforeseen consequences independently of us. Arguably it must treat number-words as proper names of objects.

(ii) *The Augustinian Picture:* At the heart of Frege's conception of language is the view that the role of linguistic expressions is to stand for entities, either directly (in the case of content) or mediately (in the case of sense and reference).[3] Around this 'insight' is constructed his function-theoretic analysis. In his eyes, Platonism with respect to functions, concepts and abstract objects is not at all an optional extra. Nor is his Platonism about senses something which we can now jettison without ado. For not only are senses the references of proper names formed by the operator 'the sense of "ξ"', they are the references of expressions in oblique contexts. Moreover Platonism about sense is his distinctive way of meeting the publicity requirement for objects of judgment. Hence a further development of Frege's theory must not only be realist with respect to truth, but also Platonist with respect to sense. Consequently, theories of language which connect sense intimately with the practice of the use of language, and view sense as internally related to use (i.e. view the use of expressions as *constitutive* of their sense, rather than externally related to it) adulterate the pure Fregean fount with waters drawn from alien wells. For such a picture secures objectivity of sense without the Augustinian model of correlation, and generates a conception of sense which does

---

3. Of course, there are exceptions (e.g. the assertion sign, or unbound variables) and complications (e.g. 'All *F*'s' does not stand for all *Fs*, nor for the class of *Fs*). This is altogether typical of any philosophical theory which is built upon the Augustinian *Urbild* (cf. Baker and Hacker, *Wittgenstein: Understanding and Meaning*, p. 34ff.).

not generally involve representation of an entity. Hence Frege would claim that such an account cannot literally employ function/argument analysis.

(iii) *Contextualism and Compositionalism:* The primacy of the name of a truth-value is, in Frege's mature theory, correlative to the compositional principle of sense. This principle is invoked, in his late writings (infra p. 381ff.) to explain the 'mystery' of our capacity to understand new sentences. A general theory, Fregean in spirit, which cleaves to this principle must associate the understanding of judgments with the form and content of the sentences expressing them (irrespective of their use!) and conceive of this understanding as derivation or calculation from senses of constituents and mode of combination. Moreover, this must be done in a way that is *compatible with variable analysis of structure.*

(iv) *Function/argument analysis:* We can no more conceive of a Fregean account of language which repudiated the idea that concept-words are function-names than we can conceive of a Cartesian metaphysics which is not dualist. Nor can we jettison his conception of a function and claim to have extended his theory. His function-theoretic apparatus carries important commitments. Since functions must have entities as values, he required that thoughts and truth-values be objects, and indirectly that sentences expressing thoughts literally be names of truth-values! Since it is an intrinsic feature of a function what its value is for any argument, the truth-value of a sentence must be determined alone by the argument and function as whose value it is presented (i.e. independently of matters of fact). Finally, since a sentence indicates its sense by presenting a truth-value as the value of a function for an argument, the connection between its sense and its reference must be internal, and derivatively so too must be the relation of the sense to the reference of any expression. If these commitments of Frege's extension of function theory to language are misguided and incoherent, then it is futile *ab initio* to seek to broaden his account into a viable semantic theory.

Apart from these general doubts and problems the introduction of the two truth-values gives rise to a significant number of quite specific puzzles. These, it is commonly admitted, are aspects of a sort of scholasticism which infects Frege's mature system. This elicits varying reactions. At one extreme it is treated as a cosmetic problem symptomatic of ageing. Astute plastic surgery, involving minor excisions and a slight face-lift will restore the appearance of youthful vigour consonant with the underlying virility of the basic ideas. At the other extreme, each paradox or apparent absurdity can be taken as the surface manifestation of a deep-rooted cancer ramifying through the entire body of his logical system.

To adjudicate between these divergent reactions requires patient unravelling of threads of complex and interwoven arguments on three subjects, viz. assertion, truth and thoughts. On these three topics he wrote a considerable amount, hence our labour is not the wholly speculative one of reconstructing a possible dinosaur from an apparent vertebra. His discussions of the three themes are partly or wholly developments of his earlier theory of content. Hence we can deepen our understanding of the mature system by tracing the continuities which Frege preserved as well as the changes he introduced under the pressure of the innovations in function/argument analysis and the splitting of content into sense and reference. His reflections on all three subjects produce weird and baffling results. The incoherences of the *Begriffsschrift* account of assertion are retained, and new ones added. Thoughts are conceived as being wholes composed of parts, but also as being the values of functions for arguments and as decomposing into function and argument. It is held to make sense to think that $2^2 = 4$ is identical with $3 > 2$, hence perfectly possible to assert this thought. Since thoughts *stand for* truth-values, truth is denied to be a property of thoughts or sentences. And truth is finally conceived to be ineffable: it is impossible really to make the assertion that one vainly tries to make by uttering the sentence 'It is true that water is wet'. All these remarks look absurd. Each such contention mislocates the boundary between sense and nonsense. The aspiring Fregean theorist must not only expand an embryonic theory to apply it to the intricate phenomena of natural language and do so within constraints determined by the basic motivation of Frege's system, he must also conceive of the 'scholastic excrescences' upon the subjects Frege discussed as excisable with a few quick cuts. But if these are manifestations of deep-rooted malady and not mere surface imperfections, the body of Frege's thought will no more survive their excision than an animal will survive removal of its vital organs.

2. Assertion

At first glance the *Basic Laws* seems to introduce far-reaching changes in the explanation and use of the assertion-sign and its constituent content-stroke. Since judgeable-content is split into sense and reference, the term 'content-stroke' is replaced by 'horizontal', and the horizontal is given an altogether different explanation.

Appearances are, however, misleading. There is no change in the general conception of what it is to make an assertion, i.e. of the relations between entertaining, judging and asserting, or of the relation between the act and the instrument of assertion. Or rather, there are no significant changes save those necessary to ensure a smooth mesh between the account of assertion and the new sense/reference distinction. These changes do not reach down to the deeper confusions about assertion in *Begriffsschrift*. The psychologistic conception of judgment and inference is retained; the view that inference is possible only among true assertions soldiers on; the dissociation of assertoric force from the predicate and the need to represent it by a special sign in concept-script is preserved. Not only is there no deep change in the philosophical analysis of assertion and judgment, but the basic algorithms of the signs '⊢────' and '────' have also 'hardly changed' (BLA, p. x). Thus '⊢──── A' is still the expression of an assertion in concept-script. The judgment-stroke '|' is prefixed to well-formed formulae of the form '──── A'. The assertion sign is still used for the representation of *rules* of inference (e.g. *modus ponens*). The horizontal '────' continues to indicate the scopes of logical operators. Hence it remains a component of all molecular sentences. Finally, the rationale for introducing these symbols into concept-script persists, namely to 'eliminate guesswork' and to provide an explicit realization of the contrast between merely expressing a thought and asserting it. The novelties are in fact minimal.

Since Frege now sharply distinguished between what a symbol expresses and what it designates, he also distinguished, parallel to the possibility of expressing a thought without asserting it, the possibility of designating a truth-value without making any assertion. A sentence in concept-script, e.g. '2 + 3 = 5', only designates a truth-value without making any assertion, or saying which truth-value is designated (BLA, §5; FC, 34). Prefixing the assertion sign to the name of a truth-value yields a compound symbol (e.g. '⊢——— 2 + 3 = 5') which does *not* designate anything, but rather makes an assertion (FC, 34n.). Concept-script stands doubly in need of a special sign to be able to assert something as true, to make 'a move in the language-game'.[4]

Like his conception of merely expressing a thought, Frege's conception of merely designating a truth-value is muddled. His fundamental mistake is to suppose that it makes sense to claim that a sentence (or the expression of a thought) *designates* anything at all. But there are further confusions. The hint that designating a truth-value is incompatible with making an assertion (FC, 34n.) threatens to prevent integration of the representation of assertion in concept-script with the sense/reference distinction. A symbol of the form '⊢——— A' does express a thought (which is asserted). But this thought, independent of human volition, 'determines' a truth-value just as the sense of any expression willy-nilly determines its reference. Therefore, there is no scope for denying that '⊢——— A' designates a truth-value unless another major thesis is also denied. Inconsistency on this point probably results from Frege's confusions about truth together with his misconceptions about expressing a thought. Had he eliminated the notion that a clause in a molecular sentence merely expresses a thought without asserting it, he would have eliminated too the notion that such an expression merely designates a truth-value without saying which it is. Had he rid himself of the idea that to make an assertion is to make the transition from a thought to the True, he would no longer have imagined that expressing a thought, which does as a matter of fact designate the True, necessarily falls short of making an assertion. Failing to clarify those matters, he fell into inconsistency.

The horizontal is treated as a function-name. It is elucidated by the stipulation that its value for the True is the True, for the False, the False; i.e. like multiplication by 1, it maps these two objects onto themselves. Although *Begriffsschrift* paraphrased the horizontal as 'the circumstance that', Frege later offered only a formal explanation, no longer searching for a natural language equivalent or analogue. Being a function-name, '———' is subject to the canon of completeness of definition. Hence he stipulated that for any object as argument which is *not* a truth-value, the horizontal-function takes the False as its value. Hence '———2' refers to, is a name of, the False (just like '2 + 1 = 4'). To this sign, of course, one 'cannot' prefix the judgment-stroke since one 'cannot' assert the False. But to the negation of '——— 2', viz '——⊤— 2', one 'can' prefix the judg-

---

4. Cf. Wittgenstein, *Philosophical Investigations*, §49.

ment-stroke ('⊢⊤— 2') since '—⊤— 2' designates the True.[5] From a formal point of view this effects some simplification in what counts as a well-formed formula.

These modifications do nothing to allay our original qualms over the intelligibility of the content-stroke. Indeed, they are exacerbated. For now the new function ——— ξ is said to be a concept (since its value for any object as argument is a truth-value). But is this not problematic? What concept is it? It might be thought that '——— ξ' stands for the concept of being identical with the True,[6] for this concept has the True as its value for the True as argument and the False as its value for any other object as its argument. A natural riposte is that there is no such thing as the concept of being identical with the True. Such expressions as 'Julius Caesar is identical with the True' and 'If the King has moved, castling is prohibited is identical with the True' are gibberish. This argument generates a dilemma: either Frege's stipulations about the reference of '——— ξ' succeed in endowing it with a sense and reference despite the lack of any equivalent expression in natural language, or the absence of such an explanation proves that '——— ξ' does not stand for a concept, despite the fact that his stipulations supply this putative function-name with a value for every non-empty name in its argument-place. This dilemma is but a special case of a perfectly general problem arising out of his function-theoretic analysis of concepts, namely, whether it is really intelligible to explain concepts, to elucidate our use of concept-words, in the normal form of specifying a function, viz. by stating what is its value for all possible arguments (supra p. 253f.).

A further difficulty consequent on construing the horizontal as a name of a concept is the requirement that, when completed by an appropriate kind of name, the result should express a thought yet not be an instrument of assertion. This perpetuates the confusions of *Begriffsschrift* and casts further doubt on the claim that '——— ξ' is a concept-word (sic!). For a concept-word is explained as an expression that yields a *sentence* when its argument-place is filled by a proper name (PW, 229). Is there anything other than a sentence which expresses a thought? Is there any such thing as an expression of a thought which cannot be used to say something? Affirmative answers are essential to justify Frege's use of the horizontal, but negative answers are implied by his characterization of thoughts (PW, 197) and our concept of expressing a thought.

The innovations in the account of assertion in the *Basic Laws* contribute to a smoother analysis of the logical connectives and to a dovetailing of the sense/reference distinction with the formal system of concept-script. But they do nothing to remedy the deep flaws in the early conception of what an assertion is and what it is to express a thought.

---

5. Indeed, by functional abstraction '——— ξ' is defined not only for all proper names as argument expressions, but for all expressions whatever (BLA, §5). If 'Φ' is any function-name whatever, '——— $\Phi(\xi)$' is a concept-word. Hence '——— ξ' is systematically ambiguous, taking proper names, concept-words or function-names as argument expressions, and transforming them into names of truth-values or into concept-words.
6. Cf. D. Bell, *Frege's Theory of Judgment*, p. 19 (Oxford University Press, Oxford, 1979).

Frege's last three published papers all discuss judgment and assertion. The exposition is independent of the formal notation of concept-script. Nevertheless these last thoughts reveal no change in the rationale for introducing the judgment-stroke and horizontal into his notation. He continued to insist that 'the logical imperfections of language' stand in the way of logical analysis of proofs (PW, 253), that it is a defect that no part of a sentence invariably carries assertoric force (PW, 252) since this produces confusions. It seduces us into thinking that proofs may proceed from mere hypotheses (CT, 553) or even from thoughts held to be false (PW, 244f.), whereas in fact inferences must be among assertions. It deludes us into construing logical connectives as taking judgments as arguments rather than contents of judgments (CT, 539f.). All these thoughts are familiar from his previous work. We would look in vain for significant recantations of fundamental theses.

There are, however, two novelties in these last writings. First, Frege made explicit his conception of the temporal anatomy of the act of assertion. Making an assertion consists in a sequence of acts. One must first grasp a thought, then one must judge it to be true, and finally one must manifest this judgment by uttering a sentence which expresses it in the appropriate form. There had been previous hints about the necessary temporal ordering of these acts (e.g. PW, 7, 185, 198), but the doctrine now became overt (N, 127; T, 22; PW, 267). He thus elaborated a mythology about sequentially ordered mental acts instead of scrutinizing the internal connections between the concepts of understanding a sentence, believing it to express a truth, and using it to make an assertion.

His second innovation[7] is the doctrine that a sentence-question and the corresponding declarative sentence express the same thought (T, 21; N, 117, 119, 129; CT, 539). An appropriately paired question and declarative sentence express different relations of a person to the same object, viz. the thought expressed by both. A sentence-question expresses a thought without judging it to be true. Frege used this claim to support his familiar thesis that it is possible to express a thought 'without laying it down as true' (T, 21). He then argued that the act of grasping a thought should *always* be distinguished from the act of judging a thought to be true (N, 119). In particular, this distinction should be applied to declarative sentences even though both acts are there so closely conjoined that it is easy to overlook their separability (T, 21 f.). Frege even suggested that his term 'thought' should be defined as the sense of a sentence-question (N, 119).

His reasoning about sentence-questions seems very naïve. From the fact that somebody expresses a thought by answering 'Yes' to the question 'Is two prime?', Frege inferred that the sentence-question itself must *contain* this thought (T, 21). Apparently every sentence-question *expresses* a thought (cf. T, 22). This is doubly absurd. First, sentence-questions are typically used to ask questions, and to describe someone as asking a question is normally incompatible with describing him as expressing a thought. It would be nonsensical to state that I expressed the

---

7. The germ of this idea is present in Lotze *Logic,* I-ii-§40; in Frege's *Nachlass* it crops up much earlier than in his publications (see PW, 7f., 138f.).

thought that the morning plane from Timbuctoo has arrived if what I did was to telephone the airport and ask 'Has the plane from Timbuctoo arrived yet?'. Second, the argument that a sentence-question must contain a thought since an elliptical answer to it expresses a thought would have immediate parallels unacceptable to Frege. He denied that an imperative expresses a thought (T, 21). Yet, if somebody says 'Please pass the butter', I might reply 'No, I won't' or 'I can't'; in each case I would express a thought, just as I would in answering 'Yes' to a sentence-question. The parallel between these utterances generates a dilemma: *either* I can express a thought by giving an elliptical response to an utterance which does not contain a thought (in which case the doctrine about sentence-questions collapses) *or* I can express a thought in this way only if the utterance to which I respond contains a thought (in which case the doctrine about orders, requests, desires, etc., is incoherent). Frege's reflections on non-declarative sentences seem irredeemably flawed.

The idea of explaining the connection between a sentence-question and the corresponding declarative sentence in terms of their expressing different relations of speakers to a common abstract object (a thought) is very alluring. It might even be thought the key to a comprehensive account of internal relations among speech-acts. But the root idea is defective by Frege's own lights. If the object of assertion is the reference of the indirect statement correctly used for reporting the assertion (in *oratio obliqua*), the object of questioning should be the reference of the indirect question correctly used for reporting the question. But typically indirect questions cannot be substituted for indirect statements *salva significatione*, nor vice versa. Hence the references must differ, and so correspondingly must the senses of the paired questions and declarative sentences.

Frege's account of assertion together with this late doctrine about sentence-questions is widely thought to be the foundation for a general theory of force. It seems to open up the vista of a systematic explanation of the use of utterances to perform speech-acts. The general programme is to single out something (the 'sense' of a sentence) which is the object of many different speech-acts. Sentences or utterances are then analyzed on the model of the symbolism '|' and '——— ξ' into two components: a force-indicator and a specification of the sense. Frege's comments on sentence-questions suggest introducing a symbol for interrogative force such as '?' and representing paired assertions and questions in the form '⊢——— A' and '?——— A'. Modern theorists have gone much further in the differentiation and analysis of 'force' and in the elaboration of formal calculi embracing imperative and erotetic inference. All of this is manifestly at odds with Frege's purposes. He would have judged the putative accomplishments of theories of force to be nonsense. Nonetheless, it would be claimed, his analysis of assertion does serve as the solid foundation of theories of which he never dreamt. Our careful inspection shows this thesis to be erroneous. His distinction between judgment (represented by '|') and content (represented by '———') is the archetype of the modern distinction between sense and force, and is one of the Pillars of Wisdom, yet excavation around his account of assertion and his use of '⊢———' bring to light nothing but sand.

3. Truth

Frege's theory is commonly, and correctly characterized as Realist. Thoughts are true or false, and their truth or falsehood is determined by how things are in reality. The Real, one might say in a moment of metaphysical abandon, determines the True and the False. How so? A sentence expressing a thought decomposes (in the simplest case) into a proper name and a concept-word. The reference of the proper name is an object, that of the concept-word is a concept. Object and concept alike belong to the domain of the objective, the mind-independent 'realm of reference'. The truth of the thought expressed by the sentence is determined by the references of the constituent expressions of the sentence expressing the thought. If the object a falls under the concept F (or, more happily, has the property F) then the sentence 'Fa' expresses a true thought. So too with any relational sentence: 'Jupiter is larger than Mars' expresses a true thought just in case Jupiter is related to Mars by the relation which is referred to by 'is larger than', which relation 'belongs to the realm of reference' just as do Jupiter and Mars (PW, 193).

This schematic account looks very like familiar correspondence theories of truth. Propositions are true if they correspond to the facts, false if they do not. So, if it is admitted that the truth of the thought expressed by a sentence is determined by the arrangement or concatenation of the references of the constituent expressions of the sentence in reality, why not take the further step which is so inviting? For plainly one may now propose that if the referents of the sub-sentential expressions are concatenated in a manner corresponding to their arrangement in the sentence, then *it is a fact* that things are in reality as the thought expressed says they are. A true proposition or thought corresponds to the facts, a false one does not. (Whether one goes farther and claims that a true proposition *names* a fact is a further matter.) At any rate, facts seem denizens of the realm of reference and it is in virtue of their obtaining or not obtaining that thoughts are true or false.

Frege did not follow this path. He denied that facts belong to the realm of reference, that true thoughts (or the sentences expressing them) refer to (or name) facts, that facts correspond to true thoughts, and that truth consists in correspondence with anything at all. Facts, he contended, simply *are* true thoughts (T, 35). The suggestion is not uncommon.[8] For it speak the following features: facts, like true thoughts or propositions, obtain independently of cognition or cogitation, are timeless ('once a fact, always a fact'), are supported by evidence, confirmed by tests and experiments, discovered by investigation, are independent of language, of being stated or expressed. Finally, it looks as if 'the thought (proposition) that p is true' is synonymous with 'That p is a fact' or 'It is a fact that p', and equally that if what a person believes is true, then what a person believes is a fact: *ergo*. . . .

---

8. Cf. C. J. Ducasse, 'Propositions, Opinions, Sentences and Facts', *Journal of Philosophy* XXXVII (1940): 701–11; R. Carnap, *Meaning and Necessity,* p. 28, and, more surprisingly, A. N. Prior, *Objects of Thought,* p. 5.

However, the proposal is not acceptable. Facts, though not denizens of the 'realm of reference', are not denizens of the 'realm of sense' either. The fact that Bishop Berkeley is buried in Christ Church chapel is surely one and the same fact as the fact that the author of *Principles of Human Knowledge* is buried in the cathedral of Oxford. But, according to Frege, the corresponding *thoughts* are altogether distinct. Facts, it seems, are 'extensional' and thoughts 'intensional', so facts cannot be identified with true thoughts. Second, Frege's characterizations of thoughts do not fit facts. Thoughts are composed of senses of expressions, but it would be bizarre to think of facts as so constructed. A true thought is a mode of presentation of the True. It is surely nonsense to suggest that a fact is a mode of presentation of a truth-value. Third, the ways in which we speak of facts and thoughts (propositions) do not mesh. If the proposition or thought that it is warm in Oxford is true, then it is a fact that it is warm. But the fact that it is warm is not a true proposition. One may say alike that one fact about Rupert is that he was rash, or that one true proposition about Rupert is that he was rash. But it is senseless to say that one fact about Rupert is the true proposition that . . . or that one true thought about Rupert is the fact that. . . .[9] True thoughts may be detailed, accurate, explicit; but while one may give detailed, accurate and explicit accounts of the facts, they themselves are not detailed, accurate and explicit. The fact that Smith whines may annoy me, but the true thought need not. Snodgrass' death may be due to the fact that he forgot to turn off the gas, but not to the true thought that he so forgot.

We try to make the thoughts we think square with, or match, the facts. Thoughts or propositions are true in virtue of the facts, they correspond to the facts. The story of a reliable witness is commonly born out by the facts, and we see this when we check his story against (the known) facts. These turns of phrase are not part of a *theory* of any kind, they do not 'commit us to an ontology of facts' any more than talk of the average man commits us to an ontology of such pseudo-objects. Nor do they commit us to a certain theory of truth, viz. the correspondence theory. What is wrong with the correspondence theory of truth is not that it is false that true propositions correspond to the facts, but that it erects this innocuous platitude, this 'grammatical statement', into a pseudo-explanatory theory accompanied by a rich ontology of facts and propositions. It is just because 'The proposition that p is true' is equivalent to 'It is a fact that p' that 'The proposition that p' is equivalent to 'The proposition that is made true by the fact that p'. And it is also because of this that Frege's claim that one cannot say that a thought is true if it corresponds to the facts (since if it is true it *is* a fact) is wholly misguided.

Frege's own arguments against the Correspondence Theory of Truth turn on the notion of correspondence and its inappropriateness in explaining what it is for a thought to be true. First, correspondence is a relation, truth is not (T, 18). But the correspondence theorist need not claim that truth is a relation, merely that the property of truth is attributable to propositions only if the relation of corre-

---

9. Cf. Rundle, *Grammar in Philosophy*, p. 324ff.

spondence obtains between them and reality. Second, correspondence may be absolute or relative to a certain respect (T, 19). Absolute or perfect correspondence, Frege claimed, is complete coincidence, which is incompatible with the correspondence-theorists' demand for two distinct entities standing in a certain relation to each other. Imperfect correspondence passes *that* test, but fails because it is a matter of degree, whereas truth does not admit of degrees. However, if the correspondence in question is relative to a certain respect, an infinite regress is generated. To establish whether it is true that *p* we would first have to establish whether it is true that the thought that *p* and reality correspond in respect *C*, and to establish whether *that* in turn were true we would first have to ... and so on *ad infinitum* (cf. PW, 128f.)

The argument is not compelling. The kind of correspondence necessary for the theorist to sustain his account is arguably precisely that the referents of constituents of a sentence be so arranged in reality (or 'in a fact') as the sentence specifies them as being, e.g. that for a sentence of the form '*Fa*' the object *a* have the property *F*, that for a sentence of the form '*aRb*' *a* stand in the relation *R* to *b*, and so on. The alleged regress is not obviously vicious, inasmuch as to establish whether '*Fa*' corresponds with reality one need not (indeed cannot) launch an independent investigation into whether '"*Fa*" corresponds to reality' corresponds to reality.

Since Frege's arguments against the Correspondence Theory of Truth are so feeble, and since his Realism comes so close in spirit to a classical correspondence theory, we might well wonder what deeper reasons there are for his repudiating the theory. The operative consideration is surely that he *did* think that sentences have a reference, *did* think that they named an entity 'in the realm of reference', but that *what* they thus stand for or name is not a fact but a truth-value. Had he argued that a true sentence stands for, corresponds to, or names a fact, then obviously it would be false that all true sentences have the same reference, since they correspond to *different* facts. This would have blocked the direct definition of the extension of a concept in terms of a course-of-values, and that in turn would threaten to rule out any logical transition from concepts to their extensions as required by the philosophical arguments of the *Foundations of Arithmetic*. Might he not have argued instead that all true sentences have the same referent, viz. reality?[10] This would differ in substance from his own account only if it left false sentences without any referent, on a par with fiction. But that suggestion would be unacceptable since it would exclude the analysis of concepts as functions; a properly defined function must have a value for every argument, whereas *ex hypothesei* the 'function' which is identified with the concept of being a prime in the analysis of the true thought that 3 is a prime would have no value for the argument 4. The True and nothing but the True would suffice for the essential purpose of forging a *logical* connection between a concept and its extension. And only if the True is accompanied by the False can concepts be exhibited as func-

---

10. Cf. C. I. Lewis, 'The Modes of Meaning', *Philosophy and Phenomenological Research* 4 (1943–44): 242.

tions and function/argument decomposition be secured as the general pattern of logical analysis. It is these considerations *au fond,* not his rather weak arguments, which preclude recognition of facts as references of sentences. They would, as it were, get in the way of truth-values. Hence it is not surprising to find them summarily ejected from the realm of reference and hardly surprising to find them reinstalled on the level of senses. 'The fact that p' has the role of a proper name; hence, unless every such proper name lacks reference, there must be some corresponding objects, and Frege apparently assumed that objects can be assigned without remainder to the realms of sense and of reference. (Compare his account of *oratio obliqua.* From the negative observation that names do not have their customary reference in indirect speech he concluded that they have their customary senses as their references in such contexts. The negative claims in both cases might be separated from the positive ones.)

Frege clearly regarded the two truth-values as indispensable for his execution of a reduction of arithmetic to logic within the framework of function/argument analysis. His introduction of the True and the False was motivated by a theory (or perhaps a vision), and these two logical objects were integrated into the foundation of his later formalization of logic. He proclaimed the discovery of these two *objects* to be as important an intellectual achievement as the discovery of a pair of new chemical elements (PW, 194)! He did not arrive at his observations about truth via a close analysis of how competent speakers explain and apply the term 'true' in everyday speech. Nonetheless, he might have arrived at important insights into the meanings of 'true' and 'false' though he set out from a different point on an expedition intended to yield different treasures. He is often claimed to have anticipated the redundancy theory of truth, and he apparently clarified the widely misunderstood connection between truth and assertion. Such claims on his behalf make his analysis of the concept of truth worthy of further consideration.

On reflection any such suggestion seems unwarranted. We shall show that his 'analysis of truth' is the incoherent product of irreconcilable tensions within his logical theory. Unless his conclusions are treated in isolation from his reasoning, they stand condemned. And no real improvement is possible in his analysis without a recantation of some fundamental theses and a removal of some basic confusions. Hence his remarks about truth are without substantial value to modern philosophers who aim to elucidate the meanings of 'true' and 'false'.

The difficulties with his account are immediate and striking. Sometimes he argued that truth is a property of thoughts (and derivatively of sentences), though a property which is simple, 'primitive', irreducible to anything else, indefinable, and peculiar in that 'predicating it is always included in predicating anything whatever' (PW, 128f.). At other times he argued that 'truth is *not* a property of sentences or thoughts, as language might lead one to suppose' (PW, 234), and he supported this conclusion by remarking that 'the sense of the word "true" is such that it does not make any essential contribution to the thought' (PW, 251f.). It seems bewildering that Frege offered identical grounds for reaching opposite conclusions. Confusion seems worse compounded when he denied that 'true' is an 'adjective in the ordinary sense' (PW, 251) or when he characterized the function of 'true' as a vain attempt 'to make the impossible possible [viz.] to allow what corresponds to the assertoric force to assume the form of a contribution to the thought' (PW, 252). He capped these muddles with the mysterious claim that assertion involves a transition from a thought to a truth-value, that the term 'true' attempts to make this transition but remains always at the level of thoughts, and that it is impossible to say that a thought designates the True (SR, 64f., BLA, §5). Frege's remarks seem opaque and gratuitous; he seems to blunder his way from one thesis to its contradictory and then back again. Any gold seems buried under mountains of dross.

Closer examination shows that these are not the ravings of a lunatic. Frege was dragged in different directions by irreconcilable demands built into his logical theory. Driven first to one extreme, then to another, he was in the thrall of creatures that he had created: at each stage in his progress he must have thought '*This* isn't how it is, yet *this* is how it *must* be'.[11] But he could renounce his servitude only by bringing his edifice of logic down on his own head. The ingredients of this dilemma are plain.

On the one hand, the True and the False are essential to his identification of the extension of a concept with the course-of-values of a function. The link between concepts and their extensions is forged by introducing the two truth-values as the referents of expressions of thoughts and by redefining concepts as functions whose values are always truth-values. This theoretical apparatus presupposes that it makes sense to specify which truth-value is assigned to particular expressions of thoughts, e.g. to assert that the referent of '$\underline{\qquad} \Phi(A)$' is the True; otherwise, it would be impossible to explain which function from objects to truth-values is the referent of the concept-word '$\Phi(\xi)$', since the only available explanation would be the patently circular one that the reference of '$\Phi(A)$' is the value of the function $\Phi(\xi)$ for the argument A. Unless it is feasible to state for which arguments the function $\Phi(\xi)$ takes the True as its value, for which arguments the False, there would be no possibility of defining the extension of a concept Φ as those arguments for which the function $\Phi(\xi)$ takes the True as its values. Moreover, this identification of an extension with the course-of-values of a function presupposes that those objects for which the function $\Phi(\xi)$ takes the True as value are just those objects which fall under the concept Φ, i.e. just those objects of which the predicate Φ is true. As Frege explained, it is true that A falls under the concept Φ if and only if the value of $\Phi(\xi)$ for A as argument is the True (FC, 30). The True is elucidated as the common referent of all true thoughts (FC, 28; PW, 255). Hence, it follows from the intelligibility (and non-triviality!) of specifying whether the True is the value of the function Φ for the argument A that it is intelligible and non-trivial to characterize the thought expressed by '$\underline{\qquad} \Phi(A)$' as being true. Truth then must be considered to be a property of thoughts. It is a genuine property, one which some thoughts have and others lack, and one which is timeless (T, 29) and independent of human thinking and rec-

---

11. Cf. Wittgenstein, *Philosophical Investigations* §112.

ognition (PW, 133). Any retreat from this position about truth would sweep away the possibility of analyzing concepts as functions and of identifying extensions of concepts with courses-of-values.

On the other hand, Frege's theoretical apparatus demands another apparently incompatible conception of truth. He had to marry the claim that certain symbols are names of truth-values with the thesis that they are expressions of thoughts. The concept of truth was the go-between. A formula in concept-script such as '⊢—— $(-1)^2 = 1$' is said to state that the True is the value of the function $\xi^2 = 1$ for the number -1 as argument, but it is also held to assert the thought that -1 falls under the concept of being a square root of 1, or equivalently that -1 is a square root of 1. But the statement that the True is the value of $\xi^2 = 1$ for the argument -1 must itself be rendered, according to the elucidation of 'the True', into the statement that it is true that -1 falls under the concept of being a square root of 1. On pain of contradiction, it must then be affirmed that asserting that it is true that -1 is a square root of 1 is identical with asserting that -1 is a square root of 1. This in turn seems to entail that in uttering a sentence of the form 'It is true that p' or 'The thought that p is true' a speaker makes the very same assertion that he does if he utters the sentence 'p'. Frege was thus driven to embrace this conclusion. He observed that 'the word "true" contributes nothing to the sense of the whole sentence in which it occurs as a predicate' (PW, 252). Hence he concluded that 'true' is devoid of sense or content, adding the clarification that it does not yield expressions without any sense, but rather that its sense is degenerate in that 'it does not make any essential contribution to the thought' (PW, 251f.). This proves that 'truth is not a property of ... thoughts, as language might lead one to suppose' (PW, 234). Moreover, the apparent vacuity of truth licenses the claim that truth is 'distinguished from all other predicates in that predicating it is always included in predicating anything whatever' (PW, 129). Any retreat from this conception of truth would apparently sweep away the possibility of identifying names of truth-values as expressions of thoughts and thereby thwart the application of Frege's calculus to the analysis of judgments and inferences.

His quandary about truth is the surface manifestation of inconsistent ideas built into the foundations of his mature system of logic. To affirm that truth is a redundant property of thoughts seems to undermine the possibility of giving a non-circular definition of any concept by stating that it takes the True as value for certain entities as arguments. But to affirm that truth is a non-vacuous property of thoughts suggests that the formula '—— $\Phi(A)$', viewed as a name of the True, would express a thought stronger than the thought that A falls under the concept Φ. Frege was suspended between two distinct centres of gravity, and caught thus like Sciron between sea and sky, he never came to rest. To the end he was unable to settle the question whether truth is a property of thoughts, resolving provisionally to call truth a property 'until something more to the point is found' (T, 21). His optimism was unwarranted, since his dilemma could be avoided only by starting again from scratch.

His struggles with the analysis of truth generated much muddled reasoning. In particular, his thesis that 'true' has a vanishing sense might be thought to rest on an argument fallacious by his own lights. In a report in indirect speech such as 'He asserted (the thought) that 2 is a prime', Frege viewed the indirect statement as an expression which designates a thought. When 'true' occurs as a grammatical predicate as in the sentence '(The thought) that 2 is a prime is true', the subject-term appears to be the same proper name of a thought. To the extent that Frege held such a noun-phrase uniformly to have the same significance, it must here be taken to *designate,* not to *express,* the thought that 2 is a prime. But, on this view, the role of the predicate 'is true' is to map a thought onto its truth-value (the True). This evidently qualifies this expression to count as a concept-word. Moreover, the concept designated by this expression is not degenerate since many objects (even thoughts) do not fall under it. The expression 'the thought that' seems to correlate names of truth-values with names of thoughts, and the predicate 'is true' apparently effects the same correlation in the reverse order. From the fact that these inverse operations cancel out when combined it cannot be inferred that 'true' is degenerate in sense. Of course, Frege's reasoning might be vindicated if he held that the noun-phrase 'the thought that 2 is a prime' expressed the thought that 2 is a prime when completed by the predicate 'is true' and hence denied that this phrase is always a proper name. But this seems to be an illegitimate application of the expression 'to express a thought' and to render problematic the identification of proper names.

A second argument is even more clearly fallacious. He often conjoined to the negative thesis that truth is not a property of thoughts the positive one that a thought is related to its truth-value as the sense of a sign to its referent (PW, 234; cf. 194). From the premise that a sentence of the form 'The thought that p is true' contains the same thought as the simple sentence 'p', he argued, 'It follows that the relation of the thought to the True may not be compared with that of subject to predicate' (SR, 64). This seems an exemplary *non sequitur*. The expression 'the True' is introduced by Frege as a proper name of an object. Consequently, how can any sentence incorporate *this expression* in such a way that it even *appears* to function as a predicate? Conversely, how do observations about sentences of the form 'The thought that p is true', where the expression 'true' occurs as the predicate, have any bearing on whether 'the True' designates an object? Is it not an illusion that there even seems to be a conflict here? Why cannot the thesis that '(is) true' ascribes a property to thoughts coexist with the claim that 'the True' designates an object which is the reference of a (true) thought? We might compare the relation of 'is true' to 'The True' with the relation of 'is worth £1' with '£1'; surely there is nothing at all paradoxical about the idea that '£1' designates the value of an object which has the property of being worth one pound. Conflict would not emerge even if Frege were to admit 'the True' as a legitimate designation for the truth-value of the thought that 2 is a prime. Though the sentence '2 is a prime is the True' would then by his standards be well-formed and express a true thought, the predicate of this sentence, on his view would be not 'the True' but 'identical with the True' (cf. CO, 43f.).

A third muddle arises out of the phrase 'passage (or transition) from a thought to a truth-value' (SR, 64f.). Though apparently opaque, this phrase can be readily explained. The basic idea is the assignment of a value to a function for a given argument. An expression such as '$e^{i\pi}$' indicates a sense (how a number is presented as the value of a function for an argument), and the stipulation '$e^{i\pi} = -1$' specifies the reference determined by this sense. Consequently, such an identity-statement effects a transition from a sense to a reference, i.e. it assigns a particular referent to a given sense. On this model, the analogue for defining a concept-word would be the assignment to an expression of the form '—— $\Phi(A)$' of a particular truth-value. If concept-script were enriched with a proper name of the True, say 'T', then '—— $\Phi(A) = T$' would attach a reference to a thought. This expression would effect a transition from a sense to a reference. But *that* Frege declared to be impossible. Two different arguments support this conclusion. The first is that only the expression of a thought may have the True as its referent. If this is to ban any such simple name of the True, the rationale for that restriction is utterly opaque; why is it impossible to bestow a name on what is alleged to be a genuine (logical) object, or alternatively, why does 'T' not express a thought? But the consequence is clear: instead of '—— $\Phi(A) = T$', the only possible expression for specifying the value of a concept would have some such form as '—— $\Phi(A) =$ —— $\Psi(B)$', and this would express merely the thought that two different thoughts agree in truth-value without saying which truth-value they shared (BLA, §5). The other argument is that '—— $\Phi(A) = T$' would express the same thought as 'It is true that A falls under the concept Φ', and that expression, by combining a subject and a predicate, 'reaches only a thought, never passes from sense to reference, from a thought to its truth-value. A truth-value cannot be part of a thought, any more than, say, the sun can' (SR, 64). This reasoning seems altogether bizarre. If '—— $\Phi(A) = T$' cannot make a transition from a sense to its reference because it expresses a thought, then *pari passu* the stipulation '$e^{i\pi} = -1$' cannot attach a referent to the sense indicated by '$e^{i\pi}$', because this too expresses a thought and the number -1 can no more be part of a thought than the sun can (cf. PMC, 157). Consequently the phrase 'transition from a sense to a reference' would have no coherent application at all. Frege did not draw that radical conclusion. Rather, he apparently thought that there was a peculiar difficulty in expressing this transition in the case of expressions of thoughts. The crucial difference between expressions of thoughts and other expressions with sense must be that the attempt to specify a reference yields merely another expression of the very same thought. Whereas '$e^{i\pi} = -1$' says something different from '$e^{i\pi}$', the expression '—— $\Phi(A) = T$' says the same thing as '—— $\Phi(A)$'. This is nonsense since '$e^{i\pi}$' says nothing. The proper conclusion, of course, is that there is nothing analogous to '$e^{i\pi} = -1$' in the case of the thought represented by '—— $\Phi(A)$'. This reflects the facts that expressions of thoughts are not names of anything and that placing such an expression on either side of the identity-sign produces nonsense (supra, p. 250). Hence there is no such thing as defining a *function* whose values are truth-values. Having launched himself decisively into nonsense at the very outset of theorizing, Frege found himself in a typhoon of troubles.

These basic logical difficulties do not exhaust his travails. He was also caught in a strong cross-current of confused ideas about the connection of assertion to truth, and he was carried hither and yon by this. His root confusion here is closely related to his misconceptions about what it is to express a thought. He based his analysis of assertion on the idea that to make an assertion is to put a thought forward *as true* or to advance a proposition *as a fact*. It then seemed evident that to say 'The thought that *p* is true' or 'The proposition that *p* is a fact' *is* to assert the thought that *p*. But, given that 'The thought that *p*' contains the whole of the thought expressed by 'The thought that *p* is true', it follows that the function of the predicate '(is) true' or '(is) a fact' is not to add anything to what is expressed (the judgeable-content or the thought), but to manifest assertoric force. That position Frege advanced in *Begriffsschrift* (BS, §3); on this early view 'true' has a role despite lacking any content. Later, he retracted this positive claim, though retaining the negative thesis that 'true' contributes nothing to the thought expressed. Saying 'The thought that *p* is true' does not necessarily constitute the assertion that *p*; hence 'true' does not carry assertoric force, but rather the form of the declarative sentence 'That p is true' is the (unreliable) vehicle of this force. Where this force is lacking, the predicate 'true' cannot reinstate it (T, 22). Hence Frege ultimately concluded that 'true' has no genuine role at all in our language (PW, 252). It represents a vain attempt 'to make the impossible possible' by incorporating assertoric force into the expression of a thought. The need to use the term 'truth' in characterizing the laws of logic he ascribed to 'imperfection of language'. The term '"true" makes only an abortive attempt to indicate the essence of logic'. Instead, he suggested, 'what logic is really concerned with is not contained in the word "true" at all but in the assertoric force with which a sentence is uttered' (PW, 252). It is noteworthy that he arrived at this confused conception only after the collapse of his logicist programme, hence only once he had no immediate need for the True and False!

Confusion is evident from the beginning of these reflections about truth, but it became more acute later. What threw Frege off balance at the outset is such expressions as 'to put a thought forward as true' or 'to advance a proposition as a fact'. These are indeed approximate synonyms of 'to assert'. They describe the performance of a basic speech-act. For this reason, the act of putting a thought forward as true cannot be identified with uttering any form of words, *a fortiori* not with uttering a sentence (or clause!) of the form 'The thought that *p* is true'. Frege initially laboured under the misconception that the concepts of truth and assertion were internally linked in this simple way. He never shook off this error and started afresh. Instead he modified his earlier view. For a time he held the act of asserting a thought to make a transition from the thought to its truth-value (SR, 65), whereas the predication of truth of a thought is powerless to do so. As a corollary, he doubted the possibility of *saying* that a thought is true (BLA, §5). Ultimately he decided that 'true' is designed for the purpose of saying that thoughts are true, but that this purpose is sabotaged by some ineluctable design-failure in our language (PW, 252). The concepts of expressing a thought and putting a thought forward as true are much more subtle than Frege ever realized.

His remarks on truth are not the beginning of wisdom about the meaning of the word 'true' and its role in our language. They spring directly from the procrustean demands imposed by his logical system and his analysis of arithmetic. They are full of important inconsistencies which are symptoms of deep-seated confusions. Though we might seem to discern in them the redundancy theory of truth, closer scrutiny reveals that these faint glimmerings are the glitter of pyrites, not of gold. His recognition that 'It is true that p' and 'p' are logically equivalent threatened to upset his function-theoretic framework of analysis and to carry in its wake a redundancy theory of the True. The whole body of his remarks is also influenced by a muddled conception of assertion which runs, like a shining vein of fool's gold, through the bedrock of his theory from *Begriffsschrift* to his last essays.

4. Thoughts

The thought, like the judgeable-content, is the object of assertion or denial. It is what we prove when we prove something, what we know when we know what is the case. The thought that $2 + 2 = 4$ is what is asserted in using the sentence '$2 + 2 = 4$' to make an assertion. In making an assertion we 'take the step from the level of the thought to the level of reference', viz. that of truth.

Ordinary language seemed to confirm that thoughts are abstract entities (at least once the errors of psychologism are swept away). In this respect, Frege believed, it is not misleading at all; it reflects the true nature of things. He exploited the conception of thoughts as objects by admitting thoughts both as the values of variables and as the values of certain functions for appropriate arguments. In expressions of the form 'The negation of ξ' or 'The negation of the negation of ξ' (N, 132f.), or 'He believed ξ' (SR, 66), the gap-holder is replaceable by thought-designations. If, in 'the negation of ξ', we substitute a thought-designation, the resultant complete expression will designate a thought, the function *the negation of* ξ taking thoughts as its values for thoughts as its arguments. Indeed, as we have noted (p. 324ff.), it is arguable that every well-formed formula of concept-script, indeed every sentence, represents a thought *as* a value of a sense-function for an argument.

We need not dwell at length upon Frege's Platonism about thoughts. It is a manifestation of a phase in the history of philosophical logic stretching from Bolzano into the twentieth century which is now mercifully in eclipse.[12] Frege's own rationale for the reification of thoughts is generally regarded as superficial. The objectivity and timelessness of truth, the possibility of public agreement and disagreement about theses or doctrines, the intelligibility of hypotheses and suppositions, do not require us to acknowledge the existence of thoughts as aethereal intermediaries between sentences and what they are about. Nor is the analysis of

---

12. For a discussion of its history, see A. N. Prior, *The Doctrine of Propositions and Terms*, ed. P. T. Geach and A.J.P. Kenny (Duckworth, London, 1976); for criticisms of the doctrine, see A. N. Prior, *Objects of Thought*.

indirect discourse or of belief-sentences as involving relations between a person and an imperceptible object seen as a plausible vindication of Platonism about thoughts. We do indeed speak of thoughts or propositions, talk of relations of compatibility, inconsistency, implication, etc. that obtain between them, describe people as entertaining, supposing or postulating them. But such talk no more involves an 'ontological commitment' than does discourse about averages.

If we follow Frege in conceiving of thoughts as abstract objects, then we may be inclined to demand that a criterion of identity be specified for them. This appears to many to be a necessary condition for the intelligibility of discourse about thoughts. Frege himself allegedly[13] recognized the legitimacy of this requirement, insisting on the need for its fulfilment in the logical analysis of number-words (FA, §62). But his publications fail to specify any general conditions for holding that two declarative sentences express the same thought. These writings contain only scattered dogmatic verdicts about the identity and difference of senses expressed by specific sentences or formulae (e.g. (BLA, §2) '$2^2 = 4$' and '$2 + 2 = 4$' differ in sense (cf. BLA, §10, p. xviif.; SR, 56f., 62, 64, 66, 73, 76f.)). There is no hint of any method for testing whether an arbitrarily selected pair of sentences express the same thought. However, some passages from the posthumous publications seem to make a stab at filling this apparent lacuna:

> It seems to me that an objective criterion is necessary for recognizing a thought again as the same, for without it logical analysis is impossible. Now it seems to me that the only possible means of deciding whether sentence A expresses the same thought as sentence B is the following. . . . If *both* the assumption that the content of A is false and that of B true *and* the assumption that the content of A is true and that of B false lead to a logical contradiction, and if this can be established without knowing whether the content of A or B is true or false, and without requiring other than purely logical laws for this purpose, then nothing can belong to the content of A as far as it is capable of being judged true or false, which does not also belong to the content of B; [and vice versa]. . . . Thus what is capable of being judged true or false in the contents of A and B is identical, and this alone is of concern to logic, and this is what I call the thought expressed by both A and B. (PMC, 70f.)[14]

Such remarks seem to bolster the case for attributing to Frege the principle that the sense of a sentence is identical with its truth-conditions (cf. BLA, §32). For he here seems to provide the appropriate correlative account of criteria of identity, viz. that two sentences express the same thought if and only if they have the same truth-conditions.

This alluring proposal to line Frege's thinking up with sophisticated modern ideas is doubly anachronistic and betrays deep misconceptions. First, the concept of truth-conditions is altogether alien to Frege (infra, p. 373ff.), the product of the *Tractatus,* not the invisible companion of the *Basic Laws.* It yields results

---

13. This involves a misinterpretation of his demand for a 'criterion of identity' for numbers (supra, p. 219ff.).
14. Cf. PMC, 67; PW, 197.

discordant with his specific judgments about identity and difference of sense; in particular, it would make all arithmetical truths and all laws of logic identical in sense! It is even incompatible with the very passages marshalled in its support from his writings. He explicitly distinguished and excluded from his considerations any sentence which 'contains a logically self-evident component part in its sense' (PMC, 70; cf. PW, 197). The only conceivable rationale for this qualification is that the conclusion of identity of sense would otherwise not follow from the premises about the relations between two sentences. Therefore, he must have thought that two sentences can meet his putative conditions for equipollence without having the same sense, provided that at least one has a logically self-evident component part in its sense. But, if the conditions for equipollence are identity of truth-conditions, it follows that two sentences can have the same truth-conditions without having the same sense! (On the assumption that any basic logical law counts as logically self-evident, this conclusion would be needed to defend the doctrine that the basic laws of logic are axioms independent of each other). The second anachronism is related to the first: the very idea that Frege adumbrated a criterion of identity for thoughts rests on the model-theoretic conception of validity according to which logical consequence turns on (semantic) relations among *sentences*. We might concede that it makes sense to consider whether the validity of every argument is left unaffected by the uniform substitution in it of one sentence for another, though we might doubt whether this constituted a 'means for deciding or judging' whether two sentences were logically equipollent (or identical in truth-conditions). But he did not share this conception at all: logical consequence depends on relations among *thoughts,* not among sentences. Consequently, for him it would be unintelligible to propose any procedure which examined logical consequences in order to determine what senses pairs of declarative sentences had! The extraction of thoughts from sentences is in principle prior to any discernment of logical relations. The interderivability of the thoughts expressed by a pair of declarative sentences is no more a *test* of identity of thoughts than correspondence with the facts is a *test* of truth (or than the *possibility* of correlating two sets one-to-one is a *test* of them having the same cardinality).[15] The fundamental question is not whether Frege outlined an effective operationalist criterion for identity of thoughts, but whether he could make sense of the suggestion that any stipulations whatever about sentences could lay down general conditions under which two sentences would express the same thought.

What then is the purpose of the passages apparently formulating 'criteria of identity' for thoughts? Presupposing that the concept of a thought is simple and unanalyzable, Frege foreswore trying to define it and instead contented himself with giving elucidations or hints about how he intended it to be understood. Logicians were to be steered towards his concept of a thought. What they understood by 'judgment', 'assertion', 'proposition', 'statement', etc., is in the target area, but it must be purified in two respects. First, it must be shorn of assertoric force; i.e. it must be conceived wholly as the object of assertion. Second, it must be purged

---

15. Cf. *Wittgenstein's Lectures, Cambridge 1932–1935,* ed. A. Ambrose, p. 164ff.

of accretions (tone or colouring) irrelevant to the cogency of inferences; otherwise the (alleged) subjectivity and epistemic privacy of colouring will plunge logicians into endless and fruitless controversies about logical questions, in particular about whether sentences express 'congruent' contents (PMC, 67). The formulations of 'criteria of identity' for thoughts discharge this second negative task of explanation (cf. PW, 138ff.). They are the direct heirs of the earlier elucidation of identity of judgeable-content in terms of equipollence in inferences (BS, §3). As is typically the case, these elucidations tacitly make use of the concepts to be explained in the phrasing of the explanation. Since the cogency of inferences depends on the relations among thoughts, it is circular to delimit thoughts by the explanation that the thought expressed by a sentence is restricted to that part of its content which affects the cogency of inferences.[16] In Frege's view, this circularity is unavoidable (cf. PW, 107; FG, 59), yet the explanation, taken with a grain of salt, may help gain a correct understanding of what a thought is. A corresponding negative elucidation is equally circular: two thoughts are different provided neither can be inferred from the other by appeal only to general logical laws, i.e. provided neither one follows from the other. As was customary, Frege rephrased this explanation as the principle: two thoughts are different provided it is possible that either one could be true without the other being true (PW, 141, 143; cf. FA, §44). No more than Mill or Lotze was he committed by this phraseology to the absurd doctrine that every necessary truth follows from any thought whatever, still less to a general account of the validity of inferences on the lines of modern logical semantics. Frege's ambitions were much more limited. He wished merely to give an informal elucidation that would obviate an anticipated misunderstanding of his concept of a thought.

The weakest possible conclusion to be drawn is that Frege failed to provide any criterion of identity for sense (thoughts). But it might be thought that this lacuna could somehow be filled in conformity with his general pronouncements about sense and his verdicts on identity and difference of sense in respect of particular utterances. The only putative criterion for which there is any textual support would be a psychological or epistemological one. According to this, two sentences have the same sense (or express the same thought) if and only if it is impossible

---

16. We are apt to convert this truism into what Frege would have viewed as a paradox. Taking validity to depend on relations among sentences, we give the non-circular explanation: two sentences express the same thought if they can be derived from each other by appeal to logical laws alone. But in Frege's view, there is no such thing as inferring a thought from itself; hence, if the sentences are interderivable by appeal only to logical laws, the derivation does *not* amount to an *inference,* and yet it is allegedly sanctioned by *logic!* We are tempted towards the claim that two thoughts are the same if each one can be inferred from the other, but this is as self-contradictory as the principle that two objects are the same if they have all their properties in common. (Mathematicians dodge this paradox by a double use of terms such as 'theorem': on the one hand Zorn's Lemma and the principle of well-ordering are described as equivalent but distinct theorems of set theory because they are provable each from the other, and on the other hand, these are described as different formulations of the same theorem because they are logically equivalent.)

for someone who understands both to believe that they differ in truth-value (cf. FC, 29). Quite apart from more general objections, this proposal raises specific difficulties. First, such a psychologistic criterion of identity would be altogether uncharacteristic of Frege's strategies. Second, his analysis of belief sentences *presupposes* an antecedent complete account of sentence-sense, since the 'that'-clause in 'A believes that *p*' is held to have as its reference the thought that *p*. Finally, it is hopelessly unclear how to settle the question of the possibility or impossibility of belief. After all, if we practise hard enough with the White Queen for half an an hour a day, we might believe as many as six impossible things before breakfast! Or is that, as Alice suspected, impossible? Other commentators have sought to fill the lacuna in Frege's account by reference to considerations going beyond anything that he openly advocated. Carnap has proposed explicating identity of sense in terms of intensional isomorphism of sentences,[17] and Dummett suggests treating the sense of an expression as a canonical method for making certain what its reference is,[18] thereby linking the sense of a sentence with a canonical procedure for demonstrating its truth-value. Rather than rushing into a detailed examination of these proposals or a mapping out of other parallel possibilities, we might more profitably reconsider the status of the question to which they are supplied as answers. What would Frege have thought of the contemporary dogma 'no entity without criteria of identity' in application to thoughts? There are two fundamental reasons of principle for supposing that he would have found this rigid demand incoherent. The first is already evident: we refuse to acknowledge anything but the specification of some relation among *sentences* as a criterion of identity for the thoughts expressed, whereas he held that the conditions for thought-identity turned on relations of logical consequence which themselves cannot be explained otherwise than as relations among thoughts. No conceivable answer that would satisfy our requirement could be acceptable to him. The second objection is equally insuperable. What he understood by a criterion of identity for numbers satisfies a demand for proof (FA, §62). A definition of a number-word ('the number of *F*'s') yields a criterion *(Kennzeichnung)* for recognizing a number as the same again only if it alone suffices to generate a *proof* of every true or false statement of identity which is formulated with the aid of this number-word (viz. every judgeable-content of the form '$A = NxFx$'). Consequently, specifying a criterion of identity is only required where it is legitimate to demand proof and only intelligible where it is possible to produce one. Neither condition is met by what Frege called 'thoughts'. If a further proof were necessary to justify a transition in a proof from one asserted sentence to another (or one formula in concept-script to another), then what these sentences express (their senses) would *ipso facto* be different, for it would be false that anybody who recognized that what one expressed was true would immediately recognize that what the other expressed

---

17. Cf. Carnap, *Meaning and Necessity*, §14ff. His suggestion, of course, does not conform with Frege's admission of the legitimacy of alternative logical analyses of a single thought.
18. Dummett, *Frege: Philosophy of Language*, p. 105.

was also true (PW, 143, 197).[19] On pain of contradiction, there is no such thing as a *proof* of identity of sense, *a fortiori* no such thing as a criterion of identity of sense which would serve as a basis for proofs of identity of sense. The fact that two sentences express the same thought must be self-evident, something that neither needs nor admits of proof.

> [H]ow does one judge whether a logical analysis is correct? We cannot prove it to be so. The most one can be certain of is that as far as the form of words goes we have the same sentence after the analysis as before. But that the thought itself also remains the same is problematic. When we think that we have given a logical analysis of a word or sign that has been in use over a long period, what we have is a complex expression the sense of whose parts is known to us. The sense of the complex expression must be yielded by that of its parts. But does it coincide with the sense of the word with the long established use? I believe that we shall only be able to assert that it does when this is self-evident. (PW, 209f.)

To feel qualms about abstract objects and hence to demand a criterion for effectively settling whether two declarative sentences express the same thought is to make a break with deep aspects of his framework of thinking. It is to deny that the concept of a thought cannot be analysed, and it is to adopt a conception of validity of inference against which Frege resolutely set his face. Modern philosophers must either abandon portions of their catechism or pass a verdict that his conception of the thought is both deeply flawed and useless for purposes of theory construction.

Disregarding the issue of criteria of identity, the glaring flaws in Frege's account of thoughts can be arranged under three heads:

First, simplicity and complexity. Viewed as the *values* of sense-functions for sense-arguments, thoughts are conceived as simple objects. Viewed as presenting a truth-value *as* the value of a function (concept) for some argument, thoughts are conceived as complex entities. This ontological singularity is directly inherited from a similar tension in the earlier theory of judgeable-content (supra p. 153f.). *Officially,* complexity is cashed in terms of function/argument decomposition. But, as we have seen (supra p. 325ff.), the gravitational pull of the part/whole pattern of analysis proved irresistible, even in the very context of function/argument decomposition (cf. PW, 255; CT, 538) and despite manifest inconsistency. This clash of divergent 'pictures' or models of analysis produces pseudo-problems (e.g. how do the parts of a thought fit together into a whole?). It thus encourages the superphysics of thoughts, a mythology of abstract entities existing in a supernatural domain and subject to their own adamantine laws of combination and dissociation.

Second, Frege's conception of thoughts exacerbates his misconceptions about assertion. Since thoughts are identity- and existence-independent of human acts

---

19. There is a tension here in Frege's position. It is not evident that, if what the formulae 'A' and 'B' each expresses are so related that contradictions can be deduced from 'A & ¬B' and 'B & ¬A', then anybody must grasp immediately that what is asserted by uttering 'A' is the same as what is asserted by uttering 'B' (Compare PMC, 69f. with PW, 197).

of assertion, they are, in a sense, externally related to assertion. He did not conceive of the thought as an abstraction from acts of assertion, but of acts of assertion as (somewhat mysteriously) *engaging* with thoughts which exist in their own right. But 'No assertion without thought asserted' (BLA, ii, §105) is not like 'No kicking a goal without a ball' but like 'No kicking a goal without scoring'. A thought is indeed what is asserted in an assertion, but that is a grammatical statement, not an instrumental one. It makes no sense to speak of an assertion made unless it makes sense to report what was asserted in *oratio obliqua*. But Frege played havoc with our notion of what can intelligibly be said to be asserted. For given the combination of compositionalism about sense and reluctance to impose any category restrictions other than those required by his functional type-theory, he must permit the 'existence' of nonsensical thoughts as well as the possibility of asserting gibberish. But a person can no more assert that the number two loves moving space (whatever words he may utter) than a painter can paint a five-sided hexagon (whatever he may draw).

Third, Frege's conception of thoughts as abstract objects brings down on his head the congenital weakness of Platonism, the inability to give a perspicuous account of the internal relations holding among concepts. One aspect of this problem concerns logical relations among assertions. The governing paradigm here is the relation of a conjunction to one of its conjuncts: the thought that A and B entails the thought that A because the first thought is a compound thought one of whose constituent parts is the second thought. Where no entailment obtains between two thoughts, Frege standardly gave the explanation that the first does not contain the second (e.g. PMC, 71; SR, 69). This leaves wholly unilluminated the validity of the inference from the thought that A to the thought that A or B. Indeed, he was drawn towards inconsistent statements: on the one hand, the thought that A or B is a compound thought one of whose constituents is A, but on the other hand, the thought that A or B must be part of the thought that A since the entailment holds. Of course, his explanations of the logical constants do specify how to determine the truth-value of a compound thought from the truth-values of its constituents, but in doing so they by-pass the issue of what thought is expressed by a molecular sentence, proceeding instead to calculate the reference of this expression directly from the references of its parts. Platonism about thoughts is a dust cloud that threatens to blanket this source of light. The other parallel problem affects the account of the foreign relations of thoughts, in particular the connection between thoughts and their truth-values. Officially senses determine references (SR, 58); hence, as a special case, thoughts determine their truth-values. How are we to interpret this claim? A thought is a self-subsistent object, a truth-value another such object. What can it mean to declare that one abstract object *determines* another? Frege's conception of this matter becomes clear from his remarks about the relations among sentences, thoughts and truth-values. A sentence is supposed to express a thought and to designate or refer to a truth-value. However, it is not really sentences that are true or false, but what they express (FC, 31; PW, 129, 253); hence strictly speaking, we should assert that it is the thought which refers to the True or the False. Though apparently anodyne, this formulation masks a deep incoherence. By conceiving of thoughts as intermediaries between sentences and truth-values, he in effect transformed thoughts into extraordinary *signs;* they have the role of signs (since they are held to have entities as their references), but unlike familiar signs they are purified of all empirical dross, and they carry out their tasks independently of human stipulations or decisions. They are, as it were, the only *real* signs. This is incoherent. Thoughts resemble other abstract objects in being imperceptible, but there is no such thing as an imperceptible sign. Correspondingly, these entities are held to signify something, but it is far from clear that there are any such objects as the True and the False to be signified. If the problem is to make intelligible the idea that different signs may signify the same thing despite their difference, then the interposition of a further ethereal sign between these signs and what is signified makes nothing clearer. But if thoughts cannot be construed as purified signs, we are plunged again into the mystery of how a thought can determine the True or the False! Platonism about thoughts thus threatens the intelligibility of some of Frege's most fundamental claims.

In fact his discussion of thoughts is no more than a 'fairy tale about symbolic processes'.[20] Thoughts or propositions are not kinds of objects and they do not have referents. Truth and falsehood are not named by sentences, nor are they the references of thoughts. To say, correctly, that thoughts (or propositions) are true or false quite independently of us is not to make claims about the objective relation between two abstract objects, a thought and a truth-value, but rather to make a claim about us and the world we talk about. To claim that the thought that the sky is blue is true independently of us is merely to claim that the sky's being blue is independent of us. The thought that I assert must be the same as the thought that you deny if we really disagree. But this does not commit us to the existence of abstract entities in a 'third realm'. It merely commits us to the existence of a grammatical articulation in our language, viz. that if I utter a sentence '*p*' assertorically, and you utter its negation assertorically, then it *follows (ceteris paribus)* that you denied what I asserted. Complexities, to be sure, there are, and refinements and qualifications are necessary. What is not necessary is that we abandon our moorings in the safe haven of good sense and sail forth with Frege upon the treacherous Platonic deeps.

It is striking that modern philosophers commonly draw the wrong moral from Frege's debacle. They either withdraw to an austere form of nominalist puritanism or indulge in an orgy of Platonizing. The first reaction is to view the source of a multitude of evils in his theory as located in his Platonist excesses. What must be done is to eschew all talk of abstract entities such as thoughts, propositions, senses, meanings, objects of assertion in favour of the more austere vocabulary of sentence-tokens, occasion-utterances, inscriptions, etc. For any abstract entities introduced into a respectable philosophical system must have a role in theory-construction, and there are no operationalist criteria available for determining the identity-conditions and hence the conditions of employment of such objects in a

---

20. Wittgenstein, *Philosophical Grammar*, p. 256.

theory. Quine and his disciples preach this new asceticism. The second reaction is to locate the flaw of Frege's system in his *naive* Platonism. What is needed is not less, but more of the same. A respectable philosophical theory requires bold and calculated postulation of abstract entities. Sophisticated Platonism will be vindicated by the explanatory powers of the resultant theory. We must introduce such abstract entities as possible worlds, sets of truth-conditions of sentences (to explain the semantic relation of logical consequence), and the descriptive-content of sentence-radicals (to provide the foundations of a theory of force).

We are watching a familiar philosophical drama. The antitheses of a powerful philosophical vision contain within themselves the hidden errors of the condemned theses. Thus the typical reactions to Cartesianism are idealism or phenomenalism on the one hand and materialism or behaviourism on the other. Both are violently anti-Cartesian. Yet they carry within themselves flawed material derived from the parent against which they are rebelling. So too here. Both the extreme nominalist and the extravagant Platonist reaction share with the Fregean strategy a common conception of philosophical understanding. They all see their endeavours as quasi-scientific theory building in pursuit of explanations of linguistic, cognitive or arithmetical phenomena. The one constructs austere, highly economic, theories, eschewing any talk of propositions, meanings or senses, and hence accusing ordinary language of sins of hypostatization. The other constructs rich, elaborate theories on the assumption that if a theory requires the hypostatization of a plethora of theoretical entities, *that* is the best vindication of their actuality. Neither strategy sees the philosophical objections to Frege's thoughts as deep, calling for a fundamental reorientation rather than ingenuity, for a different perspective instead of more technical inventiveness. What is awry with his system is not that it is a theory which fails to approximate sufficiently to the truth, and hence needs further refining, or replacement by a better theory which will come closer to the truth. It is rather that there can be no such thing as a *theory* about the bounds of sense. What is at issue with Frege's central pronouncements is not whether they are true or not, but whether they make sense.

PARALIPOMENA

13
The Wisdom of Hindsight: Sense, Understanding, and Truth-Conditions

1. The missing keystone

Our logical excavations are virtually over. There is, of course, much that we have not laid bare. New trenches could be opened to bring to light areas we have neglected. More careful scrutiny of elements which we have only briefly examined may yield noteworthy results, contributing to a more detailed and exhaustive picture of the architectonic of Frege's thought. We have said little about his analysis of oblique contexts, his conception of presupposition, his controversy with Hilbert. Our remarks on his demand for completeness of definition, his analysis of number, his elucidation of the mathematicians' notion of a function, and on his general conception of mathematics have been the absolute minimum necessary for our purposes. Many other topics have gone unmentioned. Our goal has been to reveal the philosophical foundations of Frege's thought, to elucidate the general constructional principles of his edifice, and to scrutinize the central weight-bearing members of the building. Critical evaluation of these produced damning results. The foundations are rotten, the principles unsound, and the supporting members flawed and cracked. Our investigation may seem open to the charge that

> it seems only to destroy everything interesting, that is, all that is great and important? (As it were all the buildings, leaving behind only bits of stone and rubble.)[1]

Our piecemeal critical probes have apparently brought low a noble edifice. But this may be illusory. The great vaults of a building are held together by the keystone. Instability in its absence does not derogate from the stability of the structure when it is in place. Is not the keystone to Frege's system the idea that the sense of a sentence is its truth-conditions? And have we not, by ignoring this cen-

---

1. Wittgenstein, *Philosophical Investigations*, §118.

tral thesis, drawn mistaken conclusions about the stability of his thought? Might it not be that everything great and important has actually survived our destructive battering and still soars overhead, preserved intact by its internal cohesion?

We have indeed made nothing of the notion of truth-conditions. We have not drawn attention to the passage that is cited as the *locus classicus* of a truth-conditions theory of meaning:

> not only a reference, but also a sense, appertains to all names correctly formed.... Every ... name of a truth-value *expresses* a sense, a *thought*. Namely, ... it is determined under what conditions the name refers to the True. The sense of this name—the *thought*—is the thought that these conditions are fulfilled....
>
> The names, whether simple or themselves composite, of which the name of a truth-value consists, contribute to the expression of the thought, and this contribution of the individual (component) is its sense. (BLA, §32)

This explanation of the notion of sense apparently states a generalization applicable to all declarative sentences; it identifies the sense of such a sentence with its truth-conditions; and it accords to sentences priority over subsentential expressions in the construction of a theory of sense. Taken together with his elucidations of the logical constants of his concept-script, this gloss of 'sense' seems to most philosophers to entitle Frege to be hailed as the founder of truth-conditional semantics. It supports the contention that 'Frege actually said that the truth-conditions determine the sense of a proposition'.[2] The explicit identification of sense with truth-conditions is one of the prominent theses of the *Tractatus*, which Wittgenstein, it seems, obviously learnt from Frege.

> The conception pervades the thought of Frege that the general form of explanation of the sense of a statement consists in laying down the conditions under which it is true and those under which it is false (or better: saying that it is false under all other conditions); this same conception is expressed in the *Tractatus* in the words, 'In order to be able to say that "p" is true (or false), I must have determined under what conditions I call "p" true, and this is how I determine the sense of the sentence' (4.063).[3]

The two philosophers shared

> the notion that the meaning of a statement is fixed by determining the circumstances under which it is true or false. Certainly such a notion of statement meaning was endorsed both by Frege and by the Wittgenstein of the *Tractatus*.[4]

Indeed, Frege's position[5] is the same as that expressed by the passage:

> The expression of agreement and disagreement with the truth-possibilities of elementary propositions expresses the truth-conditions of a proposition.

---

2. G.E.M. Anscombe, *An Introduction to Wittgenstein's Tractatus* (Hutchison, London, 3rd ed., 1967), p. 59.
3. Dummett, *Truth and Other Enigmas*, p. 7; cf. pp. 176, 185, 379, 384, and 452.
4. C. Wright, *Wittgenstein on the Foundations of Mathematics* (Duckworth, London, 1980), p. 9.
5. Ibid., p. 9n.2.

> A proposition is the expression of its truth-conditions.[6]

Agreement on this fundamental conception is no coincidence.

> Frege maintains that the sense of a proposition ... is fully determined by stating what necessary and sufficient truth-conditions it has. ...
> This doctrine is to be found ... , by direct derivation from Frege, in Wittgenstein's *Tractatus*.[7]

Unanimous assertion of congruence in their views and of actual historical influence licenses the use of such rhetorical questions as:

> Who else, after all, invented the idea that the meaning of a sentence is to be identified with its truth-conditions?[8]

Frege's title 'Founder of Truth-conditional Semantics' is conferred *nemine contradicente*.

This verdict has dramatic corollaries. It sanctions the attribution to Frege of a number of the guiding ideas of modern formal logic and philosophy of language, and it even does so in the face of little or no direct evidence for his holding these views. Six corollaries are particularly noteworthy:

(i) Frege's elucidations of the primitive symbols of his concept-script[9] constitute correct semantic analyses of the fundamental logical constants of the first-order predicate calculus with identity. By appeal to these explanations, he showed that his axioms express unconditional truths and that his inference rules are truth-preserving. His exposition clearly distinguishes the issue of derivability within his axiomatization of logic from the issue of unconditional truth. He inaugurated the modern era of formal logic both by drawing a clear boundary between semantics and syntax and by developing a wholly semantic conception of the logical validity of arguments.[10] A logical truth is formulated by a statement that is true in all possible conditions,[11] and an argument is valid provided that the truth-conditions common to all of the premises are contained in the truth-conditions of the conclusion.[12] Whether this relation holds is independent of whether the premises are used to make assertions and of whether what they say is true or false. Hence it is clear that 'one can *draw inferences* from a false proposition.'[13]

(ii) The correct explanations of the senses of words, together with proper analysis of sentence-structures, must deliver the sense (or meaning) of any well-formed sentence by specifying its truth-conditions. Frege provided 'the foundation for a theory of meaning': he showed how 'the meaning of a statement is determined by the meanings of the words ... of which it is composed, and by the way

---

6. Wittgenstein, *Tractatus Logico-Philosophicus*, 4.431.
7. Geach, 'Frege', in *Three Philosophers*, p. 141.
8. B. Harrison, *An Introduction to the Philosophy of Language*, p. 64f.
9. Barring, of course, the judgment-stroke and the horizontal.
10. Cf. Dummett, *Frege: Philosophy of Language*, p. xiii.
11. Provided, of course, that its only constants are logical constants.
12. Wittgenstein, *Tractatus Logico-Philosophicus*, 4.46ff, 5.11ff.
13. Ibid. 4.023.

in which these are put together to form the statement'.[14] His fundamental insight was that 'sentences are constructed in a series of stages'.[15] This enabled him to demonstrate that the standard explanations of 'everything' and 'something' suffice to determine the truth-conditions of statements of multiple generality. The conception of step-wise sentence construction

> allows the use only of the simple account of the truth-conditions of sentences containing signs of generality that is adequate for those involving only simple generality, as a general account applicable to all signs of generality, provided that the application is made only to the stage of construction at which the sign of generality in question is introduced.[16]

Frege generalized this idea in constructing a semantic theory adequate for all mathematical statements and for a wide range of statements of natural language: 'Given a basic fund of atomic sentences, all other sentences can be regarded as being formed by means of a sequence of operations'[17] of three types, and their truth-conditions can thus be specified by iterated applications of simple semantic principles. 'It belongs to the essence of a sentence that it should be able to communicate a *new* sense to us. A sentence uses old expressions to communicate a new sense.'[18] The key to the explanation of this important fact is the basis of Frege's philosophy of language: he 'was unwaveringly insistent that the sense of a sentence—or of any complex expression—is made up out of the senses of its constituent words', and hence he laid down the principle that 'the sense of a word or of any expression not a sentence can be understood only as consisting in the contribution which it makes to determining the sense of any sentence in which it may occur ... [i.e.] the conditions under which that sentence is true and the conditions under which it is false'.[19] Truth-conditional semantics for natural languages is packed economically into the slogan 'It is only in the context of a sentence that a word has a meaning'.[20]

(iii) Frege employed the notion of truth-conditions for the novel purpose of welding together formal logic and the semantic analysis of language. 'Frege's formal logic is an integral part of his theory of meaning.'[21] Any determination of the validity of a proof turns on ascertaining the truth-conditions of its premises and conclusion, and establishing their truth-conditions is a clarification of the meanings of the statements which constitute the proof. 'The validity of a proof depends upon the meanings of the statements which form the premises, conclusion and intermediate steps of that proof, and their interrelation.'[22] Frege recognized that

14. Dummett, *Frege: Philosophy of Language*, p. 2; D. Davidson, 'Truth and Meaning', *Synthese* 17 (1967), p. 304ff.
15. Dummett, *Frege: Philosophy of Language*, p. 10.
16. Ibid. p. 11.
17. Ibid. p. 16.
18. Wittgenstein, *Tractatus Logico-Philosophicus*, 4.027f.
19. Dummett, *Frege: Philosophy of Language*, p. 4f.
20. Ibid. p. 6.
21. Ibid. p. 670.
22. Ibid. p. 2.

bringing absolute rigour to the process of mathematical proof depends on providing 'the foundation of a theory of meaning'.[23] In his view, 'it is the business of *logic* to give' a semantic analysis of sentences.[24] This is a major shift from the traditional conception of logic as a codification of norms for making inferences with the goal of attaining truth or knowledge, for according to that view rules of inference could have only an external correspondence with norms for the correct use of the 'arbitrary' symbols of language. Frege transformed the traditional conception by adopting the principle that the laws of logic are held in place by the whole body of meaning-rules for expressions of natural languages. They constitute, as it were, the axis of rotation of our thoughts, judgments, and assertions,[25] for they are derived from the *essence* of linguistic symbolization. Frege saw that logical laws are to be treated as derived rules of a theory of meaning:

> logic is not a field in which *we* express what we wish with the help of signs, but rather one in which the nature of the natural and inevitable signs speaks for itself. If we know the logical syntax of any sign-language, then we have already been given all the propositions of logic.[26]

This conception is condensed into the thesis that the truths of logic are 'boundary stones set in an eternal foundation, which our thought can overflow, but never displace' (BLA, p. xvi). On Frege's view, logic obviously is something sublime which has a peculiar depth and a universal significance; it presents the *a priori* order of language, thought, and the world, and hence it must be of the purest crystal, 'the *hardest* thing there is'.[27]

(iv) The notion of truth-conditions furnishes the key to the nature of understanding and communication.

> To understand a sentence means to know what is the case if it is true.
> (One can understand it, therefore, without knowing whether it is true.)
> It is understood by anyone who understands its constituents.[28]

Frege arrived at this conception once he distinguished the thought expressed by a sentence (its sense) from the truth-value designated by it (its reference). Indeed, the impetus for his making the sense/reference distinction was his concern with giving a clear explanation of how it is possible to understand a true identity-statement without recognizing its truth. The answer is that someone can grasp the truth-conditions of a sentence, whether it formulates his own thought or another's, without knowing whether they are fulfilled. Moreover, this is possible even in the case of 'new sentences which we have never heard or thought of before, so long

---

23. Ibid. p. 1f.
24. Ibid. p. xiii (our italics).
25. Hence, if serious disagreement about logical laws were to arise, it must manifest a global disagreement about the structure of a theory of meaning. (E.g., the conflict between intuitionistic and classical logic reflects a fundamental dispute between 'realist' and 'antirealist' semantics.)
26. Wittgenstein, *Tractatus Logico-Philosophicus*, 6.124.
27. Wittgenstein, *Philosophical Investigations*, §97; cf. §§89, 96.
28. Wittgenstein, *Tractatus Logico-Philosophicus*, 4.024.

as they are composed of words which we know, put together in ways with which we are familiar'.²⁹ Obviously, 'we understand the sense of a propositional sign without its having been explained to us.'³⁰ Frege gave a clear explanation of this fact by appeal to his thesis that the sense of a sentence is compounded out of the senses of its constituents.

> This means that we understand the sentence—grasp its sense—by knowing the senses of the constituents, and, as it were, compounding them in a way that is determined by the manner in which the words themselves are put together to form the sentence. We thus derive our knowledge of the sense of any given sentence from our previous knowledge of the senses of the words that compose it. . . .³¹

It would be ludicrous to assert that understanding a sentence typically consists in a conscious calculation of its truth-conditions from its constituents and structure. Rather, 'if anyone utters a sentence and *means* or *understands* it he is operating a calculus according to definite rules',³² but these rules must be brought to light by constructing a semantic theory. The calculus of meaning-rules, the mechanism of understanding, is hidden and awaits discovery. But we know in advance of apprehending any details that the sense of an expression is what is grasped in understanding it, and hence that a theory of sense is a theory of understanding.³³

(v) Truth-conditions are assigned to sentence-types once and for all; semantic analysis abstracts from all features which vary with the circumstances in which different tokens of a sentence-type may be produced (matters of 'pragmatics'). A declarative sentence containing indexical expressions (e.g. demonstratives, personal pronouns, 'here', 'now', or tenses) may typically be used to make different assertions in different contexts. A specification of its sense will only determine what is asserted by an utterance of one of its tokens relative to the context of this utterance. Hence its sense alone does not determine the truth-value of one of its tokens. '[I]t is its sense *taken in conjunction with the context of utterance* that in general determines the reference.'³⁴ Given that the sense of an expression is a means for determining its reference, then it is natural to suppose that

> [t]oken-reflexive expressions . . . bear a sense which provides a means of determining their reference in a systematic manner from the circumstances of utterance. . . . When token-reflexive expressions are used, . . . truth or falsity will depend upon the circumstances of utterance as well as the sense of the words employed.³⁵

---

29. Dummett, *Frege: Philosophy of Language*, p. 3.
30. Wittgenstein, *Tractatus Logico-Philosophicus*, 4.02.
31. Dummett, *Frege: Philosophy of Language*, p. 4.
32. Wittgenstein, *Philosophical Investigations*, §81.
33. Dummett, *Frege: Philosophy of Language*, pp. 133, 293, 634; *Truth and Other Enigmas*, p. 122. Cf. J. H. McDowell, 'On the Sense and Reference of Proper Names' in M. Platts (ed.), *Reference, Truth and Reality* (Routledge, London, 1980), p. 143.
34. E. J. Lemmon, 'Sentences, Statements and Propositions', p. 95, in B. Williams and A. Montefiore eds., *British Analytical Philosophy* (Routledge and Kegan Paul, London, 1966).
35. Dummett, *Frege: Philosophy of Language*, p. 383.

Consequently particular utterances, not the senses of sentences, must be regarded as the primary bearers of truth-values,[36] but only the truth-conditions of type-sentences bear on the validity of arguments. A parallel abstraction concerns the uses made of different tokens of a single type-sentence in different circumstances of utterance; on one occasion an assertion may be made, in another a conjecture formulated, in yet another a supposition entertained. Such differences in use are not differences in sense, but rather differences in 'force'.[37] They are irrelevant to semantics and therefore to logic. In particular, logic is indifferent to the difference between an asserted and an unasserted proposition. (This distinction might be characterized as 'merely psychological'.[38]) A theory of sense is not an analysis of language in use, but rather a study of abstract potentialities for use.

(vi) The notion of truth-conditions can serve as the key-concept for a general theory of meaning and understanding applicable to the whole of a natural language only if it can be stretched somehow to encompass sentences not standardly assessable as true or false. In particular, some account must be supplied of how to understand non-declarative sentences (questions, imperatives, optatives, exclamations) and declarative sentences not used to make assertions (especially 'performative utterances' and expressive uses of language). The answer is supplied by extending the distinction between sense and 'force'.

> Judgment, command, and question all stand on the same level; but all have in common the propositional form, and that alone interests us. What interests logic are only the unasserted propositions.[39]

Any significant sentence must be associated with a definite 'descriptive content' or 'propositional radical' (its truth-conditions), and the precise nature of this association will depend on the form of the sentence (e.g. whether it is an imperative or interrogative or contains an explicit performative).

> We thus arrive at the distinction, originally drawn by Frege, between the *sense (Sinn)* of a sentence and the *force (Kraft)* attached to it. Those constituents of the sentence which determine its sense associate a certain state of affairs with the sentence; that feature of it which determines the force with which it is uttered fixes the conventional significance of the utterance in relation to that state of affairs. . . .
> It is difficult to see how a systematic theory of meaning for a language is possible without acknowledging the distinction between sense and force, or one closely similar.[40]

---

36. Ibid. p. 400; cf. p. 385.
37. E. Stenius, *Wittgenstein's 'Tractatus'*, chap. IX (Blackwell, Oxford, 1964); J. R. Searle, *Speech Acts*, ch. 2 (Cambridge Univ. Press, Cambridge, 1969).
38. Wittgenstein, *Notebooks 1914–1916*, p. 96. Cf. Russell, *The Principles of Mathematics*, pp. 35, 49.
39. Wittgenstein, *Notebooks 1914–1916*, p. 96. Cf. Russell, *The Principles of Mathematics*, pp. 35, 49.
40. Dummett, *Truth and Other Enigmas*, p. 449f.

Because Frege drew this distinction himself, his own analysis of the sense of number-words is not restricted to declarative sentences used to make assertions, but it explains 'their use in any utterance whose force can be described in terms of the truth-conditions of some sentence'.[41] The natural correlate of a theory of sense is a theory of force.

Frege originally declared his fundamental idea to be the analysis of judgeable-contents into function and argument (BS, Preface). Later he asserted, 'Concept and relation are the foundation-stones upon which I erect my structure' (BLA, §0). Do we now see what he did more clearly? The nearly unanimous verdict of modern commentators is that the real foundation of his thought was the insight that the sense of a sentence is its truth-conditions. He should have stated that his edifice rests on the notion of a truth-condition; he should not have left us to guess that the proper business of the logician is the semantic analysis of sentences of natural language; and he should not have put his readers off the scent by harping on the logical defects of ordinary language when all that he meant is that the task of constructing a theory of meaning is not trivial.

'No matter', it might be replied. 'Let us not pick a quarrel with Frege's *exposition* of his ideas. Having finally ascertained the true nature of his system, in large measure through the benefit of hindsight, let us instead rejoice in his invaluable legacy. The vision that he bequeathed us is subtle and sophisticated, sweeping in its scope and awesome in its power. Frege filled in little of the detail, and what he did sketch in is open to question. The importance of his work lies elsewhere. He single-handedly established the framework which modern formal logic and philosophy of language have been exploiting for half a century. He effected a revolution in philosophy. He installed a new paradigm. Previous work in formal logic now seems quaint and outmoded, and pre-Fregean investigations of language seem hopelessly simple-minded and defective. His discoveries in philosophy are comparable to relativity theory or quantum mechanics in the evolution of physics. Their primary importance is unveiling a "research programme". We should pause to admire his New Creation. And then, if we have the wit and the will, we should join in the vast enterprise of pushing back the frontiers of semantic analysis, of widening the domain governed by understood principles of logic, of laying bare the mechanism of understanding, and of supplementing the theory of sense with satisfactory theories of force and of context-dependent utterances.'

There is no doubt that this programme is firmly entrenched and that it exercises an immense influence on contemporary philosophy. There is some doubt about the value of some of its aspects and about the proper tactics to be followed (e.g. whether 'truth-conditions' should be cashed in terms of conditions of verification or in terms of verification-transcendent conditions of truth). But whatever the verdict of such evaluation may be, there is serious ground for doubt whether it has any bearing on Frege's thought at all. For there is overwhelming evidence that in every major respect the whole modern conception of truth-conditional semantics is inconsistent with his leading ideas.

---

41. Ibid. p. 106.

2. The collapse of the vault

It would have been revolutionary in 1879, or equally in 1893, to argue that the cogency of an inference depended solely on relations among the *sentences* employed in its formulation. Most nineteenth-century logicians would have found this contention incomprehensible because they held the relation between a declarative sentence and what it asserted to be external and dependent on arbitrary choice; language was conceived to be merely the instrument of thought. Frege did not dissent from this tradition, but openly embraced it. Had he wished to convey a fundamentally different conception of logic, he could not reasonably have expected to do so without laying as much stress on this innovation as he did on the replacement of subject/predicate analysis by function/argument decomposition. If he shared the modern conception of logical validity, he certainly kept it to himself.

It is salutary too to remember how modern the semantic conception of validity is. Although Russell acknowledged a profound debt to Frege 'in all questions of logical analysis', *Principia Mathematica* did not formulate a clear distinction between logical truth and provability within its axiom system, and hence it did not envisage the possibility of proofs of consistency, independence, and completeness. Wittgenstein criticized the conception of logical truth that he discerned to be common to Frege and Russell. He thought them mistaken to hold that logical truths rested on axioms which were generalizations certified to be self-evident by inspection of the primitive concepts of logic (the concepts or propositional-functions designated by the logical constants). His fundamental ideas were presented as novelties: 'the "logical constants" are not representatives' and 'one can calculate whether a sentence belongs to logic, by calculating the logical properties of the *symbol* . . . using only *rules that deal with signs*'.[42] This conception was so alien to Russell that he failed to grasp it from Wittgenstein's account.[43] He continued to think of propositions as Platonic objects, as what is believed or what is stated,[44] and hence he struggled in vain to ascertain what the property of logical truths was which Wittgenstein called the characteristic of being tautologous.[45] Russell could not appreciate that the peculiar status of logical truths turned solely on features of symbols; although he adopted Wittgenstein's view as his official creed, he persisted in thinking that logical truth was a mysterious property of Platonic entities.[46] Those philosophers who were converted to the modern seman-

---

42. Wittgenstein, *Tractatus Logico-Philosophicus*, 4.0312 and 6.126 (our italics).
43. It would be fascinating to know what Frege made of the *Tractatus* and what Wittgenstein replied to his queries and objections, but that is a lost chapter in the history of modern philosophy.
44. E.g. Russell, *Logic and Knowledge*, p. 187; though, inconsistently, he had just asserted 'A proposition is just a symbol' (ibid. p. 185).
45. B. Russell, *Introduction to Mathematical Philosophy*, p. 203ff. (Allen and Unwin, London, 1919).
46. Compare his exposition of the theory of types. He changed nothing substantial, but added as a coda 'The theory of types is really a theory of symbols', as if that were enough to meet Wittgenstein's criticism! (Russell, *Logic and Knowledge*, p. 267).

tic conception of validity were directly influenced in this matter by the *Tractatus,* not by Frege.⁴⁷ Schlick in 1925 considered logic as the study of the interconnections of judgments, and he held syllogistic logic to be capable of presenting all logical relationships among judgments.⁴⁸ By 1930 his conception was transformed: syllogistic logic was swept away in favour of modern quantification theory, and Wittgenstein was credited with the decisive innovation of showing that 'the validity of logic has nothing to do with any properties of the universe, ... [but] it is concerned ... with the equivalence of different expressions'.⁴⁹ The dawn did not break until the 1920s, and then the light increased swiftly among philosophers.

Developments within formal logic support the same conclusion. The first glimmerings of the conception of a completeness proof for a logical system appeared only early in the 1920s. The general problem was first formulated in full generality in 1928, and in 1930 Gödel proved the completeness of the first-order predicate calculus with identity as formalized in *Principia Mathematica.* Formal logicians rapidly exploited the potentialities of metalogical proofs once they had a firm grasp of the semantic conception of validity. The evolution of formal logic and of a philosophical clarification of the nature of logic should instil a certain scepticism about the claim that the modern conception is derived directly from Frege's writings.

These general doubts can be sharpened by reviewing prominent features of Frege's thought that are inconsistent with truth-conditional semantics and the modern semantic conception of validity.

(i) Inferences can be drawn only from true assertions. An array of sentences formulates a cogent inference only if the premises are severally both asserted and true. There is no such thing as drawing a conclusion from a supposition; a sentence expressing a false thought is 'logically useless' (PMC, 79; cf. 16ff., 182n.); and reference-failure disqualifies a sentence from occurring in any cogent reasoning (PW, 129f.).

(ii) Frege denied that various degenerate cases of sound arguments express cogent inferences. In some cases, he held that there was in fact no inference; e.g. each of the argument-schemata

$$\frac{P}{P} \qquad \frac{P \mathbin{\&} Q}{Q \mathbin{\&} P} \qquad \frac{P \to Q}{\neg Q \to \neg P} \qquad \frac{P}{\neg\neg P} \qquad \frac{\neg(\exists x)(Fx \mathbin{\&} Gx)}{\neg(\exists x)(Gx \mathbin{\&} Fx)}$$

fails to represent an inference-pattern because the premise and the conclusion in each case have the same sense, i.e. they are two different ways of saying the same thing (BS, §22; CT, 541ff., 548, 553ff.). In other cases he thought that there is

---

47. One clear example is Carnap, who, while paying generous tribute to his teacher Frege, explicitly attributes to Wittgenstein the semantic analysis of logical truth (*The Philosophy of Rudolf Carnap,* ed. P. A. Schilp, pp. 12f., 24f., 46f. (Open Court, Illinois, 1963)).
48. M. Schlick, *General Theory of Knowledge,* trans. A. E. Blumberg (Springer, Vienna, 1974), p. 102f.
49. M. Schlick, *Gesammelte Aufsätze* (Gerold, Vienna, 1938), pp. 33f., 224f.

no *possibility* of inference from what is expressed by the premise; e.g. the principle *ex falso quodlibet*.

$$\frac{P \,\&\, \neg P}{Q}$$ cannot express a cogent inference just because no sentence of the form $P \,\&\, \neg P$ expresses a true thought (cf. PW, 244ff.) or can be asserted without absurdity (CT, 556).

(iii) Frege never formulated truth-table definitions of the logical constants of the propositional calculus. He treated these symbols as indefinable, and the so-called definitions as mere informal elucidations. According to the *Basic Laws*, truth-tables would be incomplete even as elucidations, while the tabular explanations of *Begriffsschrift*, if treated as full explanations in terms of truth-values, are inconsistent with Frege's judgments about identity and difference of judgeable-content.

(iv) All of the logical constants of the *Basic Laws* name functions whose values are the objects the True and the False. Each is held to have a sense and a reference, and hence each is a 'representative' just as any well-defined concept-word is. Every logical law states a generalization about functions.

(v) The propositional calculus and the predicate calculus with identity are axiomatized in *Begriffsschrift* independently of the apparatus of truth-values and truth-conditions. Truth and falsity play no role even in the informal elucidations, and it would lead to inconsistency to identify judgeable-contents with truth-values. Frege's invention of his formalization of logic is therefore manifestly independent of the semantic conception of logical validity. So too, *ceteris paribus*, is the persistence of this system in the *Basic Laws*.

(vi) Concept-script is an independent language for the perspicuous formulation of inferences. The variables that occur in its formulae are not dummy expressions of natural language (sentences, names, concept-words, sentence-connectives). Hence it is in principle impossible to calculate whether a formula in concept-script expresses a logical law by appealing to rules that deal with the signs of natural language! In Frege's view, examination of the structure of the sentences used to formulate an inference cannot 'guarantee the formal correctness of thought-transitions' (SJCN, 84f.), nor, *pari passu,* can it certify whether a sentence expresses a logical truth.

Each of these six ideas is incompatible with some basic aspects of the modern semantic conception of logical validity. Together they add up to a powerful *prima facie* case against ascribing that conception to Frege, unless he failed to draw even the most elementary consequences from his alleged decisive innovation.

Similar discrepancies spring to view in comparing his conception of sense and its applications with the leading ideas of truth-conditional semantics.

(i) Frege's sole explanation of sense in terms of 'truth-conditions' is explicitly concerned only with well-formed formulae *in concept-script*. 'Every such name of a truth-value [i.e. every name correctly formed from our signs] expresses a sense, a *thought*' (BLA, §32).

(ii) The sense of any complex expression is a way of presenting an entity *as the value of a function* for some argument(s). Hence the sense of a declarative sentence must be a way of presenting the True (or the False) as the value of a concept (or relation) for some argument(s). The thought expressed by using a declarative sentence to make an assertion is that the True is the value of a particular function for some specified argument(s). The only method for ascribing a sense to a sentence makes explicit reference to the True and to a function whose value is this object. Hence to deny that sentences are names of truth-values is to undermine the basis of Frege's explanation of the sense of a sentence.

(iii) Truth and falsity are not properties of sentences or even of thoughts. Rather, the True and the False stand to thoughts as objects stand to the senses of proper names. Only derivatively, i.e. via the thought expressed, can the True or the False be called the references of declarative sentences.

(iv) The *sense* of a sentence containing indexical expressions will vary according to the context of its utterance. It cannot be said to have a complete sense at all, and different tokens produced in different contexts will typically express different thoughts (BLA, p. xvif.; T, 24). Sense is not a feature of context-dependent type-sentences.

(v) Any two sentences expressing necessary truths have identical truth-conditions, viz. each is true in every possible situation. But Frege differentiated the thoughts expressed by equations and inequations, even though he argued that the thoughts expressed, if true, were necessarily true and, if false, necessarily false. He stated explicitly that the sense of '$2^2 = 4$', differs from the sense of '$2 > 1$', '$2 + 2 = 4$', and '$2^4 = 4^2$' (FC, 29; BLA, §2).

(vi) The same thought may be decomposed in radically different ways into functions and arguments, and it can be expressed by sentences (or formulae in concept-script) having radically different constituents and structures. Alternative analyses of a single thought may involve ascent in the hierarchy of logical types (e.g. BLA, §22) or descent (e.g. BLA, §34; cf. FA, §64). The logical decomposition of a thought is not bounded by the uniqueness of the complete semantic analysis of a declarative sentence expressing it. Indeed, a thought is not literally composed of components assembled in a particular structure in the way that a sentence is composed of words put together in a definite structure allegedly uniquely characterized by its 'constructional history'—unless possibilities for alternative analyses of thoughts explicitly licensed by Frege are to be retracted and the applicability of function/argument analysis to *thoughts* is to be withdrawn.

(vii) Frege did not introduce the sense/reference distinction to resolve any puzzles about the nature of understanding.[50] He did not explicitly address the problem of how it is possible to understand an identity-statement without knowing its truth-value. His only remarks about understanding are casual and peripheral, since he obviously regarded the task of clarifying the nature of understanding as belonging to the province of psychology, not of logic.

(viii) He did not generalize the distinction between sense and force, though both

---

50. Pace Dummett, *Frege: Philosophy of Language*, p. 293, and McDowell, op. cit., p. 145.

Russell and Wittgenstein did around 1913.[51] Frege went only as far as distinguishing the assertoric force of a declarative sentence from the interrogative force of a sentence-question. Although he ascribed senses to orders and to sentences expressing desires or requests, their senses are not thoughts, and he indicated no way to relate their senses to the senses of declarative sentences and the constituents of declarative sentences. Nor is it obvious whether he would have welcomed contemporary efforts at generalization as extensions of his own theory. It is typically contended by supporters of this modern theory that every sentence, no matter whether declarative, interrogative, imperative, or optative, splits up into two elements, a sentence-radical and a mood-operator. The sentence-radical incorporates the 'descriptive content' of the sentence

> which is in general independent of whether it is being used to make an assertion or give a command; this descriptive content corresponds precisely to what Frege calls the sense of a sentence, or the thought it expresses. . . .
>
> On this view, assertoric sentences, imperatives, sentential interrogatives and optatives would all express thoughts: they would differ only in the force attaching to them—the linguistic act which was performed by uttering them. We can do various things with the expression of a thought: assert that it is true, ask whether it is true, command that it be made true, wish it were true.[52]

But it makes no sense to command a Fregean thought *to be made* true. Thoughts are timelessly and unalterably true, false or truth-valueless. If the so-called 'descriptive content' of a command corresponds precisely to the thought that the command is obeyed, then if the command *is* obeyed, the thought *is* (timelessly) true, and if not, it *is* (timelessly) false. But one cannot coherently represent an order as a requirement to make a Fregean thought true. Furthermore, Frege adamantly insisted that all inferences are from assertions to assertions. The idea of inference from unasserted thoughts to unasserted thoughts would be no more intelligible to him than the idea of inference from a contradiction! Hence thoughts could not provide the materials for constructing a logic of imperatives or a logic of questions.

These eight observations amount to a powerful *prima facie* case against ascribing to Frege the modern conception of a theory of meaning erected on the foundation of the notion of truth-conditions.[53] Had he arrived at that conception, his writings would be rife with gratuitous blunders and inexplicable lacunae.

---

51. B. Russell, 'Theory of Knowledge' (unpublished typescript at McMaster University, written in 1913), p. 196; Wittgenstein, *Notebooks 1914–1916*, p. 96.
52. Dummett, *Frege: Philosophy of Language*, pp. 305, 307.
53. The very idea that Frege thought it to be the business of logic to give a semantic analysis of sentences is a disastrous preconception for the interpretation of his philosophy. It involves a fundamental overestimation of the discontinuity of his logic from preceding conceptions and a counterpart underestimation of development within the modern tradition of analytic philosophy. It generates the illusion that Frege constructed a philosophy of language, and it alone lends support to the dubious doctrine that he regarded the philosophy of language as the foundation of the whole of philosophy (Dummett, *Truth and Other Enigmas*, p. 441f.; Sluga, *Gottlob Frege*, p. 2).

This barrage of serious doubts and difficulties can be redoubled by considerations from another quarter. One might well request an explanation of the concept of a truth-condition. What exactly is asserted by a philosopher who maintains that the meaning of a sentence is its truth-conditions? Unless this can be clearly explained, we must surely doubt whether truth-conditional semantics clarifies anything at all.

One strategy in supplying an explanation would be to appeal to various internal connections between truth, assertion, and understanding. To assert that p is to assert that it is true that p, and hence to understand what is asserted is to understand what would be the case if it were true that p. And by stipulation, 'the truth-conditions of the sentence "p"' means the same thing as 'what would be the case if it were true that p' (or perhaps 'what would be the case if "p" were true'). Unfortunately this explanation accomplishes less than nothing. It fudges the distinction between a sentence and what is asserted by uttering a sentence. Even waiving this difficulty, we would be left with a platitude. For what would be the case if it were true that p is that p! Hence, to understand what is asserted by somebody who asserts that p is to understand that p (i.e. what would be the case if it were true that p). How can the introduction of the terminology of 'truth-conditions' transform this truism into the fundamental idea of a theory of meaning?[54]

An attempt to give genuine content to the concept of a truth-condition must take a different form. One obvious strategy would be to explicate the main theses of the canonical texts of truth-conditional semantics. By far the most influential of these works (and the one to which commentators explicitly liken Frege's theory of sense) is Wittgenstein's *Tractatus*. Hardly less so is Carnap's *Meaning and Necessity*, which elaborates the general position of the *Tractatus* but adds the orthodox distinction between intension and extension that Wittgenstein sought to do without. Let us inspect the flesh that these seminal works put on the bones of 'truth-conditions'.

Fundamental to the picture theory of meaning in the *Tractatus* is the notion of an elementary proposition. Logical analysis of any proposition brings us ultimately to elementary propositions each of which consists of names in immediate combination and depicts a state of affairs *(Sachverhalt)* in virtue of the representation of objects by each of its simple names. The truth-possibilities of elementary propositions are the possibilities of existence and non-existence of states of affairs, and given that every atomic proposition is compatible with every other one and with the negation of every other one, a possible world is completely described by listing all elementary propositions and then specifying which are true, which false (what Carnap calls a 'state description'). The truth-conditions of a sentence are its agreement and disagreement with the truth-possibilities of elementary propositions. What a sentence expresses is its truth-conditions. Hence to understand a

---

54. A symptom of the vacuity of the concept of a truth-condition is that it becomes difficult, if not impossible, to conceive of any theory of meaning whatever that does not explain meaning in terms of truth-conditions (cf. Dummett, *Truth and Other Enigmas,* p. xxii).

sentence is to know what its truth-conditions are, i.e. how to determine its truth-value in any possible world (state description). The sense of any significant sentence can be specified as a truth-function of elementary propositions.[55]

This conception has dramatic consequences. It implies that there is a unique ultimate analysis of the sense of any sentence, a single representation that simultaneously makes perspicuous all of its logical relations with every other sentence (once analysed). The truth-conditions correlated with a sentence having a definite sense do not depend on facts in any way; hence the sense of a sentence cannot depend on the truth of any other sentences. Truth or falsity, however, does typically depend on the facts; the sense of a sentence does not typically determine its truth-value as a function determines its value for a given argument. But there are sentences with 'degenerate' sets of truth-conditions, viz. those whose truth-conditions include all possible worlds ('tautologies') and those whose truth-conditions are empty ('contradictions'). The truth-values of tautologies and contradictions are determined by their senses alone. Every tautology has the same sense as every other one, and its truth is consequent on the meanings of its constituents (since the meaning of a constituent can be seen as its contribution to the truth-conditions of sentences in which it occurs). The independence of elementary propositions guarantees that the only necessity is logical, i.e. that all necessary truths can be exhibited as truths of logic by appealing to the explanations of the logical constants (truth-functions). Finally, there is no such thing as a significant sentence which is not a truth-function of elementary propositions, and hence the appearance that there are non-extensional occurrences of propositions must be explained away as an illusion.[56]

The *Tractatus* employs the notion of truth-conditions to develop a semantic or model theoretic conception of the validity of arguments. Validity can be strictly defined in terms of truth-conditions. In the simplest case, the proposition that p follows from the proposition that q provided only that every possible world verifying q is also a possible world verifying p. This makes clear that the validity of an argument is independent of the truth of its premises. Wittgenstein thought that an extension of truth-tables gave a decision procedure for logical truth (and hence too for validity). He drew some obvious consequences. Logical truths are 'flat'; none are more primitive than any others. Hence a system of logical truths is not axiomatic. There are no axioms, indeed nothing needing to be certified by appeals to self-evidence. The role of explanations of logical constants is not to foster insight into the 'primitive truths' of logic, but rather to provide a rigorous means for distinguishing tautologies and contradictions from sentences with sense. A 'derivation' of a logical truth from others is merely an expedient for recognizing a tautology in a complicated case—an expedient for ascertaining something capable of direct verification via truth-tables. In this case there is the sharpest possible

---

55. Wittgenstein, *Tractatus Logico-Philosophicus*, 4.02ff., 4.2ff., 4.5ff., 5ff.; cf. Carnap, *Meaning and Necessity*.
56. Wittgenstein, *Tractatus Logico-Philosophicus*, 3.23ff., 4.05ff., 4.46ff., 5.54ff.

separation of derivability from validity, and the truth-table decision procedure allows this distinction to be drawn sharply for every argument.[57]

This whole conception of truth-conditions is far from trivial. It commanded great interest when published by Wittgenstein. But it is far from any conception identifiable in Frege's work, or even coherently ascribable to him. This is obvious. On almost every major point he demonstrably held a view antithetical to the one presented in the *Tractatus*. In particular, he did not hold that all necessary truths are identical in sense, that all necessity is logical (i.e. that all necessary truths are analytic), that all operators on sentences are truth-functional, that a logical system could be anything other than axiomatic, or that the correctness of an argument can be separated off from the truth of its premises. To the extent that the position of the *Tractatus* is treated as definitive of the concept of a truth-condition, it is evident that Frege did *not* consider the sense of a sentence to be its truth-conditions. The explanation for this fact is plain. What generates the dramatic and unprecedented conception in the *Tractatus* is the pair of ideas that relations between thoughts or assertions turn on relations among (the meanings of) *sentences* and that there is a *single set of elementary propositions* of which every significant sentence is a truth-function. Neither of these crucial ideas surfaced in Frege's thinking.

The search for a philosophical clarification of 'truth-conditions' turns up a dilemma. There is a tolerably definite explanation, for which the *locus classicus* is Wittgenstein's *Tractatus*. But the concept there explained manifestly does not fit Frege's conception of sense, even if many commentators mistakenly ascribe it to him. Abandoning the *terra cognita* of the *Tractatus*, we are left on the uncharted deep with nothing more useful for navigation than a compass according to which every direction is north; for by the phrase 'what would be true if it were the case that . . .' we are told nothing whatever. Consequently, it seems, the contention that Frege conceived of the sense of a sentence as its truth-conditions is either false or vacuous.

With this second wave of objections, the prosecution in the case of the Estate of Frege v. the Wisdom of the Twentieth Century can rest its case.

3. Understanding and thought-building-blocks

A defence of the wide-spread claim that Frege's theory of sense is a semantic analysis of declarative sentences might stress his appeal to senses of sentences to establish the possibility of mutual understanding and successful communication. The sense of a sentence is what is asserted in making an assertion, and hence it is also what is grasped by somebody who understands what is asserted. How could successful communication occur without a shared object of understanding? On Frege's view the theory of sense constitutes an explanation of understanding and communication. But, as we well know, a theory of understanding is the same thing

---

57. Ibid. 5.1ff., 6.1ff., 6.37ff.

as a theory of meaning. Does it not follow immediately that the theory of sense is a semantic theory, covering at least those sentences that express thoughts?

This *prima facie* case can be greatly strengthened by a further observation. The fundamental problem for a semantic theory is thought to be the 'creativity of language'. There are indefinitely many meaningful sentences that can be formulated in a natural language such as English. A criterion of adequacy for a theory of meaning is that it specify the meaning of *every* meaningful sentence. It must account for the possibility of understanding a well-formed sentence never previously encountered. The only apparent explanation is to relate the meaning of a sentence in a systematic way to its grammatical structure and the meanings of its constituents. The essential task of a semantic theory is to execute this programme. Against this background, it is striking that Frege actually framed the question of how it is possible to understand novel sentences; and it is even more noteworthy that he appealed to the notion of sense to provide a solution by arguing that the sense of a sentence is uniformly built up out of the senses of its constituent expressions in accord with its structure.

> It is remarkable what language can achieve. With a few sounds and combinations of sounds it is capable of expressing a huge number of thoughts, and, in particular, thoughts which have not hitherto been grasped or expressed by any man. How can it achieve so much? By virtue of the fact that thoughts have parts out of which they are built up. And these parts, these building blocks, correspond to groups of sounds, out of which the sentence expressing the thought is built up, so that the construction of the sentence out of parts of a sentence corresponds to the construction of a thought out of parts of a thought. And as we take a thought to be the sense of a sentence, so we may call a part of a thought the sense of that part of the sentence which corresponds to it. (PW, 225)

On Frege's view, the existence of thought-building-blocks is necessary to explain understanding: 'The possibility of our understanding sentences which we have never heard before rests evidently on this, that we construct the sense of a sentence out of parts that correspond to the words' (PMC, 79). This idea also explains how a language has the power to express an unsurveyable profusion of thoughts. 'What makes this possible is that a thought has parts out of which it is constructed and that these parts correspond to parts of sentences by which they are expressed' (PW, 243). These passages seem to demonstrate conclusively that Frege was aware of the fundamental problem of semantics and that he outlined a generative account as a solution to this problem—a solution the pattern of which, though perhaps not the details, meets the requirements for a proper semantic theory. What could be stronger evidence than this that the theory of sense was intended to be a theory of meaning?

Before appending 'Q.E.D.' to this argument, there are four preliminary reasons for doubting its conclusiveness. First, all four passages indicating a generative theory of understanding are late in date. The first three were written in 1914, and the fourth probably not before 1922. Even if Frege later used his theory of sense to account for the ability to understand sentences, it cannot be inferred that this was his primary purpose, or even a subsidiary one, when he first introduced the notion of sense in 1891. Second, the four passages are not independent of each other. The first three are closely connected in purpose and in detailed content, while the fourth appears directly to echo the first. It appears that he suddenly discovered the idea, repeated it for a brief period as part of a set argument, and then rediscovered it in redrafting some earlier material for publication. Third, the generative theory of understanding plays a subordinate role in all four arguments where it occurs. In the first three it supports the thesis that a thought has parts corresponding to proper names in a sentence expressing it; i.e. that a proper name must always have a sense as well as a reference. In the fourth, it supports the claim that a molecular sentence expresses a thought compounded out of thoughts. In none of these passages did Frege present the generative theory of understanding as a major plank in his theory of sense; rather, he offered it as further support for theses long familiar from his expository writings. Fourth, his interest in the problem of understanding new sentences postdates extensive discussions with Wittgenstein, for whom this problem was a primary concern. Wittgenstein was impressed by the fact that 'it belongs to the essence of a sentence that it should be able to communicate a *new* sense to us'.[58] The explanation for this fact he located in the observations 'A sentence must use old expressions to communicate a new sense. . . . It is understood by anybody who understands its constituents'.[59] He then fleshed out this schematic proposal with the picture theory of meaning.[60] It is possible that Frege, under the influence of discussions with Wittgenstein about the theory of symbolism, suddenly came to see the issue of understanding new sentences as something philosophically important, that he discovered among his reflections on sense the outline of a general solution to the problem, and that he then used this fact to support some of the more contentious details of his theory of sense. He may have tried to graft onto his own root-stock what he perceived to be a prize-worthy blossom growing on an otherwise noxious plant. The clear implication of these four observations is that we should carefully examine how central the generative theory of understanding is to Frege's theory of sense.

His proposal to appeal to senses of expressions in order to explain the understanding of sentences rests on the contention that the structure of a declarative sentence is isomorphic with the structure of the thought which it expresses. This idea was an important ingredient of the theory of sense from the very outset. Frege held a thought to be essentially complex, i.e. to consist of parts. This seemed to him a prerequisite of the possibility of objective relations among thoughts just as other logicians had considered the possibility of logic as a science of judgments to presuppose the complexity of judgments. A sentence too obviously consists of parts. In general, the sense of part of a sentence is itself part of the sense of this

---

58. Wittgenstein, *Tractatus Logico-Philosophicus,* 4.027; this notorious feature had been earlier noted by von Humboldt and others.
59. Ibid. 4.03, 4.024
60. Ibid. 4.03ff.

sentence. This thesis does not assert isomorphism. Yet Frege obviously inclined toward the position that 'the analysis of the sentence corresponds to an analysis of the thought ... [;] I should like to call this a primitive logical fact' (PMC, 142). Moreover, he advanced specific claims of isomorphism. The saturated components of sentences (names) correspond to saturated thought-components, while unsaturated sentence-components (predicates or concept-words) correspond to unsaturated elements of thoughts. To every proper name in a declarative sentence there must correspond a sense which is part of the thought expressed by the sentence. The fact that an atomic sentence appears as part of a molecular sentence (e.g. a disjunction) mirrors the fact that the thought expressed by it is part of the compound thought expressed by the molecular sentence, for only on this supposition is it obvious that a thought may be expressed outside fiction without being asserted (cf. PW, 198). Finally, just as an unsaturated part of a sentence is uniquely determined by subtracting from the sentence some saturated expressions, so too an unsaturated part of a thought is apparently determined uniquely by subtracting from the sense of a sentence the senses of some proper names (PW, 192). All of these aspects of isomorphism are prominent and often reiterated in Frege's theory of sense. He gravitated towards the idea that a declarative sentence is a model of the thought expressed.

The idea of isomorphism of sentences and thoughts, however, bears two important qualifications. The first is that 'one and the same thought can be split up in different ways and so can be seen as put together out of parts in different ways' (PW, 202). At the very least, this requires that any isomorphism be complex; i.e. to each way of articulating a sentence into component expressions there must be a corresponding way of articulating the thought into thought-components. The second qualification is a general caveat that caution must be exercised in drawing inferences from the nature of language to the nature of thoughts. 'We should not overlook the deep gulf that yet separates the level of language from that of the thought, and which imposes certain limits on the mutual correspondence of the two levels' (PW, 259).

Up to this point there is nothing startling in the general contours of Frege's thinking. It was a commonplace of traditional logic that judgments are complex entities, that they are constructed out of independently apprehended constituents (concepts), that the subject/predicate structure of a sentence roughly mirrors the decomposition of the judgment into two terms, and that the concatenation of words into complex subjects or predicates roughly reflects the composition of complex concepts out of characteristic marks *(Merkmale)*. In this respect, only his detailed tactics, not the broad strategy, differentiates Frege's part/whole analysis of thoughts from the familiar synthetic conception of judgments. On the other hand, his entire strategy of logical analysis turns on function/argument decomposition of thoughts (and truth-values), and this framework does not consort well with treating thoughts as complex entities built up out of parts. There are deep tensions. In particular, thoughts must have function/argument structure if they are to match the 'decomposition' of truth-values into concepts and objects, but typically neither a function nor its argument are parts of the value of this function for this argument. Moreover, the doctrine of the primacy of the thought demands that logical analysis start out from thoughts to arrive at thought-components, not vice versa, and it is linked with the legitimacy of radically dissimilar analyses of a single thought—an idea difficult to marry with the construction of a thought out of fixed parts. The theory of sense from its inception had an in-built instability (supra, p. 322ff.).

The emergence of the notion of a strict isomorphism between sentences and thoughts brings the stresses in the theory of sense right to the surface of Frege's thinking. The idea of thought-building-blocks is explicitly tied to the analysis of thoughts into parts; it is invoked to prove that, since the reference of a proper name cannot be *part* of a thought, there must be something else (the sense of a name) which is part of a thought. But it is also associated with analysis of thoughts into functions and arguments. Indeed, despite their manifest inconsistency, isomorphism requires both part/whole decomposition and function/argument decomposition of thoughts: the unsaturated *parts* of a thought are clearly *functions*. Although he acknowledged that neither a concept nor an object taken as its argument is part of a truth-value (PW, 255), he failed to draw the parallel conclusion about the functions and arguments resulting from decomposition of thoughts. Despite their obvious inconsistency, he also conjoined the thesis of the primacy of the thought over thought-components with the idea that a thought is constructed out of independently given building-blocks, and he failed to reconcile the demand for unfettered functional-abstraction in his analysis of generality with the conception that a thought has a definite constitution out of given components. Finally, the doctrine of isomorphism of sentences with thoughts assigns to *sentences* a definite role in the logical analysis of thoughts. In particular, it would impose restrictions on the possibility of representing a thought as the value of a function for an argument. This idea is unprecedented in Frege's thinking. Although the contextual dictum in the *Foundations* appears to assign to sentences a primary role in the doctrine of content, it really was meant to formulate the thesis of the primacy of judgeable-content over unjudgeable-content. Not only is the idea unprecedented, but also it would have seemed absurd to Frege. Thoughts are abstract objects existing independently of language and human cognition. That a given sentence expresses a thought can no more limit the range of possibilities of its representation as the value of a function than can the fact that

'$x^2 \rceil_2$ ' designates the number 4 limit the range of possibilities for representing 4

as the value of functions for different numbers of arguments of different levels. The more seriously we take Frege's suggestions of isomorphism between sentences and thoughts, the more incoherent we must judge his conception of sense to be.

To construct a theory of strict sentence/thought isomorphism would make two basic demands. First, it would require that there be a unique ultimate analysis of each thought. From this complete decomposition various partial analyses could be derived by abstracting from part of the thought-structure; and perhaps it might be allowed that functional abstraction might generate further analyses from the basic one by authorizing ascent to unsaturated thought-components at higher levels in the hierarchy of sense-functions. But every admissible analysis of a thought must be systematically derivable from its ultimate decomposition. The second presupposition must be that no two sentences with incompatible structures can express the same thought. Only if the grammatical structure assigned to each can be related to a single underlying complete grammatical analysis does it make sense to assert that they express the same thought.[61] Both those requirements are wholly alien to Frege's thinking. The first is antithetical to the fundamental thrust of function/argument analysis of thoughts. The range of possibilities for presenting any entity as the value of a function for an argument is always wide open; any level of function and functions of any number of arguments may be used. There could be no rationale for a restriction that would not threaten to block the way to the functional abstraction required to support his analysis of generality. The second requirement is equally problematic in the context of Frege's theory of sense. He maintained that pairs of sentences with radically different structures may express a single thought, even in concept-script. The generalization expressed by '$\vdash$———— $\Phi(x)$' can be decomposed into the sense of a second-level concept and the sense of a first-level concept (corresponding to the formula '$\vdash$——$\overset{a}{\smile}$—— $\Phi(\mathfrak{a})$'), but on pain of contradiction with his doctrine that free variables only indicate objects, this decomposition of the thought cannot correspond to thought-components expressed by the sentence-constituents in '$\vdash$———— $\Phi(x)$'. Similar radical grammatical asymmetries typify pairs of sentences in natural language which express the same thought, e.g. 'The concept *man* is realized' and 'There are men' (cf. CO, 49) or 'Two is a prime' and 'It is true that two is a prime'. Since neither of the basic requirements of strict sentence/thought isomorphism is fulfilled according to Frege's account of sense, it seems best to discount the importance of his building on the theory of sense a generative explanation of understanding. This has the appearance of being a late brainwave, an argument that seemed to support one of his entrenched theses and to provide a fresh *raison d'être* for the notion of sense after the collapse of his logicism. It thus appealed to him in spite of its not squaring with the most fundamental aspects of his conception of sense. The only other options would be to convict him of blatant inconsistency with the ideas informing the *Basic Laws* or to interpret him as having repudiated without notice this whole framework of thought.

This discussion might raise a doubt whether the very idea of a systematic theory of understanding based on the composition of thoughts out of the senses of sentence-constituents is not an artifact of our manufacture, not Frege's. Had twentieth-century philosophy and linguistics not made so much of the 'creativity of language', would anybody have considered the few scattered passages in his writings about thought-building-blocks to be of any real significance for interpreting his theory of sense? Did he mean to construct a *theory* of understanding? Perhaps his intentions were far more modest. After all, he glossed his thesis that a sentence is 'a mapping of a thought' with the comment: 'corresponding to the whole-part

---

61. The transformation of actives into passives would be the simplest paradigm for the identity of underlying grammatical structure beneath superficial difference.

relation of a thought and its parts we have, *by and large,* the same relations for the sentence and its parts' (PW, 255; our italics).[62] Similarly, to the thesis that we can understand novel sentences because sentences serve as images of the structures of thoughts, he added: 'To be sure, we really talk *figuratively* when we transfer the relation of whole and part to thoughts' (CT, 537; our italics). These cannot be construed as insignificant qualifications to a rigorous theory of understanding and a strictly compositional conception of the sense of a sentence. They completely subvert the claim to have constructed such a theory. The first sabotages the thesis of universal isomorphism, while the second undermines the idea that a thought is literally compounded out of thought-building-blocks. The residue seems little more than a reiteration of Frege's assertion that there is a good, though imperfect, correspondence between the articulations of sentences in natural languages (especially German!) and the structure of the corresponding formulae in concept-script (cf. CO, 45). He might merely have stated that, when translating English or German into concept-script, we do not typically proceed by translating each sentence *en bloc* into a formula, but by translating some of its constituent words or phrases into symbols standing for functions and arguments. Had he done so, would we discern in his theory of sense the least hint of a theory of understanding?

4. Mistaking a reflection for a reality

It seems incredible that anybody could reflect seriously on Frege's writings and then propose that his works outline the basic ingredients of truth-conditional semantics. His leading ideas and fundamental preconceptions are manifestly discordant with salient features of the standard accounts of truth-conditional semantics. How then is it possible that a chorus of modern philosophers should hold this radically distorted conception of Frege's thought, and use it as the basis for their detailed interpretations of his writings? Have we not stumbled here upon a hitherto unknown form of madness? The explanation is simple. Philosophers approach the study of his thought with a definite expectation or preconception about its essential nature. They already know that he was a sophisticated logician and that his thought was praised by the main early expositors of truth-conditional semantics (Wittgenstein and Carnap). Hence it is taken for granted that he must have held the modern semantic conception of logical validity. The only scope for investigation seems to be the details of how he elaborated this idea. Philosophers became the dupes of their own preconceptions like the spectators at a magician's show. The first step in Frege's performance, though executed with a fanfare in plain view of his readers, is the one that altogether escapes notice.[63]

The fundamental and explicitly avowed inspiration of his logical system was the generalization of function theory and its consequent application to the study

---

62. He never retracted the emphatic claim that the sense expressed by a sentence may be much more complex than is suggested by the sentence itself (SR, 76ff.).
63. Cf. Wittgenstein, *Philosophical Investigations,* §308.

of the laws of inference. *Begriffsschrift* introduced judgeable-contents, the fundamental 'atoms' of inference, as the values of functions for arguments. It introduced concepts as functions taking judgeable-contents as values and analysed judgeable-contents as decomposing into function and argument, concept and object. The logical connectives of the propositional calculus were viewed as functions from judgeable-contents to judgeable-contents. The result of these mysterious manoeuvres was a powerful form of representation for the presentation of valid inference, in particular in mathematical reasoning. The formal system was clearly superior to any hitherto discovered. This alone committed Frege to a particular way of looking at matters. So powerful an instrument must have been well-forged, otherwise how could it be so successful? It must give us the key to the nature of what it so triumphantly analyses, even though we may not yet fully understand how or why! And so the 'decisive movement in the conjuring trick has been made, and it was the very one that we thought quite innocent'.[64] Success ensured that function/argument analysis was firmly ensconced in the hearts and minds of post-Fregean logicians.

Mesmerized by his virtuosity, we are induced by Frege to avert our eyes from the price that had to be paid for this new tool. We do not notice his systematic violation of fundamental norms of function theory, e.g. the impossibility of independently identifying the values of these new kinds of functions for their arguments, the representation of both function and argument as constituents of the value of that function for that argument. We do not baulk at the obscurity of the fundamental notions, e.g. of a concept viewed as a function. We disregard the abuse of language involved in presenting elements of the theory, e.g. treating sentences as names and allowing them to flank the identity sign. The reason is perhaps a kind of mathematical pragmatism. In the face of such formal success, it might seem unobjectionable to us to leave the nature of these new, or newly revealed, entities undecided. 'Sometime perhaps we shall know more about them—we think'.[65] As long as an open-minded attitude is adopted towards their true nature it is surely quite innocent to introduce novel notions of functions, concepts, objects and judgeable-contents into logic.

The *Foundations* analyzed natural numbers as the extensions of certain concepts. Marrying this development to the formal system of *Begriffsschrift* did indeed require a thorough reappraisal and revised explanation of the fundamental notions. This seems to demonstrate the virtues of an undogmatic pragmatic strategy. For the further development of the theory apparently made clear the true nature of the basic notions involved in it. The *Basic Laws* was the report on these new discoveries. The True and the False emerged as previously unrecognized objects which are really the values of concepts. Contrary to the immature theory, there are no objects which serve simultaneously as the object of assertion and the

64. Ibid.
65. Ibid.

value of a concept for an argument. Other innovations and emendations followed in the wake of these, but the overall development appeared to vindicate the first insights into the function-theoretic forms and norms of judgment and valid inference.

Our sustained and willing suspension of disbelief seems amazing. Concepts are not functions and concept-words (predicates) are not function-names. Applicatives are not names of second-level functions and sentences are not names of the values of functions for arguments. Logical connectives are not function-names and truth and falsehood are not logical objects. It may seem incredible to assert this. For if it were true, would it not be altogether extraordinary that Frege's formal logic 'works', that it reveals the formal logic of generality, that it is *the* method for representing patterns of reasoning? Is this alone not sufficient to demonstrate that he was right, in broad outline, if not in detail? Does the success of his logical system not show that the forms of thought and language alike have, *au fond,* a function-theoretic structure?

So to think is to think confusedly. The forms and norms of function-theoretic logic are not reflections of the structure of reality, as Frege thought. Nor are they revelations of the hidden structure of natural language, let alone of the real forms of human thought as modern philosophers are inclined to think. Frege made no new discoveries about the hitherto hidden nature of reality, thought or language. Rather *he invented a new form of representation,* and projected his function-theoretic forms onto the nature of what they represented. He took the norms of his function-theoretic notation as reflections of what he represented by its means. Nor was this unnatural, since whatever can be represented in this form of representation must first be *transformed* into the norms of representation that govern its structure.

Give or take a few paradoxes, which Frege held merely to reveal the defectiveness of natural language, most sentences of natural language seem amenable to translation into concept-script. And given a certain self-denial in employment of higher-level functions, most formulae of concept-script can be found a rough equivalent in natural language. This, coupled with the fact that inferences *are* (sometimes opaquely) presented in natural language may lead us, as it led Frege, to think that language itself has a crude function-theoretic structure, i.e. that predicates (properly parsed) are function-names, singular referring expressions are names of arguments, and sentences are names of values of functions for arguments. For were this not so, how could language do defectively what concept-script does impeccably? In the belief that concept-script reflects accurately the true forms of thoughts, Frege proceeded to measure natural language against the yardstick of concept-script in the belief that he was holding it up against the nature of things. Not surprisingly he found it different, and saw this difference as a demonstration of the logical defects of language. We too follow this strategy, distinguishing where necessary the grammatical form of a declarative sentence from the logical form of the statement made by its utterance. In a sapient remark which has outraged some contemporary philosophical logicians Wittgenstein pointed out that

> 'Mathematical logic' has completely deformed the thinking of mathematicians and of philosophers, by setting up a superficial interpretation of the forms of our everyday language as an analysis of the structures of facts. Of course in this it has only continued to build on the Aristotelian logic.[66]

This observation merits careful reflection.

Frege's concept-script is simply an alternative form of representation to natural language. It allows us to present certain inferences in a manner more readily surveyable and more mechanically checkable than their normal representation in ordinary language. But it does this only at a price, for it can do so only if we impose the relevant function-theoretic forms on the arguments we wish to represent. We are all trained to do so, and by and large find it unproblematic. Yet important features of our practice should be noted.

(i) There are some kinds of sentences which cannot readily be mapped into the requisite forms of the predicate calculus, e.g. sentences with dummy subject terms, such as 'It is raining' and 'It is time to go'.

(ii) There are certain kinds of linguistic forms which have no direct correlate in a function-theoretic calculus. Many verbs are paradigmatic examples, since their multiple polyadicity violates the requirement that functions have a unique number of arguments. Consequently trivial forms of inference dependent upon this feature can only be represented in the calculus via drastic transformation involving quantification over actions and events.

(iii) Familiar forms of inference cannot be captured at all by the standard apparatus of the logical calculus, e.g. relations of exclusion between determinates of determinables. Conversely, the (modern) calculus generates forms of valid inference which no one (including Frege) would antecedently have admitted as inferences at all, e.g. '$p \vdash p$', '$p \;\&\sim p \vdash q$'.

(iv) Concepts which differ profoundly from each other are given brisk Procrustean treatment in order to fit them into the moulds of the predicate calculus. Sortal nouns, mass nouns, predicative adjectives, attributive adjectives are treated alike as 'predicates' and presented uniformly as properties of an object. We thus impose a uniform logico-syntactical form upon concepts (expressions) which differ *toto mundo* from each other.

(v) Although the representation of generality by means of second-level functions is a powerful tool for the analysis of complex arguments involving multiple generality, its achievements are purchased at the cost of gross distortions. First, the existential quantifier is a form into which we squeeze such logically diverse expressions as 'there is', 'something is', 'exists', 'a so-and-so', 'some', 'one so-and-so', etc. Second, we do not distinguish, in our representation of generality by means of the universal quantifier, between logically distinct kinds of generality, e.g. 'all primary colours', 'all students at St. John's', 'all men', 'all natural numbers' are presented in the same form. Far from highlighting logical or conceptual

---

66. Wittgenstein, *Remarks on the Foundations of Mathematics*, p. 300. Dummett characterizes this remark as 'plainly silly' ('Wittgenstein's Philosophy of Mathematics', p. 171, in *Truth and Other Enigmas*).

differences, we obliterate them. Third, by representing generality in the form of second-level functions and concepts in the form of first-level functions we both distort language and thought and also invite nonsense. The statement 'Something moved' invites the sensible question 'What is it that moved?'. But translating 'A person moved' into the form '$(\exists x)(Px \ \& \ Mx)$' (viz. 'There is an x(something) such that x(it) is a person and x(it) moved') not only misleadingly represents being a person as a property of something, but it also invites the senseless question 'What is the thing that is a person?'[67]

(vi) We represent the expressions 'and', 'not', 'or', 'if . . . then . . .', etc. as truth-functional sentential operators, although they all have many familiar uses in natural language in which they do not fulfil this regimented role.

Many more examples could be produced, and the ones just listed could be elaborated. They are not presented here as novelties, merely as reminders. We are not suggesting that modern function-theoretic logic should be abandoned, or that quantification theory is an illegitimate form for presentation of certain kinds of inference. Nor are we implying that further refinements of the calculus may not enable it to give a less distorted presentation of aspects of language, thought and inference which are, in its present state, mangled. The point is to draw attention to the nature of the formal calculus devised by Frege and developed by his successors, and to make clear its limited philosophical significance.

Frege invented an ingenious formal calculus by generalizing and developing the abstract forms of the mathematical theory of functions. For certain restricted purposes it proved far more flexible and powerful than the syllogistic which it replaced. But like syllogistic (and indeed any future formal calculus that might be devised) it has no special metaphysical justification—for there is no such thing. It is just another form of representation devised for special purposes[68] in the study of valid inference. And like syllogistic[69] it requires that we project the indefinitely variegated forms of expressing our ordinary concepts into the limited, regimented forms of the calculus. Where syllogistic required the transformation of all premises of an argument into the forms of the theory of classes, quantification theory demands a similar transformation into the forms of function theory. This can

---

67. Cf. Wittgenstein, *Lectures on the Foundations of Mathematics, Cambridge 1939*, ed. C. Diamond, pp. 167, 263f., 268 (Harvester Press, Sussex, 1976), and *Philosophical Grammar*, pp. 202ff., 265ff.
68. Although it is worthwhile asking oneself what exactly these are, and discovering how limited their philosophical interest is!
69. Wittgenstein's remark that modern mathematical logic has, in its superficial interpretations of the forms of natural language, 'only continued to build on the Aristotelian logic' finds at least part of its rationale here. So too his comment that Frege's notions of concept and object are 'the same as subject and predicate' (*Philosophical Grammar*, p. 205) must not be read (ludicrously) as a jejune manifestation of ignorance of Frege's thought, but as a penetrating criticism. Frege's categories of concept and object are the same as the traditional categories of subject and predicate in so far as they serve 'as a projection of countless different logical forms'. Furthermore the function/argument (concept/object) notation represents (atomic) judgments uniformly as attributions of predicates to undifferentiated (featureless) subjects. Hence it can rightly be said to be a 'sublimation' of the subject/predicate form (*Philosophical Grammar*, p. 265).

always be done, *somehow or other*. But it can only be done, in the case of function-theoretic logic *as* in the case of syllogistic, by employing a multitude of *different methods of projection* in order to transform the many heterogeneous forms of expression in natural language, whose uses constitute the logical structure of our concepts, into the few function-theoretic forms of quantification theory. If we then look to quantification theory to discover the 'real logical forms' of our thought, we mistake a deliberately distorted reflection for the object reflected. Wittgenstein illuminates this idea by a powerful analogy:

> Suppose we were set the task of projecting figures of various shapes on a given plane I into a plane II. We could then fix a method of projection (say orthogonal projection) and carry out the mapping in accordance with it. We could also easily make inferences from the representations on plane II about the figures on plane I. But we could also adopt another procedure: we might decide that the representations in the second plane should all be circles, no matter what the copied figures in the first plane might be. . . . That is, different figures on I are mapped onto II by different methods of projection. In order in this case to construe the circles in II as representations of the figures in I, I shall have to give the method of projection for each circle; the mere fact that a figure in I is represented as a circle in II by itself tells us nothing about the shape of the figure copied. That an image in II is a circle is just the established norm of our mapping.[70]

The warning is not idle, since the error is widespread, and gross philosophical confusions stem from it. It is not the circles on the plane of projection, the forms of our logical calculus, which will tell us anything about what is projected (whether it be conceived of as the constituents of reality or of the 'depth of grammar' of language) but the diverse methods of projection. To reveal these we must clarify for ourselves the structures of the concepts, *manifest in the different uses of the expressions expressing them,* which we project into the predicate calculus. Ludicrous consequences ensue if this is not understood. Philosophers seek to *explain,* and *make clearer* the inference from 'A shut the door with a bang' to 'A shut the door' by translating it into the form 'There is an event x such that x was a shutting of the door, x was done by A, and x occurred with a bang' and invoking conjunction elimination. This is meant 'to help elicit in a perspicuous and general form the understanding of logical grammar we all have that constitutes (part of) our grasp of our native tongue'.[71] But this 'explanation' does no more than cast clouds of obscurity over the blindingly obvious. There is no unclarity about such kinds of inference and no problem about them which the predicate calculus could possibly solve. They are transparently valid, and it is because of this that they set a problem *for the predicate calculus,* i.e. of finding a function-theoretic form in which to present them—on the dubious assumption that it is desirable (*l'art pour l'art* as it were) to find methods of projection for as many forms possible! Even

---

70. Wittgenstein, *Philosophical Grammar,* p. 204f.
71. D. Davidson, 'The Logical Form of Action Sentences', in N. Rescher, ed., *The Logic of Decision and Action* p. 115 (Univ. of Pittsburgh Press, Pittsburgh, 1967).

greater absurdity follows if we think that what is to be found on the plane of projection provides a logico-metaphysical proof of what exists on the projected plane. Thus translation of action- or event-sentences into the forms of the predicate calculus is sometimes thought to be a 'proof' that events really exist (sic!), or a demonstration that we need to 'postulate' them (as if they would not go on occurring (not 'existing'!) if we did not 'postulate' them).

A formal calculus is of very limited philosophical significance and has limited philosophical use. Certainly there are arguments the inferential structure of which is difficult to survey, and a well-designed calculus may make clear something difficult to grasp in the structure of natural language. This is one merit of an alternative notation (comparable perhaps to the computational advantages available in mathematics from switching from one notation to another). Furthermore, the forms of the Fregean (or Russellian) function-theoretic calculus, just because they involve different norms of representation, provide a useful, and occasionally illuminating, contrast to natural language. Where philosophical puzzles generated by our ordinary forms of representation hold us in a vice, it is sometimes helpful to contrast these with an artificial alternative form. Although the predicate calculus does not 'prove' that existence is a second-level concept (and so 'demonstrate' a fallacy in the Ontological Argument for the existence of God), it may help to free the log-jam generated by the deceptive superficial similarity between the grammatical predicate 'exists' and other grammatical predicates. Of course, the price we pay for this potential illumination is high, for we all too often delude ourselves into thinking that we have solved a philosophical problem by such a translation, whereas all we have done is transpose it onto another plane.

Beyond these humble, but not insignificant, achievements there is but meagre philosophical harvest to be reaped, although a multitude of fresh pests to eradicate. At any rate, one may wisely immunize oneself with a strong dose of scepticism against deep metaphysical, ontological or psycho-linguistic truths which are alleged to flow from the invention of a formal calculus.

> For we can avoid ineptness or emptiness in our assertions only by presenting the model as what it is, as an object of comparison—as, so to speak, a measuring-rod; not as a preconceived idea to which reality *must* correspond. (The dogmatism into which we fall so easily in doing philosophy.)[72]

A double dogmatism is now born of quantification theory. On the one hand there is a stubborn determination to mould all sound inference to this pattern, and to find here the essence of language, thought and reality. On the other hand there is a refusal to countenance any alternative philosophical corona around this central sun. This, in particular, debars one from recognizing that different metaphysical and psycholinguistic theses accompanied Frege's vision of his similar theory. Hence one fails to see that the surrounding philosophical *gas* (ours as well as his) is optional, disputable, and typically disreputable.

---

72. Wittgenstein, *Philosophical Investigations*, §131.

5. In the end lies a fresh beginning

The desire and need to impose an order upon the past is not merely a matter of idle curiosity; it is also a quest for self-knowledge. A clearer representation of the past helps us to gain a sharper view of the present. When we study the remote past we strive to understand not only actual episodes, but also the *possibilities* for thought and action which confronted people in previous times. The pictures the historian paints provide a measure, and a contrast, against which to comprehend possibilities currently available as well as the reasons and causes of present impossibilities. They give us a firmer grasp of the historically conditioned nature of what is humanly possible, not merely with respect to forms of social and economic organization or structures of political power, but also with respect to the goals, values or forms of life which inform these, and which confer meaning on human lives.

In the case of intellectual history, the importance of grasping the nature of the limits of what is possible for thought is paramount. Without an understanding of the range of questions available in a culture with respect to a given problematic subject, and an insight into the nature of the limitations conditioning the questions that could intelligibly be raised, we cannot hope to comprehend the appeal of the answers that were offered. Equally, without understanding the limits restricting the ranges of possible *intelligible* answers to these questions in the past, we will fail to grasp correctly the very questions themselves. Although the accessibility of 'possible worlds' may be a philosophers' fantasy, the accessibility of ideas, the limits of historically possible thoughts, is not.

When the object of investigation in the history of thought is not temporally and culturally remote, finding a correct perspective is learning something new about current ideas, and this not merely by way of contrast and heightened intellectual self-consciousness. To know how our ideas evolved is to see them afresh, in a different light. For we can then hope to see how forms of thought which are 'natural' to us become possible. Genetic analysis is not conceptual analysis. But knowledge of the origins of our ideas gives us a new perspective; above all, it shakes the compelling power, the mesmerizing character, of our intellectual paradigms. This encourages us to question what we normally take for granted, to examine critically what typically goes unnoticed because it is so pervasive. The merit of such questioning is not its destructive force. Paradigms of conception, explanation, and understanding are not such frail creatures as to crumble immediately under any critical examination, although they may do so if the circumstances are ripe. Its merit is rather that it enables us to see these paradigms from the correct logical point of view, to see them for what they are, not adamantine, sempiternal structures reflecting the transcendent nature of things, but our own historically conditioned intellectual constructions in terms of which we strive to make the world and ourselves intelligible to ourselves.

The importance of this for the history of philosophy is obvious. For although philosophy is not an explanatory science, being rather a description of our forms of representation, philosophers struggle with the gossamer webs of their contentious subject-matter equipped with a no less historically conditioned framework of concepts and ideals of understanding than those of others. That historical consciousness is essential for history of philosophy is particularly evident from brief reflection on the extent to which the power of a now rejected form of philosophical thought is difficult to recapture. The New Way of Ideas, originating with Descartes and enthusiastically embraced by Locke and his successors, was apprehended as a great liberating innovation, opening up fresh vistas for the analysis of thought and experience. It dominated European philosophy for more than two centuries. Yet it is now commonly thought to be *baffling* how a so 'transparently' unsatisfactory conception could do so. After all, a first-year undergraduate can be taught how to tear yawning holes in this ill-woven fabric. That we perceive this pattern of philosophical analysis to be flimsy and threadbare is important. If it could take in Descartes and his rationalist successors, Locke and his empiricist heirs, then the fact that we do not or cannot feel its force tells us more about *ourselves* than about them. The lessons we should derive from this reflection are manifold. We will not have understood correctly the classical Way of Ideas until we cease to be baffled by its original irresistible appeal. We will not have fulfilled our task as historians of philosophy until we have made clear its compelling force in its historical context. Crucially important though this is, we must go considerably farther. It is not enough that we make clear the appeal of a given philosophical strategy and framework of thought. For if we can say of it that it is definitely misguided, rests on misconceptions and misunderstandings, it is also incumbent upon us to explain why what is so obvious to us was so totally inaccessible to our predecessors. After all, it is not because we are so much cleverer than Descartes or Locke that we can see the way through some of the mazes in which they got lost. We must make clear why the very questions they asked led them astray, how the presuppositions underlying the questions constituted a field of force deflecting their answers. We must formulate the questions which they failed to ask (in philosophy the right question is worth ever so much more than a correct answer); and we must also show what stood in the way of raising the correct questions.

The latter point already makes clear the extent to which doing history of philosophy involves doing philosophy. One cannot wholly separate the understanding of the work of a past philosopher from grasping the cogency of his argument (any more than we can understand a personal history without grasping the reasons, motives and goals of its protagonist). And one cannot wholly separate grasping the cogency of argument on a philosophical subject from the task of clarifying the very subject oneself.

Philosophers contemptuous of history of philosophy view the subject as a form of intellectual antiquarianism. Historians of ideas who believe in a form of transgenerational cultural solipsism can see no way to bring the thought of past philosophers to bear on contemporary philosophizing. Both are mistaken—in more ways than one. Although the presuppositions underlying philosophical questions, as well as the range of possible answers to them, may change over time thus affecting the content of the questions, these questions are not typically thereby rendered obsolete. The central questions of philosophy do not so much shift as shimmer. In each generation they are reflected afresh, in novel forms, in mirrors of differing degrees of concavity or convexity, containing different flaws and surface blemishes. Questions about the nature of thought and its relation to reality, about the relation of subjective experience to its objects, about the nature of self-consciousness or the relation of will to action, about doubt and certainty, truth and existence, reason and desire, constitute, if not perennial,[73] at least fairly permanent foci of philosophical concern. These questions are rooted in our thought, experience and language. Though they present themselves in different forms to successive generations, though they are viewed from different perspectives, in the context of changing presuppositions, new knowledge, varying paradigms, they constitute persistent problematic areas in the structure of thought and language of our culture, and perhaps of any human culture that engages in philosophical reflection.

Nor is it true that the answers given by past generations to such problems are of mere antiquarian interest. Philosophical criticism of past philosophical analyses is not like scientific criticism of obsolete scientific theories. Great philosophical pictures have the lasting interest and fascination which they possess in part at least because they represent clear articulations of persistent philosophical temptations to illusion. Given the structure of our discourse about the mental, the asymmetry between first- and third-person psychological utterances, the fossilized metaphors of introspection by now naturally embedded in our language, the forms and phenomenology of perceptual experience, dreaming, and self-consciousness, then the seductive powers of a Cartesian conception of the mind are likely to prove as permanent as these conceptual forms. Hence each generation must immunize itself afresh against such intellectual diseases, and may do so by studying their paradigmatic exemplification in the thought of a great philosopher who unwittingly succumbed. Although the specific forms which such intellectual ills assume vary with time, we are best equipped to discern our own confusions about the nature of our conceptual structures by scrutinizing past attempts to clarify similar, if not identical, structures.

The errors of our predecessors often find structural *analogues* in contemporary philosophical confusions. Cartesian mind/body dualism, for example, is mirrored (on a different plane) by current central-state materialist brain/body dualism. The distorted conceptions of language, thought and reality which characterized philosophical thought in relation to syllogistic, distortions consequent upon misapprehensions of the nature of a formal calculus for representation of valid inference, find powerful analogues in the no less (but differently) distorted conceptions which inform modern reflection associated with function-theoretic calculi. Consequently analysis of past error contributes to the identification of our own intellectual maladies, for the bacillus is often a mutant from a common stock.

It would, however, be altogether misleading to represent the history of philosophy as merely the study of error and illusion, as if we were (as we might fondly hope) the first generation to see glimmers of light, so that only with us philosophy came of age. It is also the study of insight and illumination. Painfully achieved

---

73. Cf. Q. Skinner, 'Meaning and Understanding in the History of Ideas', p. 50 (supra p. 10).

insights into the non-uniformity of the forms of human knowledge, e.g. distinctions between *a priori* and empirical knowledge, between scientific and historical knowledge, or between empirical knowledge and apprehension of value, are a precious legacy. Distinctions which we take for granted, e.g. between logical connections and causal connections, between questions of fact and questions of right, between genetic and conceptual investigations, are a hard-won heritage. By studying how these were achieved we can arrive at a deeper apprehension of their significance, a firmer grasp of what precisely was gained in terms of clarity of thought and potential fruitfulness. The results of the endeavours of our predecessors, even where inadequate and mistaken, often contain a vision of the truth seen through a glass darkly. They frequently captured a baffling conceptual articulation in a distorted manner, and their partial insights may both point us in the right direction, as well as warn us of associated forms of error. Nor is this legacy idle, for each such fundamental philosophical distinction throws up fresh problems with which we must grapple, and our often extended applications of these distinctions produce new confusions, requiring in turn novel clarification of conceptual boundary lines.

The study of past philosophical thought is not only a philosophical exercise in its own right. It exposes us to weakened strains of philosophical diseases which may give us partial immunity to more virulent contemporary strains. And it may provide us with fragmentary sketches of the labyrinthine forms of human thought through which we are struggling to find our way. Above all, a sharpened awareness of the once unnoticed flaws in past styles of philosophical thought should force us to examine the forms of philosophical analysis which we currently employ to discover whence comes their force and fascination, and what are their limitations.

Much of the apparatus of contemporary philosophical logic and philosophy of language is commonly attributed to Frege. One purpose of our investigation has been to conduct such logical excavations as would reveal the real historical foundations of some of those aspects of modern philosophy. Our logical 'dig' has brought much to light that has not generally been acknowledged. With this in view, we can perhaps now see Frege's work correctly in its nineteenth-century context. His purposes were not ours. He was not elaborating the depth-grammar of every possible language. He was not constructing a theory of meaning for a natural language, and he was not investigating the nature of our capacity to understand language. What was merely instrumental for him seems to us his greatest achievement, namely his complete formalization of the predicate calculus. This magnificent formal system was not invented to account for the logical structure of language; indeed, had language been constructed from such a logical blueprint (PW, 269), it would hardly have been necessary to invent this artificial language, the concept-script. Its purpose was to represent the objective, language-independent, structure of thoughts, and that in order to reveal the logical foundations of arithmetic. What was central to him, namely to set mathematics on secure foundations and to demonstrate that numbers are logical objects, is peripheral to us.

We learn little from reading Frege as if it were only *per accidens* that he is not our contemporary. So to read his works both impoverishes and distorts his thought. Moreover, instead of gaining a perspicuous representation of a *different* style of thought from ours, a strategy of logico-philosophical analysis *against* which to match ours, we obtain only a broken reflection of our own ideas. There is, as we have tried to show, a substantial legacy of typical nineteenth-century philosophical thought about logic at the foundations of Frege's system. The decisive premises he never questioned seriously would one and all be repudiated by late twentieth-century philosophers and logicians. He thought that logic is concerned with relations between objects of judgment, that objects of judgment are Platonic entities (the only alternative he entertained to this conception was psychologism). He conceived of language as externally related to thought. What grammar is to language, logic is to thought; the two are, he supposed, *toto caelo* distinct. The laws of logic he envisaged as being ultimate transcendent laws of thoughts, not empirical or even transcendental laws of human thinking. The nature of the acts or processes of judging, thinking and understanding are, in his view, as irrelevant to the study of the laws of thoughts as they are to the study of the laws of number. This misplaced anti-psychologism allowed him to take for granted, and build on, crumbling foundations.

Frege's achievement was, as one should expect, largely formal. He pursued his elaborate and sophisticated logicist thesis to its bitter end. While the bare idea of logicism had long been in the air, he was the first to give it substance and to attempt its realization. From his failure much is to be learnt. (It is also worth noting that even if it had not produced contradictions, it would not have shown what he wanted, viz. the unity of logic and arithmetic.) His formal system of logic was a triumph. Negatively, this invention put an end to age-old controversies concerning metaphysical issues lying at the foundations of the old logic (e.g. the 'reality' of relations). But, as the sparks fly upwards, it produced its own harvest of philosophical and metaphysical confusions which plague us to this day. These are likely to remain with us as long as we continue to employ a function-theoretic form of representation for analyzing inferences. Until a novel paradigm becomes entrenched, each generation will have to immunize itself afresh against the philosophical diseases that flourish in this formal culture, to fight itself free from the intellectual illusions produced by this powerful hallucinogenic and so come to see aright the nature and value of this system of logic.

Just as the shaky foundations of syllogistic did not irreparably mar the formal system itself, so too the rickety philosophical substructure of Frege's formal system leaves the formal structure intact. This is not a coincidence. Yet it would be wrong to approach his propaedeutic to his formal theory in the hope of finding a modern semantic account of validity or an explanation of the logical connectives in the standard modern form. He was not, and could not consistently have been, the inventor of truth-conditional semantics. Nor did he produce any account of logical truth acceptable to us, even though his conception was one of the two principal alternatives envisaged by most nineteenth century thinkers.

Since Frege conceived of the business of logic as the study of the relations between and modes of decomposition of Platonic entities, his concern with natural language, its structure and limits, is incidental and at best indirect. Far from connecting the notion of meaning with that of the use of language, his Platonism is simply inconsistent with the basic ideas involved in the dictum 'the meaning of a word is its use in the language', and his concepts of content and sense alike are neither identical with nor isomorphic with our notion of linguistic meaning. If his basic analytical tools, sense and reference, function and argument, were the fundamental tools for philosophy of language, this would be fortuitous. But these very instruments, taken as Frege designed them, are, we have argued, deeply flawed and have dubious application for purposes of philosophy of language. To point this out is not to confer an accolade upon modern machine-tooling in philosophy; our instruments differ from his, but they are not therefore less flawed.

Frege's conceptions were different from ours. The strengths and weaknesses of his ideas do not bear *directly* on current wisdom (nor vice-versa). But in so far as extensive excavations around the foundations of his thought revealed unsuspected defects, this strongly suggests the need for a similar critical investigation into the presuppositions of modern truth-conditional semantics. Although his ideas are only indirectly responsible for what his successors have made of them, nevertheless, at a high enough level of generality, there are important affinities which we crudely sketched and denominated 'Neo-Pythagoreanism'. Though contemporary philosophers have, in various respects, gone farther than Frege, and also psychologized his conception in ways he never envisaged and would hardly have welcomed, the kinship in *Weltanschauung* is undeniable. As long as we read him as a contemporary, and not as the nineteenth-century mathematical logician he was, we will be impressed by the affinities. But what one takes for granted in common with the philosopher one examines is unlikely itself to get examined. If we stand back and read Frege in his true light, we can become aware of the different nature and sources of his Pythagoreanism. Subjecting it to critical scrutiny, we can see its flaws and confusions. This in turn may give us fresh awareness of what we take for granted, and induce us to examine the presuppositions of our own philosophical conceptions and paradigms of understanding. Though as intangible as that insight into the past which is the most important product of archaeology, this gain in knowledge would be a truly rich reward, far outweighing any feeling of disappointment at having brought to light so little buried treasure in the course of these extensive logical excavations.

Index

(Most citations from, references to, and critical discussions of secondary literature on Frege are to be found in footnotes.)

Abbe, E., 12
Abstractionism, 7, 56, 57, 58, 60
Act/object distinction, 40, 107. *See also* Judging; Judgment, object of
Aesthetics, 9, 43
Algebra, logical, 78f., 81, 89, 109f., 123. *See also* Boolean algebra
Ambiguity, 69, 131. *See also* Type-ambiguity (of concept-script)
Analytic truth, 34, 178
Analytic philosophy, 7. *See also* Philosophy of language
Ancestral (of a relation), 13, 35
'And'/'but', 20, 336
Annahme, 84f.
Anscombe, G. E. M., 366
Anti-Cartesianism, 7
Anti-geneticism, 5f.
Anti-psychologism, 6, 8, 37, 39, 67, 88, 200, 202, 280
   about judgeable-content, 46ff.
   about the laws of logic, 41ff.
   about unjudgeable-content, 49ff.
Aristotle, 11, 15, 21, 130, 155

Assertion, 19, 36, 37, 39, 40, 50, 55, 57, 60, 77ff., 85ff., 107, 339ff., 358f., 367
   distinguished from predication, 18, 20, 79f.
   object of, 29, 104ff., 278, 343
Assertion-sign, 29, 78ff., 83ff.
   incoherence of, 96ff.
   paraphrase of, 91ff.
Assertoric force, 18, 29, 78ff., 90
Auffassungsweise, 138f., 159, 166, 168, 173, 239
Augustinian picture of language, 23, 24, 56, 61f., 172, 192, 267, 337
Austin, J. L., 7, 216, 317
Axioms, 229

Bacon, F., 25
Baker, G. P., 24, 26, 60, 199, 337
Basic Laws of Arithmetic, 3, 17, 19, 20, 103, 117ff., 230, 233ff., 240, 265ff., 272, 274, 294
Baumann, J. J., 51

399

Begriffsschrift, 3, 17ff., 21, 29, 34, 35, 50, 79, 84ff., 91ff., 102, 104, 106, 114ff., 119, 124f., 127f., 133f., 142, 145, 146, 160, 162, 171, 178, 181, 188, 205ff., 218, 220f., 230, 233ff., 240, 269f., 274
Bell, D., 341
Bennett, J., 4
Bentham, J., 58
Berkeley, G., 51, 52, 57, 58, 66, 322, 331
Berlin, I., 7, 22
Bestimmungsweise, 138f., 146ff., 151, 159, 168ff., 221, 226, 229, 280, 285
Bivalence, 6, 271
Black, M., 114
Bolzano, B., 6, 40, 42, 59, 77, 78, 326, 353
Boole, G., 12, 13, 14, 18, 22, 41, 42, 47, 73, 74, 81, 83, 85, 89, 108, 110, 117, 122, 125, 133, 185
Boolean algebra, 7, 11, 12, 13, 14. See also Algebra, logical
Bynum, T. W., 6

Calculus, formal, 390f.
Cantor, G., 12, 222
Carnap, R., 24, 49, 291, 329, 344, 357, 374, 379, 386
Cartesianism, 7, 25, 28, 59ff., 361
Categories, 143
Cauchy, A. L., 33
Central-state materialism, 19, 26
Church, A., 244, 291, 305, 312f., 329
Coleridge, S. T., 6
Colouring, 47f., 54, 278f., 336
Compositionalism, 9, 28, 338
Concepts, 5, 49, 56, 57, 58, 65, 105, 140, 143, 170ff., 183, 241, 252ff.
  as functions, 10, 11, 16, 21, 24, 153f., 170ff., 252ff., 263f., 266, 270
  as references of concept-words, 21, 189, 246
  extension of, 180, 253, 266, 268ff., 276
  first-level, 56, 171ff.
  identity of, 180
  paradox of, 247f., 254f.
  second-level, 56ff., 69, 181ff., 216
  specification of, 281
  unsaturatedness of, 189, 264

Concept-formation, 56ff., 60, 134, 171, 178ff., 182, 214
Concept/object distinction, 74, 135, 266
Concept-word, 170ff., 183, 256, 283, 289, 335f. See also Predicate
Content, 53, 68
  conceptual, 33ff., 105f., 128, 130, 133, 138f., 202, 212
  identity of, 163
  judgeable-, See Judgeable-content
  unjudgeable-, See Unjudgeable-content
Content-stroke, 83, 85f., 92. See also Horizontal
  paraphrase of, 91ff.
Context-dependence, 130f. See also Indexical
Contextual definition, 6, 8, 203, 216, 224, 227ff.
Contextual dictum, 5, 19, 134, 135, 194ff., 230. See also Heuristic Maxim; Restrictive Condition; Sufficiency Principle
  in *Foundations,* 199ff., 215ff.
Contextualism, 20, 194ff., 215, 217ff., 227, 230, 293ff., 338
  in *Begriffsschrift,* 205ff.
Copernicus, N., 11
Copula, 77ff., 256
Count-statements, 56, 69, 187, 216. See also Number, analysis of
Course-of-values, 241f., 252, 253, 272
Cousin, V., 42

Däniken, Erik von, 26
Davidson, D., 258, 368, 391
Decompositional analysis, 15, 105, 135ff., 145ff., 159ff., 221
Dedekind, J. W. R., 11, 12, 13, 146
Definite article, 335. See also Descriptions, theory of
Definition, 115, 159f.
  completeness of, 116, 140, 153, 214, 247, 253, 258, 292
  contextual. See Contextual definition
  explicit, 196, 198, 228ff.
  formal, 159, 176ff., 305f.
  implicit, 8, 229
  incomplete, 124
  piecemeal, 124, 229

INDEX

Democritus, 17
De Morgan, A., 12, 13, 14, 38, 72, 73
Depth grammar, 5, 22, 23, 24
Descartes, R., 5, 6, 10, 12, 15, 21, 28, 39, 47, 55, 57, 58, 271, 331, 394
Descriptions, theory of, 228
Descriptive content, 377
Direction, identity of, 161, 230
Dirichlet, P. G. L., 14, 312
Dray, W., 7
Ducasse, C. J., 344
Dummett, M. A. E., 5, 6, 8, 11, 13, 15, 19, 22, 62, 72, 106, 115, 124, 125, 129, 131, 158, 163, 170, 195, 216, 223, 236, 240, 244, 251, 255, 288, 293, 306, 308, 320, 331, 332, 357, 366, 367, 368, 370, 371, 372, 376, 377, 378, 389

Elucidation, 115, 119f., 127
Equinumerosity, 161, 218f., 222f., 268
Equipollence, 8, 126. *See also* Thought, criterion of identity for
Equivalence class, 161
Equivalence relation 320f.
Erdmann, B., 41, 43, 44f., 50
Ethics, 9, 43
Euclidean geometry, 23, 70, 120, 121, 229
Expression of thought, 98ff., 103, 342f.

Fact, 91, 344ff.
Fiction, 80, 99f., 129, 291f., 316. *See also* Reference-failure
Fischer, K., 6
Fodor, J., 272
Force, 107, 279. *See also* Sense/force distinction
  theory of, 100, 343
Formal concept 184
Formalism, 8, 29, 50
Foundations of Arithmetic, 3, 17, 35, 102, 106, 195, 197ff., 215ff., 230, 233ff.
Frege, G.
  achievement of, 397f.
  as a mathematician, 16ff.
  education of, 6f., 11f., 26
  his mature phase, 233ff., 239f., 266ff., 278
  influences on, 7, 11
  linearity of his thought, 21, 230
  misinterpretation of, 25ff., 239ff., 386ff.
Freudian psychoanalysis, 21
Function, 5, 14, 18, 136ff., 147, 153, 180, 242ff., 263
  as second-level concept, 69
  calculability of, 9, 312
  composition of, 185
  extensionality of, 111
  identity of, 189
  linguistic, 19, 172
  partial, 124
  second-level, 176, 181ff., 215, 247
  specification of, 137, 149ff., 174ff., 214, 267
Functional abstraction, 141f, 152, 157, 160ff., 166, 173, 182ff., 189, 192, 267, 310ff.
'Function and Concept', 237ff., 241ff., 262, 266
Function/argument analysis, 9, 24, 28, 76, 78, 135, 136ff., 158, 166ff., 191, 197, 221, 338
Function-name, 257ff.
Function/operation, 111
Function-theoretic logic, 8, 9, 13, 14, 22, 23, 241ff., 271ff.
Function theory, 14f., 18, 23, 71, 136ff., 143ff., 157ff., 160, 171, 192, 259, 312
Furth, M., 306

Galileo, G., 25
Geach, P. T., 5, 18, 19, 58, 79, 109, 124, 169, 172, 255, 332, 367
Gedankenbausteinen, 325, 381ff.
Generality, 179, 181ff., 208ff., 213
  multiple, 181f., 186
Gibbon, E., 66
Gödel, K., 17, 374
Grammar, 67ff., 74, 76, 107, 239, 256, 261, 263, 265
  philosophical, 333ff.
Grammatical/logical form, 23, 67ff., 77f.
Gregory, D. F., 13

Hacker, P. M. S., 24, 26, 60, 199, 337
Hamilton, Sir William, 18, 42, 64, 68, 72

Hamilton, Sir William Rowan, 14
Hardy, G. H., 176
Harrison, B., 18, 367
Hart, H. L. A., 7, 9
Haydn, J., 4
Hegel, G. W. E., 23, 36
Herschel, J., 12
Heuristic Maxim, 200f., 204, 212, 215, 293, 297f.
Hilbert, D., 114, 121, 229, 293, 365
History of ideas, method in, 3ff., 27, 28f., 233ff., 393ff.
Hobbes, T., 63, 64
Hodges, W., 112, 216
Horizontal, the, 340f.
Hume, D., 39, 51, 55, 331
Huntingdon, E. V., 12
Husserl, E., 12, 41, 50, 249
Hypothesis, 38, 39

Idea, 5, 46ff., 61, 64, 65, 66
Idealism, 47, 51, 54, 57f.
Identity, 50, 56, 57, 128, 150, 187, 213, 217, 220f., 223f., 229f., 245, 276, 280f., 284, 286, 289, 304f., 320
  contingent, 18
  criterion of, 251
Image, 63
Imperative, 20, 343
Indefinables, 318
Indexicals, 36, 130f., 335, 376
Indicative mood, 77ff.
Indirect statement, 321
Inference, 37ff., 60, 155, 374
Instantiation, 179
Intension/extension, 24, 315
Internal relation, 100, 126
Isomorphism. *See also*
    Gedankenbausteinen; Grammatical/logical form
  intensional, 357
  of thought and language, 73f., 76, 239, 248, 316ff., 382ff.

Jevons, W. S., 6, 7, 12, 40, 42, 73
Jourdain, P. E. B., 19, 21

Judgeable-content, 19, 35ff., 49, 50, 52, 53, 68, 69, 72, 78, 84, 86, 89, 102, 104ff., 135, 136ff., 145ff., 169, 171, 271, 274, 279
  alternative analysis of, 139, 142, 151, 154ff., 166f., 173f., 195, 214, 221f., 298
  anti-psychologism about, 46ff.
  as object, 145ff., 153
  criteria of identity for, 125ff.
  decomposition of, 133ff., 145ff., 164f., 191ff., 195ff., 315, 328
  primacy of, 105, 122, 133f., 142, 201, 206, 230, 294f.
  relation to meaning, 128ff.
  specification of, 152f.
  synthetic conception of, 135, 139ff., 158
Judging (act of), 36, 39, 63, 77, 80, 87ff., 104
Judgment, 36, 37, 38, 40, 59, 85
  Cartesian conception of, 39, 77, 80
  content of, 46, 59, 63
  grounds of, 38
  object of, 36, 40, 46, 49, 59, 61, 87ff.
  primacy of, 21, 127, 194
  subject of, 40
  vehicle of, 36
Judgment-stroke, 83f., 86ff.
  paraphrase of, 91ff.

Kant, I., 6, 7, 10, 12, 15, 28, 41, 55, 64, 108, 331
Kaplan, D., 132, 291, 313
Kelsen, H., 7
Kenny, A. J. P., 47, 258, 331
Kerry, B., 17, 74, 247, 269
Keynes, J. N., 40, 43
Kneale, W. and M., 13, 15, 123
Knowledge, Cartesian conception of, 122
Knowledge, logical source of, 121
Korselt, A. R., 293
Kossak, E., 222
Krug, W. T., 6, 42

Language, 67
  as a calculus, 22, 23, 26, 28
  creativity of, 381ff.

function/argument structure of, 22, 24, 25
ideal, 70f.
logical analysis of, 23, 335ff.
natural, defects of, 34, 66, 68ff., 75f., 81, 97, 162, 288, 388
relation to thinking, 68ff.
Legal philosophy, 9
Leibniz, G., 5, 7, 71, 110, 134
Leibniz's Law, 168, 220
Lemmon, E. J., 131, 370
Lewis, C. I., 346
Locke, J., 10, 15, 47, 52, 57, 58, 64, 394
Logic, 37, 38, 40, 43
axiomatization of, 121ff.
Frege's conception of, 121, 355
laws of, 37f., 41ff., 50, 55, 70, 73, 107, 121
mathematization of, 11ff.
relation to language, 67ff., 260ff.
subject matter of, 37ff., 107ff., 121, 132, 134, 144, 156, 192, 261ff., 369
Logical connectives, 12, 92, 111, 114ff., 122f., 145, 189, 266, 373, 375
Logical form, 23, 391. *See also* Grammatical/logical form
Logical truth, 367, 373, 379
Logicism, 3, 7, 8, 17, 27, 35, 163, 268
Lotze, H., 6, 7, 36, 42, 47, 63, 65, 68, 71, 125, 130, 285, 326, 342, 356
Lucretius, 17

McColl, H., 13
McDowell, J. H., 370, 376
Mansel, H. L., 42, 64, 130
Marty, A., 196
Marxist theory of history, 21
Mathematical induction, 3, 13, 17, 34, 35
Meaning, 134, 197
theory of, 24, 28, 29, 132, 370, 374ff., 381
Meinong, A., 40
Merkmale, 57, 134, 177f.
Metalogic, 374. *See also* Model theory; Validity, semantic conception of
Mill, J. S., 37, 40, 51, 58, 59, 68, 69, 77, 89, 108, 110, 125, 130, 156, 170, 265, 356

Modality, 130, 143
Mode of presentation, 285, 289, 292, 300ff., 317f. *See also* Sense
Model theory, 6, 355
Modus ponens, 89, 95, 100f., 108, 109, 119, 155, 163
Mozart, W. A., 4

Necessary truth, 41, 55
Necessity. *See* Modality
Normative science, 43, 107f.
logic as, 37, 43
Nozick, R., 7
Number, 29, 56, 67, 74, 224ff.
analysis of, 217ff. *See also* Logicism
as objects, 204
criterion of identity for, 127, 357
definition of, 17, 161
Humean definition of, 223ff.
Number-words, 50, 56, 202ff., 216ff., 221f., 224ff.

Object, 5, 16, 49, 56, 57, 137, 138, 143, 153, 164ff., 183, 250ff. *See also* Judgeable-content as object; Number as object; Platonism; Thought as object
Objectivity, 50
Oblique context, 290. *See also* Oratio obliqua
Occam's razor, 144
Ontological reduction, 105
Ontology, 173
Oratio obliqua, 94, 104, 276f., 343, 347

Part/whole relation, 135, 138, 145, 174, 191ff., 318ff., 325ff.
Peacock, G., 13, 14
Peano, G., 12, 147, 186, 288
Peirce, C. S., 11, 12, 13, 15, 40, 43, 147, 188
Philo, 15
Philosophy, 273
Frege's conception of, 7, 8, 26f.
nature of, 7, 11, 27, 67, 72, 393ff.
of language, 7, 8, 17, 29, 67, 334f.

Plato, 11, 21, 25, 28, 67
Platonism, 6, 28, 54, 55, 57, 59ff., 66ff., 73, 75, 103, 123, 143, 149, 314, 319f., 333f., 337, 354, 359f., 398
  Frege's, 7f., 16, 23
Port Royal Logic, 18, 47, 64, 108, 125, 130
Pragmatics, 370
Predicate, 77ff., 255ff.
Predicate calculus. *See* Quantification theory
Primary/secondary propositions, 108ff., 122, 133, 134, 185f.
Prior, A. N., 88, 280, 322, 344, 353
Private language, 46
Proof, 33ff., 38, 70f., 152, 167, 288, 306, 321, 369
  indirect, 38
Proper names, 53, 164ff., 203, 250f., 284, 288, 300f., 335
  complex, 281
  ordinary, 8
Property, 57, 138, 139, 143, 255
Proposition, 36, 88. *See also* Judgeable-content; Thought
  terms of, 77f., 105
Propositional calculus, 110ff., 114ff., 118ff.
Propositional function, 15, 19, 137, 206, 207, 264
Psychologism, 8, 41f., 44, 47, 49, 51f., 60, 67, 68, 73
Pythagoras, 25
Pythagoreanism, 22, 28, 398

Quantification theory, 5, 11, 13, 14f., 23f., 35, 106, 145, 181, 184ff., 213
Quantifier, 181ff.
Quine, W. V., 216, 361

Rawls, J., 7
Realism, 55. *See also* Idealism
  in semantics, 20
Reductionism, 143f.
Reference. *See also* Sense/reference distinction; Truth-values
  determination of, 10, 304, 312ff., 359
  principles of, 290ff.
  realm of, 316f., 323. *See also* Fiction
reference-failure, 69f., 80f., 129f., 281, 290ff.
Relation, 49, 55
Representation, forms of, 388ff.
Restrictive Condition, 201, 203f., 212f., 215, 221, 293, 295f.
Rule-following, 43
Rundle, B. B., 88, 345
Russell, B., 12, 15, 17, 19, 20, 26, 27, 40, 50, 74, 77, 78, 130, 137, 144, 147, 175, 186, 206, 207, 211, 228, 264, 287, 308, 330, 334, 371, 373, 377
Russell's paradox, 17
Ryle, G., 7, 27, 66, 88

Schlick, M., 374
Schering, 12
Schleiermacher, F. D. E., 43
Schirn, M., 161, 163, 223
Schloemilch, O., 50
Scholz, H., 163, 223, 225
Schröder, E., 12, 15, 36, 72, 81, 89, 110, 133, 180, 222
Scope (of logical operator), 185
Searle, J. R., 371
Semantic analysis, 144, 154ff., 162, 195, 261, 367, 369f., 377
Semantics, 216, 368, 381
Semantic value, 106
Sense, 19, 54, 57, 61, 71, 286, 300ff., 314ff. *See also* Mode of presentation
  ambiguity of, 310ff., 317
  and functional abstraction, 310ff.
  as intermediary between symbol and reference, 322
  as object, 302f.
  as thought-constituent, 318ff.
  determines reference, 10, 304, 312ff., 359
  identity of, 304f., 321
  of simple expression, 307f., 314, 318
  principles of, 290ff.
  realm of, 316f., 323
  reification of, 314ff.
  relation to meaning, 129
  specification of, 320f.
  theory of, 103
  without reference, 309f. *See also* Reference-failure

'(On) Sense and Reference', 248, 286f.
Sense/force distinction, 20f., 24, 28, 110, 278ff., 282ff., 290ff., 371f., 376f.
Sense-function, 325
Sense/reference distinction, 11, 21, 24, 28f., 53, 55, 106, 237ff., 316, 369
Sentence, 36, 38, 50, 68, 336
    assimilation to proper name, 9, 19, 21, 124f., 247, 250, 266, 287, 289, 376
    function/argument analysis of, 15, 24
    new, 381
    non-declarative, 336
    primacy of, 19, 134, 194f., 197, 216, 230
    sense of, 370
    step-wise construction of, 15, 368
Sentence-question, 129, 342f.
Sentence-radical, 378
Sheffer, H. M., 12
Sherlock Holmes, 9
Skinner, Q., 10, 395
Sluga, H., 7, 15, 65, 125
Solipsism, 51
Sorites paradox, 122, 213
Speech-acts, theory of, 5. *See also* Sense/force distinction
Spencer, H., 6, 42
Stenius, E., 371
Strawson, P. F., 130, 131, 317
Stumpf, C., 196
Subject/predicate, 56, 77ff., 135, 136, 257, 260, 263f.
    form in logic, 77ff.
Subordination (of concepts), 184
Sufficiency Principle, 201ff., 213ff., 219f., 222, 225, 293, 296f.
Syllogistic logic, 13, 23, 78f., 156, 390
Syntax, 173. *See also* Grammar
Syntax/semantics distinction, 6, 367

Tautology, 379. *See also* Logical truth
Thinking, 44, 60, 63ff., 73
Thomae, J., 29
Thought, 10, 36f., 40, 53, 59, 63, 67f., 103, 131, 276, 278ff., 283f., 286, 315f., 353ff., 377
    alternative presentation of, 321f.

alternative analysis of, 298, 328f., 376, 383ff.
as ideal sign, 360
as object, 279, 353f., 359
as value of function, 281, 323ff., 358, 383ff.
complexity of, 322ff.
compound, 327
constituents of, 322ff. *See also Gedankenbausteinen*
criteria of identity for, 8, 354ff.
designation of, 280
expression of, 280
laws of, 42ff.
part/whole analysis of, 325ff., 358, 383ff.
primacy of, 294f., 384
structure of, 75
Thought/language distinction, 68f.
Token-reflexives, 370. *See also* Indexicals
Tolstoy, L., 22
Tone, *See* Colouring
Trendelenburg, A., 7, 65
True, the, 245. *See also* Truth-values
Truth, 43, 46, 49, 106, 116, 121, 273, 344ff.
    correspondence theory of, 344ff.
    laws of, 36
    redundancy theory of, 262, 349ff.
Truth-conditional semantics, 24, 28, 366ff., 373ff., 378
Truth-conditions, 6, 18, 20, 134, 354f., 365ff., 375f., 378ff.
Truth-function, 117f.
Truth-tables, 12, 114f., 118, 145
Truth-values, 16, 21, 107, 111, 145, 237, 243ff., 249, 262, 272f., 275f., 283, 287, 301f., 347f., 369f., 376
Type-ambiguity (of concept-script), 162, 166f., 189
Type-theory, 124, 141, 142, 183, 188f., 192

Underdetermination of functions, 140, 174
Understanding, 19, 23, 54, 60f., 279, 288, 376, 380ff.
    generative theory of, 20, 23, 368, 370, 380ff.

Unjudgeable-content, 35, 36, 49ff., 104, 128, 133ff., 164ff., 192. *See also* Concept; Object
    anti-psychologism about, 49ff.
Unsaturated expression, 264. *See also* Concept-word; Predicate

Vagueness, 70, 129, 253, 292
Valid inference, 43, 55, 60
Validity, semantic conception of, 9, 37, 70, 109, 155, 261, 355, 367ff., 373ff., 379
van Heijenoort, J., 114, 236
Variable, free, 106, 112, 116, 124, 138, 159, 176, 179, 183, 186, 205ff.
Venn, J., 12
von Humboldt, W., 382

Waismann, F., 208, 260
Watts, I., 77
Weierstrass, K. T. W., 33
Whately, R., 37, 72, 77
Whewell, W., 12
White, A., 88
Wittgenstein, L., 5, 7, 22, 24, 25, 39, 44f., 60, 95, 121, 122, 123, 124, 216, 251, 254, 260, 290, 317, 326, 348, 360, 371, 373, 374, 377, 386, 387, 389, 390, 391, 392
    Tractatus Logico-Philosophicus, 6, 73, 111, 119, 136, 140f., 184, 199, 288, 292, 366, 367, 368, 370, 378, 379, 380, 382
Wölfflin, H., 5
Wright, C., 366
Wundt, W., 43